Writing Better Technical Articles

Writing Better Technical Articles

Harley Bjelland

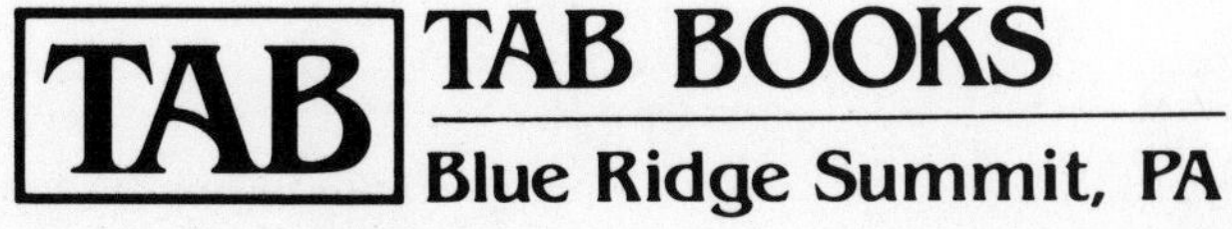
TAB BOOKS
Blue Ridge Summit, PA

Trademarks

CompuServe	H&R Block
Delphi	
Dialog	Lockheed Missles
GEnie	General Electric
IBM®	International Business Machines Corp.®
Knowledge Index	Lockheed Missles
Mead Data Central	Mead Corp
Newsnet	
Orbit	SDC/Orbit
Videolog	Schweber Electronics
Vu/Text	
Wilsonline	H.W. Wilson
WordStar®	
WordPerfect	
XYWrite™	
Yellow Pages	Dun & Bradstreet

FIRST EDITION
FIRST PRINTING

Library of Congress Cataloging-in-Publication Data

Bjelland, Harley.
Writing better technical articles / by Harley Bjelland.
p. cm.
ISBN 0-8306-8439-5 ISBN 0-8306-3439-8 (pbk.)
1. Technical writing. I. Title.
T11.B54 1990
808′.0666—dc20 89-49448
CIP

TAB BOOKS offers software for sale. For information and a catalog, please contact TAB Software Department, Blue Ridge Summit, PA 17294-0850.

Questions regarding the content of this book should be addressed to:

Reader Inquiry Branch
TAB BOOKS
Blue Ridge Summit, PA 17294-0214

Acquisitions Editor: Roland S. Phelps
Technical Editor: David M. Gauthier
Production: Katherine Brown
Book Design: Jaclyn J. Boone

Contents

Appendices

Acknowledgments

My wife, Tokiko (AKA Dorrie), has been infinitely patient during the long inspiration, gestation, and perspiration periods required for this book. And she served well as my "guinea pig" in making my text more useable.

Many thanks to Roland Phelps, Electronics Acquisitions Editor at TAB BOOKS who helped me along the way and who believed in the book.

Floyd Ashburn, owner of Compatibility + Plus in Springfield, Oregon, supplied me with an excellent IBM XT and with much sage advice and assistance along the way. Thanks Floyd.

I'm grateful for the excellent assistance the people at the Eugene, Oregon, library gave me in finding obscure facts that I did not even know existed.

And thanks to Thor, my computer, who took so much of the drudgery out of writing and made it a creative experience.

To the light of my life,
my wife,
Tokiko.

Introduction

This book is targeted first at showing how to write better technical articles, since this is the easiest and best way for a technical professional to become published for the first time. However, all of the techniques taught in this book apply to all types of technical writing, including books, memos, technical manuals, proposals, letters, etc. So the book can be used as a basic text applying to all forms of technical writing.

This book was designed to be read from beginning to end. The chapters form a series of activities which show you, step-by-step, how to get published. Each chapter builds and expands on the material in the previous chapter(s).

To use this book, first read the Table of Contents. This gives you a quick outline of the entire book. Next, browse through the book, look at the headlines, the section headings and the visuals to get a better "picture" of the book's contents and the manner in which information is presented.

Finally, start with Chapter 1 and read through all of the other chapters, in order. Study the examples and use the figures and formulas provided to create your own letters/memos and reports. Use the Precedent Sort to organize your writing.

Above all, be an active reader. When you read, keep a marking pen nearby and underline or highlight the points you especially want to remember. This activity reinforces the information in your mind and stores it in your memory bank for later recall. And it helps you locate the information later.

Don't wait until you have read the entire book to start practicing the writing techniques taught here. Remember the aphorism, "What I read, I forget. What I see (visuals), I remember. What I do, I understand."

As you read through the book, try writing short paragraphs, a memo, a query letter. Practice organizing topics for an article. Make a sketch of something and write a description of how it works. Don't worry, no one is going to read it, this is just practice. Visit the library and get acquainted with the marvelous references available.

This book was not written just to be *read.* This book was written to be *used.* So, use it well and you'll find that writing of things technical will indeed become easy, and rewarding in many ways.

Although the title of this book is *Writing Better Technical Articles,* all of the principles taught will also help you write better memos, letters, reports, proposals, manuals—all of the day-to-day writing you have to do in your profession.

In science, agriculture, bacteriology, chemistry, forestry, engineering, mining, medicine, metallurgy, physics, or any other technical field, writing is essential in most stages of every important project if a person does not want to remain anonymous.

Chapter 1 gives an overview of your first step, to select a good subject, one that suits both you and your audience. If you start out in the wrong direction, you'll waste a lot of valuable time and effort. Chapter 2 shows you how to get started right with a good subject.

Once you've selected a good subject, you should perform your brief, preliminary research, then write a query letter to the periodical(s) of your choice. Chapter 3 covers this important aspect of writing. This step is very important. You will save a lot of your valuable time and effort if you query a periodical before you write the article so you can become familiar with that periodical's specific editorial requirements and their need for articles. Some periodicals are totally staff-written, so it would be a waste of your time to try to publish in such periodicals.

After an editor's go-ahead, complete your in-depth research as discussed in chapter 4. Some of the information you need may be in the library, so this chapter, with appendix A, gives you a comprehensive, up-to-date refresher course in how to use a library.

The most important single key to written communication is:
Organization

An article, properly organized, is half written. In chapter 5 you'll learn how to take your raw data and your visuals, and organize them using the *Precedent Sort.* The precedent sort is a unique technique, a "formula" that I have adapted, tested, and refined over many years and used in all types of publications, ranging from simple memos to full-length books.

Briefly the precedent sort shows you how to create and organize a random group of topics so that they form a logical, smooth-flowing, coherent, well-written narrative. The precedent sort helps you develop the proper and logical sequence to use to effectively communicate your ideas to others.

For the next step, add your visuals, as described in chapter 6. It's also very important to plan and sketch your visuals before you write so your illustrations form an integral part of your text, instead of being tacked on as afterthoughts. Visuals can simplify your writing since each picture can save hundreds of words. Proper visuals also make technical concepts much easier to understand than words do.

Once your visuals have been properly integrated with your outline, it's time to write your first draft. Chapter 7 gives an efficient, relatively painless, orderly method of doing so.

One of the most important uses of the modern personal computer is its application as a word processor. It has revolutionized the way we write. No longer do we have to write, rewrite, and retype time and time again to come up with a perfect manuscript. Word processing programs store all the text we input so we need to edit only those portions that need changing to arrive at a complete draft.

And most word processing programs operate with companion spelling programs, which you can expand to include the special words or jargon of your technical profession to a supplemental electronic dictionary. In addition, a thesaurus is available as an adjunct to many word processing programs to help select alternate words to better express your ideas. Some special programs check the grammatical aspects of your writing, including sentence length, percentage of long words, overusage of the passive voice, etc.

The age of computer-assisted writing is indeed here and promises to do even more in the future to take much of the drudgery out of this important task of communicating. Chapter 7 describes how to effectively use the important capabilities of modern word processors and auxiliary programs to enhance, simplify, and make your writing easier.

Chapter 8 shows you how to add a beginning and an end to your article, and how to prepare your manuscript for submittal to a periodical. Also included is a comprehensive summary of the survey I have made of the editorial requirements, both current and future, of over 75 periodicals in all branches of engineering and science.

For the last step, you mail your manuscript off to the editor. But don't sit and wait for your acceptance check to come in and for your article to appear in print. Start researching another article immediately; before you realize it, you'll have a half dozen articles ready to list on your resumé and you'll be on your way to writing a full-length book.

Chapter 9 covers the revolution that is occurring in on-line systems and shows how to research the thousands of databases located throughout the United States and in many foreign countries. This exploding field promises to radically change the manner in which we do research. Printed books may soon be on the way out since a single 256-megabyte optical storage disk can store the equivalent of 300 to 400 books. With a personal computer, a modem, and a telephone line, you can research gigantic databases from your home or office.

There are millions of pages of information in the thousands of information repositories located throughout the world.

Chapter 10 is a bonus chapter—a post-graduate course for writers who are serious about improving their writing abilities even more. You'll learn about readability principles and techniques that make your writing easier to read and understand, and to increase the chances of your getting published.

1

Why You Should Write: Publish and Flourish

Your knowing is nothing, unless others know you know.

Persius

Forbes magazine asked a number of successful corporate executives what people should learn to help prepare them for careers. Their answer:

Teach them to write better.

The primary goals of this book are to meet the *Forbes'* recommendation and to dispel the notion that engineers and scientists cannot write. This book is going to teach technical people how to write better and with less effort.

This book will help overcome the overwhelming and largely unnecessary emphasis on style, spelling, grammar, and the Roman-numeral outlines that discouraged you in your high school and college English classes. This practical text will teach you how to write more proficiently using new effective writing formulas and techniques that have never before been applied to technical writing.

THE ANONYMOUS ENGINEER

Kevin Hansen worked diligently as a design engineer for a number of years. Few members of upper management knew his name. No one criticized or complimented him for his accomplishments. Quiet and unassuming, he felt he was just an anonymous engineer in a big group of engineers.

One day while browsing in the public library, Kevin delved into the topic of visual aids, remembering how inadequate visuals usually were at technical conferences. His interest grew when he discovered that no single book or article covered the subject adequately, so he researched the topic, took notes, outlined and wrote an article on the proper use of visual aids. It was accepted by a periodical, and he received a check and a copy of the periodical a short time later.

The next day Kevin's supervisor called him into his office and complimented him on his article. That afternoon the vice president of marketing called Kevin and sought his permission to reprint the article and distribute it to all of their employees.

During the next two weeks Kevin received calls from four different companies asking his permission to reprint the article and distribute it to all their employees.

Suddenly the anonymous engineer became known, not only in his own company, but throughout his profession.

This could happen to you once you break out of your indecision, take the plunge, and write an article for publication. It can change your professional and personal life, give you more confidence, and reward you in many other ways. Most of all it gives you a special inner satisfaction of self-accomplishment that few in your profession earn.

Writing is not a dull, unnecessary task to be endured. Writing is an important part of your work that can do much to advance your professional career. If you wish to be really successful, you must learn to write well. Writing is often the only tangible result of your efforts. Engineering/scientific writing is a skill to be learned, a professional tool that is as important as your technical knowledge and experience. Technical people who write well are readily known and recognized by their supervisors and contemporaries. Otherwise their ideas, no matter how good, may be ignored and overlooked. And being ignored and overlooked for one's honest efforts is enough to discourage any creative person. Recognition, however, leads to advancement, new assignments, and salary increases.

BUT FEW ARE PUBLISHED

Only 1 percent of all degreed engineers ever have an article published. This book shows you how to join this elite group of professionals.

Writing of technical matters always pays off. When you've published a number of good articles, a few scientific papers, or a solid book, it can make the difference between a routine career and an outstanding one. As an author, you're certain to gain professional recognition.

Studies have shown that most of the knowledge scientists and engineers learned in college becomes obsolete only four to seven years after they graduate, unless they continue their education. Obviously most professionals can't afford to take a sabbatical every few years to learn about the new developments in their profession, so they must learn about them by reading periodicals and books and by attending seminars. In some technical fields, as much as one-third of a workday must be devoted to keeping up with the ever-increasing knowledge of their specialty. Thus, all engineers and scientists incur an increasing obligation to write well so they can tell of their own developments and pass on this information to their peers.

If you're an engineer, a scientist, a technician, or an executive and have long wanted to have a scientific or technical article published, but you don't know where to begin or are afraid to start, this book will show you the direction. Techniques in this text have been developed to give you the impetus and confidence to direct you through all the necessary steps, in the proper order, to accomplish this all-important, career-advancing task.

NEW WRITING FORMULAS AND TECHNIQUES

The writing formulas and techniques described in this book will show you how to outline, organize, research, illustrate, and write all kinds of scientific and technical articles—and get them published! Perhaps you have some technical innovations or ideas you want to share with others in your profession. This book shows you how to make a national, even a worldwide audience aware of your accomplishments.

If you're in research or teaching, you know the value of being published for obtaining grants, tenure, and promotions. The more you publish, the more impressive are your credentials. In some academic institutions, you publish or perish.

EARN MONEY FROM WRITING

If you would like to earn money by selling your technical knowledge, this book will teach you not only how to write, but also how to market your articles. There are more than 6,000 technical, business, academic, and trade publications in the United States. These publications serve every conceivable occupation; they publish hundreds of thousands of technical articles and papers each year. It is indeed a fantastic, hungry, growing market.

If you're ready for that ultimate in technical writing—a book—all of the principles taught in this manual can be directly applied to that prestigious and profitable undertaking. A series of well-written articles that is accepted within the scientific and engineering communities is a pathway leading to a full-length book.

WHY WRITING IS DIFFICULT

Two of the biggest problems most writers face are:

1. What to write first.
2. What to write next.

The big, blank, white page staring up at you can paralyze even the most confident writer. What to write first often discourages many potential writers. Even though they bravely break down this initial barrier and write something down, they panic again about what to write next, and next, and next. They wonder if they're including all of the information, in the proper order, and. . . .

In spite of these psychological barriers, people's creative minds literally burst with thousands of ideas. These ideas come to their minds in a random order, however—a cacophony of creativity, seemingly unrelated to each other, but inescapably dictated by the complex interrelationships of each writer's background, loves, hates, experiences of all types, as well as cultural, ethnic, and educational backgrounds.

So, the problem is not how to come up with ideas, but how to organize ideas, how to use them effectively, and how to write them down in some meaningful order.

That's one of the main thrusts of this book: to show you how to organize your ideas so they form the basis of a cogent, lucid, and logical presentation.

Organization is a vital key to
unlocking your ideas and putting them in writing

THE PRECEDENT SORT

Order is essential to our lives. During every waking moment our minds are busy trying to draw order from confusion, trying to impose some pattern on the flood of stimuli flowing toward our senses. Without order, we live in chaos. In the Bible, Job speaks of death, not only as a land of darkness, but as a "land without any order." And order is essential when you read these sentences. Without organization you cannot comprehend what is written. It becomes a meaningless jumble. Order is an absolute essential to our comprehension.

Most people hate to organize simply because no logical method has been available. Yet they acknowledge that a comprehensive outline is essential in order to write a clear, concise article. Proper organization provides focus, direction, and impact for your writing. Facts don't usually speak for themselves; they speak only when related to the main topic of your article. Because the significance of facts grows out of their relationship to your topic, a logical organization is vital for readers to properly interpret and understand what is written.

What you need is a logical, painless method to organize your material, a method as easy to apply as the mathematical formulas that govern so much of engineering and science. And that's what you're going to learn in this book: how to write formulas.

You are not going to be bored by illogical, antiquated, hard-to-use methods containing Roman numerals, indentation, subordination, incantation, and other outdated ideas replete with confusion that were force-fed to you in high school and college. Rather you're going to learn a modern organizing method based on up-to-date computer technology, called the *precedent sort.* The precedent sort is easy to use, effective, and universal. It's like putting numbers in a simple formula and voila: the outline emerges!

Starting with a comprehensive outline, you will find it much easier to write and convert your ideas, your topics, into words and to string them all together to compose sentences and paragraphs until, before you realize it, you've written an entire article and you're ready to start your next one.

PEER FEAR

Another problem facing would-be writers is peer fear. The fear of writing a substandard article has deterred many engineers and scientists from being published. But you can overcome that barrier by following the procedures taught in this book to research, organize, and write your first article. You will learn what it takes to produce a publishable article and so gain confidence. Once you are published, your technical expertise will become known, you will become a recognized expert in your specific field, and others will seek your counsel. You'll meet new people, establish new contacts, and absorb more knowledge in talking with others.

BUT I DON'T KNOW HOW TO WRITE!

This book is all about ridding you of that mental block, showing you how writing can be easy. This text takes much of the mystery out of writing of things technical. Included in this practical manual are a compilation of pragmatic writing techniques, "tricks of the trade," tips that I have developed, distilled, tuned, and refined in over 25 years of scientific and engineering writing. My results were three nonfiction books; over thirty articles in national magazines; hundreds of technical manuals, reports, proposals; and thousands of letters and memos.

And don't think you need an enormous vocabulary to write. Studies have shown that a vocabulary of a mere 1,000 words covers about 85 percent of a writer's requirements on ordinary subjects. When you are writing on a specific technical subject, these basic words will, of course, be supplemented by the special words of your profession.

Most books on writing spend a considerable amount of time on grammar, but devote very little to the practical aspect of technical writing: that vital aspect of how to communicate ideas to others.

This text is different. You'll learn how to communicate effectively with words, using what I call ***invisible writing techniques.*** You'll learn writing formulas—guidelines that will help you create and organize your ideas. Then you'll learn techniques that will help you convert your organized notes into sentences, paragraphs, and a completed article in which you successfully communicate your ideas without the reader realizing how you accomplished it—a clever, silent seduction of sorts.

With the confidence and abilities you'll learn from this book, you'll overcome the unfortunate hang-ups and fears instilled by your high school and college English teachers. You'll find that writing of things technical can be easy, fun, and one of the most rewarding and stimulating activities you can participate in.

BEING A WRITER HAS ITS ADVANTAGES

Once you start writing using the techniques in this book and begin putting your ideas down on paper, you'll find that it has helped you in many ways:

- Writing your ideas down on paper will help clarify and organize your thoughts.
- You'll learn more about any subject you choose when you research, outline, and write your article.
- Once they are written down in cold type, you can study and evaluate your ideas with a critical, perceptive eye.
- Often your written documents are all that upper management sees of your work. Write well and you'll be recognized and rewarded for it.
- You're contributing needed knowledge to your field—helping others to learn.
- You might find, like so many people, that you can express your thoughts better in writing than in speaking. When you write, you can revise your material again and again until you get it just right. You can't do that when you speak. Once said, it's said, or as the Russian proverb tells us:

  A spoken word is not a sparrow.
Once it flies out, you can't catch it.

- You'll soon discover it's easy to compose other types of important communications, such as simple memos, letters, reports, and articles. The techniques revealed in this book work for all types of writing.
- You'll help your company's business and improve your own chances of promotion.
- A list of publications looks impressive on your resumé.
- The technical person who writes well is looked upon as a leader in his or her profession.

- At a technical conference you may reach, at most, 100 people. In a technical periodical, you can reach tens of thousands of people.
- You'll earn an inner satisfaction, a pride in what you've accomplished. This, many feel, is the most important reward of all.

THE IMPORTANCE OF CLEAR WRITING

Most people recognize the names of Charles Darwin and Albert Einstein. Not only were their discoveries important; these men also were outstanding writers, able to communicate about complex things in simple terms that both scientists and laypeople could understand. Many brilliant scientists have died in obscurity, their discoveries unknown because they wrote leaden prose, foggy sentences, disorganized dissertations that no one could comprehend. It was almost as if they spoke a foreign language that no one understood. Their ideas had to be rediscovered by others because they did not properly document and report their original discoveries.

An example of leaden prose is the average Ph.D. thesis about which J. Frank Dobie wrote in his *A Texan in England*:

> The average Ph.D. thesis is nothing but a transference of bones from one graveyard to another.

Much technical writing will never be read because it is nothing but leaden prose, a rearrangement of old bones.

To illustrate how leaden a bad writer's prose can become, even in everyday life, read this:

> "In accordance with our previous agreement, it is hereby requested that you cease delivery from the period beginning Tuesday, May 19, up to, but not including Thursday, May 21, and thereafter, until further written notice is received from the undersigned."

In simple terms, this is a note to the milkman, written by a government bureaucrat, that should have read:

> "Please skip Wednesday."

Which version would you prefer to write, and to read?

An excellent example of effective brevity that was composed over a thousand years ago is the best trip report ever written:

> "Veni, vidi, vici."

This trip report by Julius Caesar translates to:

> "I came, I saw, I conquered."

So, to put it graphically, this book teaches you how to take the lead out of your prose.

STEPS TO FOLLOW

This book is organized according to the steps you would logically follow in writing an article as diagrammed in FIG. 1-1. Most of your time will be spent on research, organization, writing, and revision. Experts recommend devoting the following percentages of your time to these major functions:

40% Research and organization
30% Writing
30% Revision

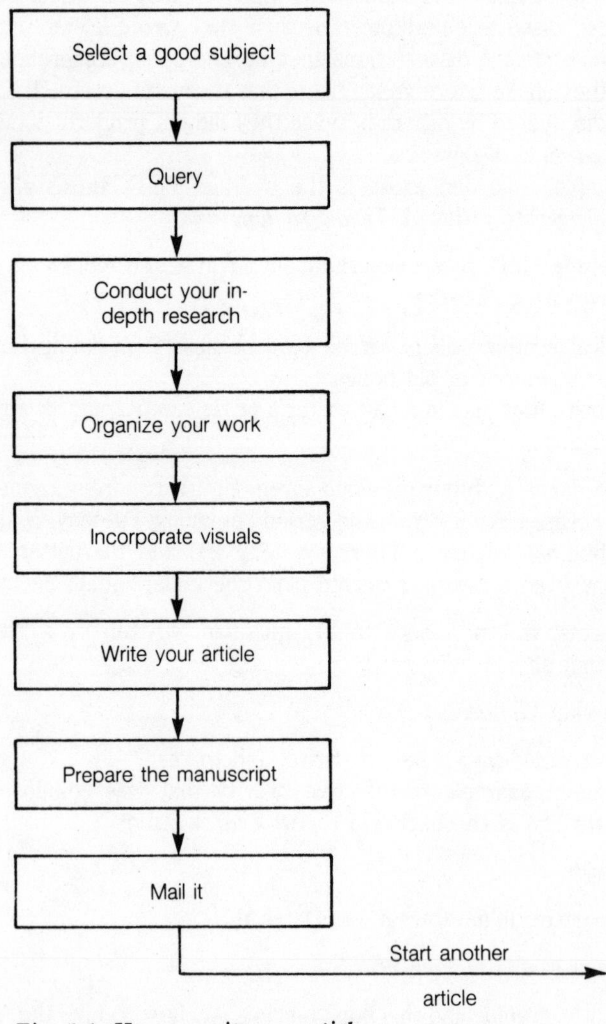

Fig. 1-1. How to write an article.

This breakdown illustrates the importance of thorough research and organization before you start writing. If you begin to write before you organize your material, it's like pouring concrete before setting up the forms. To tear apart the formless sentences and paragraphs you have written is exhausting and exasperating labor.

These percentages obviously can vary, depending on the subject you've chosen. However, if your writing takes up more than one-third of your total time, you're not spending enough time in researching and organizing, and your revision will be laborious and require excessive time.

EXERCISES

1. List 10 subjects you'd like to research, write about, and publish in a technical periodical.

2. List 10 technical periodicals you'd like to write an article for.

3. Why do you want to learn to write better?

4. Find an example of well-written technical material and a contrasting example of poorly written technical material. Check which of the following characteristics separate them into good and bad.

    Organization
    Long sentences
    Long paragraphs
    Long, complex words
    Excessive use of jargon
    Subject matter

5. Choose a highly technical article about a subject in your field of specialty. Select a half dozen consecutive paragraphs and rewrite them for a layperson. Translate the technical jargon so it can be understood by nontechnical readers.

2

How to Select a Publishable Subject

Curiosity is one of the permanent and certain characteristics of a vigorous mind.

Samuel Johnson

Whether you have already chosen a subject or whether you need to select one, the material in this chapter will benefit you because the single most important decision you are going to make is to choose a subject. Select a good subject and your article will blossom and be easy to write. Select an inappropriate subject, and writing it will be a drudge; publishing it will be impossible.

This chapter will take you, step by step as diagrammed in FIG. 2-1 from defining what makes a good subject, through the final steps of selecting a subject and creating a working title.

START NOW

Don't wait until you've finished this chapter before you start thinking of subjects. As you read through this chapter, and as you go through your daily activities, keep a pen or pencil and paper beside you to jot down a number of possible topics you're going to consider. Keep a folder to file them in. Call this your Idea File.

Topics in this chapter will jog your memory and help you create dozens of suitable topics for articles. Generally the best subject to consider first is your work, something you know well, or a topic that interests you. It's exclusively yours and you should have enthusiasm for it. But don't limit your subjects to

How to Select a Publishable Subject

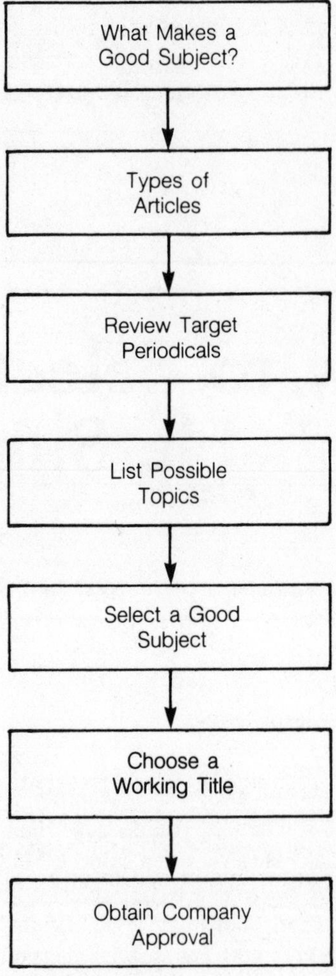

Fig. 2-1. How to select a good subject.

your work. Consider a wide variety of topics—let your imagination roam free and unfettered.

Creativity

Writing is a creative art. Scientific tests regard "facility in writing" as a basic index of creative aptitude. As a writer you are not simply going to copy down phrases, sentences, and ideas that others have already composed. Instead, you are going to review and evaluate what other writers have done with similar topics to see how their writings can influence, enhance, and supplement what you are creating.

Finally, you'll organize all of the information into a comprehensive whole, then write and revise your article until you are satisfied with it.

This process begins with that indispensable ingredient that exists in all people, in varying degrees:

creativity.

Imitate

Because you learn by imitation, you can enhance your creativity by consciously cultivating some of the attributes that characterize creativity. A creative person is:

- Unconventional
- Original
- Curious
- Sensitive
- Dissatisfied
- Open Minded
- Image Creating
- Fluent
- Persistent/Motivated
- In possession of a sense of humor

This might seem like a large number of characteristics to have, but most people possess these attributes to varying degrees. Creativity was first manifested in us when we were children—always curious, always asking questions, always trying new things. But this creativity was often stifled by parents and teachers who judged us harshly and impatiently, telling us not to ask such silly questions or do such stupid things.

Undoubtedly some of these cautions served a useful purpose, but in many instances it would have been better if we had learned the consequences ourselves.

As adults we became so conditioned to accept the status quo that we became afraid to suggest improvements for fear that our peers would judge us, laugh at us, or poke fun at us. There are thousands of examples of creative people being laughed at, from Noah to Fulton to modern-day researchers. But these creative people endured the criticism and carried their ideas to fruition.

That creative urge is still there in all of us, lying dormant, waiting to be nourished, eager to be used again. Curiosity is like a muscle: it must be exercised to be of full use.

And now you are going to have an excellent opportunity to exercise your "creativity muscles" and put them to work by writing an article that will surprise even you by its excellence. Review the characteristics that creative people possess. Ponder these priorities; practice and cultivate them consciously so they can become ingrained in you and become a part of your subconscious. Practice writing down ideas and thoughts, no matter how outlandish they may seem. Only you will be reading them. Ideas feed on ideas, so what might have seemed like a totally ridiculous idea can give birth to that great, revolutionary idea that can change your life.

Unconventional Creative people employ unusual problem-solving techniques and give uncommon answers to questions in the process of generating unusual solutions to problems. They balk at pressure to conform to the norm. These individuals are not afraid to try something different, to take a chance.

A creative person does not fit into a mold. The creative person is not necessarily the one who has the best memory. Studies have shown that people with exceptional memories are not creative, but imitative. As an example, an idiot savant can have a photographic memory, but is in no way creative.

The basic definition of creativity is the ability to come up with something new, something different. So a truly creative person is one who thinks differently, who sees objects and situations in a different light.

Original Creative people are always searching for something new and are not hampered by stereotyped solutions. Creative people can analyze existing systems, see room for improvement, and combine unusual connections and combinations to arrive at a solution. Creative people are receptive to new ideas and can rearrange basic elements in new combinations to form a new whole.

Sensitive Responding strongly to their senses of hearing, touch, taste, sensitivity to color, shapes, and textures also distinguish the creative individual. They have a greater feel for things and events that surround them. They visualize concepts better than most people.

Curious A predominant characteristic of creative people is that they have never totally lost their childhood wonder, the curiosity of their earlier years. They become intrigued by problems that puzzle them, and ponder how these problems can be solved. They also wonder about things, events, and processes. What caused them? Can the results be changed? Can it be done in a different and better way?

Dissatisfied Creative people are dissatisfied with the way things are —dissatisfied enough to change them. They don't accept situations as being inevitable, unchangeable. These people are restless, yet they cherish the opportunity to relax and switch their minds over to something different. This discontentment does not disturb them to the point of rejection, but to the point of wanting to improve it. They know there are better ways to accomplish things.

Open Minded A prime requirement of creativity is the ability to react to stimuli without prejudice, without shutting data out. Creative people open their minds to all ideas, are not bothered by the NIH (not invented here—If I didn't invent it, it's no good) factor, which prejudices so many decisions. An ability to restrain critical judgment during the creative or idea-generation process, and openness to all approaches distinguishes the creative person. Don't toss out ideas before they can be considered, evaluated, and tested. Don't believe that the obvious way is the best way. Be receptive to unusual ideas. Creative people have a talent for ambiguity and can work in situations where no clear direction exists. They can find their way in the dark without a candle.

Image Creator Creative people think in images, not words. They tend to be daydreamers and can visualize an idea taking shape. They believe it's best to solve most of the problem with images, even though their initial imagery may be foggy and ill-defined. They resist the temptation to articulate, to reduce their solution to words too soon, since words can put restrictions around an incomplete solution.

Fluent Possessing fluency, creative individuals can generate fifty uses for a paper clip. They are fluent not only in their own field, but in many related fields. Fertility of ideas also characterizes these individuals. They can generate a large number of ideas in a short time. Ideas spawn other ideas, loosening the gates of imagination so that a large number of potential solutions emerge. As a result of this chaining effect, an outstanding solution usually will result.

Flexibility The ability to adapt and adjust to new and changing situations is also important. Creative individuals can abandon old ways of thinking and initiate different directions, different problem-solving approaches. They can generate a number of different kinds of ideas and are able to break away from conventional methods of solution.

Persistent/Motivated Creative people don't work an eight-hour day; they do not function to a schedule. They tend to be totally immersed in their work and fiercely determined to succeed. Not overly discouraged by failure, they realize that some failures are necessary in learning to walk the path to finish the work they have started. They are driven and persistent. These are probably their most important and strongest characteristics. More energetic, they have a strong desire to create, and they welcome confrontation. They are not discouraged by the unknown; they accept it as a challenge. Creative people have a huge capacity to take pains. They are stubborn—determined to see their project through to completion.

Sense of Humor As George Orwell said, *"Every joke is a tiny act of rebellion."* This is characterized in creative individuals by their ability to laugh at life's foibles. Some experts claim that computers will never be able to do creative thinking because computers have no sense of humor. It is a healthy sense of humor that creative people have, not a degrading, cynical one.

KINDS OF ARTICLES

Some of the basic kinds of articles that appear in technical periodicals are described here. These are not strict divisions, and often an article can cover more than one of these kinds. This list is not in any way to be considered restrictive. Use it to jog your memory in your search for topics.

Design Procedures/Ideas This kind of article shows your reader how to design a device, plant, or product, or how to devise a procedure. It can describe the requirements to be met, alternative methods to meet them, design steps, how to check the results, special considerations, future modifications, and projected results.

Process Description A specific process is detailed in this kind of article. The process might be semimanual or automated. The raw materials and equipment required are covered, along with reliability, efficiency, and step-by-step operating procedures, including time, temperature, and pressure. Such articles also compare the advantages and disadvantages of the process with competitive processes.

Technical Descriptions In this kind of article, a description is given of the general manner in which some device performs its functions and might include a general description of the device, its operating theory, design considerations, lists of major parts, reliability, and related factors.

New Products Consider this popular kind of article carefully so that it does not become a thinly disguised product promotion piece. Stress the general use of the device, process, or product; construction data; methods of using, operating, and applying it in various situations; and anticipated results.

Management Techniques A wide variety of topics are available for this field. Many deal with the methods and means of supervising, scheduling, and motivating employees. Since people form the most important asset in any industry, this kind of article, if it is written to cover very general management problems, is always in high demand.

Computerization The use of computers is revolutionizing all industries, so articles on how computers are being used are always needed. The list of potential applications is virtually unlimited.

Product Surveys This popular topic provides an unbiased comparison of competing companies' products, processes, and devices. Write this one carefully, or you might alienate one or more companies that advertise in the periodical.

Plant Descriptions This kind of article covers the functions of the plant, considerations in location and size, equipment installed, plant capacity, plant flow, operating details, personnel facilities, and any unusual design features, such as degree of automation, energy management, and environmental design. This kind of article usually proceeds from the outside in, describing the overall forest before detailing the trees.

Mathematical/Computer/Graphical Solutions A need always exists to provide better, easier, and more accurate solutions to a wide range of design problems. Solutions can be purely mathematical, by computer, or by graphical means. Covered are calculation procedures, equations, assumptions, limitations, and one or two examples of the procedure.

How Tos The scope of this kind of article is virtually unlimited. A how-to article can show you most anything related to any aspect of your profession. They can cover how to schedule a project, how to write specifications, how to save time in a process, all the way to a book on how to write better technical articles. How-to articles will always be in high demand.

Miscellaneous The miscellaneous category covers a wide range of possible topics, such as:

- Opinion articles—Opinion articles are difficult to sell because most technical people like to separate personal opinions from their work.
- Plans for future projects—Before projects start, or as a progress report after a project is underway.
- Reporting what others have done—You might want to write about some special project underway at your company that others are working on. Often your co-workers are unable or unequipped to write about their accomplishments and need a technical reporter to write about it for them.
- Humor—Unfortunately the demand for humor articles is not too great, giving some credence to the stereotyped serious periodical editors. I once wrote a humorous satire claiming the word man-month was a sexist phrase. The article was published in a national magazine and I received many letters and calls from entertained readers and an invitation to speak at an electronics convention. However, the publisher apparently wasn't impressed and wouldn't consider any more humorous articles for his staid technical periodical.

JOT DOWN YOUR IDEAS

As you read through the various sections of this chapter, you might find that certain topics you've listed in your idea file are not suitable for a variety of reasons. If so, cross them out. But don't be too critical at this point; save the final culling out for later.

As you read and as you conduct your daily activities, continue to write down any more topics that occur to you. By the time you reach the end of this chapter, your list should be extensive and you should be ready to evaluate the topics and select the ones for your first few articles. Next, let's see what criteria a good subject must meet.

Benefit Your Reader

A primary requirement for all of your articles is that your topic benefits your readers. If it does not, it simply will not be published. Most technical articles are written to inform readers of new and future developments: how things work or how to accomplish certain design tasks or processes. Technical articles are written *by* technical professionals *for* technical professionals. All topics must be slanted so technical people can profit from reading them. A few articles are written to inform, but most are published to educate readers, to teach them more about their profession. And the more readers you can attract, the greater the likelihood your article will be published.

Always keep this question foremost in your mind when you select a subject:

"How will my article benefit my readers?"

If you can't come up with a solid answer, select another subject.

Be Topical

Technical periodicals usually cover topical subjects; that is, topics that are currently in use, recently developed, etc.

Technical books, however, cover topics that are of a more archival or long-lasting interest—subjects that are not out of date by the time the book is published. In recent years the division between the two has blurred somewhat; however, the content of most technical periodicals is still principally topical.

The reasons for this difference in coverage are twofold: First, a periodical can be written, printed, and distributed to a list of subscribers in less than a month's time, before the new gets old. A book, however, often takes a year or more from start to finish. Secondly, and perhaps more importantly, periodicals derive most of their revenue from advertising. Many technical periodicals are given free to the reader, the publishing and distribution costs being covered by the advertisers. Technical periodicals must attract new readers to every issue so their subscribers will keep reading, and hopefully responding to, the latest ads.

However, a book earns its revenue from the sale of the book alone. It is purchased solely for its contents. Readers expect topical subjects in technical periodicals, so if you want your articles to be published, *be topical.*

Be Unique

Uniqueness covers the superlatives of the professions:

- Newest
- Biggest
- Smallest
- Fastest
- Cheapest

If your topic covers a product, a development, or a process that falls under one or more of these categories, you have a built-in readership because one of the main functions of technical periodicals is to inform technical personnel about how to make things smaller, less expensive, faster, etc. You can fulfill this need if you discuss such a characteristic. If applicable, use one of these superlatives in the title of your article.

Still, uniqueness isn't an absolute requirement. If you have a new slant, a new approach to what has been published before, it's also an excellent candidate for an article.

Interest a Wide Readership

This might seem like stating the obvious, but so many people become so involved in their work that they don blinders and completely ignore this simple fact: the broader the appeal of your topic, the more likely that your article will be published.

Don't write about the design you completed using an A29BXT4 integrated circuit in your automatic bean counter for the accounting department. Broaden the appeal of your design; consider a host of other general counting applications, such as counting lima beans, jelly beans, cars, people, railroad cars, etc.

When you designed your counter, you might have concentrated on designing for an accounting bean. But unless your design is useful for dozens of other applications, few will read your article. Your readers are primarily interested in their own application, not yours.

Something You Know or Can Research

Ideally your subject should be something you know a lot about and can write about with authority. Your readers will recognize this authority and respect you for it. If you're the only person to your knowledge working on some subject and it has wide appeal, you greatly enhance your chances of publication.

However, you can also write about topics you know little or nothing about if you follow these words of advice from chapter 4:

Research the literature

If you find a topic that intrigues you and that you'd like to learn more about (your interest is vital in this approach), you can devote the necessary time to research your subject. When you thoroughly research a topic you learn a lot about it and become an expert through osmosis. Then you can write about it with authority, even though most of your knowledge was acquired from the writings of others. This is research, not plagiarism; because you're reading, critically evaluating, reorganizing, and updating the work of many others, incorporating your own ideas and slant on the topic.

To illustrate, I searched for a suitable topic for a technical article, but my work at the time wasn't a suitable topic. So, after doing a little preliminary research, I chose "Visual Aids for Technical Talks" as my topic. It was a subject that I felt had not been adequately covered in periodicals and one that I felt most readers could benefit from. Sooner or later most technical people have to give a speech requiring visual aids, whether it's a proposal presentation, a request for a grant, a progress report to management, or a technical talk.

Even though I had given a few technical talks at conferences in my professional career, I could in no way consider myself an expert on visual aids. So I went to work, thoroughly researched the subject, reviewed books and periodicals on the topic, took notes, etc.

Then, using the techniques described in this book, I wrote a query letter, received a positive response, and wrote the article. It was accepted the first time out. After it was published, I received a number of letters and calls praising the article and its practical content. Some people asked permission to reprint it as a guideline for use by their companies. In spite of the mostly library-acquired expertise on the subject, I hadn't originated any new information. I had merely researched, adapted, organized, and rewritten existing information, adding my own ideas and slant for the article.

You can do the same and discover dozens of topics you're intrigued by. Research, evaluate, organize, query, and write about them and you can also become a published expert.

And don't think you need anything of earthshaking significance to write about. Most articles you see in technical journals barely nudge the Richter scale. They are simply everyday topics that the author has researched and written about competently, in a manner that others can read, understand, and profit from.

The Periodicals You Want to Write For

This should also be so obvious that it shouldn't have to be mentioned, yet technical editors are constantly plagued with articles that are totally unsuited to the type of material they publish. It should be obvious that you don't send a technical article on the design of lasers to *Readers' Digest,* nor one on getting along with your mother-in-law to *Scientific American* (unless she's a nuclear physicist). Make sure there is a market for your article or you are wasting a lot of valuable time.

REVIEW TARGET PERIODICALS

Now that you have a pretty good idea of the kinds of topics that make for good copy, the next step is to review some of the periodicals in which you'd like to be published. A review form is provided at the end of this chapter to record this information. Xerox it so you'll have a number of blank forms to fill out when you're evaluating periodicals.

First, don't choose only one periodical to write for because the odds might be too much against you. The idiosyncracies of the editors, the advertisers, and the number of the editor's relatives on the staff all work against you. Select at least three or four periodicals in your field of expertise, periodicals you like to read. (That's a good standard: if you like to read their articles, there is a good chance that you can be published in them.)

You can probably find back issues of periodicals in your company or public library, or from your local pack-rat co-worker who saves old issues of everything. A representative list of periodicals is included in appendix C.

Even though its primary purpose is to list advertising rates. *Business Publications Rates and Data,* published by Standard Rate and Data Service of Skokie, Illinois, is an excellent source for checking on periodicals that carry advertising. SRDS lists the following:

- The publisher and editorial staff
- The publisher's editorial profile—the types of articles published
- Circulation—paid and nonpaid
- Business analysis of circulation—types of products, services covered
- Breakdown of the readership by job title

Usually a periodical's last issue of each year lists all of the articles published in that year. Check one of these last issues and review the titles. You'll get a quick, capsule view of what types of topics are popular in the periodicals of your choice. And you'll also get a taste for the kind of titles used in the periodicals. You might even want to make your own checklist to see how many of each kind of article discussed earlier in this chapter were published in each periodical.

Go through Back Issues

Next, go through three or four back issues of the periodicals you want to write for, page by page.

Although it may seem like a lot of work at first to review the periodicals you'd like to be published in, this effort will help you pick the right subject, the best title, the correct style, and will practically guarantee your article's acceptance. Like in most of life's endeavors, the more effort you put into it, the more rewarding the results.

What to Look for

Check the cover first. Are photographs and illustrations used? Are titles of some of the key articles listed on the front cover?

Review the table of contents. What types of articles were published, were they how-tos, informational, product surveys, design procedures. What is the average title length? What types of titles are used?

The next important point of contact should be the *masthead* of the periodical that should be displayed on one of the first few pages. The masthead will list the editor and the editorial and advertising staffs. It will tell you who to address your query letter to. Usually this will be the editor; however, if the periodical has special department editors; address your query letter to that special editor. Check the masthead against the table of contents to see how many of the articles were written by a staff member. If most articles are staff-written, your chances of publication in that periodical are poor. Concentrate on those periodicals that publish mostly free-lanced articles.

If you find an article written by the editor, read it carefully. It will give you a good idea of the type, slant, structure, language, and subject matter the editor favors.

Read the letters to the editor. They'll indicate how earlier articles were received and which drew the most positive or negative responses. If the editor has written an editorial, read it carefully. It might provide additional information on the types of articles the periodical is seeking.

Review the Ads

It's the advertising revenue that pays for the free (or low-cost) subscriptions to the periodical. And the advertisers, who want their advertising revenue invested wisely, are experts at pinpointing the types of people who read and respond to their ads by obtaining readership profiles from the periodicals.

A careful analysis of a periodical's ads can show you:

- What the typical reader of the periodical is like (interests, education, profession).
- The types of subjects the editor is seeking, since the articles' contents frequently complement the ads.

Often there is a close correlation between the types of articles that are published and the ads. Advertising agencies prepare lists of articles that will be featured in forthcoming issues and use this promotional material to interest companies to advertise in these specific issues. So you see the importance of studying the ads and using this information in selecting a topic for your article.

To analyze the ads, determine:

- What types of products are advertised: components, products, services, systems, books?
- What type of people and products are depicted in the ads: young, old,

professionals, blue collar? How are they dressed? What are they doing in the ad?

And don't forget the mail-order ads that are usually in the back section of the periodical. They also give some clues as to the types of products readers are interested in.

One important piece of advice about ads: if you want your article published, never antagonize an advertiser in your article! The reason should be obvious.

Study the Visuals

Review the visuals. Are photographs used? Black and white or color? What kind of captions are used on the visuals? Are line drawings, cartoons, graphs, and curves used? Is the art complex? How many visuals per article?

If photographs are used, you should determine in your query if the publisher prefers transparencies, 8-X-10-inch glossies, or contact sheets. Polaroids or low-quality prints are generally not acceptable.

Style

Study the articles to see what styles the writers use. What type of vocabulary is utilized—long words or short words, considerable jargon or none?

What is the sentence length? Are a variety of sentence lengths used? Are verbs active or passive? Are descriptive words used? Are contractions used? Is first person used: I, you, or they?

One method of determining the educational level of readers is through Gunning's Fog Index. To compute it:

1. Take a sample of 100 words of continuous writing and measure the sample with a ruler to see how many column inches the 100 words take up. Make a note of this number.
2. Count the number of words per sentence in the 100-word sample, but treat independent clauses (the parts of a sentence separated by a semicolon) as separate sentences. Calculate the average sentence length.
3. Count the number of words of three or more syllables in the 100-word sample. Don't count capitalized words, easy combinations like bookkeeper, or verbs with three syllables ending in -es or -ed (for example, *edited*).
4. Add the average sentence length in Step 2 to the number of polysyllable words per 100 words. Multiply this figure by 0.4.
5. Pick three or four more random samples of continuous writing and

repeat this procedure. Average the result. This will give you the number of years of schooling a reader would need to read the article with ease and understanding.

Following are examples of Gunning's Fog Index for the world's best seller and for some popular magazines:

The Bible	6–7
Reader's Digest	8
Time	10
Atlantic Monthly	12

CHOOSE A TOPIC

By now you should have an idea file, bulging with at least ten or twenty potential article topics. Next, sit down in a quiet place with a pen or pencil and read over your list. As you read, more new ideas and new slants will occur to you. Write them down, no matter how impractical they seem. You can weed out the bad ones later.

Visualize yourself in the reader's shoes. What would *you* like to see in the periodicals that you and your co-workers read? Write these ideas down, using just a few words to describe your topic.

Finally, talk to your co-workers. Ask them what they would like to read about, what topics haven't been covered, and what they like or dislike about articles in current periodicals. Note down these additional ideas.

Now that you have a long list of potential articles, it's time to narrow your list down to three: one primary topic and two backup topics just in case your primary topic doesn't work out. Carefully evaluate the topics and cross off the ones that do not meet the criteria established earlier in this chapter.

SELECT WORKING TITLES

With three topics selected, it's time to pick working titles for all three. The title is an extremely important part of your article—it's usually the first thing a reader sees. Your title will be read by thousands of people, but probably only a few hundred will read your entire article. If you create a just-right title when you write your query, your chances of being published will be much enhanced.

Because of the mountains of literature available, and with more being printed and circulated every month, your reader is pressed for time and will first skim your title to determine whether or not to read on. So your title must be good enough to hook your reader. Investing a lot of time and effort in creating a grabbing title will result in big dividends.

A descriptive title is also important so people using on-line data banks (see chapter 9) can key in the proper words to locate your article after it has been published.

When you performed your periodical survey, you probably noted how often certain words appeared in titles:

- How to
- New
- Unique
- Smallest
- Easy
- Fastest

A good title is actually a summary of an article, but it can't be too long—no more than six to eight words. Use the active tense in your title; avoid jargon. Use words that are familiar, specific, common, and short.

Types of Titles

A title that hooks gives the reader a little information and entices the reader on. Sometimes a title need do no more than pique a reader's curiosity—a curiosity that can only be satisfied by reading your entire article.

Four types of titles are commonly used Label, Question, Imperative, Statement.

A *label* title can consist of one or more words and is a name or label that describes the contents of your article; for example:

- Consulting
- Brainstorming
- Computer Gold

The label title is the type of title most often used.

A *question* title should lure the reader into the body of your article. For example:

- *Ideas: Where Do You Find Them?*
- *Is Your Chemical Process Foolproof?*

When properly used, question titles can titillate your readers enough so they must read your article to find the answer to the question posed.

An *imperative* title urges your reader to action and is often used in how-to articles:

- *Check Your Unit's Reliability Under Operating Conditions*
- *How to Design a Suspension Bridge*

This is also an excellent type of title to use for informative articles.

A *statement* title is a short sentence that summarizes your article; for example:

- *New Computer Techniques Improve Reliability of Communication*

When generating titles, avoid, if possible:

- Articles (the, an, a).
- Numbers and complex symbols. Technical titles should be informative, not mysterious.
- Long titles.
- Vagueness—Remember that your title might be the only information given on your article in abstracting journals and indexing services.
- Unnecessary words, such as: *Study of, Investigation of, A Final Report On.*
- Passive voice.

Titling Techniques

When people start to read a periodical, they might first look at the front cover, then the table of contents, then start flipping through the pages until a title or a visual catches their attention. If their interest is tweaked enough, they'll stop and begin to read your article. Your title or visuals have then accomplished one of their missions.

You can use a number of techniques to further enhance the saleability of your title, to make it stand up and be noticed, even in staid, conservative periodicals.

- Alliteration—*Computer Commuter*
- Parody—*What a Difference a Computer Makes*
- Play on Words—*Only Your QC Department Knows*
- Rhymes—*Spy in the Sky*
- Coined Words—*Euphemania*
- Paradox—*Spend Dollars and Save Millions*

Write down a number of possible titles for each article and select a working title that meets these criteria.

OBTAIN COMPANY APPROVAL

Most companies have regulations that their employees must obtain permission before they publish articles, even if they are going to write about something not connected with their work. Before you query publishers, check your company's policies on this matter. Can you submit material directly to publications, or must you go through your firm's public relations or advertising departments? Is it necessary to have the material screened by your legal department for patent or confidentiality conflicts?

PERIODICAL ANALYSIS

Periodical Title ______________________________

Date (Mo/Yr) __________ How often Pub.? __________ Free or Sub __________

What's on cover? ______________________________

Editor's name ______________________________

Special Editors ______________________________

Editorial subject ______________________________

Table of Contents—No. of articles ______________________________

How many are staff written? __________ Free lanced? __________

Ads ______________________________

____

____

____

____

____

____

Comments ______________________________

____

____

____

Fig. 2-2. Periodical analysis form.

ARTICLE ANALYSIS

(Use a separate sheet for each article)

Article title ______________________________

Author ______________________________

Column inches per 100 words ______________ No. of inches ______________

No. of words ______________ Type of article ______________

Avg. Sent. Length ______________ Avg. Para Length ______________

Visuals: Quantity ______________ Types ______________

Photos ______________ Types ______________

Special visuals or special effects ______________________________

Comments ______________________________

____

____

____

____

Fig. 2-3. Article analysis form.

EXERCISES

1. Generate at least twenty uses for a paper clip.

2. Narrow your list of subjects down to three: one prime and two backups.

3. Narrow the list of periodicals you'd like to be published in down to three.

4. Review your periodicals and narrow your list to one. Keep the other two as backups in case your first choice doesn't work out.

5. For the selected periodical, analyze at least two issues of the periodical using the analysis form in FIG. 2-2 at the end of this chapter.

6. In your selected periodical, analyze at least three articles using the form in FIG. 2-3.

3

Query

He has half the deed done
who has made a beginning.

Horace

Two basic approaches to selling an idea for publication are:

1. Query first, then write the article.
2. Write the article, then send in the completed article.

The first method is by far the most practical, since you need invest only a small amount of your time in preliminary research and in writing a query letter (a brief proposal) to sell your idea. If the editor likes your concept, he or she gives you a tentative commitment that the periodical will buy your article if it lives up to their expectations.

Thus, you sell your article before you write it. That's what most professional writers do, and that's how most editors prefer to work with writers.

IT'S IMPORTANT TO SELL YOUR EDITOR

Before we get too deep into developing a query, let's take a look at the most important individual you will be dealing with in your quest for publication: the editor. Whether this person is an editor-in-chief, a managing editor, an associate editor, or an assistant editor, this is the individual you have to sell your article to. True, the editor may have technical experts on the periodical's staff to "referee" or review your article, but the editor is going to make that first important decision whether to even consider your query, and then, whether to publish you.

More than any one individual, the editor knows what a reader wants because, if he or she doesn't satisfy the reader, the publication ceases to exist.

The Editor's Job

An editor's job is not an easy one. I know whereof I speak because I served as an editor on two technical journals for a number of years. Writing columns, prodding authors, editing papers, corresponding, attending meetings, scheduling, budgeting, attending conferences, and a dozen other activities fill an editor's day to overflowing. Still, it is a rewarding job to guide scientists and engineers through all of the steps necessary to achieve publication.

Technical editors come from both sides of the spectrum: technical and editorial. A constant debate exists over which source contributes the best editors. Although I came from the technical side (I worked as a scientist and as an engineer for many years), my experience has been that excellent editors can come from either profession. As usual in any profession, it's the individual's ability and determination that count, much more than a background or previous occupation.

Editor's don't have a monopoly on technical articles and must keep their pipeline full of articles in various stages of completion to fill their periodicals. An editor might have to deal with as many as five to ten times more authors than the periodical can possibly publish and may end up with four to nine authors with bruised egos for each article published.

Most editorial offices operate on a panic schedule. An enormous amount of work must be accomplished in a limited time in dealing with advertisers, printers, authors, mailing lists, computer typesetters, electronic paste-up, illustrations, etc. to produce a periodical that must always be completed and mailed on a fixed calendar schedule.

So, before you write your query, consider the harried editors you're writing to and make that query short, succinct, exciting, and enticing enough to make that editor smile and conclude, "I just have to publish this article."

Why Query First?

In addition to saving your and the editor's valuable time, a query will benefit you in the following ways:

1. You will be able to determine if the subject matter of your article is suitable for that particular publication. Requirements vary considerably from time to time, and a query is a quick way to find out the current editorial needs.

2. If the periodical is entirely staff written, a query will quickly establish this fact and save your effort.
3. You only have to wait a week or two for a reply to a query. However, a periodical may take two to three months, or more, to evaluate a completed article and render a decision.
4. The editor, in his or her response, will send you the periodical's editorial guidelines regarding length, format, types and number of illustrations desired, etc., so you'll have all of this information available before you invest your valuable time in researching and writing a full-length article.
5. You'll have established some rapport with an editor that will serve you well for this and future submittals. Editors prefer repeaters; they cultivate writers who can continue to deliver good articles.
6. It's the professional way of conducting your writing business.

MAGAZINES TO QUERY

A large number of representative periodicals are listed in Appendix B. Other sources have comprehensive lists of technical publications and technical associations in all fields of technical activity. They not only list the periodical, they also provide information about the topics the periodical covers. Some of these references are:

- *Encyclopedia of Associations*—Gale Research Agency
- *Scientific, Engineering, and Medical Societies Publications in Print* —James Kyed and James Matarazzo
- *Standard Periodical Directory*
- *Ulrich's International Periodical Dictionary*—R. R. Bowker

If your library doesn't have the periodicals you want to review, your friendly librarian can probably arrange to borrow a few back issues from other libraries.

SIMULTANEOUS QUERIES?

Suppose you're not certain that a specific periodical will be receptive to your query, so you decide to write to a number of different periodicals at the same time, with the same article proposal, on the chance that at least one will buy your idea.

In a word: don't!

It's possible that more than one editor might like your idea and ask for a completed article. Then you're in trouble. You'll have to renege on your original promise to some of the editors. Editors who have been turned down have long memories and may be biased against you for future articles.

Play it safe and submit a query letter to only one periodical at a time. If your proposal is rejected, immediately submit your query to another periodical . . . until you find a receptive editor.

VANITY PUBLISHING?

Instead of having the periodical pay for your article, some editors actually have the audacity to insist that the author pay to be published in their periodical or journal. These appropriately named, ***vanity publishers*** are an insult to the profession. Such ego publications are not worth the paper they're printed on.

If an editor replies to your query with, "We'd love to publish your article; however, our editorial policy requires that an author must pay $100 a page for the privilege of being published in our periodical . . . ," simply reply, "No thanks," and send your query to an editor who is buying, or who will at least publish it at no charge to you.

Stay away from vanity publishing. If you're going to invest your valuable time and effort in researching and writing an article, you shouldn't have to pay to have it published.

Some periodicals, often technical journals, operate on a limited budget. Because they are the official journals of nonprofit technical societies, understandably they can't pay for articles. If you feel that exposure in these periodicals will enhance your career, this type of periodical may be your best choice. But absolutely refuse to pay to be published!

TELEPHONE OR WRITTEN QUERY?

You can contact an editor by telephone or by letter. (In the future you may be able to send the editor your query by modem. See chapter 9.) If you believe that you can effectively sell your article on the telephone, do so.

Although a telephone query has the advantage of giving you a quick answer to your query, I believe it's best to send an editor a written query. A written query gives you time to plan what you're going to say, and to say it the way you want to. Also, it's less of an imposition on a busy editor's time if he or she receives a written query that can be read and pondered at leisure and discussed with co-workers before a decision is made. Often such decisions must be made during staff conferences, involving a number of decision makers, so a written query would serve best.

HOW MUCH PRELIMINARY RESEARCH IS NEEDED?

The amount of preliminary research you need to do before you write a query depends on how much you know about your subject before you query. If you're writing on a topic that you've been working on for some time, you need only search your brain and organize this information to write an effective query.

However, if you're writing on a subject about which your knowledge is limited, you're going to have to search the brains of others in enough depth to write a query that can convince an editor that you can write an authoritative, full-length article and deliver it on schedule. For this preliminary search, it's best to select one or more sources that are authoritative and recent to make sure your information is accurate and up to date.

Shortly I'll present a typical query letter so you can see what it contains and gauge how much preliminary research you must do to sell your article to an editor.

Before we cover a query letter, however, you're going to be introduced to the fabulous four-part formula. This simple, but powerful formula should form a basic part of everything you write, whether it's a short memo, a report, an article, a chapter in a book, whatever . . . Once learned, you'll find that the four-part formula simplifies your writing, as well as makes it more saleable, more attractive to a reader, and easier and more enjoyable to write.

FOUR-PART FORMULA

This old and well-proven formula for effective writing is based on the fundamental way in which people react. The four-part formula, described in Walter S. Campbell's excellent book, *Writing Non-fiction,* is valid for everything you write and is, simply:

- HEY!
- YOU!
- SEE?
- SO!

HEY! The first part of the formula requires that you catch the reader's attention by using some interesting phrases or attention-getting statement, such as:

"Sex is America's second favorite sport. Sex is also America's favorite sport."

YOU! For this part of the formula, convince the reader that what you said in your opening and what you are going to say in the rest of your article, affects, and is of interest to him or her. For example:

"Do you realize that if it weren't for sex, you wouldn't be here today?"

SEE? Now that you have your reader's attention, you are ready to write the *body,* or main part, of your article. In this section you present your facts, your story to hold your reader's interest. To show the reader how you prove your thesis, demonstrate the points you want to make and deliver the message you have to give, such as:

"Among the million and one advantages of sex are . . ."

SO! In the final part of your article you convince your reader that he or she has profited from reading your communication. Leave your reader with some closing idea, thought, conclusion, summary, or a call to action of some kind:

"So, go out and get sexed. It'll clean the cobwebs out of your brain."

HOW TO USE THE FORMULA

Whether you realize it or not, you use the four-part formula many times a day. For example, suppose you see your co-worker, Clyde, walking down the hall at work:

"Clyde!" you call out. (HEY!)

Turning, Clyde walks over to you. "Yes?" he asks, frowning and tugging at his ear.

"How'd you like to go to the ball game tonight?" (YOU!)

"Who's playing?" Clyde responds, leaning against the wall.

"The Dodgers. I have an extra ticket. Pizzica is pitching and Sather is catching. Should be a super game!" (SEE!)

"Sounds great," Clyde smiles. "Okay."

"Pick you up at seven," you say. (SO!)

There! You've automatically used the four-part formula, without even having to think about it. It's a completely natural formula, based on the way people react to a situation. See how effective it is?

In commercial advertising, the four-part formula is also used extensively:

> "Our Hartford automobile tires are triple-steel-belted (HEY!) to keep you (YOU!) safe on the roads. Tested on rough, washboard roads and high-speed freeways, Hartford tires show no wear after five million miles (SEE?). They're on sale at your local dealer. Buy a set today. (SO!)

And to illustrate the use of the four-part formula in writing a simple memo:

TO: Jennifer Johnson
FROM: Quentin Czarnecki
SUBJECT: Lab tests on Solvent X-23 (HEY!)

The lab tests you (YOU!) performed last Friday were not documented properly. We need your complete report so we can distribute copies to Research. (SEE?)

Please complete the report by Friday, January 13, or you'll be pulled off the project. (SO!)

When you start using the HEY! YOU! SEE? SO! formula consciously, you'll find that you will create more interesting writing that is effective, concise, and easier, almost formulalike, to write.

REQUIREMENTS OF A QUERY

A successful query must:

1. Be brief. A maximum of one page. Editors are busy people and are prejudiced against writers who can't state their case briefly.
2. Hook the editor. You must convince the editor that your idea is a good one that will benefit the readers.
3. Summarize your article. Give enough information so the editor can make an intelligent evaluation of your idea.
4. Establish why you are the person to write the article. Outline your unique qualifications.
5. Include your proposed title and specify the tentative length, number of illustrations, etc.
6. Assure the editor that you've cleared the article subject matter with your company management, if that is required where you work.
7. Tell the editor how long after go-ahead you'll deliver the completed article.
8. Be typed, single-spaced, neat, error-free.
9. Address the editor or editorial staff member by name. Don't use "Dear Sir"—it's a tipoff that you haven't even bothered to look at a copy of their periodical to obtain the correct addressee.
10. Ask for a copy of their writer's guidelines.
11. Enclose a self-addressed, stamped envelope (SASE) for the reply.

AT LAST, A SAMPLE QUERY

Since one of the fundamental adages about writing is "Write about what you know best," I'm going to illustrate in FIG. 3-1 with a sample query to sell an editor on publishing an article about helping authors organize their material when they write for technical periodicals.

Query

31 February 1999

Wayne Brent, Editor
Advanced Technology
1313 Idaho Street
Waterloo, TX 77079

Dear Mr. Brent,

1. Engineers and scientists, who love order and precision in things, hate to outline because so far no logical system has been devised for outlining. Yet, everyone acknowledges that a good outline is essential to write a good article. And you, as an editor can recognize when one of your authors has used an effective outline to produce an excellent article. (HEY! YOU!)

2. Scientific professionals love to use formulas. All they have to do is select the right formula, input the proper values, turn the crank, and come out with the correct answer. (HEY!)

3. So, why not a formula for outlining? (HEY!)

4. That's what I've been developing over the past few years, an easy-to-use formula for outlining technical documents, ranging from simple memos, to proposals, to technical articles . . . even to books. I've named this formula the precedent sort. It promises to modernize many aspects of writing of things technical. (SEE?)

5. I'd like to write an article for you which I've titled HOW TO ORGANIZE TECHNICAL WRITING. I can cover the topic in about 2,000 words, complete with examples of how to use the precedent sort to organize technical material. I'll use four or five illustrations. This will be the first time this technique will appear in public print, so I know your readers will be interested in, and profit from, this topic. (SEE?)

6. As for my credentials, I'm a full-time free-lance writer, working in both the technical and commercial fields. I have a B.S. in Engineering and have worked as an electronic engineer in both design and program management, so I am very familiar with engineering writing requirements. I've had three non-fiction books published and have authored over two hundred technical manuals, proposals, plus numerous technical articles published in national periodicals. (SEE?)

7. Please send me a copy of your writer's guidelines. I'm enclosing a SASE.

8. I can deliver a completed article one month after your go-ahead. I'll look forward to your reaction to this exclusive submission. (SO!)

Sincerely,

Harley Bjelland
Free-Lance Writer
P.O. Box 1776
Any City, MN 54321

Fig. 3-1. Query letter.

Use of the Four-Part Formula

The paragraphs in the figure have been numbered simply for reference. They shouldn't be numbered in your query.

This query follows the basic four-part formula. The HEY! and YOU! parts are somewhat intermingled, as they usually are. By the end of the third paragraph, the editor should be hooked, wondering if there really could be a formula for such a subjective task as outlining.

Paragraphs four through six cover the SEE?. The last paragraph covers the SO!, making a commitment and asking the editor to respond. Note that the editor is being informed that this is an exclusive submission, a requirement some periodicals insist on.

Analysis of Sample Query

Let's go back to the Requirements of a Query to see how the sample query meets the criteria established.

1. Brief. The query is about one page long.
2. Hook. The editor should be hooked, his or her interest aroused, by the end of the third paragraph.
4. Credentials are established in paragraphs four and six.
5. The title, length, and number of illustrations are proposed in paragraph five.
6. It's not necessary for me in this case to obtain company clearance since I'm a free-lance writer.
7. A copy of the periodical's Writer's Guidelines is requested and a SASE is enclosed.
8. The last paragraph makes a commitment to deliver in 1 month.
9. The letter is typed, single-spaced, neat, and free of errors.

SOME MECHANICS

To help with the preliminary planning of your article, here are some basic recommendations on length, etc.

The specific periodical you're aiming for usually has certain length requirements, although these are often flexible. Most articles range from 1,000 to 3,000 words; about 2,000 to 2,500 is the average.

One thousand words is about four pages of double-spaced, typewritten material, so your typed draft should be between four and 12 pages long.

As far as illustrations are concerned, use at least one or two for even the shortest article, and perhaps five or six (count tables as illustrations) for longer articles. For some periodicals you need not worry about supplying finished artwork since they usually redraw everything you submit to meet their own specifications. Some periodicals, often professional society journals, require camera-ready copy. Your editor should answer this question in response to your query. Photographs can be reduced by your publisher as required.

A typical three-column, 8½-by-11-inch page of a periodical contains about 1,000 words, without illustrations, so you can estimate the number of finished pages for your article. If you use the recommended two visuals per periodical page, they'll occupy about one-third of the printed page, leaving about 700 words per page when two visuals are used. You can expect to receive from $25 to $100 per page, up to as much as several hundred dollars if you have an outstanding article.

EDITOR'S RESPONSE

For a query, you should hear from an editor in two to four weeks. If you haven't heard after about a month, write a short, polite note to jog your editor's memory, asking if a decision has been made on your article.

Your editor's response can be one of three types:

1. Your query has done its job and you're given a go-ahead to write a full-length article.
2. Your proposed article is accepted conditionally; that is, the editor makes some suggestions for modifying your article (e.g., length, slant, etc.) before you submit your completed article.
3. Your proposal is rejected. Unfortunately this is a cold fact of the writing profession. Don't take it personally. There are many reasons for rejecting an article that have nothing to do with the quality of the article.

If you receive one of the first two responses, go to work: accomplish your in-depth research and write your article. If you receive the third response, before you get too depressed, quickly write another query to another periodical and mail it immediately! That's why I recommended that you have at least two back-up periodicals in mind in case your first choice doesn't work out. There's nothing like the renewed hope of being accepted the next time out to assuage the disappointment generated by a rejection.

EXERCISES

1. Narrow your list of subjects down to one.

2. Conduct enough preliminary research to draft your query.
3. Draft a query letter to the periodical of your choice.

4

How to Research the Literature

Knowledge is of two kinds.
We know a subject ourselves,
or we know where we can find
information upon it.

Samuel Johnson

To paraphrase Samuel Johnson a bit, the next best thing to knowledge is knowing where to acquire knowledge. And the best repositories of knowledge are libraries and commercial data banks. A phenomenal amount of information is available in these storehouses of learning, but sometimes the sheer magnitude of them makes it difficult to find what you're looking for. This chapter will show you what to look for and how to find it in the most expeditious manner.

The first time you research the literature for your article, it should be a cursory look, preliminary research, just to make sure that the subject you have chosen has not been overdone and to determine if enough literature exists to form the basis for a full-length article. This first look will give you enough information to write a query letter to interest an editor. After you receive your go-ahead from your editor, you can conduct your in-depth research to collect the information needed to write your full-length article.

The principles discussed in this chapter will help you in both searches. The major difference is that you will probably spend less than an hour on your preliminary research and perhaps a few hours on your in-depth research.

WHAT RESEARCHERS NEED TO KNOW

Any researcher needs to know two things:

1. Where to look.
2. How long to look.

This chapter, along with appendix A, shows you where to look. The "how long" is up to you.

When you're about to write an article and want to know what others have done in a similar vein, you need to research the literature. Knowing where to look and what to look for can save you much valuable time.

Each year, skilled librarians and abstracters spend hundreds of thousands of hours categorizing, cataloging, indexing, referencing, and abstracting the thousands of periodicals and books that are published that year. This chapter provides some paths to follow to help you locate the specific answers you need in the books and periodicals in this veritable mountain of information.

Books are the most important reference for research, but they are often out of date. Periodicals contain the most up-to-date information, but they generally do not cover subjects in adequate depth.

A LITERATURE SEARCH

The reasons for conducting a literature search are many and varied:

1. When you review established works, you're often learning from the best brains in the field.
2. Often you'll get new ideas, new slants on your article as you review other articles. Your creativity will be stimulated when you read how someone else attacked a similar problem.
3. In today's increasingly competitive technological world, you must be aware of, and hopefully leap-frog, what your rivals have done and have written about.
4. Another aphorism, "Two heads are better than one," applies. When you read how someone else handled a problem, you're reaping the results of two great minds working on the same problem: yours and the author's.
5. Billions of dollars have been invested in the projects that spawned those articles and books. Why not take advantage of this enormous investment?

A tremendous amount of information is available in those mountains of literature. And it's free. All it takes is a little time and effort, and some help in finding the proper direction for searching.

HOW TO GET STARTED

Your first step should be to that fount of knowledge, the library—either your company library or a university, college, or good-sized public library. (In chapter 9, on-line data banks, which are rapidly becoming a major source of information, will be covered.) If you have trouble locating the right library, the *American Library Directory* (ALD), Jacques Press/R. R. Bowker Co. (2 volumes) lists over 35,000 U.S. and Canadian libraries of all kinds: public, academic, company, association, etc. The directory is arranged alphabetically by state and province, and summarizes each library's holdings.

The library you select might not have on hand all of the documents you're seeking, but they generally have most of the basic reference books, abstracts, and indexes you'll need. If they don't have the book, periodical, or report on hand, they can probably order it for you through an interlibrary loan.

Start by checking the card catalog, which indexes all the holdings of that particular library. Many of the more modern libraries are in the process of converting to a computerized system where you can access the library's holdings from a computer terminal. If you have a computer at home, you can access some of these public libraries from your home or office.

No matter which system you use, the basics of finding information are still the same, so let's start with an old-fashioned manual card catalog, then we'll cover a typical electronic card catalog.

THE CARD CATALOG

Arranged alphabetically, manual card catalogs have an Author card, a Title card, and one or more Subject cards for each holding in that library. The card catalog indexes books, reports, pamphlets, periodicals, handbooks, transactions, etc.

An Author card is illustrated in FIG. 4-1. The articles *a, the,* and *an* are ignored in alphabetizing. Two basic methods are used in alphabetizing: word by word, and letter by letter.

The word by word system	The letter by letter system
New York	Newark
Newark	New York

Delivery Decimal System

Many libraries use the Dewey Decimal System, in which books are catalogued from 000 to 999, according to subject:

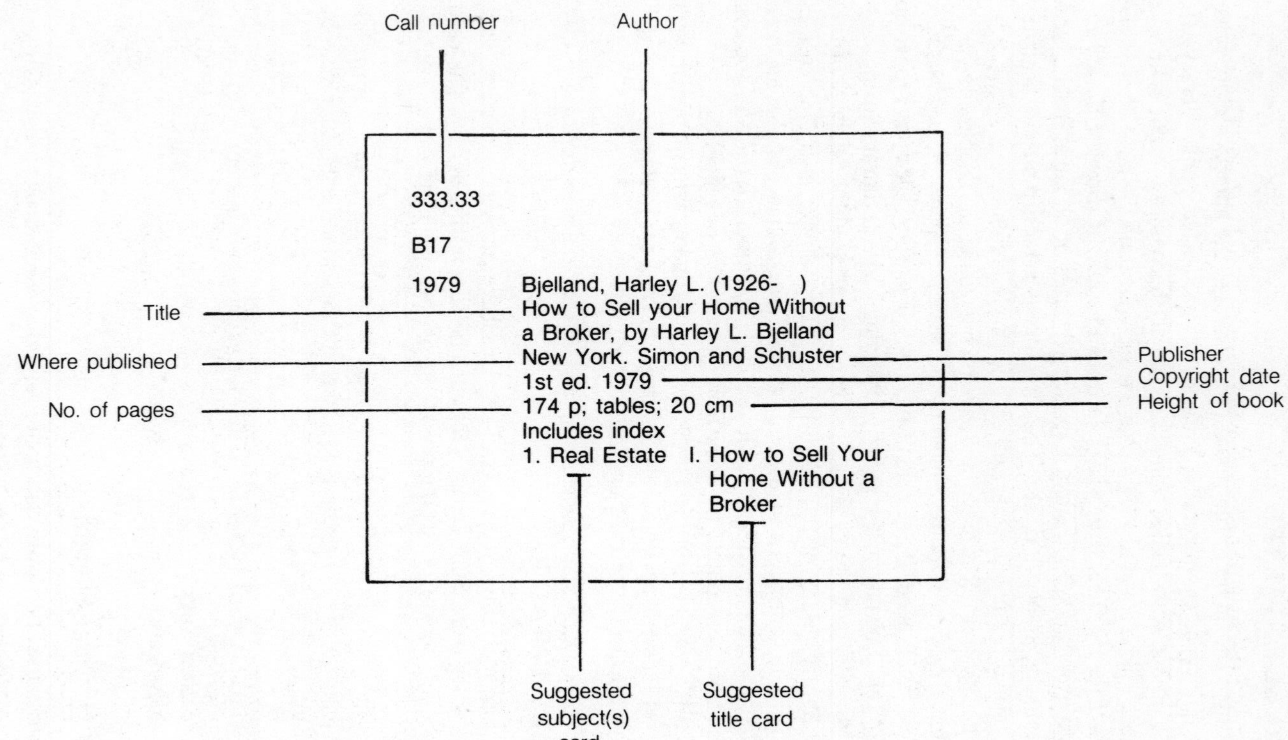

Fig. 4-1. Author card.

The Library of Congress system is also further subdivided, for example:

T	Technology
TA	Engineering. Civil engineering
TC	Hydraulic engineering
TD	Environmental technology. Sanitary engineering
TE	Highway Engineering. Roads and pavements
TF	Railroad engineering and operation, including street railways and subways
TG	Bridge engineering
TH	Building construction
TJ	Mechanical engineering
TK	Electrical engineering. Electronics. Nuclear engineering
TL	Motor vehicles
TN	Mining engineering
TP	Chemical technology
TR	Photography
TS	Manufactures
TT	Handicrafts. Arts and crafts.
TX	Home economics

How to Use the Card Catalog

You can search for your topic by any combination of subject, author, and title. When you locate something of interest, be sure to write down the *full call number,* the *title,* and the *author* of the book or periodical in your index card notes. This not only helps you locate the specific book on the shelves, it also gives you a general location to browse through on the nearby shelves for related books on the same subject. Also, if the book you want has been checked out, the librarian will need this information to reserve it for you.

ELECTRONIC CARD CATALOGS

Libraries are gradually shifting over to a computerized card catalog. Many are converting only their more recent acquisitions and are gradually adding their older acquisitions to their computer database. So the total conversion may take a number of years to be completed, but eventually the manual card catalog will be mostly a museum piece.

Computerization of the library's holdings offers a number of features not practical in a manual card system. You can usually find what you're looking for in a shorter time, and you can search by author, title, subject, and call number. Also, you can also search using wild cards when you know only part of a title, name, or subject. In addition, the computerized system can tell you if the book

is checked out, and if so, when it is due back—excellent features also not practical in a manual system.

As an example of the use of the computerized filing system, this is the procedure I used to call the Eugene, Oregon, library to obtain their on-line card catalog. My inputs are in boldface.

After my modem dialed the library's number, I received a message:

1200 connect

that told me that the connection had been made from my home to the library and that I would be conversing with the library at a 1200 baud rate. I hit my Enter key and the following series of conversations then took place:

USER NAME: **EPLTEL**
PASSWORD: (I TYPED IN THE PASSWORD)
WELCOME TO THE EUGENE PUBLIC LIBRARY VAX 11/750
YOU MAY FIND MATERIAL IN FIVE DIFFERENT WAYS:

BY TITLE	EXAMPLE	TI-HAWAII
BY AUTHOR AND TITLE	EXAMPLE	AT=MICHENER/HAWAII
BY SUBJECT	EXAMPLE	SU=HAWAII
BY CALL NUMBER	EXAMPLE	CD=919.69 C226

PLEASE TYPE IN YOUR INQUIRY AND PRESS THE RETURN KEY.

I typed in **SU=TECHNICAL WRITING**

THERE ARE 12 TITLES IN THE CATALOG FOR SUBJECT TECHNICAL WRITING. BELOW ARE LISTED THE FIRST 10 (To conserve space I've listed only three.)

1. Technical Writing (12)
2. See also English language—Technical English
3. See also Science news (2)
4. Etc.

To continue, SELECT ONE of the CHOICES BELOW.

I then input **1** as my choice and hit my **ENTER** key. I then received the following message:

1. 657.1 884	Crouch, William A Guide to Technical
2. 808.066 D661	Dodds, Robert H. Writing for technical Wri
3. 808.066658 H367h	Helgeson, Donald Handbook for Writing Pro

Press 1 to see more TITLES
Press 2 to display PREVIOUS SCREEN
Press 3 to display FULL BIBLIOGRAPHIC data for one of the above titles.
Press 4 to display LOCATION and AVAILABILITY of one of the above titles.
Press 5 to TERMINATE this inquiry.

I then input a 3 and a 1 and received the following message:

808.066658 Helgeson, Donald V.
H367h Handbook for writing technical proposals that win contracts/
Donald V. Helgeson..
Englewood Cliffs, N.J. : Prentice-Hall, c1985
xxii, 218 p. ; 25 cm.
Includes Index

Next, I wanted to check if the book was available and hit 2 to display the previous screen then I hit the 4 key and the following message appeared:

LIBRARY	LOCATION	VOLUME	AVAILABILITY
EUGENE	SHELVES		ON LOAN DUE BACK 06 MAY

Seeing that the book was checked out and that I had saved a trip to the library, I then logged off.

Whichever system is available to you, browse through either or both the computer or the manual card catalog for your first effort. It's a good place to start.

But the card catalog is only a beginning. Next broaden your search into other areas of interest for additional information.

OTHER SOURCES

To expand your search, consider all possible documents, even those not in your library. These documents may be classified as shown in FIG. 4-2.

General References

For a general, broad picture of a topic, to help orient yourself at the start of a literature search, general reference books are often a great help. Usually these books do not have up-to-date information, but they do give you good background material. They also list additional references to check. General reference books are listed in appendix A.

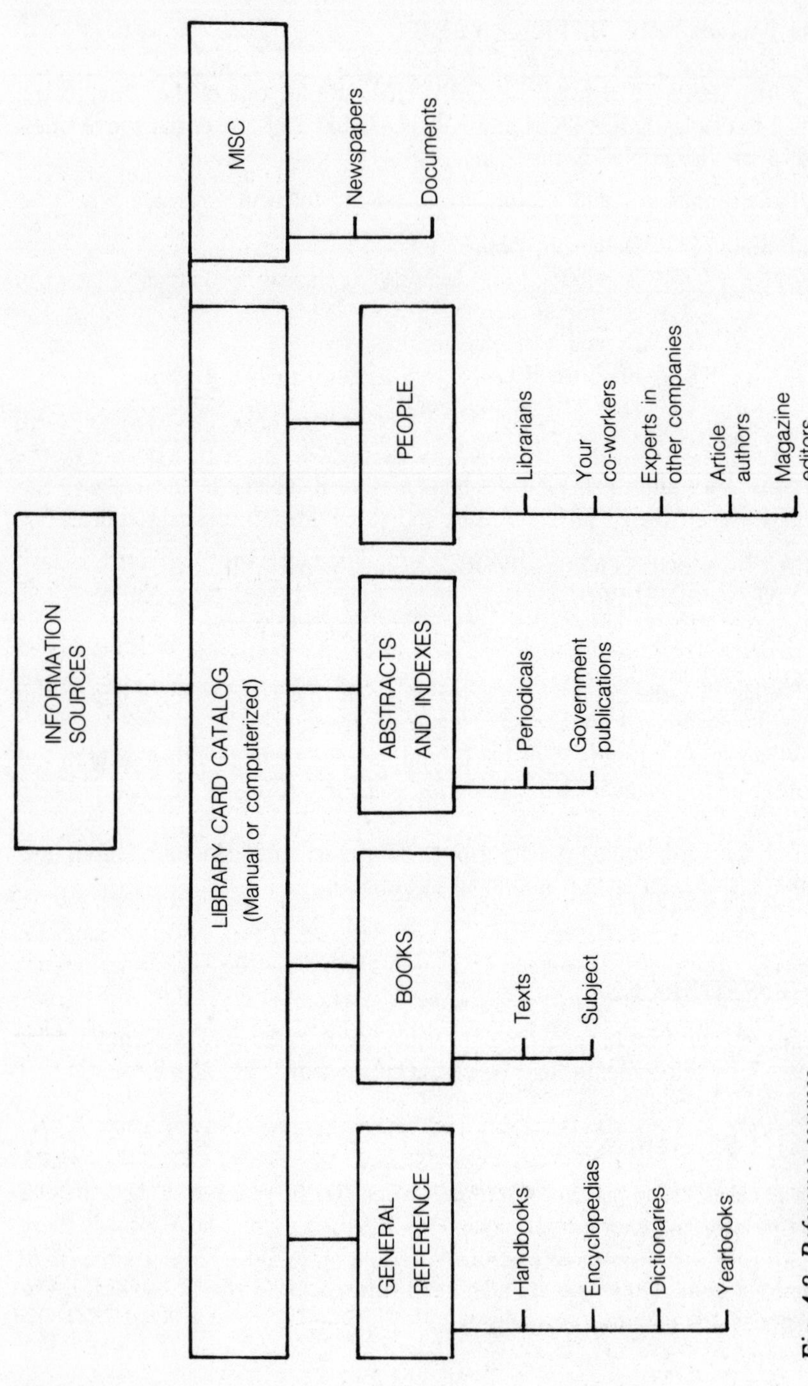

Fig. 4-2. Reference sources.

BIBLIOGRAPHIC REFERENCES

To list bibliographical references in your article, use the following format:

Books
Lastname, firstName. *title of book.* New York: Simon and Schuster, 1979.
Periodicals
Lastname, firstname. article title. *"periodical title."* 29 (June 1999): 1-4. (29 is volume number, and 1-4 are the pages the article is included on.)

Books on Reference Sources

If you have no idea where to start looking for references, you can begin with the following sources. They are listed alphabetically by title, followed by the publisher and years published.

A World Bibliography of Bibliographies, (1965-) Besterman. 5v. Includes 117,000 items, grouped under 16,000 headings and subheadings. A classified bibliography of separately published bibliographies of books, manuscripts, and patent abridgements. International in scope.

Bibliographic Index (1938-) Wilson. A cumulative bibliography of bibliographies. An alphabetical subject arrangement of separately published bibliographies, and of bibliographies included in books and periodicals. About 2,600 periodicals are examined regularly.

A Brief Guide to Sources of Scientific and Technical Information, (1980) Information Resources Press. A concise, selective guide to sources of information for the engineer and scientist. Lists principal science libraries in the United States.

Government Reference Books, (1970-) Libraries Unlimited. Annotated list of bibliographies, directories, dictionaries, statistical works, handbooks, almanacs, and similar reference sources published by the U.S. government.

Guide to Reference Books, Eugene P. Sheehy. Chicago; (1986-) American Library Association. Subject, title, and author index of sources in all fields. Special section on pure and applied sciences. An excellent, comprehensive guide. The best!

Guide to Reference Material-Vol. 1: Science and Technology A. J. Walford (ed.); (1973-77) The Library Association, London 3v. Excellent, comprehensive lists of reference material in science and technology. Worldwide listing, Emphasis on English-language materials.

How to Find Out About Engineering, (1972) Pergammon Press. 271 p. Guide to various sources in engineering and its many branches. Covers bibliographies, encyclopedias, dictionaries, use of libraries, and standard reference sources. Name and subject indices.

New York Times Guide to Reference Materials, (1985) Popular Library.

Shows where to look first when all you have to work with is a name. Covers a wide range of references; also has excellent sections on how to find information.

Science and Engineering Literature: A Guide to Reference Sources, (1980) Libraries Unlimited. General guide to sources of science and information fields. Over 1,200 reference sources covering basic reference books in general science, mathematics, physics, chemistry, computers, astronomy, geology, biology, engineering, and medicine.

HOW TO TAKE NOTES

Now that you know where to locate your information, you should adopt a good note-taking procedure. The method I use is described here.

The best medium on which to record your research notes is index cards. One side is normally lined; the other side is blank. They're stiff and easy to manipulate later when you organize them. They're available in 3-×-5, 4-×-6, and 5-×-8-inch sizes. Choose a size that's big enough to contain your notes and stick with that size. The 3-×-5 size is most commonly used.

Some general guidelines to observe when you take notes are:

1. Choose either the ruled or unruled side of the cards for writing your notes. (I prefer using the unruled side for my notes since I occasionally incorporate a simple figure or illustration along with my notes.) Whichever side you choose, be consistent—write your notes on only one side of the card.
2. Put only one idea, one topic, on a card. That way they'll be easier to organize later into topics and subtopics using the precedent sort. An *idea* or *topic* can be defined as any small amount of information that will not have to be broken up so that the parts can be placed at separate points in your outline. At the top of each card (see FIG. 4-3) put

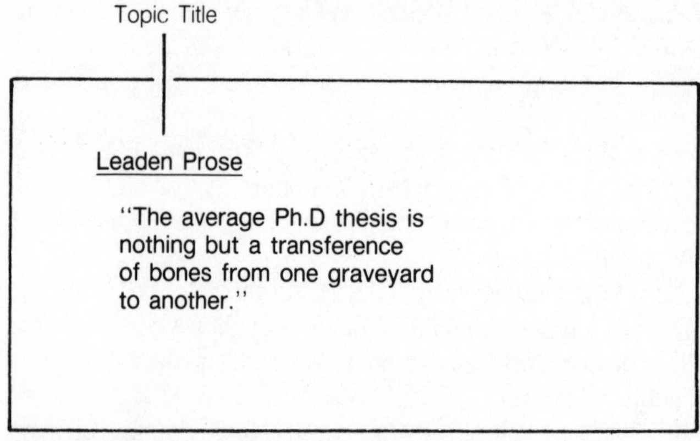

Fig. 4-3. Index card—topic title and notes.

a title, or topic, or subject, or heading or whatever you'll need for writing your notes and reviewing them later.

3. Use the opposite side of the card (see FIG. 4-4) to record bibliographic information. If the note is your own idea, record this fact on the bibliographic side. If you don't, when you've accumulated a lot of notes, you might later have difficulty remembering which ideas came from you and which came from the references.

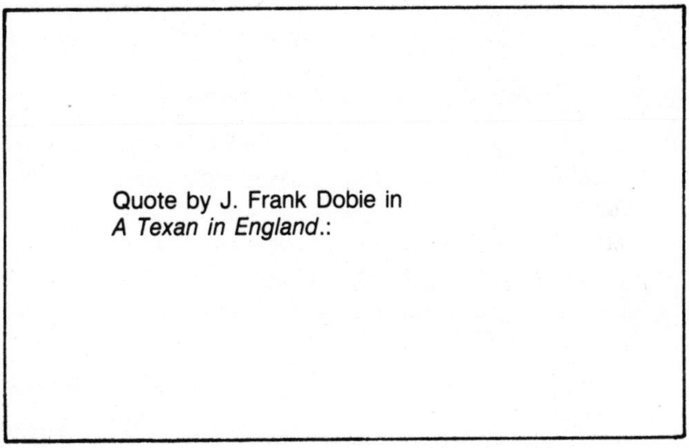

Quote by J. Frank Dobie in
A Texan in England.:

Fig. 4-4. Index card—bibliographic data.

4. Take many more notes than you need. Later you can prune them down to the essentials.
5. Be careful of using too much shorthand; you might forget what you meant by it. Record enough information to save yourself a trip back to the library.
6. Each card should have the following information as a minimum:
   a. The book or periodical reference
   b. Topic or subtopic title
   c. Your notes

7. Read through your references carefully to make sure you understand them, then digest and rewrite what you've read in your own words. Write down key ideas, words, and phrases.

When you do your research, you'll be taking three kinds of notes:

- Quotations
- Paraphrases
- Personal comments

Copy verbatim only if your reference has said something unique. You'll find that for 90 percent of the time you take notes, you will be paraphrasing the information. When it's in your own words, it suits your purposes better than the quotation, and you can usually shorten it.

If you must use a quotation, such as from an expert or well-known person, put quote marks around it, even if you paraphrase it later. Copy it exactly, including punctuation, spelling, capitalization, and paragraphing. If you omit part of the quotation that does not apply to your topic, indicate the omitted part with ellipses (. . .).

HOW TO REVIEW BOOKS QUICKLY

You're not going to have time to completely review all of the books you'd like to research, so there are some shortcuts that will give you enough information for your article:

1. Skim the book. Read the chapter titles and the bold headings, and look at the illustrations. They'll give you a quick summary of the book's contents.
2. Read the title page; the book may also have a subtitle. Note the author and his or her qualifications. Check the publication date to see how recently it's been published or if it's only a reprint of an earlier edition. Note the publisher. The book is more likely to be an authoritative reference if the publisher specializes in that type of book.
3. Read the foreword, preface, and introduction; they often give the purpose and intent of the book.
4. Review the table of contents; it outlines the book's contents.
5. Read the opening paragraphs of each chapter.
6. Check the index for topics of interest and see how many pages are devoted to these topics. If you find some topics that pique you, look them up, skim them, and make any notes of topics that interest you.
7. Check the bibliography. It will reveal the author's sources and whether he or she is using up-to-date information.

EXERCISES

1. Look up the periodicals of your choice in either Gales Research Cos. *Book Review Index* or *Ulrich's International Periodical Directory.* Record any pertinent information.

2. Review at least six books or articles in periodicals that cover a similar topic to the one you have chosen.

3. For these references, critique the published material. Why is your proposed article going to be better than those already published? Is yours more recent? Does yours provide a better coverage? Does it cover items not included in the other articles?

5

Organize Your Material Properly

Order and simplification are the first steps toward the mastery of a subject . . .

Thomas Mann

In a major city, a four-story, red-brick, office building had to be moved to a vacant lot twelve blocks away to make way for a freeway. Three movers were invited to a conference held by the building's owners. They presented their credentials to be awarded the job.

Ace Movers, the biggest of the three, won first chance. Ace sent their engineers and mechanics to the location and they immediately tried to lift the building with jacks and move it to its new site. Although they labored long and hard, the jacks were not sturdy enough, and broke each time they were used to try to elevate the four-story structure. So this method had to be abandoned.

Modern Movers, the next largest in size and the second choice, decided that thinking hard was better than brute force. They hired a group of college professors and computer consultants who were full of ideas and good intentions. They concentrated deeply on ways to move the building. They computed the forces needed to move the structure using large blimps to lift it and carry it away. But it would have taken 10,000 giant blimps and wire harnesses, which would cost too much. As their computers spewed out calculation after calculation, the learned men also considered helicopters and dozens of other methods to lift it up all at once. But their calculating, thinking, and talking exhausted them, and they finally theorized that it was absolutely impossible to move the building.

SIMPLIFY

Walter Bell, of Bell Movers, the tiniest of the moving companies, had the next chance. Walter told his engineers to "Simplify." No one really knew what he meant by that, but he called his few employees together and accompanied them to the building. They walked around it, measured it, calculated, and came up with a plan. Their detailed plan was to start at the very top, carry one brick at a time to the new location, and reassemble the building according to the detailed sketches they had made. It was a very simple solution and it took many trips to move the building, but before long the building glistened in its new location in the middle of the formerly vacant lot. Bell and all of his employees earned special bonuses for solving the problem so effectively.

Writing an article, or a story, or a book is a similar project. It's too big a job to jump into without planning. It's much too practical a job to only think about, much less to try to do by good intentions alone. Yet, by reducing it to a detailed and well-thought-out plan, a series of small steps, and doing each of these steps, one at a time, even a small person can do a big job.

That is precisely what a good outline, a logical ordering of your topics and ideas, accomplishes for you. Organization reduces a huge, seemingly insurmountable project into a series of small, interrelated steps, each of which can be done one at a time and be joined together, until the total writing project is completed.

BRAINSTORMING

By this time you should have accumulated a number of notes pertaining to your topic. Now that you have built up your confidence by amassing a lot of background information, it's time to proceed to the next vital step: brainstorming.

Brainstorming is a process where you think about your topic uncritically, without giving any thought to organization or judging an idea's worth. You open the floodgates of your mind and write down every idea you have about your subject, in whatever random order they occur to you. Let the ideas gush out in any form, in any order. Don't try to evaluate your material at this point. Creative thinking requires a positive attitude, so write down all of your ideas, knowing for certain that some of them will be good.

Judging too early is the big enemy of brainstorming and must be avoided or it might cause you to cut off the flow of ideas or disregard promising ideas. You can produce ten times as many ideas as you could if you paused to judge each one when it occurs. Defer your judgment or you might evaluate and discard ideas that do not seem good, but which, through association with your ideas, could give birth to new ideas.

Evaluate and organize your ideas later. To brainstorm, you must let your brain roam free, unfettered, unbiased, unorganized, uncensored.

To start your brainstorming session, get the set of note cards you've accumulated from your library research and procure an additional supply of blank note cards. Start reading through your note cards and let your mind do a little creative wandering as you read them. This activates your mental muscles, setting the marvelous process of idea association to work.

Each time you read one of your note cards, one or two more ideas or thoughts will pop into your mind. Jot them down on separate blank cards, one idea or thought per card. A huge quantity of ideas is needed because quantity helps breed quality. Build new ideas and modify the ideas you've already written down. Write down all your thoughts, no matter how wild and impractical they seem. You are writing them down for only yourself to see, so don't be embarrassed by what you write.

Asking yourself questions is often an excellent way to get started because questions are the creative acts of the mind. Read through these questions slowly and see which you can apply to your topics.

Ask yourself, "Can I . . .

- Use a different shape, a different form?
- Make it faster, less expensive, smaller, lighter?
- Add new features to make it more versatile?
- Combine it with something else?
- Reverse it or change the sequence?
- Find new applications for it?
- Use different ingredients to obtain new properties?
- Adapt ideas from a related product or process?
- Split it up into smaller parts?
- Use a different process?
- Merge old ideas?

Brainstorming is not a five-minute task. It might take an hour or two, perhaps longer if you have a complex subject to discuss. But this is prime time. It gives you the basis, the framework for your article—the points, the unique ideas, the thoughts you are going to cover. This is truly the most creative part of your writing; it is the task that can make your article an outstanding one.

Go through your old and new cards a few more times. More and more ideas will occur. Write them down, uncritically, and you'll be amazed at how the stack of cards has grown into a full deck.

When you feel your creative juices have finally run their course, set all of your notes aside and do something else for a few hours. Take a walk, shoot some baskets, hit a few golf balls, do something physical to rest your mind and let your subconscious go to work. If you can spare the time, sleep on it. Next you're going to be ready to form the skeleton of your article, to organize it, and to make order out of chaos.

ORGANIZATION

The most important single key to effective written communication is good organization! A document properly researched and organized is half written.

In a survey of technical writers—professionals who earn their living from writing—only 5 percent said they used no outline for their writing. Most likely some of the inept technical writing you have encountered was written by this 5 percent.

Most writers agree that one of the most difficult tasks is organizing their work. In a recent survey 1000 engineers were asked what bothered them the most about writing. Leading the list of dozens of complaints at 28 percent was organizing and outlining. Some people equate creating an outline to having a dentist perform a root canal. But this book promises the novocaine to take the pain out of outlining, and to make it into an enjoyable, logical, and profitable enterprise.

It's difficult to overemphasize the importance of a well-constructed, comprehensive outline. The more time you spend on creating and revising your outline, the easier your article will be to write. While your ideas are still in outline form, you can review, evaluate, modify, correct, rearrange, even totally reconstruct them much more readily than a written draft of an article. You can easily spot missing or inconsistent topics in an outline and correct them. And you can spot and eliminate topics that aren't necessary. An outline gives you a track to run on. It's a beacon that lights the way when the going gets rough.

Once you start putting words on paper, it is extremely difficult to totally rearrange your ideas and still maintain continuity. A completed rough draft almost sets the article's contents and order in concrete, requiring a jackhammer effort to change it. Detailed outlines encourage forethoughts, rather than afterthoughts.

This chapter takes that very difficult, even onerous task of organizing and shows you how to organize your thoughts according to sequence, importance, problem/solution, or whatever order you elect to use. Introduced in this chapter is the precedent sort (p-sort), a revolutionary, new technique that performs this vital task of organizing, making it almost formulalike in its simplicity, precision, and ease of use.

You'll find that once your ideas are properly organized into an outline, your words, sentences, paragraphs, and pages fall into place effortlessly. When you write, it's essential to know what you're going to say next. The precedent sort, or p-sort, gives you this direction.

The P-Sort

A comprehensive outline makes your article easier to write because once you know that you've covered all of the important points you need not worry about being distracted by thinking about your other topics. You can concentrate your efforts totally on the topic you're writing about.

The p-sort shows you how to establish this direction. It works for articles, for chapters in a novel or a nonfiction book, in technical writing, in a memo, in speeches, in all types of writing.

If someone hands you a poorly organized report, it's as if you received a dozen beakers of chemicals that comprise one of your company's products, with no directions on how to process or mix them. All the chemicals (facts) are there, but they are of no use unless they are processed (organized) properly.

That's what the p-sort does for you. It provides the directions to assemble the vital parts of your articles in the proper order; to form a unified useful product that other people can understand.

Basic Writing Steps

Two basic steps in writing, whether it's for an article, a letter, a proposal, a report, whatever, are:

1. Research and organize your ideas
2. Write and revise

Many people skip the vital first step, dive right in, and start writing. Invariably they end up thrashing about, drowning in a rambling, circumlocutious document that is difficult to write and even more difficult to read and comprehend. All of the facts might be there, but unless they are arranged in a logical order, they can't make their way from the published paper into the reader's mind.

Our brains simply are not geared to think in a logical, deductive, chronological order when we're creating something. Our minds hop about, generating random ideas loosely connected by some inexplicable, associative, magical mental process. Each individual thought might be a brilliant idea by itself, but unless they're strung together into some logical, orderly pattern, the result is a meaningless jumble to anyone else.

To demonstrate this phenomenon, try this experiment. Sit down and write everything that comes into your mind for about five minutes. Then read it back and see how your mind skipped wildly about and how little sense your unorganized thoughts would make to some stranger trying to read and understand what you just wrote down.

OUTLINES

Proper organization of your material before you write:

- Makes writing easier.
- Lets you focus all of your efforts on only one topic at a time.
- Helps you include all of the points you want to make in a logical sequence.
- Enables you to start at any point in your article since you know what is going to be included in the entire document.

A good outline is a strong skeleton that you can later flesh out with words to create a well-formed, complete body. If the skeleton is not assembled properly, the body will not operate smoothly, but in jerky, disconnected movements.

Most outlining techniques you have used have given you a lot of confusing theory, replete with Roman numerals, but with very little practical help on this important endeavor. You've probably read about, and have suffered through, the Inverted Pyramid, the Suspense Formula, the way to indent and count by Roman numerals, ad nauseam, but no technique has shown you a simple way to organize and join your topics together.

Over the past twenty-five years I have evaluated, developed, and tested numerous methods of organizing, and have adapted and modified the Bubble Sort method used in computer programming to organize material for writing. I've named this method the Precedent Sort, or p-sort. It's a systematic, logical way to organize a large group of randomly generated subjects into an orderly whole. It shows how to gather the scattered bones and put together a properly constructed skeleton.

Topic vs. Sentence Outlines

Before the precedent sort is covered, it's important to discuss the two basic categories of outline forms that are used:

- Topic outline
- Sentence Outline

A *topic outline* lists the topic to be covered in each specific part of an article. This type of outline can be used when the specific meaning of each topic is clearly understood. For example, a topic outline would use the word *operation.*

A *sentence outline* provides a more thorough definition of what the topic covers. For example, instead of the word *operation,* a sentence outline would form a complete thought such as, "How does it operate in a severe heat environment?" A sentence outline clarifies the topic and makes the final writing easier and faster. Often the sentence outline can form the logical topic sentences for paragraphs, or the key statements for entire strings of paragraphs.

Both topic and sentence forms can be combined effectively in developing an outline. Use whichever best suits your purposes.

Ordering Sequences

Before you begin to organize your material, you need to decide what type of ordering sequence you will use. Outlines can be sequenced in a number of possible ways:

- Chronological
- Spatial
- Increasing detail
- Major divisions
- Inductive/Deductive
- Known to unknown (analogy)
- Increasing or decreasing importance
- Cause/Effect
- Comparison/Contrast (Advantages/Disadvantages)
- Literary or suspense

You'll be using one or a combination of these sequences in writing your article.

A *chronologically ordered* outline reports a series of events by arranging the topics in the sequence in which they occurred. You write about the event or the topic that occurred first, then second, etc., until you have reached the last event or topic. This type of outline is suitable for trip reports, processes and operating procedures, progress reports, lab reports, analyses, etc. (Chronological ordering was used to arrange most of the chapters of this book in the order you should normally follow in writing an article.) Chronological ordering, however, does not provide proper emphasis because the most important parts of an article might be buried in the middle. Also, there is a tendency to include unnecessary information to make sure the article covers everything.

Spatial ordering organizes topics in geographical, or physical order. It may be used in describing a plant, sales and marketing reports by geographical areas, or a company organization by divisions and departments. This type of outline is also relatively easy to write, however, it also does not place proper emphasis on the most important topics. In addition, information of dubious value might have to be included for completeness.

An *order of decreasing detail* is used in journalism. Newspaper articles begin with the general information everyone wants to know. Subsequent paragraphs add more detail until the writer runs out of facts or the editor runs out of patience and limits the space allocated to the article. A news story has to be written in this manner so it can be chopped off after any paragraph to be squeezed into the newspaper space with the article still complete and the gist of the story still included. This type of outline can be used to present information of value to a broad technical audience of readers, with varying levels of interest and backgrounds, such as in engineering or market research reports with a wide circulation.

The *major divisions order* is used for a topic that easily divides itself up into natural and obvious parts. For example, a proposal or lab report is often divided into the specific major steps dictated by the standards of the company involved. A description of an organization naturally follows this order since it is based on a company organization chart, describing the functions of the president down to the lower levels, and the various divisions and departments of a company.

The purpose of the *inductive/deductive order* is to convince the reader that the conclusions in the report can be arrived at by logical induction or deduction. The inductive order begins with specific observations and leads to a general conclusion. You present evidence that you believe will support your conclusion. For example, you begin by saying that ABC Company's chemical products are inferior, then add that their petroleum products are inferior, then continue to add more negative evidence, and conclude that ABC should be banned from all manufacturing. Specific instances of the inferiority of some of the company's products lead to a general ban of all the company's products.

The deductive order is the opposite and proceeds from general observations to a specific conclusion. You state a general proposition or viewpoint, then present evidence to support it. For example, you might start out by stating that all foreign automobiles are unreliable, then proceed to point out examples where this unreliability has been demonstrated so that the overwhelming weight of the evidence you present adds up to confirm your proposition.

When the material involves controversy, the inductive, or facts-to-conclusion, method is more convincing than the deductive. If you have a difficult selling job, use the inductive order. Induction leads the reader gently to your conclusion. The evidence is presented to gradually lead to a concluding statement, and can be accepted at face value by even a doubting reader. It makes the writer look unbiased, although you really aren't. You simply don't present evidence to counter your proposition and conclusion.

On the other hand, the deductive order risks alienating the reader who, resisting the initial statement, may view even the strongest evidence in a negative light. This order helps readers get to the point quickly, however, and saves time for top managers by putting the recommendations up front. With the recommendations incorporated at the beginning of the document, the rest of the document is used to justify the conclusions.

The known to unknown order is used in documents such as technical manuals when explaining new equipment or a theory of operation. The writer starts with facts the reader understands, then bridges to another fact that can be deduced easily from the first fact, and continues to link facts until the reader is informed of, and understands, the unknown. Often an analogy will help with this method. This method is often called the *professorial,* or *teaching,* method.

Items or the topic can also be arranged in *order of increasing or decreasing importance,* depending on the desired effect. An order of increasing importance, typical of mystery stories and scientific reports, presents the least important information first, then builds to a conclusion by presenting the most important information—the results—last. If done properly, this method can capture and keep a reader's interest all the way to the end. But it is also difficult to use because few topics lend themselves to an order of increasing emphasis.

An order of decreasing importance is typical of reports that start with a long abstract, or summary of conclusions and recommendations. This order puts the most important facts or data first, to give the reader a quick review of points, then continues down the importance ladder to the bottom rung, the least important. This order is preferred by managers who want to read the results first. However, because of the decreasing level of interest, many readers will drop off on the way to the end of the article.

Cause/effect is used to explain what forces or events produced particular results. And, if the effect is not a desired one, such as in a problem/solution, the conclusion is a recommendation of what should be done to prevent a recurrence of the wrong result. It is also used for proposals and feasibility studies.

Comparison/contrast compares or contrasts two or more items, such as leasing costs versus purchasing costs or the performance of one company's products against another's, using an advantages/disadvantages comparison. This type of order is repetitious, however, unless comparisons and contrasts are done efficiently through the use of tables and graphs, rather than text.

The *literary* or *suspense* formula reveals a little information at a time and builds up a suspense to a hopefully unexpected and entertaining climax at the end of the article. Humor in nonfiction uses this order effectively.

WHICH ONE SHOULD YOU USE?

Depending on the purpose of your article, you can organize your material in one or more specific ways:

- *To inform, document, or entertain*
  - Chronological
  - Spatial
  - Major Divisions
  - Increasing Detail
  - Increasing or Decreasing Importance
  - Literary or Suspense

- *To convince or to get action*
    - Inductive or Deductive
    - Problem/Solution

- *To explain or to get understanding*
    - Known to Unknown
    - Comparison/Contrast

THE PRECEDENT SORT

Art is nothing without form.
Gustave Flaubert

Without some methodology, it's difficult to evaluate twenty to fifty or more randomly listed topics simultaneously and decide, by trial and error, the order in which they should be arranged. It's like the kid in the candy store, who said, "It's easy to decide if I want to buy either chocolate drops or peanut brittle, but when they add jelly beans, salt water taffy, and caramels, I get all confused."

The precedent sort (p-sort) puts order into this task. It breaks down a complex organizing task into a series of one-on-one decisions and shows you how to compare topics or ideas, one pair at a time, and select the topic or idea that has precedence; that is, the one which should come first. You continue until all comparisons are made and *violà,* your organization is complete.

Now that you have ten to fifty topics listed on your index cards, you can see the difficulty in trying to sort them into some meaningful order, unless some system is available to assist you.

Principle of Fewness

The Principle of Fewness states: the fewer items that are presented as a group, the easier they are to evaluate and understand. So the task is to reduce this group to as small a group as possible, one that can be comprehended and dealt with easily. Enter the p-sort. The p-sort optimizes the Principle of Fewness by reducing the number to be compared to a minimum: two. You compare only two items at a time, evaluating one against the other to determine which you should cover first in your article.

An Example The p-sort is best illustrated by a simple example. This example will use numbers to illustrate the principle. Next an example of writing a technical article will illustrate the application of the p-sort to a practical writing problem. I recommend that you take a set of 3-×-5 cards, write the listed numbers on them, and stack them in the order shown here. That way you'll obtain some good experience with the technique.

Your set of five 3-×-5 cards should each have a number on it and be arranged in the following order:

Card A	1941
Card B	187
Card C	1001
Card D	456
Card E	26

Note that it is not necessary for you to identify the individual index cards with letters as used in this example. I used the letters only so I can refer to them in the text when explaining the p-sort process.

Your task is to arrange these numbers in ascending order. You probably could do this quicker by sorting the index cards by hand, but this example has been made short and simple to illustrate the basic principles of the p-sort.

First, place all of the cards faceup in a stack and pick up the first card, Card A. With your other hand, pick up Card B and compare the two, as shown in FIG. 5-1.

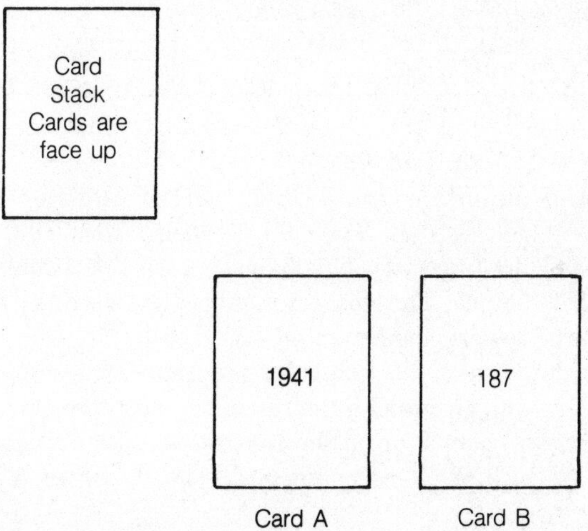

Fig. 5-1. First comparison.

Card B has a smaller number than Card A, so place Card A facedown in a new pile we'll call the Unordered pile and hold Card B. You are going to work your way through the entire stack a number of times. If the next card you pick up is lower in value than the one you're holding, you'll place the higher numbered card facedown on the Unordered pile and continue to hold the lower-numbered card.

To continue the first round, pick up the next card from the top of the stack and compare Card B with Card C, as shown in FIG. 5-2. Since 187 is smaller than 1001, continue holding Card B. Place Card C facedown in the Unordered pile on top of Card A.

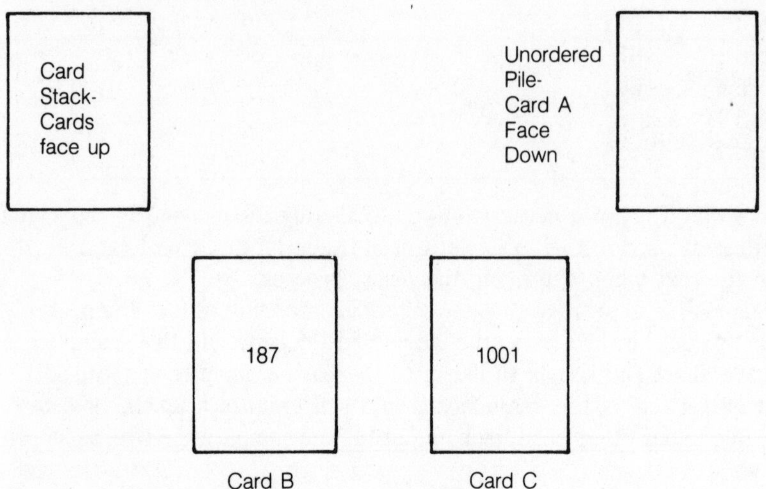

Fig. 5-2. Second comparison of first round.

Pick up the next card from the stack and compare Card B with Card D. Again, 187 is smaller than 456, so place Card D facedown on the Unordered pile and continue to hold Card B.

To complete the first round, pick up the last card from the stack and compare Card B with Card E. Since 26 is smaller than 187, place Card B facedown on the Unordered pile. Finally place Card E facedown in a new pile we'll call the Ordered pile. The smallest number (26) has now "bubbled up" to the top and is the lowest number of all the cards.

This procedure is akin to a computer programming sorting routine called the *bubble sort.* As you go through the remaining unsorted cards, successively smaller numbers will bubble up to the top and be placed facedown on top of Card E in the ordered pile. One number will join the ordered pile after each round of comparisons.

To start the second round, turn the unordered pile faceup, pick up the first card, Card A, and compare it with Card C, as shown in FIG. 5-3. Since 1001 is smaller than 1941, place Card A facedown on a new Unordered pile and pick up Card C.

Compare Card C with Card D, as shown in FIG. 5-4. Since 456 is smaller than 1001, place Card C facedown on the Unordered pile and hold Card D. Continue the process until you have compared all the cards. At the end of the second round, you should be holding Card B. Place Card B facedown on top of Card E on the Ordered pile. So far two cards, Card E-26 and Card B-187, have bubbled up to the top.

To start the next round, turn the Unordered pile faceup again and repeat the procedure until you have sorted all the cards. For five cards (n=5), this should require n-1, or 5-1 = 4 rounds of comparison.

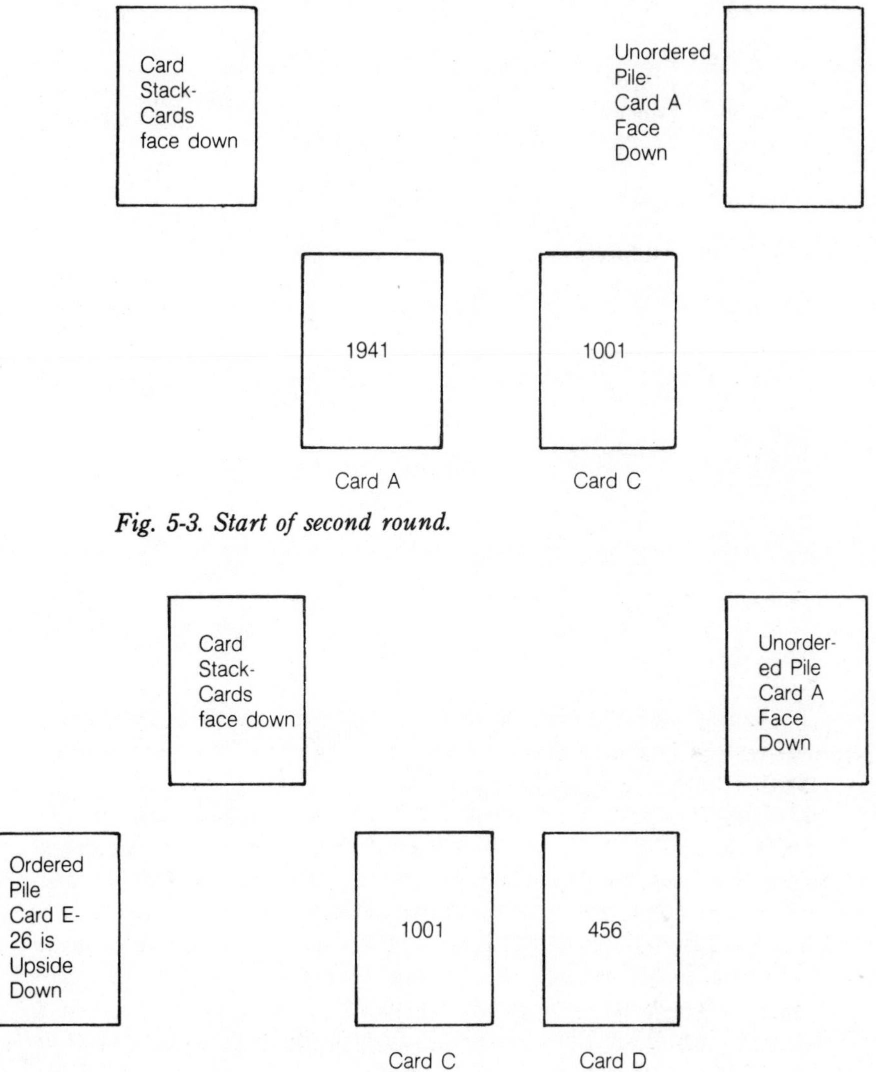

Fig. 5-3. Start of second round.

Fig. 5-4. Second comparison, Round 2.

The final order is shown, left to right, in FIG. 5-5.

A More Practical Example For the next example, an inductive order of organizing will be used, arranging the topics for an article in a periodical to convince the reader to buy your product. Assume you are going to write an article for a periodical about a new device you have just developed for your company. The topics you want to cover include the following. These were written down in a random order and will be organized using the p-sort.

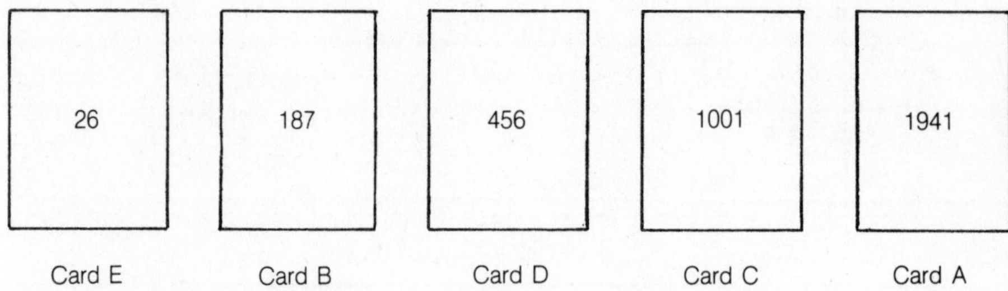

Fig. 5-5. Final sorted order.

Topic A	Applications
Topic B	Price
Topic C	How does it work?
Topic D	Reliability
Topic E	Future developments of product
Topic F	Self test
Topic G	Background of product
Topic H	Field test results
Topic I	Models available
Topic J	Modularity

For more practice on the p-sort, write down each topic from FIG. 5-6 on a separate 3-X-5 index card. You'll end up with ten cards, each with a single topic written on each.

To organize the topics in the order you're going to use, compare the topics listed in FIG. 5-5 one pair at a time, and decide which topic you should logically discuss first in your article. To start the procedure, put all of the cards faceup in a stack, then pick up and hold the first card, Topic A—Applications. Pick up Card B from the stack and compare the Topic A card with the Topic B card. Logically Topic A—Applications should be covered before Topic B—Price can be established because price depends on the application, so keep holding the Topic A card and place the Topic B card facedown to start an Unordered pile.

Pick up the next card on the stack and compare the Topic A card with the Topic C card. Logically, Topic C—How does it work? should come before Topic A—Applications can be understood, so place the Topic A card facedown on top of the B card on the Unordered pile and hold the Topic C Card.

To continue, pick the next card off the stack and compare the Topic C card with the Topic D card. Logically, Topic C—How does it work? should be known before Topic D—Reliability can be established, so place the Topic D card facedown on the Unordered pile and hold the Topic C card.

Continue this process, always placing the card bearing the topic that should come later facedown in the Unordered pile and holding the card with the topic that should be covered first. When you reach the end of the stack, you should have Topic G—Background of product in your hand. All the other cards will be facedown in the Unordered pile. So, at the end of this round, place the Topic G—Background of product card facedown in a new pile, the Ordered pile. Topic G has bubbled up to the top and has become the first subject that will be discussed in your article.

To start the second round of comparisons, turn the Unordered pile faceup and pick up the first card, Topic B—Price. Pick up the next card from the stack and compare Topic B with Topic A. You should conclude that Topic A must be discussed before Topic B can be established, so place the Topic B card facedown to start a new Unordered Pile and continue to hold Card A.

Next, compare Card A with Card D. You should conclude that Topic A should be covered before Topic D, since reliability depends on the application. So place the Topic D card facedown on the Unordered pile and continue to hold Card A.

Continue going through the stack. By the time you reach the end of the second round of comparisons, you should be holding the Topic C card. Place this card facedown on top of the Topic G card on the Ordered pile. Now you have two topics, Card G and Card C, that have bubbled up to the top to become the first two topics you will discuss in your article.

The first-to-last order I came up with after all of the topics had been compared and bubbled up to the top follows:

Topic G	Background of product
Topic C	How does it work?
Topic A	Applications
Topic J	Modularity
Topic F	Self test
Topic H	Field test results
Topic D	Reliability
Topic I	Models available
Topic B	Price
Topic E	Future developments of product

Note that although this is a "formula," you must still make subjective judgments in all the pairings, and your opinion of what should be covered first might differ from mine. So you see that this technique has room for individual creativity and that's precisely why it is so universal in its applications.

The basic principle of the p-sort is to put blinders on—to compare only two topics at a time, then concentrate and select the one topic you should cover first in your article. Thus the method of organization is broken down from the mind-boggling, complex problem of simultaneously trying to compare and organize twenty to fifty topics, to one of comparing only two at a time.

Spend all of the time you need in making the comparisons—minutes, hours, or longer if needed. If your organization doesn't come out the way you like it after going through all the cards, start over and repeat the process as many times as you need to come up with an order that satisfies you. You'll know you have the right order when you can go through all the cards and not need to change the order of any of them.

You can use the same technique to organize:

- Chapters in a book
- Topics within each chapter
- Events in short story
- Topics in a technical manual or report
- Events in a novel

Filtering Your Ideas

Now that you have all your cards in order, you're in a position to filter out unnecessary information. Up to now you have deliberately restrained or suppressed your critical judgment and evaluation of your ideas. Now it's time to be critical, to make judgments, to filter your notes.

To prepare to judge, find a quiet corner and go through your note cards, one by one. As you read through them, it will soon become obvious that some of the notes don't belong in your article. If, as you review them, you are suddenly jolted by a note that doesn't seem to belong, set that card aside and continue reviewing. Remember, you were encouraged to take more notes than you needed, so don't be afraid to filter the surplus ones out. Continue going through your note cards, setting aside any that don't belong.

Go through your note cards again and see if they read better this time. If you again have a jolt in continuity, perhaps you should put one or more of the cards you discarded back in your Ordered stack. If you feel something is missing, make a note of it on one of your blank cards and insert it where it's needed.

Repeat this process a few times until the notes form the pattern you want—until they're right, with no jolts in continuity, no missing steps.

Remember, organizing is not a static process; it's a dynamic one. So continue to make any changes you feel are necessary at any time. Your index cards are ideally suited for this procedure. You can move them around, add to or subtract from them, and totally reorder them until you are completely satisfied with your organization.

Outline Formats

Most people with whom I've discussed outlines have very unpleasant memories of the Roman-numeral outlines that were force-fed them in high school and college. So we're not going to dredge up those unpleasant memories. We're going to devise a simple outline procedure that you can adapt to your purposes. This technique uses no Roman numerals whatsoever.

By this time you should have your note cards arranged in the proper order, so it's time to place them all in a single row, glance through them to see how they look, then arrange them according to topics, subtopics, etc.

Line up all your cards in one or more vertical rows. Then, by simply moving some of them to the right and left, arrange them so that the subtopic cards are physically indented from the topics and that the sub-subtopics are physically indented further from the subtopics, etc. This will give a physical outline of how they all line up and are subordinated.

The best method is to study each card in order and decide if it has the same *weight,* or the same importance, as the topic above it. If the lower card has the same importance, don't move it. If it is a subject of less importance than the one above it, physically move it to the right. If a topic is of more importance than the one above it, physically move the topic card to the left. This physical method of indenting your topics gives you an excellent overall picture of your entire outline.

The system I use at this point, once I have my cards laid out in a physical outline, is to type these topic titles into my word processor, using the following coding and indentations. I copied only the topic titles from my cards.

ORGANIZE YOUR MATERIAL

<u>Movers and Shakers (Anecdote)</u>

<u>Brainstorming</u>

-Uncritical thinking
-Review note cards

--Add new ideas as they occur
--Modify existing ideas as required

-Not a five-minute task
-Continue until creative juices cease to flow

<u>Organization</u>

-Single, most important key to good writing
-Difficult task

--28% say most difficult of all writing tasks

-Time spent in organizing saves writing time

--Less confusing
--Spot items left out
--Spot surplus items
--Encourage forethoughts, rather than afterthoughts
--Simplify writing

So, for an article in which the topics and subtopics are limited, a simple code suffices. I underline a major topic, indent and use one dash for a subtopic, and indent further and use two dashes for a sub-subtopic. It's a simple but effective way to outline.

At this point in your article you can transfer the topic titles from the top of your note cards into your word processor or to a few sheets of paper, complete with the proper subordination. Don't copy all of the notes on the index card; copy just the topic title at the top of your card. Now you have what we'll call a *working outline* of your article completed.

Next?

The p-sort lets you order any number of topics (you can expand the list of topics up to perhaps fifty or more) in any desired sequence, which makes your writing flow logically and easily from one topic to the next. It's much easier to concentrate on and compare only two topics at a time and decide which comes first, rather than trying to evaluate fifty randomly arranged topics at once and trying to arrange them by trial and error.

You can use this valuable technique to organize the subjects in a memo, the topics in a long technical report, or the subjects in a proposal. It's a powerful technique. The more you use it, the easier it becomes.

Guidelines in Using the P-Sort

Some important guidelines to observe when using the p-sort:

1. Proceed logically from the top to the bottom of the Unordered pile. Don't jump around; your results will be confusing and the comparisons will be more difficult.
2. Don't decide that it's not important which topic comes first in a pairing. Make a judgment for each comparison, weighing all of the factors as best you can. If all else fails, just guess, but you must make a decision or the process won't come out right. If they're that close, it probably doesn't matter which comes first. Besides, if you change your mind on the order later, it's easy to modify.
3. Before starting to make your comparisons by pairs, make sure you understand precisely what each topic is and means. This is very important. If you're not sure, write out a statement or two, or an entire paragraph, on the card to accurately define each topic before you start your comparisons. There is no restriction on what you can write on a card, so define it so that you know exactly what it means.
4. To organize a huge list of items, say fifty or more, first sort the index cards into smaller groups of related items. Then take each group, one at a time, and order each group by itself using the p-sort. Finally, use a master p-sort to organize the order of the groups.

A Caution

Don't conclude that the p-sort is a mechanical, computerlike way of organizing topics because it isn't. You still must use your judgment, your creativity, just as you do when you use all formulas. The p-sort simply helps you break down a huge, perplexing problem into a series of simple, orderly, logical decisions, then helps you organize the results.

Also, the outline you have created here need not be cast in iron. If there are compelling reasons to change the order around, do so. It's a lot easier to work with and manipulate an outline after it's "almost organized," than when you're trying to organize fifty randomly listed topics.

OTHER APPLICATIONS OF THE P-SORT

You can use the p-sort for many other applications such as how to:

1. Schedule tasks, itemizing the order in which jobs must be accomplished to reach a goal in a logical, efficient manner.
2. Rank the topics in a sales brochure, listing the most important features first.
3. Arrange a large group of items in numerical, alphabetical, or chronological order.
4. Assign jobs. Use two precedent sorts for this application: one set to rank your employees in order of their competence, the second to rank the jobs to be done in their order of difficulty. Finally, assign your best employee to your hardest job, your second best to the second-most difficult task, etc.
5. List your own jobs to be done in your daily life, in a required time or spatial sequence, with the hardest jobs first, or in whatever order you choose.
6. Arrange ideas that must be sold or justified in order of difficulty, such as in a proposal where you should spend the most time discussing difficult problems that are to be solved.
7. Arrange a big decision into a series of little decisions that must be settled before the big decision must be made.
8. Judge contests, ranging from speeches (ranking the speakers) to a cake-baking competition, to beauty contests.

The next time you have a document, an article, a book, a report, or a proposal to organize, use the p-sort. You'll find it to be a capital idea.

EXERCISES

1. Why should you outline your articles before you write?

2. Get a set of 3-X-5 cards (or whatever size you prefer) and use the p-sort to alphabetize this list:

   Chemist
   Mechanical Engineer
   Architect
   Executive
   Scientist
   Writer
   Technician
   Civil Engineer
   Physician
   Programmer

3. How many passes did it take to organize the group in Question 2?

4. Organize the topics of your article using the p-sort.

6

Visuals

A picture shows me at a glance
what it takes dozens of pages
of a book to expound.

Turgenev

Good visuals can make or break an article. People understand and remember images and pictures, not abstract words. Visuals can clarify images too complex to be conveyed by language. Imagine the impossibility of fully describing a Van Gogh painting in words alone.

It's important to tentatively plan what visuals you're going to use *before* you write your text so that the visuals form an integral part of your article, rather than an appendage tacked on as an afterthought. Visuals must mesh with your text, like two gears that drive a machine. They must work in concert, each dependent on the other, to describe an object, a process, or a concept.

Because readers generally do not study visuals in detail, the text accompanying the visual should include enough information to help the reader understand the significance of the visual.

You can modify, add to, or delete visuals later when you write to a plan. It's much easier to modify a plan than to work with no plan at all.

WHY USE VISUALS?

Visuals—a term encompassing line drawings, graphs, photographs, pie charts, curves, and tables—serve many functions in an article:

1. They are often used to explain difficult concepts that are impossible to put into words. With text each word is read in series, one word after the other, and it takes many words to convey a thought or to create a complete "picture" in a reader's mind. With a visual, the whole picture is available at once, in parallel, and the reader can see how the various components of the picture relate to each other in size, in position, and in function.
2. Visuals provide eye relief, a needed rest for the eyes, by breaking up the solid, imposing text of an article with white space and a different media of presenting information. This pause helps readers better absorb the content of your article.
3. Technical people are accustomed to visuals and spend much of their professional life working with engineering drawings, curves, graphs, tabular data, etc.
4. Technical subjects generally deal with concrete things, which can usually be better explained with visuals.
5. Visuals increase reader retention of your article since visuals stick in your reader's mind much longer than words. It's been proven that we generally remember only 10 percent of what we read. However, if a visual accompanies the text, we remember 30 percent of a written message. That's a 300 percent increase!
6. Everyone who sees well-drawn visuals sees the same "picture." But words are abstractions and mean different things to different people. If you use the simple word *tree,* one person might visualize a huge, spreading maple tree, another might think of a Christmas tree, and a third might picture a scrub oak.
7. Less effort is required to understand visuals. With text your reader is forced to follow the writer's sequence of words, thoughts, and the manner in which the writer composed them. With a visual, a reader can start anywhere he or she likes.
8. Visuals save words.

SYNERGISM

When visuals are used effectively, along with the proper text, the phenomenon of synergism results. *Synergism* is the combined impact of separate inputs that, when effectively combined, increase each other's effectiveness. The total effect is greater than the sum of their individual effects; in other words one plus one equals three in this case. Your material can be three times more effective when visuals and text are properly combined. To achieve synergism, the visuals must be tied to the text so the two complement each other. If they don't, the theory "a picture is worth a thousand words" does not apply.

GENERAL CONSIDERATIONS

Once you have your article fairly well outlined, it's time to plan your visuals. Although the number of visuals can vary considerably from one type of article to another, a rough guideline is to use two visuals for each 1,000-word periodical page. Each visual typically replaces about 150 words of text, resulting in about 700 words of text and two illustrations per periodical page.

Note that all visuals, except tables, are generally assigned consecutive figure numbers and are referenced in the texts.

Visuals for technical periodicals are printed in widths ranging from about 2 inches to about 6 inches. Most visuals will be only one column wide. Publications use either two- or three-column widths. A three-column width is about 2 inches; a two-column width is about 3¼ inches. So make sure the visuals you choose will still be legible if they are reduced to 2 to 3¼ inches in width.

Each visual should carry only one idea, in simple form. Complex visuals defeat their own purpose. Each visual must also have a specific title that tells what the illustration is all about. But don't use more than ten words in your title.

Avoid color visuals. They're expensive to compose and convert for use, and most technical periodicals refuse to use them.

Later, when you're writing the text of your article and are unable to describe something, ask yourself, "Could I do this better with a visual?"

TYPES OF VISUALS

Following are the kinds of visuals commonly used in articles, ranked in order of the acceptability, usefulness, and ability to present the maximum information:

1. Line drawings
2. Exploded views
3. Photographs
4. Gazintas
5. Drawing trees, Organizational charts, Flowcharts
6. Graphs, Curves
7. Bar charts, Pie charts
8. Tables
9. Cartoons

The applications, advantages, and disadvantages of each of these visual types will be covered in the sections that follow.

Line Drawings

A line drawing is simply an artist's concept of an object constructed with lines, as shown in FIG. 6-1. The line drawing is the most useful of all visuals when you have to show how an object looks, what its innards are composed of, how it works, etc.

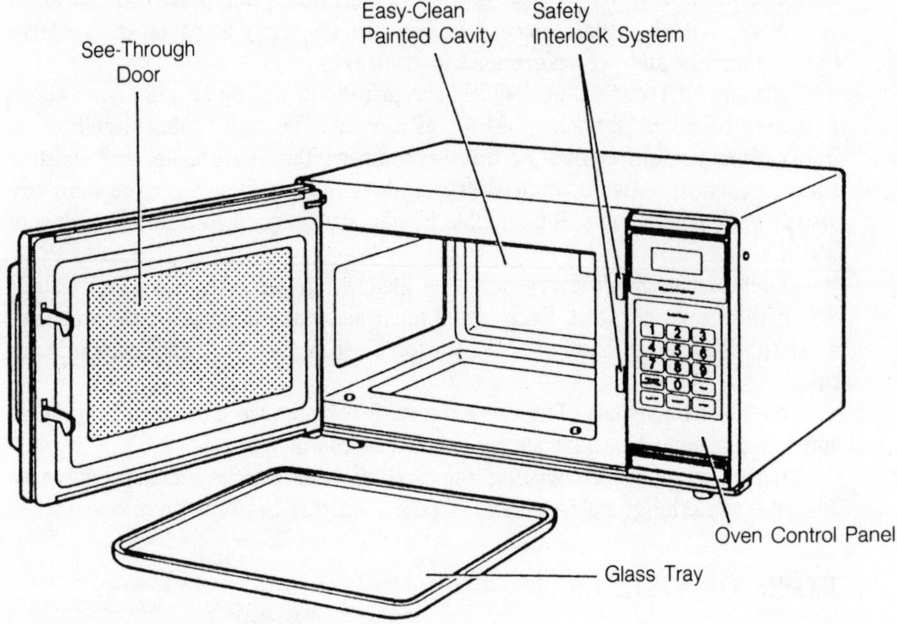

Fig. 6-1. Line drawing.

Line drawings can be restricted to include only what the originator wants to, omitting extraneous details. Cutaways can show the inner workings of the device.

Although line drawings are expensive to make, they can be created from engineering drawings, from the object itself, from photos of the object, or completely from the artist's creative imagination. Also, a wide variety of artist's aids are available from art stores and large stationery stores to help create the simpler visuals. Such items as templates, paste-down lettering, graphic material, and stencils considerably simplify the construction of line drawings.

If you intend to use line drawings in your article, however, check with your editor first. It's possible that the periodical does not want to invest the money required for a complex line drawing, and you may have to supply camera-ready artwork for such an illustration.

Exploded Views

Exploded views are so-called because they appear to be a stop-action photograph taken of an object when it is exploding apart. As you can see from FIG. 6-2, the exploded view is an excellent visual when you need to illustrate the inner workings of an object. However, exploded views are complex, and very costly to draw. Exploded views can illustrate the basic components that make up an assembly and can show how something is assembled or disassembled.

Photographs

When it comes to illustrating precisely how an object looks, nothing is better than the realism of a photograph, as depicted in FIG. 6-3. Photographs show exact details, but this feature could be a disadvantage if there are some components you don't want to show. Photographs give much greater credibility than a drawing because the reader knows the object being photographed actually exists and was not constructed from some artist's imagination. Photos provide drama and informational impact. They usually cost much less than a line drawing or an exploded view, too; however, they can only show surface views of objects. They can't show parts not visible from the exterior as an exploded view can.

To be photographed, an object must be available and in a position where it can be photographed, either in a studio or on site. Photos can be touched up to eliminate some clutter and extraneous information, but only to a limited extent.

Polaroid photos and photos taken with a low-quality camera are marginal for publication. Most editors refuse to use them. Periodicals generally prefer 8-×-10 glossies. They must be mailed in a protective envelope since a creased photo is useless. Never use paper clips on photographs because they can damage the glossy surface. Some editors will accept negatives; however, you may not be able to get them back.

To illustrate size on a photograph, use a ruler, a hand, or another familiar object in the photo. Make sure all photos are sharply focused, are properly cropped to get rid of distracting objects, and have plain, uncluttered backgrounds that do not draw attention from the object you're depicting. Because most visuals in an article are only one or two columns wide, don't expect to use up too much of a periodical page for your photos.

Gazintas

Gazintas, derived from the expression "goes into," are visuals that show the vertical and/or horizontal hierarchy of an object, an idea, an article, an organization, etc. The two types of drawings included in the Gazinta category are drawing trees, and block diagrams.

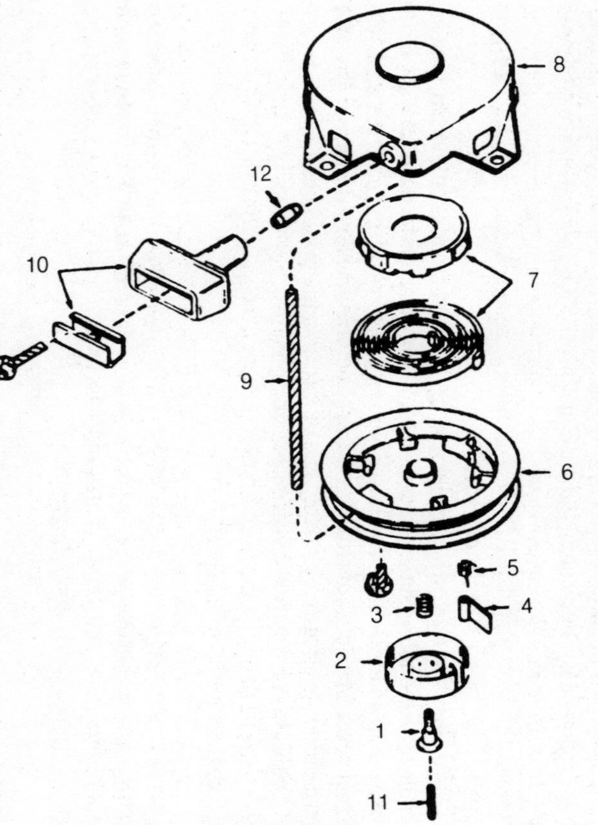

Fig. 6-2. Exploded view.

Ref. No.	Part No.	Part Name
	590420A	Starter, Rewind
1	590409A	Screw, Retainer
2	590410	Retainer, R.H.
3	590411	Spring, Brake
4	590148	Dog, Starter
5	590412	Spring, R.H. dog
6	590413A	Pulley
7	590414	Spring & Keeper Assy.
8	590536	Housing Assy., Starter (Incl. No. 1)
9	590636	Rope, Starter
10	590387	Handle Assy., Starter
11	590549	Pin, Centering
12	34986	Stop & Staple, Rope (Not used on all models)

(COURTESY HEWLETT-PACKARD)

Fig. 6-3. Photograph.

Drawing Trees

Drawing trees, as represented in FIG. 6-4, show the subassemblies that make up an assembly and the assemblies that make up an object. (A company's organizational chart is a form of drawing tree.) The tree in FIG. 6-4 is expanded only partially for purposes of illustration.

This house (top level) is composed of walls, a foundation, a roof, windows, and doors. The walls (Level 1), in turn, are subdivided into their Level 2 parts; studs, insulation, stucco, and siding.

In a drawing tree, all subassemblies of the same relative importance (or size) are on the same vertical level. As you progress from the bottom to the top of the tree, the subassemblies become larger, more complex, and fewer in number, until finally the top assembly, "house," is a single entity composed of all the subassemblies listed below it.

You can use this same technique to diagram the components of a jet plane, a book, or an article. A drawing tree gives a quick representation of the parts

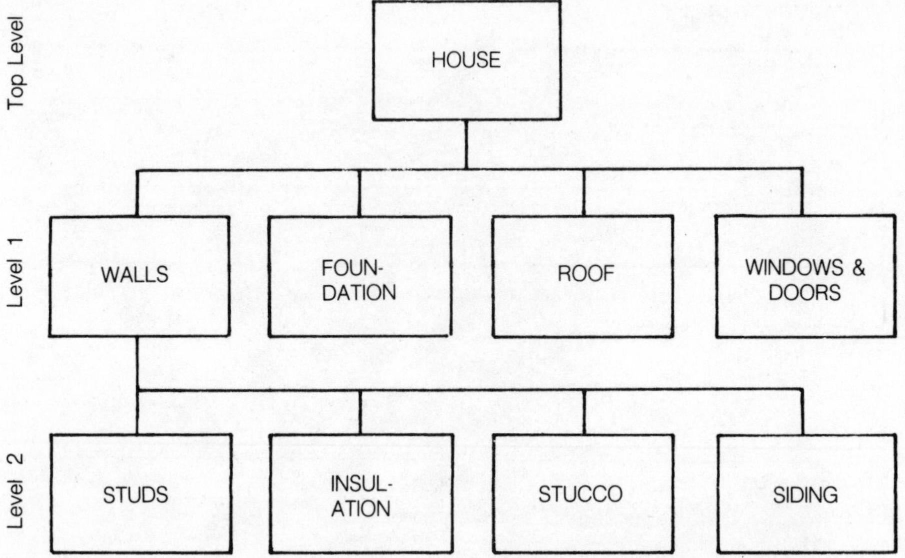

Fig. 6-4. Drawing tree of house.

that make up the whole, their relationship to each other, and the way they are assembled.

Block Diagrams

Another important form of the gazinta is the block diagram. In engineering and science, as in life, most things are not static—they move. From a speeding jet plane to a microcomputer, to a complex chemical reaction, most engineering and science is based on *dynamics,* or objects and materials in action. The block diagram is one of the most often used visuals to show how things react with each other.

Figure 6-5 is a simplified block diagram of an automobile. Turning the key (dotted line denotes an electrical connection) activates a relay on the starter, which in turn applies full battery voltage (dotted line for electrical interconnection) to the starter, rotating the starter motor. The starter is mechanically connected to the engine (a solid line indicates a mechanical link) and cranks the engine. The engine pumps gas from the gas tank via a gas line (double solid line indicates a pipe), igniting the gas, and the engine begins to run.

The engine, mechanically connected to the alternator, rotates the alternator. The alternator is electrically connected to the battery and begins to recharge the battery. Simultaneously the engine is mechanically connected to the transmission. When the transmission is activated, it engages the wheels, which are then turned and the car moves forward.

The block diagram is a very versatile visual for showing the interactions

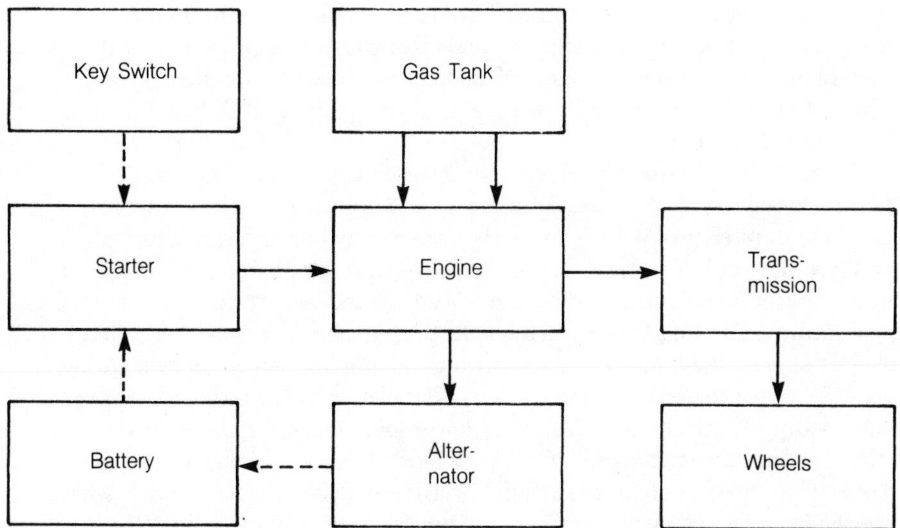

Fig. 6-5. Block diagram of an automobile.

among a number of items, whether these items are electrical, mechanical, chemical, or a combination of the three.

A pictorial form that is easy to comprehend, the block diagram shows how things interact, without showing the details of the individual components. For example, FIG. 6-5 does not show the details of how the starter uses the electrical power it receives from the battery. At this level of detail, it's necessary only to show the overall operation of the starter. Lower-level block diagrams would show the details of how the starter functions. It's confusing if you mix different levels of detail in a given block diagram.

One very important advantage of using block diagrams in your articles is that they are very easy to draw. Anyone with a straightedge can draw a presentable block diagram, and many of the simpler ones can be constructed using a word processor. Block diagrams are a powerful visual for explaining how things react, both internally and to external forces.

Guidelines for Block Diagrams

Over a period of twenty-five years, I have created hundreds of block diagrams and consider them the most useful of all visuals for any type of technical writing. Based on this experience, I have developed a number of rules and guidelines to use when creating block diagrams. They are:

Limit the number of blocks Use no more than 8 to 10 blocks in any single block diagram. Nothing can overwhelm and discourage your reader more than to be suddenly confronted with a maze of 20 to 50 blocks on one diagram, all with criss-crossing lines. The mind simply cannot grasp such a huge amount of confusing information at one time. Break your diagrams into groups of 10 blocks or less, giving your reader drawings that can be comprehended without a migraine. Also, since your editor will undoubtedly reduce your drawing to one that will fit in one or two columns on a magazine page simple block diagrams are a must for articles.

Use 8½-X-11-inch paper Use 8½-X-11-inch paper or smaller because it disciplines you to use no more than 8 to 10 blocks in a diagram.

Use conventional flow Use the conventional direction of information or signal flow of left to right, top to bottom—the same direction you use to read all printed information. Use arrowheads to indicate direction.

Reduce the Fog factor To reduce the fog factor, or confusion, in block diagrams, keep the number of lines down to an absolute minimum by showing only the major actions and interactions. Detailed interfaces can be covered much more effectively in tables, or in lower-level block diagrams.

Use functional names Give each block a short, functional name of no more than three or four meaningful words—a label that describes what specific function(s) it performs. Don't use nondescriptive titles such as Valve A. Call it e.g., Flow Control Valve, so its function is easily understood.

Use proper labels It's extremely important to make sure that the inputs and outputs that interact between separate block diagrams use precisely the same name. It's very confusing if you designate a flow input as Purge Air in one diagram and Cleanout Air in another. Your reader shouldn't have to stop and try to figure out if they're the same. Use the exact, same name for both inputs, and use the exact same name when you refer to them in the accompanying text.

Don't mix Levels Keep the functional level the same within a specific block diagram. Your reader's mind has been geared to think along a specific functional level for each diagram and it's confusing if you indicate, for example, a spark plug on an overall diagram of an entire automobile.

Alternate linkages/interfaces Indicate the primary flow with solid lines, and secondary flow within the same diagram with another code, such as dashed lines. Use different types of coded lines to distinguish other types of flow.

Text Guidelines

When you write the text that accompanies the block diagrams, the following guidelines will help:

Use Identical names Use the identical names in the text that you use in your block diagrams. I'm a fanatic about this because I've found it to be one of the single most confusing deficiencies in technical writing. When you call a block a CPU one time in the text, a computer the next time, and a processor the third time, you're planting unnecessary confusion. Your reader has a difficult enough time trying to comprehend technical concepts without having to pause and try to figure out if the writer is referring to the same part of a computer. The same guideline holds when you label input and output flow from one diagram to another. Designate them by the same name in the accompanying text to avoid confusion.

Define Terms Define abbreviations and terms the first time you use them. For example, "The *R*andom *A*ccess *M*emory (RAM), provides storage for lost bits."

Follow the Flow Direction When you write the text that describes the functions of the blocks, use an order of presentation that follows the direction of flow. First describe the input, then how the various blocks massage and control this flow. Carry the flow from left to right, all the way through until it is finally terminated, stored, or output.

Graphs

If you need to show how one or more continuous (or extrapolated) variable(s) change with respect to another, or if you need to show trends or minimums/maximums, you can best illustrate them with a graph.

A graph makes it much easier for the reader to grasp information than if it were printed in a table. A graph is a picture, rather than a set of numbers. Most graphs are two-dimensional, having a horizontal, or X-, axis and a vertical, or Y-, axis. Three-dimensional (X-Y-Z) graphs are used occasionally to illustrate more complex relationships.

Line graphs are accurate, but their data-presentation capability isn't as precise as for tables. However, communicating trends, relationships, etc. is often more important than showing exact data points.

To emphasize a curve or trend, choose a vertical scale such that the slope of the curve will be 45 degrees or greater. The idea of movement or trend is best emphasized by the steepness of a line and minimized by the flatness of a line. If your graph contains only one curve, you can add another dimension by coding or cross-hatching the area under the curve. If you use more than three lines, however, the graph can become cluttered and confusing and difficult to interpret.

When constructing a graph, label all coordinates simply. Use a minimum number of grid lines and plotted lines. Label curves directly, rather than using a key. Don't run the curves to the ends of the grids; leave a 10 to 15 percent space all around.

Line Graphs

One of the most popular of all graphs is the line graph illustrated in FIG. 6-6. It's ideal for showing a dramatic side-by-side comparison of two variables: the independent on the X-axis, and the dependent on the Y-axis.

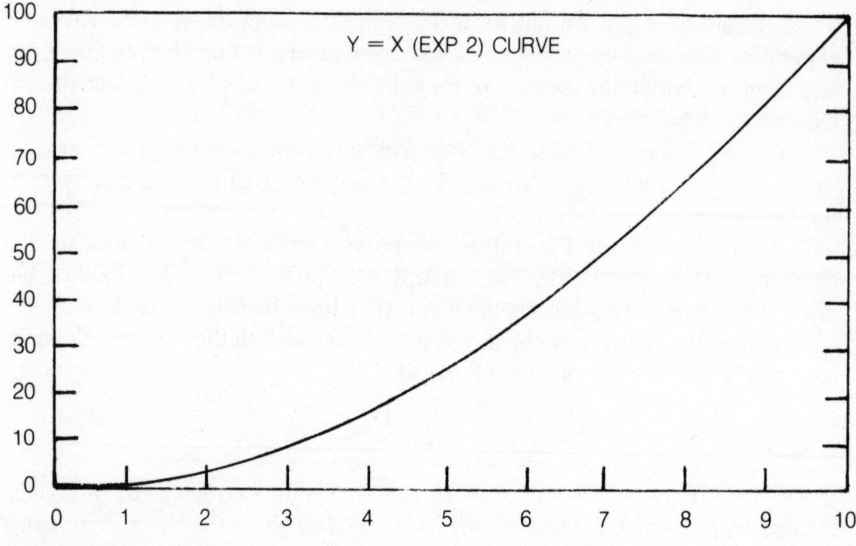

Fig. 6-6. Line graph.

Pie Charts

Pie charts provide an easy-to-understand and dramatic picture of the parts that constitute the whole, as depicted in FIG. 6-7. A pie chart gives a quick look and comparison of an item's parts. It shows proportions and the relative size of related quantities. But a pie chart is not accurate; you can't tell the difference between 29 percent and 32 percent, for example. If accuracy is required, label the percentages on or near each slice.

Don't cut the pie up into more than six slices, though. Additionally, make the smallest slice of the pie greater than 2 percent to maintain effectiveness.

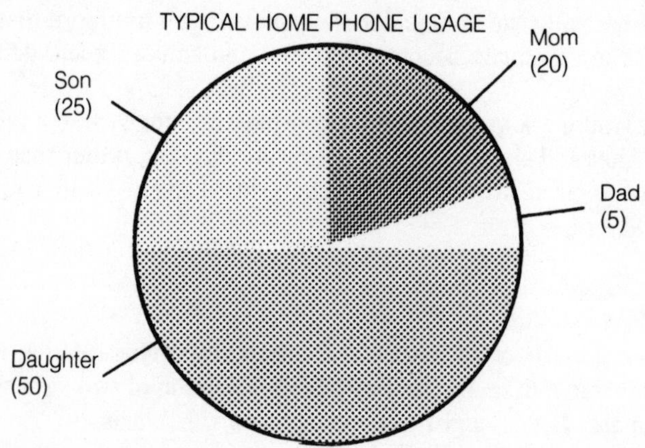

Fig. 6-7. Pie chart.

For emphasis, you can shade individual elements, or cross-hatch them to code the various parts. Make sure you label the various slices with their categories, as well as the percentages of the whole they represent. Also check to make sure all of the slices add up to 100 percent.

Bar Charts

Bar charts are a pictorial method of displaying tabular data when no mathematical relationship exists among the variables. These charts are particularly useful when you need to dramatically illustrate relationships between different sets of data. They give a quick picture and are easier to make comparisons with than pie charts. However, if the bars are too short or too long, they lose their effectiveness.

As illustrated in FIG. 6-8, it's best to separate the individual bars for clarity.

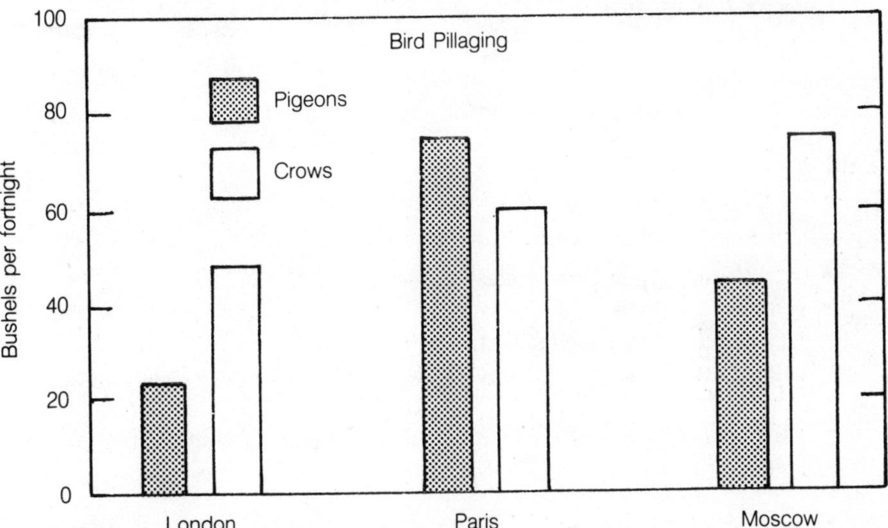

Fig. 6-8. Bar graph.

Use bars of the same width and the same spacing so they are attractive and easy to understand at first glance. You can plot the bars vertically or horizontally. Use vertical bars if you're plotting the type of data that people expect to see vertically, such as temperature, weight, etc. Horizontal bars are suitable for distance, time, speed, etc. You can subdivide the bars by shading or cross-hatching to add another dimension of information (FIG. 6-9).

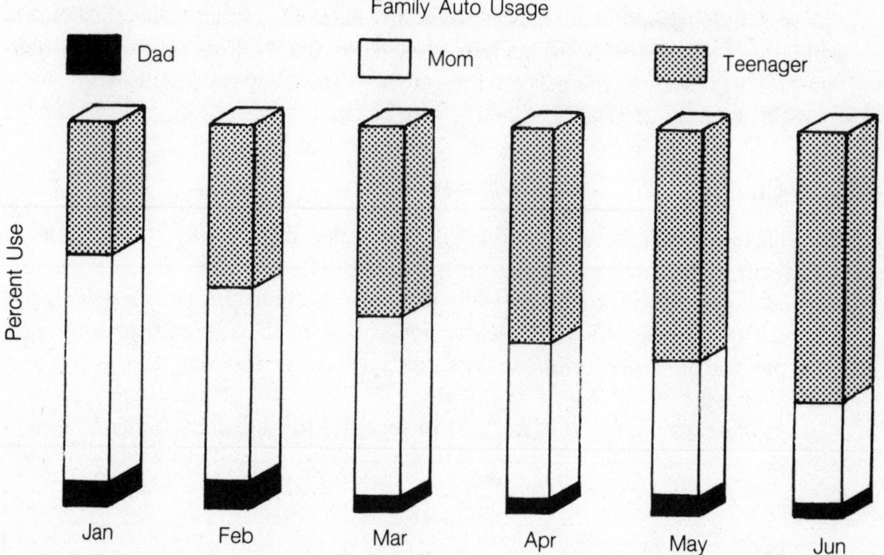

Fig. 6-9. Enhanced bar graph.

Tables

Use a table when:

- It will make a point more clearly than text can.
- You want to discuss the data in the table and draw conclusions from this data in the accompanying text.
- Putting the data in the text would take at least three times as much page space.
- The data in the table should not be duplicated in the accompanying text.
- The data in the table must also be accurate.

If there's no other way to present your information or if you have a lot of data to present and you want to show how they relate to each other, you probably can present it better in a graph. A table is less desirable than a graph because your readers must try to formulate a picture in their minds based on the numbers in the table. But if you need to present numerical data with a high degree of accuracy, a graph won't work. Instead, use a table. Make sure you discuss the figures in the table in the accompanying text to help with their interpretation.

Informal Table

An informal table, because it has no lines or ruled forms, can be included in the middle of text. It can be used to list numbers, products, etc. in vertical columns for visual clarity and quick reference. An informal table must be brief and simple. It is not identified by a table number and normally does not have a title. As an example of an informal table:

The percentage concentration of various chemicals in smog is:

Oxygen	4%
Nitrogen	6%
Carbon Dioxide	90%

The data in the informal table can be discussed in the text that follows it.

Formal Tables

A formal table requires a table number, a table name, and a minimum of internal and external lines to simplify reading the data. Columns and subcolumns should have headings that identify the nature of the data. Where needed, include units of measure, such as ounces, length in feet, percentages, etc. in column headings.

The use of a formal table, as shown in TABLE 6-1, eliminates the cumbersome repetition of units (ft/sec/sec), and the tabular arrangement helps your reader compare data easily and quickly. Often your readers won't be able to grasp the significant differences among quantities when they're presented in paragraph form, whereas in an aligned table the differences stand out.

As a guideline, when four or more sets of data are to be presented, it's best to use a formal table.

Table 6-1 Table Format.

Table Name			
		Multiple Column Caption	
Stub Caption	**Column Caption**	**Subcaption**	**Subcaption**
Heading 1	Data	More data	And more
Heading 2	Data	More data	And more
Heading 3	Data	More data	And more
Heading 4	Data	More data	And more

To prepare effective formal tables:

- Leave a lot of white space.
- Align columns on the decimal point.
- Clearly label all columns—don't over abbreviate and don't use too many columns or rows.
- Use decimals, not fractions.
- Try to keep the table width to the width of one column in the periodical you're writing for.

EXERCISES

1. Create and sketch at least two visuals per periodical page for your article.

2. Integrate your visuals by organizing them with the text to be used.

7

Write the Text

Words are the greatest things ever invented.

Sherwood Anderson

After you've completed your outline and your visuals are roughed out, you're ready to write your rough draft of the body of the document. And it should be a *rough* draft because you need not be concerned about the finer points of grammar at this time. For your rough draft, don't be overanxious about the mechanics of writing. Be more concerned with the technical content. Concentrate on getting the facts down, in the proper order.

When you begin writing, you'll discover that a significant advantage of using an outline is that you will be writing about, and concentrating on, only one topic at a time. Your mind need not wander from one topic to another, worrying and wondering what is coming next. You can mentally set aside all the other topics in your article for the moment and focus your total faculties on the one topic you're writing about.

The "meat" of most documents is the body, which fleshes out the skeleton of the outline. In this chapter, we'll cover a method I've developed in writing hundreds of reports, proposals, and technical articles, and three nonfiction books. This technique will take you through the steps of organizing, writing, and revising your documents. After you've tried these proven techniques, you'll never need another. It makes fleshing out the body of a document easy, fast, thorough, efficient, and gratifying.

WORD PROCESSORS

Whether you choose to write longhand, type, dictate, or use a word processing program to compose your article is your decision. If you have a choice, a word processing program is by far the best. Time and again it has been shown that writing with a word processor improves the quality of an author's writing. Also more and more publishers are requiring that articles be submitted on computer diskettes.

A word processor automates your writing, making it much easier. Unlike a typewriter, you don't have to worry about hitting the carriage return at the end of a line because a carriage return is automatic in a word processor. No matter how fast or how slow you type, the word processor can keep up with you. Text is automatically scrolled so you can keep typing and watch the material you input appear on the screen instantly. When you're through inputting your information, you can go back and rearrange sentences or paragraphs, as well as delete or add words, without the necessity of retyping the entire document as you would if you used a typewriter.

Many auxiliary programs are available to help you come up with near-perfect text, including a spelling program, a thesaurus, word counters, and programs to check sentence length and the use and abuse of jargon. They can also check the use of too many passive verbs, the overuse of pretentious words, etc.

Researchers have learned that people who once dreaded writing have become much more positive about writing once they have learned to use a word processor. Another important fact that researchers have discovered is that our short-term memory lasts about five seconds. A word processor can record your ideas much faster than you can input them, so your fingers, and not the computer, are the speed limit in capturing your ideas.

Besides, word processors are the wave of the future. The days of the handwritten draft being handed to a secretary for retyping are nearly gone. Soon, the only typewriters you'll be able to find will be in museums. Before long most communications will be delivered as soft copy via modems, and heavy, multivolumed instruction manuals will be delivered on small, computer diskettes to be displayed on a video screen.

Computer Knowledge Not Needed

You don't have to know anything about computers to use a word processor any more than you need to know the theory of television reception to operate a television set. All you need to learn are a few simple English commands.

When you first start using a word processor and are learning its commands, you're using the rational/logical part of your brain. And when you're using the rational/logical part of your brain, the creative/intuitive part tends to be inhibited.

However, once you have learned the basic commands, you'll find that the operation of a word processor is *transparent.* In other words, you can type and perform all the necessary commands and not be conscious of it, just as you do when you drive an automobile or walk. This is the point of reward, when the operation of the word processor becomes automatic and your creative/intuitive function begins to assert itself. Once you reach this stage, you are truly accomplishing creative writing. This is the magic time when writing becomes exciting, a pleasure.

Some important things to remember when you're using a word processor:

- When you start feeling comfortable using a word processor, it becomes very easy to stop and make corrections when you're inputting. However, when you're doing your creative drafts, strongly resist this impulse. When writing creatively you are only concerned with ideas, not grammar or spelling. All you should do in the early stages of writing is get your ideas down, and not be restricted by matters of form that can be attended to during rewrites.
- If your word processor doesn't have an automatic Save feature every few minutes, be sure to perform one yourself. Pause in your writing every ten to fifteen minutes and save what you've written, just in case you lose power.

Word Processor Capabilities

The word processor program you use should have the following capabilities as a minimum:

Help menu A Help menu should be available at all times, regardless of the writing mode you're in, and should summarize the major functions of all of the key commands.

Insert mode The Insert mode lets you add information in any section of a document. The remainder of the material moves to the right to make room for the added information. This capability lets you start with any topic you choose and add any amount of information, in any part of your document, and lets you repeatedly switch back and forth between topics and still keep everything in order.

Delete You should be able to delete a letter, delete a word, delete to the end of a line, and delete entire blocks of text.

Block Move The Block Move command is used to move large sections of text from one location to another. It is often referred to as *electronic cut and paste.*

Search The Search command locates any part of your writing for revision and correction. You should be able to search *globally;* that is, search for a specific phrase throughout the entire document.

Search and replace This command allows you to correct spelling errors or to change words in your text. You should also be able to search and replace phrases globally.

Underline The Underline command enables you to highlight words, phrases, and titles in your article.

Boldface Boldface, like Underline, enables you to highlight words and phrases. Use it for headings and subheadings.

Tab Set Tab Set is used for setting indents, table formats, etc.

Centering This command enables you to center figure titles, tables, etc.

Right Justify Use this command if you need even left and right margins on your printouts.

Line Length This command enables you to set the line length to meet the special requirements of periodical publishers.

Graphics Some of the more advanced word processors have a Graphics feature, which enables you to draw simple graphics (block diagrams, flowcharts, tables, etc.) on the screen and print them out using a dot-matrix or laser printer.

Print The Print command enables you to output your text on paper. Usually a dot-matrix printer operating in a *N*ear *L*etter *Q*uality (NLQ) mode is acceptable. Some periodicals prefer letter-quality printers. For graphics, a laser printer is desirable.

Print Spooling With Print Spooling, you can print one document when you're working on a different document. This feature is highly desirable.

Cancel Print You should be able to cancel a print in progress if you make a mistake or change your mind.

Disk Storage Some periodicals might require that your article be submitted on a diskette, in addition to a printed copy, to save the publisher the added burden of manually inputting your prose into their computer.

It would be nice if your word processor also had these features:

Speller A spelling program is handy to correct your misspelled words. The speller should also be able to catch double words and to add words to a special dictionary to store some of the jargon of your profession. Your speller should be able to correct words *in context;* that is, as they appear in your text.

Thesaurus A thesaurus helps you find synonyms for words you might tend to repeat often.

Macros Macros provide shortcuts, such as designating a special key to type in a complicated word, to set up a format, etc. A macro is a simple computer command that combines and reduces a word processor command requiring a number of programmed strokes to one or two key strokes.

Word Count Some spellers have the ability to count the words in your text. This saves a lot of work.

Diskette Storage Capability

One "byte" represents a single character—it might be a letter, a symbol, or a space. An average word length is 6 characters, so a 150K (150,000-byte) diskette can hold about 25,000 words. An average double-spaced page contains about 250 words, so even a low-capacity diskette can hold 100 pages of writing. You'll never have to worry about running out of storage space on a new diskette when you're writing an article.

MECHANICS OF WRITING THE ROUGH DRAFT

The first step in drafting an article is to gather all your index cards, notes, lab notebook, data sheets, and visuals. It's very important to make a concerted effort and spend an adequate amount of time planning what you're going to write, rather than to just sit down and write what comes into your head.

Not only is an unplanned article difficult to write, the result is often a rambling, confusing document that few people will bother to read. Also, when you start writing, you'll find that the words tend to set hard, as if in concrete. Once written down they are difficult to edit, rearrange, add to, or modify your article.

Before you start the actual writing, plan your article in a considerable amount of detail. Once planned, writing is much easier. The flow of ideas and conclusions will be arranged in a logical fashion.

Know Your Audience

Before you start to write a technical article, remember that you're writing for a specific audience. So, pause a moment and consider your readers. Are they:

- Technically inclined?
- Familiar with the background subject of your article?
- Busy, with little time to read long articles?
- Of varying backgrounds—some scientific, some executive?

Remember, your reader is not forced to read your article. It's up to you to organize and write it so it's interesting enough to be read.

When you write an article about something you've accomplished, it's only natural that, since you've been deeply immersed in your work for so long, you'll want to write about every little detail. But don't assume your readers will be interested in all these details. They need only enough information to answer the questions, "What is this all about?" "Why should I read this?" Readers will read for the article's contents, not for the joy of reading your prose.

Follow the steps listed here in writing the body of your article. It's a proven formula that will help you come up with a well-written article with a minimum of effort.

List Your Topics

To start, look over all your file card notes and visuals and review the working outline you made in chapter 5. Find a quiet place where you won't be disturbed.

Write the Text

Pick a time of day when your mental equipment works the best. If you're a morning bird, like me, choose a nice, quiet morning to work. Use a deserted office, a quiet corner of the company library or lunchroom, or an empty conference room. Sit down and concentrate. Review, read, and reread your index card notes and the other notes you've accumulated. Study your working outline in great detail. Use earplugs if you need them.

This is the single most important step in the preparation of your document, so spend at least a few hours reading, reviewing, questioning, checking, and double-checking your references and your notes. You want this information to sink deeply into your mind.

Take a Break

The next step is also very important. Put your notes away and do something else for a few hours, something totally unrelated to your report. This gestation period is needed so that the material you just studied can settle deep in your mind and that marvelous mental mechanism, your subconscious, can go to work on the material to sort, evaluate, classify, organize, expand . . .

It's ideal if you can use this time to engage in some sport or physical activity, such as jogging, bicycling, hitting a golf or tennis ball, swimming, etc. Don't engage in any work that requires deep concentration or any taxing exercise of your mentals (except for, perhaps, a crossword puzzle). And don't even consciously think about your article. Put it out of your conscious mind and concentrate on a physical activity, talk to friends, or go to a movie.

Just relax for a few hours.

Write As Fast As You Can

After you've been away from your material for a few hours, find another quiet place and get your equipment ready. Sit down in front of your word processor, your typewriter, or your dictating machine, or get a supply of pencils or pens and loose-leaf paper.

Take out your working outline if you typed it on a separate sheet and tack it on the wall. Or, if you're using a word processor, display your outline on the video screen. Also tack up the visuals you're going to use so you can refer to them. Only your visuals and outline should be visible now.

Don't consult your detailed notes under any circumstances at this point! This is very important! Seal your detailed note cards in an envelope if you must, but do not look at them now. The problem with writing directly from notes is that you'll end up simply copying the notes directly into your text. You'll be using the exact words from your notes, instead of putting all your thoughts and ideas into your own words. Working directly from notes stifles creative writing. It's better to put your creativity to work and write using only your working outline and visuals to guide the words from your subconscious mind to the keys.

Next, using either your typed working outline or the outline you input in your word processor, review the topics and decide which of those listed in your working outline will be the easiest to write first. Then start writing whatever comes into your head, referring to the applicable visuals as you write. This way you'll be writing a series of short, easily handled topics, rather than one long, continuous one.

Write or type as fast as you can. This is important! Don't let anything slow down your creative process. Double- or triple-space your writing or typing. You'll need this space later for corrections, additions, etc. Don't pay any attention to grammar, spelling, sentence structure, variety, or length. Use incomplete sentences if you must. Just get the words and ideas down.

Thoughts are associative. One thought leads to another. You must allow your thoughts to flow freely and use the style and vocabulary that are natural to you.

If you're stuck for a technical detail, a name, or a number, leave a blank space or a question mark. You can fill in the details later.

Don't let anything sidetrack you in this highly creative process. You'll find that your enthusiasm will grow once you get over the initial hurdle of starting. You'll be amazed and pleased at how quickly thoughts come to you. Your hands will not be able to keep up with your brain. Words, thoughts, and sentences will pour out of you in a logical sequence, forming a smooth-flowing narrative that you'll be proud to have authored.

As you finish each topic or subtopic, cross it off your outline. If you're writing or typing, use a new sheet of paper whenever you switch to the next topic. Write the topic name and page number in the corner as a precaution in case your sheets get mixed up. Here again the beauty of the word processor is demonstrated, because all you need to do is switch to any part of your topic outline and continue writing. All the text will simply move over to make room for what you are adding.

If you get stuck on a particular topic, switch to another one. Chances are that you'll soon get unstuck and be able to go back and finish that topic later.

The subconscious mind is probably one of the most fantastic creations in existence. It makes a supercomputer look like a windup toy. You'll be amazed at how easily you can turn out good writing, fast and almost effortlessly, with your subconscious mind running in high gear.

Check Against Your Detailed Notes

Soon after you've finished writing your rough draft, unseal the envelope with your detailed notes in it. Read through your notes again to see if you've left anything out. You probably have, so fill in the blanks left open in your important first draft.

If you're writing by hand or typing and you have left out quite a bit so you have to add a page between pages 2 and 3, number it 2a. That's the beauty of using individual sheets of paper: you can easily add, subtract, or reorganize as required. In a word processor you can easily insert whatever information you left out and whatever corrections you need to make. If you're dictating, you'll have to name the topics you are modifying and input your comments verbally or you'll have to have a rough draft typed and then mark it up.

Review

Before you start the next review, pause for a few hours (or a few days if your deadline permits it) to let your writing cool off. "Write in haste, polish at leisure" is good advice.

Then, put on your editorial hat and go through what you've written, from beginning to end. Check the sentence length, structure, grammar, and punctuation. Pause to check any spelling you're unsure of. If your word processor has a spelling program, use it. The speller will catch many of the mistakes you'd miss on a visual check.

I find that writing of things technical (and writing in general) takes two distinct and separate mental processes. First you write creatively, using the creative/intuitive part of your brain and concentrating on getting the proper technical information down, in the correct order, in your own individual style. Next, you put the logical/rational part of your brain to work and become an editor, critically blue-penciling as you read your material objectively, checking for proper grammar, spelling errors, etc. You cannot accomplish these two separate and distinct processes at the same time because, if you edit when you write your initial draft, your creative processes will be stifled. Your mind will have to continually shift gears back and forth. This mind-jerking strips your mental gears and ruins the writing process because creativity must exist without any bounds or rules. Creativity can't be tied down or restricted by grammar.

When you have your technical writing hat on, you're doing some pretty heavy creative writing. You're writing what has probably never been written before. It's coming out of your unique mind and can only be written by you.

Editing, however, is less of a creative task, but it is a vital one. Editing applies certain established, well-proven rules that have become accepted over the years. It's an extremely important task because misspelled words, awkward sentence structure, and improper grammar will completely distract from, and distort, the important things you're trying to say. Whenever you read a document replete with misspellings and errors, your mind is sidetracked and you begin to lose interest and confidence in the accuracy of the writer's technical information.

For this step, read your article through, from beginning to end, blue pencil (or pen) in hand, and make any needed corrections you find. If you're working with a word processor, you can accomplish the process on the video screen by scrolling through the text or using a printed copy. Write notes to yourself on the screen or on the printed copy about what to do later. Don't interrupt your reading too long during this process. However, read the entire draft through, at a leisurely pace.

When you come to a sentence you can't understand, just imagine how difficult it would be to understand for someone who knows little or nothing about your topic and who is reading it for the first time. Make a note to clarify it, but don't stop to correct it at this time. Wait until the next step to make the necessary revisions.

Revise

A few hours, or a day or so later, take the necessary time to do the detailed rewrite and make the additions and corrections required. Put the spelling checker to work if you have one. Use the electronic or manual Cut-and-Paste function to restructure your article as you noted during your review. Because second and third thoughts are often clearer than first thoughts, revision can be the difference between an excellent article and a mediocre one.

And revision is much easier when you're word processing. If you're working with a word processor, it's best to perform this revision step with a printout of your entire article because all of your errors will show up better on paper and you can easily jump back and forth to check and verify things in a printout.

Review the Article

Set your article aside for a few more days. Then read it through, from beginning to end, and make another set of corrections.

Check your sentence lengths. Your average sentence length should be less than 20 words; 17 is ideal. For a quick visual check, a typewritten copy usually has 10 to 12 words per line, so your average sentence length should be less than two typewritten lines. Check for long words; they are usually abstract words. Replace them with concrete words whenever possible.

BEAUTIFY YOUR BODY

Now that you have your document drafted, it's time to beautify the body: to add bullets, listings, headings, etc., to make your document easier to read and understand. Experts agree that graphic relief is needed to create white space and to make large blocks of text more readable and more attractive. I call these graphic reliefs *literary cosmetics.*

Typical literary cosmetics you can use are:

- Paragraphs
- Bullets

- Listings
- Headings/Subheadings
- Underlines

Paragraphs

Go through your material and check that your paragraphs are not too long. Keep most of them under about ten lines on a typewritten page (that's about 100 words). A final typed page should have at least three or four paragraphs per page.

But don't overdo it. Too many short paragraphs distort the effect you're trying to create: lots of white space, variety, a break for the reader's eyes.

A reader sees a paragraph as a certain amount of information he or she must gulp down in one swallow, so make them small swallows. Short paragraphs throw more *light* or understanding on your page, making it easier to read. Paragraphs also help your reader group and understand topics better.

Bullets

Bullets are the symbols used to set a list of information off from the text to emphasize certain points you're going to make. For example:

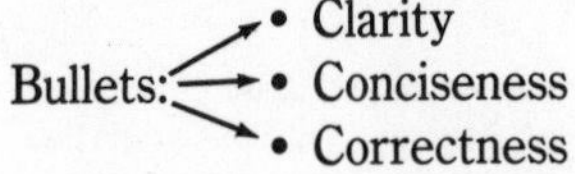
Bullets:
• Clarity
• Conciseness
• Correctness

Bullets are an additional effective way to give the reader's eyes and mind a break when you enumerate and emphasize certain items. But don't overdo them, either, or they will lose their effectiveness.

Numbered Listings

A listing uses numbers or letters to list things in sequence and set them off for the reader as individual, important topics. For example:

1. Grammar
2. Sentence structure
3. Spelling

A listing is a quick way to summarize points for your reader. It makes the items stand out. Lists take more space than sentences, but they do emphasize the data they present.

Headings

Headings provide a visible outline of your article. They also form a summary of, and introduce, the material below them.

Headings show the subjects you're covering and the order in which you're covering them. A reader can obtain a quick summary of a properly written article by reading the headings; noting the order in which they're arranged and the amount of space allocated to each. Headings can whet a reader's appetite and entice him or her into reading your entire article.

You need about one heading per two typewritten pages of 500 words to allow more light on the page. Headings should be less than five or six words long; two- or three-word headings are common. When your article is published, your publisher will usually convert your headings into boldface type so they'll stand out even more.

Another advantage of headings is that they save creating a lot of transitional phrases by using headings to bridge two topics coming up next. A heading helps you jump from one aspect of a subject to another without such much-abused transitions as:

- On the other hand . . .
- Meanwhile, back at the ranch . . .

Underlining

Underlining is used in the text of an article to call attention to a specific word or phase you want to emphasize. You're calling the reader's attention to these words.

But like most good things, underlining can also be overdone. Use them judiciously.

EXERCISES

1. Draft your entire article.
2. Review and revise your article at least three times.

8

Prepare Your Manuscript

A whole is that which has
a beginning, middle, and end.

Aristotle

Now that you have a well-formed body, it's time to put a head (lead), a foot (conclusion), a hat (Title), and a kerchief (subtitle) on your masterpiece for your readers. Each of these sections forms a significant part of your article. The relative attention-catching ability of each of these sections and the order in which the sections are read are illustrated in FIG. 8-1. The larger the area of the rectangle, the greater the stopping power—the greater the ability to arrest and hold a reader's eye, to pique your reader's interest.

As stated earlier in this book, the rising degrees of involvement for your article occur when the reader:

- Stops to read the title and subtitle. HEY! YOU!
- Looks at your visuals. HEY! YOU!
- Reads the first paragraph. HEY! YOU!
- Reads the body. SEE?
- Reads the conclusion. SO!

TITLE

If you haven't already chosen a good title, now is the time to work on it and select one that will grab the reader. The title, or *headline,* in huge, bold print is the first thing most potential readers will see. It must be precise and clear, telling them what they are about to invest their valuable time in. If the title doesn't hook them, they'll simply turn the page and your article will go unread.

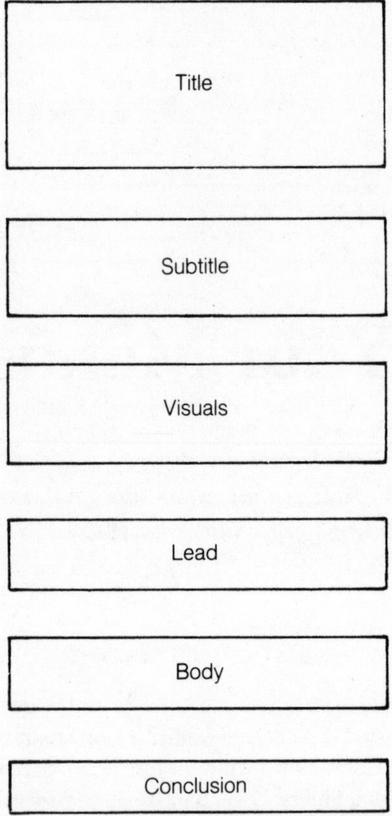

Fig. 8-1. Attention-getting values of parts of an article.

If you're not satisfied that your title will entice readers into your article, review the advice given in chapter 2. Come up with a number of possible titles (use no more than about eight to ten hard-hitting words) and work until you get a good one, an attention-grabber.

In one of my articles, the title was "Potent Visual Aids Get Your Message Across," in bold, black, half-inch letters. It told the reader how the article would help him or her use visual aids effectively. The title summarized the article in only seven words.

Once your readers have read the title and it has generated some interest, they'll read the *subtitle.* A subtitle is an expansion or clarification of the title, often in bold print slightly smaller in size than the title. A subtitle can be a sentence or two, also in bold print (typically ¼-inch letters), such as was used for my article.

Nervous about Making That Required Speech?
Visual Aids Not Only Focus Attention Away from You;

They Make for a More Effective Oral Presentation.
But Be Sure to Create and Use Them Properly.

Note also how the HEY! and YOU! parts of the four-part formula have been used. The title is basically a HEY! component that quickly and effectively informs the reader of the subject of the article. The subtitle uses a mixture of HEY! and YOU! and even includes the important word *you* in the subtitle.

Next, using your title as your guide, write a short sentence or two to use as a subtitle, expanding on the theme of your article and involving the reader, the *you,* in your subtitle.

VISUALS

If your title and subtitle have intrigued your reader, he or she will next glance at your visuals. Usually more people will glance at your visuals than will read your text. For this reason it's important to write the body so that a visual is included on the first page of your article. Review your visuals and make sure an interesting one is incorporated in the first 500 words of the body.

LEAD

Next, pull your readers into the body of your article using the *lead.* Some commonly used leads follow:

- The *summary* lead tells succinctly what the article is about. This popular type of lead is the easiest to write. It might be one or two paragraphs in length and list some of the main facts in your article. This type of lead can also serve as an abstract of your article.
- The *question* lead piques the reader's curiosity by asking a question the reader would like to know the answer to.
- The *problem/solution* lead is appropriate when your article discusses the solution of a basic problem.
- A *description* lead can describe in the first sentence the object, technique, or process your article covers. You should be able to use words that evoke concrete images for this type of lead.
- A *comparison* lead is ideal when you are analyzing or comparing two or more processes or equipment.
- The *news* lead is useful when you're announcing or describing the first installation of a unit, a new product, or a new process.
- In a *statistical* lead, you begin by comparing statistics that are shocking or very different from the norm.
- An *anecdotal* lead begins with a mini-story concerning people, and will illustrate and introduce the topic of the article. This important, human-interest lead is, regrettably, used very little in technical journals. (I used it in this book in chapters 1 and 5.)

- A *historical* lead begins by giving the origin or background of a product, service, or discovery.

A lead performs three basic functions:

- It serves as a transition from the title and subtitle into the body of your article.
- It summarizes your article and hooks your readers, promising them more information of interest to them if they'll continue to read.
- It shows them how they will directly benefit from reading your article.

As an example of an opening paragraph, consider this one from my article on visual aids:

> Sooner or later you're going to have to give a talk, perhaps at a program-review meeting or a technical conference, in a discussion with a customer on a proposal's technical aspects, or as part of a presentation to management. And, when you do, you'll discover that visual aids, in conjunction with your oral presentation:
>
> - Provide an outline of your talk for you and your audience to follow.
> - Help keep the audience's attention.

This first paragraph pulls the readers into the article by reminding them that sooner or later they're going to have to give a speech. This will appeal to almost everyone involved with any kind of technical work. Then it jumps right into the usefulness of visuals in such a presentation. Notice the YOU! component is used again to convince readers that they will benefit from reading the article.

The most important sentence in any article is the first one; it's a do or die situation. If it doesn't do (entice the reader to read your second sentence), your article just died.

The body you wrote in chapter 7 now needs a lead to pull your readers into the main part of your presentation. Write an opening paragraph that expands on your subject, one that complements the title and subtitle and hooks your reader into your article.

THE BODY

If you've done a good job with your title, subtitle, and first paragraph, you should now have the readers hooked into reading the rest of your dissertation, so the body will take care of itself.

CONCLUSION

For the conclusion, you can reword what you said in your first paragraph and leave the reader with some satisfactory thought or call to action. For example, in my visual aids article I concluded with:

> Use a lot of visuals the next time you have to give a speech. You'll find that they will do much of the talking for you.

As you can see, usually one paragraph is enough for a conclusion, but don't be afraid to use more if needed.

A conclusion can make recommendations for further developments of your subject, or it can summarize the results of your research to re-emphasize them. It can also encourage your reader to take some action. So, add a last paragraph to your article that summarizes what it's all about.

MECHANICS OF HARDCOPY SUBMITTAL

When you prepare the final draft of your text for submittal to your publisher by hard (printed) copy:

- Use a good quality of paper, typed on one side, and double-spaced.
- Leave plenty of space all around to frame the article: about one-inch margins on both sides, the top, and the bottom.
- Type your name on the top of each page (if you're using a word processor, add a header specifying your name on each page) and number each page consecutively, as "Page 2 of 8."
- Make sure the draft is neat and error free.
- Use either pica or elite type fonts. Don't use script or fancy font. And use either a laser printer or a near-letter-quality printer. Some editors are opposed to regular dot-matrix printing.

As an example of what a first page should look like, check FIG. 8-2.

For your visuals, follow these guidelines:

- Put each visual on a separate sheet of paper. Most periodicals will redraw them, so they need not be of drafting quality.
- Do not send originals; send a good-quality copy, unless an original is required.
- Make sure all your figure letters and numbers are clear and unambiguous. Define any special terms used.
- Caption each visual with a title and a figure number.
- Use standard drafting practices and standard symbology on your visuals.

Prepare Your Manuscript

Your Name
Address
City, State, ZIP
Phone Number
Social Security Number

About XXXX words

(Leave about a half of the first page blank)

TITLE OF ARTICLE
by
Your Name

Text begins here . . .

Fig. 8-2. Title page format.

Some special precautions to observe when you submit photographs are:

- Glossy 8-X-10-inch photographs are preferred; 5-X-7-inch glossies are acceptable. Mail them using stiff cardboard or some other protector in the envelope.
- Do not fold, paper-clip, or staple photographs.
- Stamp or print *Do Not Bend* on the mailing envelope.
- Write the captions and figure numbers on the backs of the photographs using a self-adhesive label, or use a transparent overlay to label the photo. Don't write on the back with a pen; it might make the photo unusable.

MECHANICS OF SOFTCOPY SUBMITTAL

Although most periodicals will still accept hardcopy submissions (typewritten copy), more and more are encouraging a softcopy (computer diskette or modem transfer) submittal, with a hardcopy. In the future most periodicals will insist on softcopy because it saves them the time and expense of retyping your manuscript. All they have to do is edit your article and format it for their specific printing requirements.

After the periodical has edited and reformatted it, they'll probably send you a diskette or printed copy of their edited version for your review and approval. All of the chapters of this book were delivered on computer diskettes to the publisher, TAB BOOKS.

Another reason that softcopy submittal will be required in the future is so publishers can send your article by modem or on a diskette to data banks. The data banks can then store this electronic manuscript so that it can be accessed by on-line systems, as discussed in chapter 9.

MAGAZINE SURVEY

To help define current and future requirements for softcopy, I queried 75 representative periodicals. The periodicals covered a wide range, from small presses to large presses, and a broad variety of scientific and engineering disciplines. The results of this comprehensive survey follow.

An overwhelming 90 percent of the periodicals either required or preferred that a prospective author query before writing an article. This saves both the editor's and writer's time and effort.

About 60 percent of those replying said they either require now, or will in the future require, that articles be submitted on computer diskettes. Nearly all editors said they still will accept hardcopy in lieu of softcopy, but this will likely change in the future.

The preferred format for softcopy was 5¼-inch diskettes; some replied that 3½-inch diskettes were also acceptable. The overwhelming choice for type of computer was IBM compatible. The ASCII format was most requested, with a scattering of WordStar, WordPerfect and XYWrite acceptable as word processing programs.

To be safe, then, the best bet is to use an IBM compatible and save your article on diskette in ASCII format. Some periodicals stated that submitting an article in a preferred format would result in a larger honorarium for the author.

Pay ranges varied considerably. About one-fourth of those responding do not pay for articles. Most of these were professional society publications, which is understandable. None of the periodicals stated the author would have to pay to be published. The pay rate ranged from $40 to $250 per page; the highest rate quoted was $2,000 for a feature article. Most pay rates were from $50 to $100 per page.

Article lengths ranged from 500 to over 4,000 words, or from about 2 to 16 pages of double-spaced copy. The average length requested was from 1,000 to 2,000 words.

Graphics are as important as text to all of the periodicals, and most of the replies stated that one to two visuals should be used for every two or three typed pages of text.

MAIL IT

Before you mail your article, make a photocopy of your hardcopy and a backup copy of your diskette, just in case something gets lost or damaged in the mail, and so you'll be able to answer any questions your editor has. If you mail a diskette, use one of the special diskette mailers you can purchase in any stationery store. Mail it First Class, with *Do Not Bend* printed on the outside.

WHAT'S NEXT?

Within four to six weeks you should hear if your article has been accepted or rejected. You may get one of three possible replies from your editor:

- "Congratulations. Your article has been accepted and will be published within the next four to six months. About six weeks prior to publication we will send you an edited version of your article for you to review and approve. At that time we will send you a check for $________. When your article is published, we will send you a copy of that issue, plus an additional copy from which you might wish to order reprints. . . ."
- "We like your article but we feel it needs some revision before we can publish it. We have assigned Hubert Humbert III as your editor, and he will be contacting you about these revisions. . . ."
- "We are sorry to have to inform you that your article does not meet our editorial requirements, so we will not be able to publish it. . . ."

If you receive one of the first two responses, enjoy. You're about to become a published author. And when you get the galley proofs to review (or a computer diskette if you submitted a softcopy), don't try to rewrite the entire article. The periodical has too much invested in editing and illustrating your article to make major changes. Review the proofs for technical accuracy. Leave the editing to the professional editors. After all, they've purchased it and can pretty much do with it what they want.

If you receive the third response, don't despair. You have tried and you are going to try again. Take heart from the appropriate words of Theodore Roosevelt:

> "Far better it is to dare mighty things, to win glorious triumphs, even though checkered by failure, than to take rank with those poor spirits who neither enjoy much nor suffer much because they live in the gray twilight that knows not victory or defeat."

If your article has gone through the entire cycle of query, approval, and writing, chances are excellent that another periodical will buy it, so query the next one on your list. Send just the query, not the completed article, to the next periodical. It takes an editor too long to respond to a complete article. Also one editor's format and requirements might be quite different from another's, so you might have to do some rewriting before another periodical will publish it. And don't inform the editor that your article has already been written and rejected.

COPYRIGHT

Most articles are copyrighted by the periodical that publishes them. However, most periodicals will let the copyright revert back to you after the initial publication, if you so request.

EXERCISES

1. Prepare your article for submittal:

   a. Compose a subtitle.

   b. Write a lead.

   c. Write a conclusion.

9

On-Line Systems

Man is a tool-using animal.

Thomas Carlyle

A revolution is underway that will radically change the look and operation of our public and private libraries, the manner in which our newspapers and books exist, and the general way in which we obtain information. This revolution in on-line systems has been brought about by a number of factors.

Information processing is growing so rapidly that it is impossible for a person to keep up with the advances in his or her profession by reading current periodicals and journals. For example, chemical and biological abstracts now exceeds a quarter of a million per year, making a manual search an impossibility. Professionals need help in locating this information quickly and efficiently.

The low cost and wide availability of the personal computer, coupled with the continually decreasing cost of digital storage, have arrived in time to become the solution to this dilemma. Taking advantage of these phenomenal advances, many on-line vendors have created on-line data banks. *On-line* means you can connect with and obtain information from these remotely located databases through the keyboard of your personal computer and via telephone lines. Data banks are the repositories of this information.

ELECTRONIC LIBRARIES

On-line databases (electronic libraries) have met this demand for ways and means of making these searches feasible and fast. Modern on-line system vendors allow you to tap into gigantic databases, search millions of documents in seconds, and view the result on your home or office personal computer. Suddenly any fact you need is as close as your telephone.

As an example, a doctoral literature search that might have taken six months a few years ago can now be done in less than ten minutes, with far more thorough results and much less cost. A manual search can cost from 10 to 100 times as much as a computer search. Also, some manual searches are a physical impossibility because access to the needed books and periodicals is not possible.

As you conduct research for your current and future articles, acquaint yourself with the use of the databases, the electronic libraries that you can access now that are as close as your telephone.

WHAT IS A DATABASE?

A *database* is an organized collection of information that a computer can search as a unit. A database can be a catalog of manufacturer's products, a series of books or periodicals, or a computer-readable version of a set of encyclopedias. It can include virtually any information that can be converted to digital form and stored for retrieval by a computer. Many databases are collections of abstracts and bibliographies from periodicals. Some are specific abstracts collected by specialty organizations. The variety, depth, and diversity of available databases are mind boggling.

Databases are sold to, or generated by, on-line vendors, who in turn sell access time to anyone wanting to search for information. Your computer, your modem, and your telephone give you this access.

Basic reasons for using a database when researching an article are:

- They help you quickly locate those few important bytes of information you need in that megabillion-mountain of data.
- You obtain instantaneous answers.
- Vendor's databases are much larger than public and private libraries.
- They can be accessed from virtually any place that has a telephone.
- Information is up to date.
- Specialized databases exist for virtually every profession.
- Information is available when you want it. You don't have to worry that someone has checked out the book or periodical you need to consult.
- The information is comprehensive, so you can research to any depth you need.
- You can query from your home or office. No longer do you have to waste time commuting, fighting city traffic and parking problems.

EQUIPMENT NEEDED

The basic equipment you need to go on-line and let your fingers do the talking is:

- Personal computer
- Modem
- Communications software program
- Printer
- Telephone

The basic hookup required is shown in FIG. 9-1.

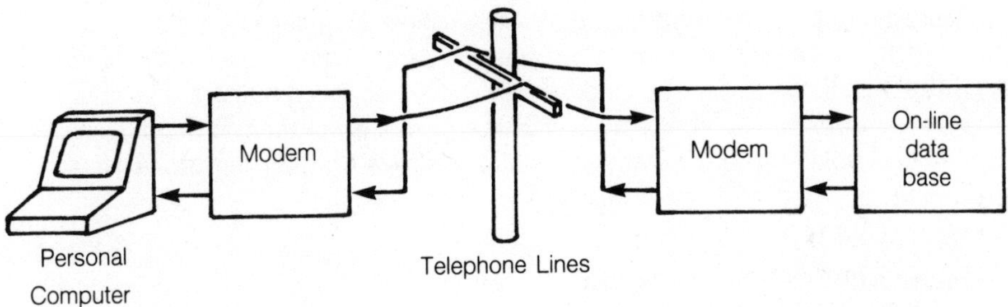

Fig. 9-1. Basic on-line setup.

Personal Computer

As far as types of personal computers are concerned, some vendors' on-line systems will work with all major brands. Many require that you use an IBM or IBM clone. You'll have to check with the specific vendors to see what computer protocols they'll accept.

Whichever computer you are going to use must have a serial RS-232C port; that is, a serial or asynchronous communication card. Some computers come equipped with a serial card; others require that you add one. A color monitor is not required, but if you want to display graphics on your monitor, make sure your computer has that capability.

Modem

A modem must be added to most computer systems. Two basic types of modems are in current use: circuit-board or stand-alone. (A third type—the acoustic modem, which contains rubber cups that clamp over your phone headset—is also available but is rapidly going out of use.)

The circuit-board modem conveniently fits inside your computer. The stand-alone modem is in a separate chassis that connects to your computer via a cable. The basic capabilities of the two types of modems are virtually the same. However, the circuit-board modem does use some power from your computer, and it cannot be transported easily so you can work on a different computer.

The stand-alone modem can easily be moved from one computer to another, but often it is inconvenient to mount and hook into your system. The stand-alone also has LEDs that help you check on the progress of your modem. Stand-alone modems usually cost more than internal modems.

Regardless of the type of modem you use, make sure your modem is, first of all, Hayes compatible. Hayes is the de facto standard for modems; fortunately most modems are Hayes compatible.

As far as speed is concerned, 300, 1200, and 2400 bits per second (bps) are in common use. You often pay a higher price for faster speed, but even if 1200 bps costs twice as much as 300 bps, you can receive data at four times the speed, so it can be a bargain if you are calling long-distance.

When you choose a modem, make sure it has the capability of working at all three speeds. That way you can choose the specific speed that satisfies your requirements, as well as those of the varied vendors. A speaker on the modem card is desirable so you can get some audio feedback on the progress of your call.

Telecommunications Program

Software is needed to tell your computer and modem what to do. A wide variety of telecommunications programs are available, ranging from free to a couple hundred dollars.

Some features to look for in a telecommunications program are:

- *Auto-Dial* The ability to dial a number or series of numbers that the computer has stored in memory, rather than having to dial your number manually.
- *Auto Re-dial* The ability to automatically redial a number if the line is busy.
- *Duplex* Both full-duplex and half-duplex capability.
- *Automated Signon* The ability to send the network number, account number, and password to the host automatically.
- *Break key* The ability to stop the host from whatever its doing and wait for another command. This is very useful when you discover that you've asked for the wrong information.
- *Data Capture* The ability to record on disk the information sent by the host so you can review it at your leisure.
- *Printout* The ability to print out data as it comes into your terminal.
- *Text File Transmission* The ability to type your message off-line, then go on-line to send your message out.
- *Word Processing* A word processing program so you can compose messages.

TYPES OF DATABASES

Two basic types of databases that vendors have available are:

1. *Full text* This type of database has the entire text of articles for you to read.
2. *Bibliographic* This type provides bibliographic references and summaries of printed publications. Most databases are of this type, but sometimes they also contain abstracts of the information you're seeking.

Some of the databases are general purpose and cover a wide range of subjects. Others specialize in various fields. A large number of science and technology databases is available.

Representative databases specializing in science and technology are:

- SCI Search Vast and expensive, but has neither abstracts nor descriptions.
- Compendex Has over a million records dating back to 1970, with extensive abstracts and a broad coverage of physics, computers, and all fields of engineering. It covers about 4,500 journals.
- National Technical Information Service (NTIS) The bibliographic database of the U.S. Department of Commerce. It contains technical reports generated by U.S. federal agencies and their contractors. These reports go beyond bibliographic citations; they also contain information on ordering complete copies of the cited reports by mail.
- Computer Database More than 530 different publications dating back to 1983 cover computer-related hardware and software.
- Inspec Focuses on physics, applied physics, electronics, electrical engineering, computers, and control engineering. More than 2,300 journals are scanned regularly, and over 330 are completely abstracted.
- Ismec The mechanical engineering equivalent to Inspec, this database has information dating back to 1973, but it only has abstracts before 1982.
- Engineering Literature Index Articles from 3,500 worldwide journals.

When you deal with these databases, you can generally obtain a full-text version of the information you're looking for, either as a printout at your terminal or as a photocopy by mail.

ON-LINE COSTS

The costs to use on-line systems vary from about $3 per hour to more than $200 per hour. In addition, most vendors have a one-time sign-up fee. Check with the individual vendors listed in appendix C to determine their requirements.

When you're first learning to search an on-line system, use the lower and less expensive 300 bps rate. After you've acquired some skill, switch to the

1200 or 2400 rates. If your budget is limited, there are some excellent after-hours databases to use: the Knowledge Index in the DIALOG system, Genie, and BRS After Dark have discount rates and an excellent variety of databases.

QUESTIONS TO ASK A POTENTIAL ON-LINE VENDOR

Before you select a vendor, seek answers to these questions:

1. Does the system have abstracts, bibliographies, or full text?
2. What does the service charge? Are there extra charges for 1200 and 2400 bps?
3. Is hardcopy available?
4. Is the coverage comprehensive and up to date?
5. Can you search on a time basis (e.g., 1985–1987)? Can you use Boolean logic in a search to minimize your hits?
6. Does the vendor have local telephone access?

HOW TO GO ON-LINE

Assuming you have your computer on and your telecommunications program installed, then:

1. Set up the proper parameters so you can communicate with the host; for example, set the baud rate, word length, parity, etc.
2. Dial the number, either manually or with the telecommunications program. Your speaker will sound the notes as they are dialed, then you'll get a short ring. Finally your computer will give you a message, for example: 1200 CONNECT to inform you that you have been connected to the remote modem at a baud rate of 1200.
3. Hit your Return or Enter key. If you don't get a response, hit the Return key again.
4. Follow the vendor's log-on procedures. You might have to give a password and account number.
5. Conduct your search and obtain whatever data you need.
6. Log-off using EXIT, BYE, or whatever your host requires.
7. Switch off your modem and go back to working on your article. Incorporate the data you just found.

If you opt to join one of the major services, they usually have training services available on the necessary procedures.

How to Search Once You Are In

You'll generally start your search with a key word, but be careful to make it a key word that is not too broad in its scope so it will minimize the number of "hits" your title will generate. *Hits* refers to the number of possible bibliographic references for the key word you have chosen. If this number is too large—for example, 100, when all you expected was 10—narrow your search parameters with carefully chosen defining words and syntax.

For example, suppose you want to locate a computer drawing program for the IBM/XT. If you start out with the key word *draw,* you're going to be overwhelmed with hits. So you should narrow it down by specifying as many of the following restrictors as you can:

- Publication title
- Article type
- Author
- Date of publication
- Combination of words, such as *drawing* and *IBM/XT*

Once you've located your reference, some vendors allow you to have the abstract displayed on your screen. Other vendors will permit the entire article to display, and a few vendors will download the article to your computer. Most of the vendors let you order a printed copy that can be mailed to you for a nominal fee.

MAJOR ON-LINE VENDORS

Some of the major on-line vendor services are listed in this section; their addresses and telephone numbers are given in appendix C. Phone numbers and addresses do change, so if you are unable to obtain a response at the number listed, dial the 800-line directory assistance service at 1-800-555-1212.

Because of the lack of space, only a few representative major vendors are listed here. For a thorough coverage of on-line vendors, check one of the following references:

Computer Readable Data Bases by Martha Williams. Chicago, Ill.: American Library Association, 1985. 2 volumes.

Data Base Directory 1984/85- . White Plains, N.Y.: Knowledge Industry Publications (1984-). Annual.

Datapro Directory of On-Line Services. Delran, N.J.: Datapro Research Corporation (1985-). 2 volumes, loose-leaf.

Directory of On-Line Data Bases, v. 1, no. 1- , Call 1979- . Santa Monica, Calif.: Cuadra Associates (1979-). Quarterly.

Some of the largest of the database vendors are listed alphabetically, along with the types of databases they offer.

On-Line Systems

Bibliographic Retrieval Services (BRS) Contains over 150 databases, is easy to use, and covers life sciences, medicine, physical sciences, applied sciences, and engineering.

BRS After Dark A low-cost system available during off hours and contains over 100 databases.

Chemical Abstracts Provides access to global scientific information by indexing 12,000 journals from 150 countries, plus patents from 27 nations. Their CA file has over 11 million abstracts. Chemical Abstracts specialties are: bioscience and medicine, chemistry, computer science, engineering, environment and energy, materials and construction, math, physics, and patents. Graphics are available from their databases.

CompuServe Owned by H & R Block, CompuServe offers a variety and broad scope of nearly 500 distinct topics and services. Investment services, reference libraries in medicine, science, and law and other disciplines are available, plus encyclopedia, news, weather, and sports databases as well.

DELPHI Covers home banking, computer periodicals, news headlines, encyclopedias, electronic messaging, business and finance, and library and computer clubs.

DIALOG This largest supplier of word-oriented databases is a division of Lockheed Missiles and Space Division. Most references are bibliographic. DIALOG has more than 100 million items of information stored. Magazine and journal articles can be searched and ordered on-line; document delivery takes from a few days to a couple of weeks. More than 320 databases are available, including chemistry, computer technology, science and technology, medicine, and biosciences. They reference more than 60,000 paper publications, and index more than 700.

Dow Jones Owned by Dow Jones Inc., and has exclusive access to the full text of the *Wall Street Journal.* Databases include financial markets, financial information, and the *Academic American Encyclopedia.*

GEnie A product of General Electric Information Services and CompuServe's closest competitor, this user-friendly service is growing rapidly and is now second to CompuServe in the number of subscribers. GEnie has databases in computing and high-technology news, educational resources, financial data, and product databases. GEnie also has a low-cost, after-hours service.

Knowledge Index This service is a subsidiary of DIALOG and is an after-hours, discount service. Most of Knowledge Index's databases are bibliographic. There are 75 different databases in 13 categories that cover chemistry, computers, electronics, medicine, drugs, science, and technology. Best for beginners.

Mead Data Central A division of Mead Corporation. Mead's original service was Lexis, a full-text database for lawyers. They also have Nexis, a full-text news service and Mead handles the *New York Times.* Lexpat, another of Mead's databases, has indexed more than 750,000 patents since 1975. Medis-Medical science is another of Mead's databases.

NEWSNET NEWSNET carries about 300 newsletters grouped into 40 different industry groups, wide-ranging subject areas, and 10 leading wire services. Over 70 percent of NEWSNET's coverage is exclusive. It covers computers, management, metals and mining, research and development, telecommunications, chemistry, and energy. Many databases are in the field of computers.

ORBIT SDC/ORBIT Owned by Burroughs, this service is bibliographic, with over 100 databases and more than 75 million citations. Specialties include patents, materials, science, chemistry, energy and the environment, engineering, and electronics.

VIDEOLOG A subsidiary of Schweber Electronics, VIDEOLOG is basically an electronic catalog of components, manufacturers, and products. You can search for more than 750,000 components for both military and commercial applications. Components are from 700 manufacturers and 14,000 suppliers of products in more than 2,000 categories. A price quote can be obtained and the product can be ordered on-line, making a paper transfer unnecessary. A paper copy will be provided for your records if needed. Good graphics permit a display of response curves, outline dimensions, etc. VIDEOLOG also has electronic industry and new-product release news.

VU/TEXT Provides full-text versions of 40 regional newspapers (36 are exclusive), plus selected articles from more than 180 regional business journals and newspapers. VU/TEXT also has the 20-volume *Academic American Encyclopedia* with more than 32,000 articles.

Wilsonline H.W. Wilson produces printed indexes for libraries and has about 20 bibliographic databases, including the *Readers' Guide and Business Periodicals Index.*

Yellow Pages Dun and Bradstreet's Electronic Yellow Pages is an electronic collection of more than 5,000 yellow pages from cities throughout the United States. It lists 6½ million entries.

DIGITAL DIPLOMAS

As libraries are computerized, it's only natural that universities and colleges follow. Some of the pioneers in these "electronic universities" offer on-line digital diplomas to people who cannot attend regular classes on campus, but who can go to class in their home using a PC and modem.

These classes have no schedule, and student/faculty conferencing can be held at mutually convenient times from the comfort of your home. Classes can be taken for credits toward degrees, or they can be audited for professional advancement.

The American Open University of the New York Institute of Technology offers a B.S. in Business Administration, General Studies, or the Behavioral Sciences. And you don't even have to visit their campus; all of your classes can be taken in your home. There is a one-time admission and matriculation fee of $150 and a charge of $85 per credit hour, which is about half what a resident student of the New York Institute of Technology pays.

Another electronic university, Connect.ed, offers a series of courses to fulfill the requirements for a 39-credit degree program leading to an M.A. in Media Studies. Connect.ed courses carry graduate and undergraduate credit from the New York School for Social Research. Courses can be taken at any time, 24 hours a day, 7 days a week, from any place in the world equipped with a PC, modem, and telephone. Graduate tuition costs $948 per course (for 3-credit courses) plus a $60 registration fee. Undergraduate (general credit) and noncredit tuition is $888 per course, and the registration fee is $20. Tuition covers all necessary charges except phone-line costs.

Nova University of Fort Lauderdale, Florida, offers masters degrees in Information Systems and Electronic Education, and Doctors of Arts in Information Sciences and Information Systems. Students must show up at the Fort Lauderdale campus twice a year for one-week institutes. Annual tuition is $4,500, and they recommend that a student budget an additional $2,500 per year for miscellaneous expenses.

DEVELOPMENTS

Some of the hardware developments that are making these dramatic changes in information access possible are discussed here.

Magnetic floppy diskettes with an optical servo can increase data storage 10 times—to 25 megabytes—over a conventional floppy disk. And 20 MB disks are economical and have become common in personal computers. A 20 MB disk has enough storage to hold 150 books of 500 pages each.

But the inventive people who develop these products have many more hardware innovations coming up. A 256 MB magneto-optical storage disk now available can hold 100,000 pages of text. That's enough for a small home library of 400 books.

Another promising development is called WORM—Write Once, Read Many (times). This is a potentially low-cost digital paper that can be used in place of a floppy disk and store as much as 700 MB. When used as digital tape, a 12-inch reel can store 1,000 gigabytes—enough to fill one billion sheets of paper with an average access time of 28 seconds. The cost is projected to be as low as ½ cent per megabyte, or one-tenth the cost of paper tape. And since it costs so little, digital paper can be a throw-away item.

An estimated less than 2 percent of the information mankind has generated is coded and stored in any kind of computer format. We are only seeing a small part of what lies ahead in the profession of information retrieval. New technologies are going to change our lives in many new and unexpected ways.

EXERCISES

1. Which on-line services would you query to obtain more information for your article topic?

2. If you have access to an on-line service at work, your college, your public library, or your home, contact one of the low-cost services and research your topic.

10

Good Writing

The greatest merit of style, of course,
is to have words disappear into thoughts.

Nathaniel Hawthorne

By now many of you will have concluded that writing of things technical can be fun and that you'd like to learn more about this fascinating and rewarding aspect of your profession and improve your writing even more. This chapter will show you techniques to use that will add a higher polish to your writing and instill in you the confidence to tackle a huge article for a first-rate periodical, perhaps even to start working toward the ultimate goal of most professionals: a book. This chapter gives you a graduate course in writing of things technical. You'll learn what good writing is and how to achieve it.

WHAT IS GOOD WRITING?

Good writing is writing that effectively communicates what the writer has to say to the reader, with a minimum of effort on the reader's part. Simple. You efficiently communicate something to someone.

Good writing uses many of the techniques you've learned throughout your life, plus what you've learned in this book and will learn in this chapter. Good writing has many ingredients:

- Logically organize the material
- Use good grammar
- Check spelling
- Use short sentences and paragraphs (vary their lengths)
- Above all, don't use big words!

Good writing is "invisible." By that I mean good writing helps your reader easily understand what you've written, but does not make obvious what techniques you've used to attain this effect. Properly used, these techniques are totally transparent to your readers.

HOW WE READ

First let's take a look at the basic way people read. The more mental energy it takes to dig meaning out of words, sentences, and paragraphs, the less energy readers have left to do something with what they learn and the more likely they will quit in exhaustion.

When we read technical material, we use energy in three basic ways:

1. To understand the words.
2. To understand the way the words relate to each other to form pictures and thoughts.
3. To understand the technical concepts described by the words and the accompanying visuals.

When we read fiction and nonfiction, we use mental energy only for the first two. Technical writing, however, adds a third and more difficult aspect: comprehending the technical concepts.

If you use short, simple, concrete words, you make the first step easy. If you use a variety of short, well-constructed sentences, you make the second task effortless. That eliminates two-thirds of the effort of reading, so your readers can concentrate all of their energy on understanding the technical concepts.

How do you do this? With good writing.

HOW WE UNDERSTAND

The basic element of writing, the basic symbol that paints a picture in a reader's mind, is the word. But the 500 most-used English words have a total of 14,000 dictionary definitions—that's an average of 28 meanings per word. And words can form radically different shades of meaning in different people's minds. For example, *house* can form a picture ranging from a tiny, rundown shack, to a condo, to a 50-room mansion. So you can see the importance of using the proper word.

Groups of words (phrases and sentences) modify and clarify the intrinsic meaning of the words they contain. Words should be used to sharpen the focus of the picture they are to create. *Man* gives you an out-of-focus, fuzzy, indefinite picture. *Fat man* brings the picture a little more in focus. *Short, swarthy, grossly fat man* sharply focuses the image the words are to convey.

The more specific the word, the more concrete, sharper, and more in focus the picture.

WORD ORDER

It's not only the words you use, but also the word order that is important. Words in your sentences must be arranged so that they can mean only what is intended. In the English language, changing the word order changes the meaning:

> Man bites snake.
> Snake bites man.

The proper word order is needed to paint the exact picture you want.

Another important aspect of helping your readers understand is consecutiveness: things happen in a given order.

> Add one liter of water to the solution, shake vigorously for 10 seconds, then add one gram of KC1.

Using the proper order makes an idea easy to follow.

Still, words are not substitutes for thoughts. You must group words together to form a sentence. But no sentence, regardless of the words you use, is any better than the thought behind it.

A sentence should carry only a single meaning. The basic pattern of a sentence is:

Subject *Verb* *Object*

Thing - - - ->Doings - - - ->Result

Girl - - - ->Throws - - - ->Basketball

Using the proper words in this basic SVO order makes your writing easier to understand because it's the sequence in which things happen. First you see the girl, then the throwing motion, then the basketball, the object being thrown.

If you say "The basketball was thrown by the girl," it's a backward construction. It jerks your mind out of gear. It's not the order in which the event occurred. First your mind focuses on *the basketball,* then on the passive *was thrown,* and finally on *the girl,* the most important part of the sentence. You have to unscramble it in your mind and put it back in the proper order before you can understand it. And this wastes your mental energy.

To make your sentence give a more complete picture, add a few more descriptive words: "The tall, blond girl lofted the basketball into the net."

You must arrange words in the proper grammatical form to carry the right meaning. An estimated 80 percent of poor writing stems from poor grammar.

In a study of 20 top writers (10 fiction and 10 nonfiction), over 75 percent of all the sentences in their writing used a subject-verb-object order: "Man bites dog." So, to be understood, it's best to use the common SVO order in most of your sentences.

We read in terms of words and we understand words within the limits of our vocabulary. Unless the message is contained within the area of the common vocabulary shared by the writer and the reader, as shown in FIG. 10-1, there is no communication. In addition, the words in the message must mean the same to both writer and reader.

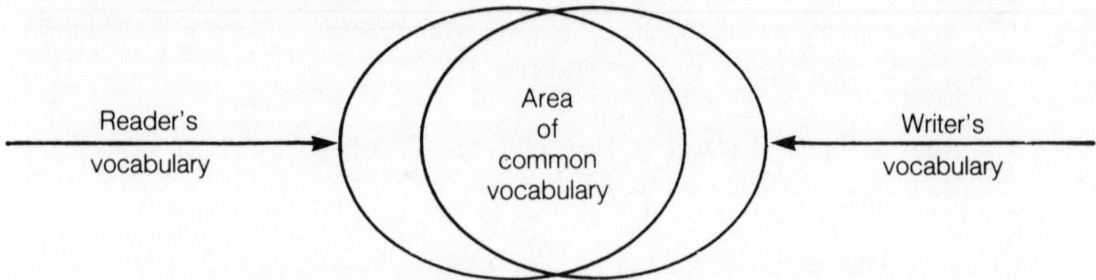

Fig. 10-1. Writer's/reader's vocabularies.

You don't need a large vocabulary for effective communication. An estimated 1,000 words cover about 85 percent of a writer's requirements on all ordinary subjects. Add some of the special words of your profession and you can certainly do most of your writing without having to resort to long, abstract, indefinite words. If you use as many of the 1,000 basic words as you can in place of complex words or jargon, your writing will be that much easier to understand.

GRAMMAR

You must use proper grammar to effectively communicate something to someone, not only to avoid misunderstanding, but also because your reader will notice improper grammar. Grammatical errors distract your reader's attention from the primary purpose of your writing: communicating technical information.

It's no grave sin to misuse a few *whos* or *whoms* because few people understand, or even care about, these finer points of grammar. But if you make obvious mistakes—such as using the wrong tense of verbs, incorrect order of words, or other serious basic errors in grammatical construction—your reader's mind stops understanding the intent of your article and starts looking for more errors in grammar. This distracts the reader's concentration on the technical content.

One of the best general guidelines for technical writing is to write as if you're taking your reader into your confidence. Try to create meaningful comparisons with things your readers are already familiar with. We learn best by analogy, by comparison, by contrast with things we are already familiar with.

SPELLING

When your document is sprinkled with misspelled words, your reader's mind is again diverted from understanding the technical content. Also, when people read a document full of typos, errors, and misspelled words, they begin to distrust the accuracy of the technical content, the visuals—everything associated with the document. With the variety of modern electronic dictionaries and with the excellent spelling programs now available in most word processors, spelling errors should soon become a thing of the past.

Of the approximately 20,000 commonly used words, only 1 percent are consistently misspelled. And of these 200, the 50 that cause the most trouble are listed, left to right, in order of difficulty:

grammar
achieve
separate
existence
assistant
dependent
accommodate
commitment
indispensable
harass
antiquated
vacuum
oscillate
dissension
battalion
inadvertent
inoculate

argument
anoint
desirable
pronunciation
repetition
irresistible
occurrence
embarrass
liaison
perseverance
insistent
ridiculous
tyrannous
connoisseur
prerogative
genealogy
dilettante

surprise
definitely
development
occasion
privilege
consensus
conscience
allotted
proceed
ecstasy
exhilarate
nickel
drunkenness
sacrilegious
iridescent
vilify

If you can study and be extra careful with these words, you'll have a good start.

Concrete Words Are Best

Concrete words paint a more understandable picture than abstract words. But how do you identify concrete words? To test for concreteness, think whether you can:

- Feel it
- See it
- Hear it
- Taste it
- Smell it

Most importantly of all, does the word invoke one, and only one, clear picture in your and the reader's minds?

Set your words in concrete and you'll have a solid foundation for your article.

SHORT IS USUALLY BETTER

Short sentences and paragraphs have been proven to be effective ways to communicate ideas. A reader tries to comprehend an entire sentence in one gulp, looking ahead to the period (the rest stop) before pausing.

Avoid long, convoluted sentences. By the time your readers get to the end of such a sentence, they have forgotten what was at the beginning. Put the main message at the front of the sentence. If your statement has to be qualified, do that in the next sentence.

Paragraphs

A paragraph is a sentence or group of sentences that express and develop one major idea. A paragraph groups together sentences that confirm the same topic and combine to form a thought-unit. The end of a paragraph provides a longer pause for the reader to catch his or her breath. The reader's mind is geared to reading and understanding a single idea, feeling that once a paragraph is read, the idea has been completely presented and he or she deserves a momentary pause before tackling the next idea or paragraph.

Use paragraphs of varied length. You can use this technique to highlight the more important ideas with longer paragraphs, contrasting them with shorter paragraphs that convey less important ideas.

Experience has shown that readers of technical material can grasp material most readily when it is presented in units of 75 to 200 words. With an average of 10 words per typed line, that's a range of 7.5 to 20 lines to allow per paragraph. Again, don't take this guideline as gospel and insist that all paragraphs fit between these limits. These numbers are average numbers, a method to check your writing along the way. Use some one-sentence paragraphs and some much longer ones. Vary their lengths for variety.

The most effective way to introduce the idea that the paragraph is to cover is to use a topic sentence as the first sentence in the paragraph. The topic sentence should contain these essential elements:

- The writer's theme or idea
- His or her viewpoint
- The topic of the paragraph

For example, consider this topic sentence: "Children's television cartoon programs are harmful." The topic sentence conveys the theme that children's television cartoons will be discussed. The viewpoint is subjective, the author's opinion. The topic is the harm that television cartoons can do to children.

A topic sentence indicates the direction an idea is to take. It helps guide the reader to the central purpose of the paragraph. It leads the reader through the development of a topic so that he or she reaches the same conclusion the writer intended. The topic sentence isn't always placed at the beginning; it can be at the end if the paragraph is developed by induction. If a writer considers a point vital, he or she might emphasize the idea at both the beginning and the end of a paragraph. Wherever it is placed, the topic sentence should dominate the paragraph, and the rest of the paragraph should serve mainly to develop it.

The topic sentence is often followed by one or more subtopic sentences that further develop, clarify, and prove the thesis of the topic sentence. Facts, statistics, or quotes (if experts can be included as part of the proof) form the subtopics. Without the proof, the reader will wonder what is the basis for the idea. The last sentence in the paragraph can be a conclusion, a call to action, or a transitional phrase to link to the next topic or idea.

Transitions

Transitions are the bridges that link paragraphs. They provide continuity so the reader can move from one topic to another without feeling a jolt in continuity. Transitions connect what has been said with what is about to be said.

Some more commonly used transitions are:

- Time *Later, the next time, finally, next, concurrently, after.*
- Place *In the next room, farther down the hall, at our plant in Texas.*
- Piling up of detail *And, also, furthermore, in addition, moreover, besides.*
- Contrast *But, however, though, although, nonetheless, yet.*
- Illustration *For example, to illustrate, in particular, for instance.*
- Cause/Effect *Thus, therefore, in conclusion, as a consequence, as a result, consequently.*
- Comparison *In a similar way, likewise, similarly, here again, consequently.*
- Concession *Although, even though, since, though.*
- Summary *To sum up, in brief, in short.*
- Repetition *In other words, that is, as has been stated.*

- Numbered or lettered steps *Used in a procedure.*
- Repetition of and references to preceding ideas, key words, or phrases.

PUNCTUATION

Punctuation is the written equivalent of the rhythm of our speech. It takes the place of the rise and fall of our voice, and the pauses and emphasis in our speech. The major function of punctuation is to make writing clearer and easier to read. Punctuation helps express, question, emphasize, surprise, and conclude.

When we speak, our listeners learn as much from the way we speak as they learn from the words we speak. We gesture, screw up our face, raise or lower our voice, pause, speak fast or slow, speak high or low, use our hands, raise our eyebrows, and use our facial expressions to add emphasis to words.

Comma About half of the total number of punctuation marks used in writing are commas. We use commas to separate words or phrases to avoid confusion and ambiguity. It's the main device by which a grouping of words, phrases, and clauses is indicated. The comma gives a short pause to illustrate and emphasize this separation.

Period The period signifies the end of a sentence. It is equivalent to a longer pause by a speaker. The period used in abbreviations and acronyms is rapidly disappearing from printed matter because it confuses readers. Readers mistakenly conclude that they've reached the end of the sentence. It's better to spell out the words, rather than use abbreviations, except for some of the more commonly used ones, such as *etc.* You can also use the abbreviation or acronym without the periods, as in *USA*.

In spite of the disappearance of periods in abbreviations, technical writing experts agree that the period is not used enough in technical writing. They believe it should be used more often to end sentences faster. Sentences should be shorter.

Semicolon The semicolon is sort of a half-period. It gives a pause of a length between the comma and the period. The semicolon is optional; you don't have to use it. In fact, most great writers seldom, if ever, use it. Some writers have gone an entire lifetime without getting acquainted with the semicolon.

Colon The colon introduces a list of details or an explanation. It is equivalent to a gesture or a lift of the eyebrow in a speaker.

Exclamation Point The exclamation point is the equivalent of a speaker emphasizing a point with a raised voice or a gesture.

Question Mark The question mark takes the place of the voice being raised at the end of a question.

In spite of all of the confusing and sometimes vague rules about punctuation, the best guideline for proper punctuation is common sense. Forget what the grammarians say about periods, commas, and quotes. If you feel that putting a stop in a specific place in your sentence will clarify it for your reader, do so. If you find you're using too many commas in your sentences, perhaps your sentences are too long and confusing. Shorten or reshape your sentences to make them easier to understand.

Like all tools, the use of punctuation marks can be overdone. Too many stops produce a jerky style, hindering your reader from understanding your writing.

THE CURSE OF BIG WORDS

Bertrand Russell stated in his *How I Write,* "Big men write little words, little men write big words."

Technical ideas are difficult enough to understand without a writer complicating them even more by trying to show off his or her vocabulary. You should write to *express,* not to *impress.* Even the most complex scientific ideas can be presented by simple (three-syllable or less) words. Some of the most famous and brilliant scientists of all time, such as Charles Darwin, Louis Pasteur, and Marie Curie, explained their complex concepts in words so simple that the public could understand and appreciate the significance of their discoveries. Undoubtedly their clear writing contributed as much to their fame as did their discoveries.

It takes a little mind (with a dictionary and a thesaurus) to write with big words. It takes a big mind to express complex thoughts and ideas in short words.

DON'T OVERDO IT

If overdone and abused, these techniques lose their effectiveness. If you use only one or two sentences in every paragraph throughout your document, or ten-word "Dick and Jane and Spot" sentences, you again distract your reader from understanding the technical message you want to convey. Your task is to use the techniques of good writing to communicate smoothly, without your reader becoming aware of the techniques you're using. The techniques of good writing should be invisible. Your reader shouldn't know what you're doing to him or her, nor how you're doing it. They should just be able to enjoy your writing, and learn from and understand what you've written.

RETENTION CURVE

One extremely important aspect of the manner in which a person's mind works is illustrated in FIG. 10-2. This curve is valid for all people, of all cultures, regardless of education. It's a fundamental fact of human nature.

Stated simply, the curve illustrates that The mind remembers best what it experiences, hears, and reads first and last. This fascinating fact holds true for

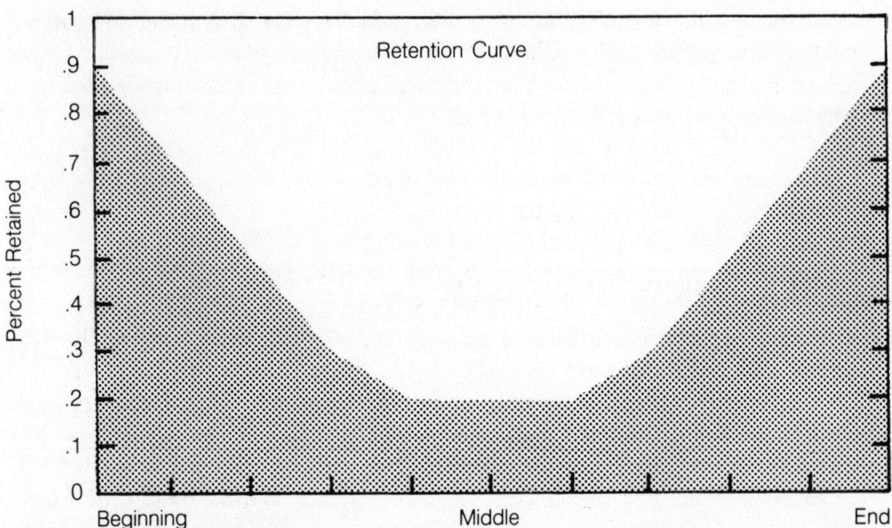

Fig. 10-2. Retention curve.

a book, a movie, a speech, a poem, an article, your life's experiences . . . whatever you see, hear, or read.

That's why the HEY!, YOU!, and SO! sections of your document are so important. Your reader will best remember the beginning and ending sections, so the summary and conclusion of your article are worth your best efforts.

The same phenomenon holds true for both sentences and paragraphs. The first and last parts of sentences and paragraphs are best remembered by your readers. That's why the topic sentence (the first sentence in your paragraph, which introduces the paragraph's contents) is so important. And the final sentence of your paragraph should be a sort of conclusion. Your readers should have the feeling of being introduced to a concept in the first sentence and should end the paragraph by understanding that concept.

This important aspect of the human mind is used in many ways, even by astute politicians. It is an established fact that on election ballots, when a person's name is in the middle of two opponents' names, that person will lose several percentage points off his or her total vote. That is true, regardless of the candidate's qualifications. People simply vote more for the names at the top and bottom of a listing.

You can take advantage of the retention curve in many ways. Keep it in mind when you want to emphasize one or more points. Put the most important points in the first part of your writing, in your first paragraphs, in the first sentences of your paragraphs, and at the beginnings and endings of your sentences. And save some of your best writing for the conclusion.

You can also use the phenomenon of the retention curve to conceal information, bad data, meager results, etc. Bury bad information, which you must include for completeness, in the middle of the document. A strong opening paragraph often will negate the bad information in the middle.

RULES OF THUMB

There are a number of rules of thumb that you'll find useful in writing to help you figure the number of words, the number of pages, etc. It takes about two pages of handwritten material to equal about one page of typed, double-spaced text. This can vary a lot depending on a person's handwriting style. If you write your material long-hand, you should experiment and make a note of its typed equivalent for future reference to properly gauge your writing length. Normal handwriting, double-spaced, includes about 125 words per page. A normal, double-spaced, typewritten page contains about 250 words, that is, about 25 lines of 10 words per line.

When you write an article for a technical periodical, add up the number of double-spaced, typewritten pages, photos, drawings, graphs, tables, etc. that you have generated. Then divide by three to get the number of periodical pages (technical periodicals usually pay by the number of pages).

Also, an article should have about one visual per finished page (or for each two or three typewritten pages) of text so the article will have some white space around it to make the layout more attractive.

Clichés to Avoid

Avoid clichés in writing of things technical. I'm including some of the most often abused ones here:

What the writer said	*What the writer should have said*
The purpose of this report is . . .	This report describes . . .
This report is submitted to . . .	This report sums up . . .
In order to conduct . . .	To conduct . . .
(*In order* can be left out of *all* writing.)	
It is recommended that the formula should be . . .	The formula should be . . .
(Watch out for *It . . . that* constructions; they waste a lot of words).	
It may be said that copper is . . .	Copper is . . .
It is interesting to note that the welded corners . . .	Welded corners will . . .
In accordance with . . .	Per . . .
As defined in . . .	Per . . .
Shall have the capability of converting . . .	Shall convert . . .

The generator shall be tested per . . .	Test the generator per . . .
Due to the fact that . . .	Since . . .
Demonstrates that there is . . .	Shows . . .
During such time . . .	While . . .
If the developments are such that . . .	If . . .
In consideration of the fact that . . .	Since . . ./Because . . .
Make as approximation as to how . . .	estimate . . .
Reduced to basic essentials . . .	simplified . . .
The purpose of this report is . . .	This report shows . . .
Would seem to suggest . . .	Suggests . . .
For the purpose of . . .	For . . . to . . .
Is designed to be . . .	Is . . .
In close proximity . . .	Close to; near . . .
Subsequent to . . .	After . . .
In the event that . . .	If . . .
In order to . . .	To . . .
Involves the use of . . .	Employs . . . uses . . .

IN CONCLUSION

In engineering and science, feedback helps stabilize a device. In our personal relationships it also helps stabilize and develop rapport among people. In writing, an author needs feedback from the readers to make future editions of a book better.

I'd like to hear your comments on how you liked the book, what you did and did not like, what I communicated well, and what I failed to communicate properly. Just write to me, in care of the publisher, and I'll use your feedback to improve future editions of this book.

Thank you. I hope to see all of you in print, soon.

EXERCISES

1. Calculate the fog index of your article (See chapter 2).

2. Calculate the fog index of a *Reader's Digest* or *Wall Street Journal* article.

3. Calculate the fog index of a leading scientific journal in your profession.

Appendix A
Reference Sources

The following categories of reference books are listed alphabetically in this appendix

- Abstracts and indexes
- Almanacs (yearbooks)
- Biographical dictionaries
- Books
- Companies
- Dictionaries, general
- Dictionaries, scientific
- Encyclopedias, general
- Encyclopedias, scientific
- Government publications
- Handbooks
- People
- Periodicals

For the reference books that follow, the title is listed first, followed by the author (if one is listed), then the publisher. A date such as (1941–) means the book was first published in 1941 and is still being published.

ABSTRACTS AND INDEXES

In this section you'll find the most up to date periodical, report, and document references. All of these abstracts and indexes, plus many more, are available on the various on-line systems.

Abstracts and Indexes in Science and Technology, 2nd ed. Metuchen, NJ: Dolores B. Owen, Scarecrow Press, 1985. Gives descriptions of 223 abstracts and indexes arranged in 11 subject categories, including on-line databases.

Applied Science and Technology Index, Bronx, NY: H. W. Wilson. (1958–). Published monthly, except August, with quarterly and annual accumulations. From 1913 to 1957, this index was known as the *Industrial Arts Index.* Arranged alphabetically by subject only. Lists articles from over 300 English-language periodicals in applied science and technology. Excellent source for the latest articles on new and old science and technology.

Astronomy and Astrophysics Abstracts, New York: Springer-Verlag. (1969–) Semiannual. Subject and author index to periodicals throughout the world printed in English, with some in French and German.

Biological and Agricultural Index, Bronx, NY: H. W. Wilson. (1964–) Alphabetical subject index to approximately 200 English-language periodicals in biological and agricultural sciences. Also has book reviews.

Biological Abstracts, Philadelphia, PA: BIOSIS. (1926–). Worldwide reporting of research in life sciences. Principal abstracting journal for biology. Contains abstracts, author index, biosystematic index, generic index, and subject index.

Business Periodicals Index, Bronx, NY: H. W. Wilson (1958–). Indexes by subject, about 275 English-language periodicals in all fields of business. Index is issued monthly and accumulated into annual volumes.

Chemical Abstracts, New York: American Chemical Society (1907–). Subject and author index to abstracts from more than 14,000 periodicals in more than 50 languages. Known as the "key to the world's chemical literature," *Chemical Abstracts* is the principal source for titles and abbreviations in the physical sciences, life sciences, and engineering.

Chemical Titles, Columbus, OH: Chemical Abstract Service (1960–). Author and keyword computer-produced index to 700 periodicals covering pure and applied chemistry and chemical engineering.

Computer and Control Abstracts: Science Abstracts, Series C, London: Institute of Electrical Engineers, and New York: Institute of Electrical and Electronic Engineers. Monthly indexes and abstracts international electrotechnology literature by author and subject. Separate index to bibliographies, books, reports, conference proceedings, and patents.

Computer Abstracts, Technical Information Co. (1957). Offers abstracts of books, periodical articles, conference proceedings, U.S. Government reports, and patents in classified arrangements. Annual and subject indexes.

Engineering Index, New York: Engineering Information Inc. (1884–). Monthly with annual accumulations. The basic English-language abstracting service. Abstracts over 2,700 professional and technical journals, as well as reports and proceedings published in 20 or more languages. A general index to engineering literature, arranged by subject, with an author index.

General Science Index (GSI), Bronx, NY: H. W. Wilson (1978–). Indexes more than 100 general science periodicals not completely covered by other abstracts and indexes in astronomy, chemistry, electricity, mathematics, physics, etc.

Index Medicus, Washington, DC: Washington National Library of Medicine (1960–). Contains subject, name, and bibliography of medical reviews. Over 2,000 periodicals are indexed, either completely or selectively.

International Aerospace Abstracts (IAA), Washington, DC: American Institute of Aeronautics and Astronautics. (1961–). Semi-monthly, with annual accumulations. Includes books, periodicals, conference papers, and translations. Indexed by subject and author. Companion service to *Scientific and Technical Aerospace Abstracts,* which covers "unpublished" material on the same topic.

Mathematical Reviews, Providence, RI: American Mathematical Society (1940–). Subject index to mathematical periodicals and books. Arranged by broad subject, with abstracts for most entries. Comprehensive coverage of the pure and applied mathematics literature.

Metals Abstracts, Metals Abstracts Trust (1968–). Author index to abstracts from 1,000 periodicals throughout the world, covering all aspects of the science and practice of metallurgy and related fields.

New York Times Index, New York: New York Times (1913–). Indexes all articles published in the *New York Times,* preserved on microfilm and carefully indexed by subject alphabetically. Semi-monthly.

Nuclear Science Abstracts—Energy Research Abstracts, Oak Ridge, TN: U.S. ERDA (1948–1976). Covers technical reports of ERDA and its contractors in nuclear science and technology. Has detailed author and subject index. Covers books, articles, and papers on nuclear science.

Physics Abstracts, Science Abstracts, Series A, Institute of Electrical Engineering, London; and Institute of Electrical and Electronic Engineering, New York (1898–). Monthly. Separate index to bibliographies, books, conferences, patents, and reports.

Reader's Guide to Periodical Literature, Bronx, NY: H. W. Wilson (1900–). Best-known popular periodical index. Author and subject index to general interest periodicals (limited, however—only covers about 174 of the thousands of magazines in circulation), including a few scientific periodicals published in the United States. From 1802–1906, it was titled *Poole's Index to Periodical Literature.*

Science Citation Index, Philadelphia, PA: Philadelphia Institute for Scientific Information (1961–). Lists the reference or coded author with his or her work for over 3,800 journals and monographic series. International. Provides access to books, papers, articles, and reports being cited in currently published papers.

Scientific and Technical Aerospace Abstracts (STAR), Washington, DC: U. S. Government Printing Office (1963–). Comprehensive abstracts worldwide of unpublished reports on the sciences and technology of space and aeronautics, especially NASA reports. Indexed by subject, report number, accession number, and individual and/or corporate authors.

Technical Abstracts Bulletin (TAB), Washington, D.C.: Defense Documentation Center (1953–). Abstracts and announces classified and unclassified or limited-distribution documents produced by the Department of Defense and its contractors and acquired by DDC.

U. S. Government Research and Development Reports, Washington, D.C.: Clearinghouse for Federal Scientific and Technical Information (1946–). Abstracting and announcement bulletin covering reports, including progress reports of R&D under government auspices.

ALMANACS

Statistics, summaries, general interest facts, updated every year. A surprising amount of information in each single volume.

Information Please Almanac, New York: Simon and Schuster (1947–). More legible and easier to use than the *World Almanac,* but its coverage is not as complete.

McGraw-Hill Yearbook of Science and Technology, New York: McGraw-Hill, annual. Reviews the past year's works in science and technology. Supplements the McGraw-Hill *Encyclopedia of Science and Technology.*

World Almanac and Book of Facts, New York: World Telegram (1868–). Annual. Often cited as the best-selling American reference work, as well as the most comprehensive almanac and most frequently useful.

BIOGRAPHICAL DICTIONARIES

Biographies are useful when you need to look up names and affiliations of experts in your field you want to consult with or write about.

American Men and Women of Science, New York: R. R. Bowker, 7 volumes (1982). Brief biographies of over 130,000 living U.S. and Canadian scientists, including positions held, education, area of specialty, etc.

Biography Index, Bronx, NY: H. W. Wilson, Quarterly (1947–). Name and profession index to biographical material in 1,500 periodicals and books. The most comprehensive index in the field. Arranged alphabetically by the name of the subject of the biography.

Chamber's Biographical Dictionary, New York: St. Martin's (1969). A good dictionary covering the great of all nations, both living and dead.

Dictionary of Scientific Biography, New York: Scribner (1970–80). 16 volumes. Comprehensive, covers all periods from antiquity to present and includes only deceased scientists. Entries give place and date of birth and death, and brief summaries of the individual's contributions to science.

Who's Who in America, Wilonette, IL: Marquis (1899–). Biennial. Best-known and most useful general dictionary of contemporary biography. People included from all fields, science, education, business, etc.

World Who's Who in Technology Today, J. Dick (1984). 5v. Covers electronics and computer science, physics and optics, chemistry and biotechnology, mechanical and civil engineering, energy and earth sciences.

BOOKS

If the information you're searching for is in a book, a number of potential sources exist for locating that book, in addition to using an electronic or manual card catalog.

Book Review Digest, Bronx, NY: H. W. Wilson (1905–). Gives digests of book reviews taken from some 75 American and English general interest magazines. Arranged by title, it carries title and subject indexes.

Book Review Index, Detroit, MI: Gale Research Co. (1965–). Author index to reviews of books in more than 450 general interest magazines.

Books in Print, New York: R. R. Bowker (1948–). Annual. Lists the books from some 3,600 publishers that are in print and can be purchased. Has author, title, and subject listings in separate volumes.

A Brief Guide to Sources of Scientific and Technical Information, Saul Herner, Arlington, VA: Information Resources Press. Guide to major sources of technical information, intended for the engineer and scientist. Indexed.

Cumulative Book Index, Bronx, NY: H. W. Wilson (1898–). Lists by subject and author English-language books in print of general interest, including the sciences. Arranged by author, title, and subject.

New Technical Books, New York: New York Public Library (1915–). Classed subject arrangement includes table of contents and annotations for each book. Subject emphasis is on the pure and applied sciences, mathematics, engineering, industrial technology, and related disciplines.

Technical Book Review Index, Washington, DC: Special Libraries Association (1935–). Monthly except July and August. Index to book reviews appearing in scientific, technical, and trade journals. Provides brief quotations from reviews. Best book review index for scientific books. Arranged by author of book reviewed.

COMPANIES

Sometimes you might need to contact specific companies in your field, or in a related field, and need their names and addresses. Try the excellent directories listed here.

Standard and Poor's Register of Corporations, Executives, and Industries, New York: Standard and Poor's (1928–). 3v. Listings of over 45,000 companies. This is the "who's who" of American companies. One volume is an alphabetical listing by company name; a second volume is a directory of executives and board members; the third volume is a set of indexes to the first volume.

Thomas' Register of American Manufacturers, New York: Thomas Publishing Co. (1906–). Annual. Massive listing. Provides a comprehensive, detailed guide to the full range of products manufactured in the United States. Lists manufacturers' names, addresses, products, and trade names.

The Yellow Pages of Your Telephone Directory. Not only for your immediate city, but telephone directories of many other cities are kept in your local library.

DICTIONARIES, GENERAL

For general definitions (good for crossword puzzles, too), these are the best:

- *A Dictionary of American English*
- *Funk and Wagnalls*
- *Oxford English Dictionary.* A classic! *Warning:* you can easily get captivated when using this fascinating dictionary. It is the most authoritative dictionary of the English language and gives the *etymology,* or history, of the words.
- *The Random House Dictionary of the English Language*
- *Roget's International Thesaurus*
- *Webster's 3rd New International Dictionary,* Merriam-Webster Inc. 2,600 pages. Can be used for years.

DICTIONARIES, SCIENTIFIC

For quick definitions of terms that might puzzle you as you search through other literature, check these.

Aviation/Space Dictionary, The, Larry Reithmaier, Fall River, MA: Aero (1989). Definitions related to aircraft armament, power plants, airline operations, and air traffic control. A good, basic aerospace reference work.

Basic Dictionary of Science, New York: Macmillan (1966). A dictionary of 25,000 terms, followed by lists of abbreviations used in science. Chemical elements with their atomic weights and numbers. Written for those with little or no scientific background.

Concise Chemical and Technical Dictionary, New York: Chemical Publishing (1974). About 90,000 entries covering trademark products, chemicals, drugs, and terms.

Dictionary of Computers, Data Processing, and Telecommunications, New York: John Wiley (1984). Definitions of some 10,000 words and phrases. Has appendix of Spanish and French equivalents.

Thesaurus of Engineering and Scientific Terms, Engineers Joint Council (1967). 690 pp. Has over 20,000 entries of engineering and related scientific terms and their relationships.

IEEE Standard Dictionary of Electrical and Electronic Terms, Institute of Elect. and Electronic Engr. New York: John Wiley and Sons, Inc. Excellent dictionary, an official source of definitions taken from standards of IEEE, ANSI, and IEC.

McGraw-Hill Dictionary of Scientific and Technical Terms, New York: McGraw-Hill (1984). 1781 pp. Over 98,500 terms with 115,000 definitions from technology and science are defined clearly and concisely.

Modern Dictionary of Electronics, New York: McGraw-Hill (1984). Brief descriptions of several thousand terms in current use. Has pronunciation guide and defines widely used symbols.

ENCYCLOPEDIAS, GENERAL INTEREST

If you know little or nothing about your topic, a general encyclopedia is often a good place to start. Occasionally they have excellent summaries on what you're looking for and, more importantly, good references to other documents having greater depth. Usually they're well cross-referenced. Encyclopedias are usually more up to date than books because they're revised more often.

Academic American, Danbury, CT: Grolier, Inc. 21 v. Falls somewhere between the *World Book* and the *Britannica* in scope and depth of treatment. The full text of the encyclopedia is available on-line through a number of commercial vendors. Intended for students in junior high, high school, and college, as well as inquisitive adults.

Colliers Encyclopedia, New York: Macmillan. Emphasis on simple explanations. Strong in contemporary science. Third in size to *Americana* and *Britannica,* but the most current, best indexed, and easiest to read of the three. Style is popular, clear, and concise.

Consise Columbia Encyclopedia, The, New York: Columbia University Press. One volume, capsule size for quick reference. Contains more separate entries than most English-language encyclopedias.

Encyclopedia Americana, Danbury, CT: Grolier, Inc. 30v. Full and scholarly. Strong on American topics, especially strong in science and technology. Good, comprehensive encyclopedia for general use.

Encyclopaedia Britannica, Chicago, IL: Encyclopedia Britannica. Full and scholarly. Good coverage of both British and American topics. The most famous encyclopedia in English. Long, detailed articles on many and diverse subjects.

Funk and Wagnalls New Encyclopedia, New York, Funk and Wagnalls. 27 v. Serves general family needs. Provides brief background on a wide variety of topics and is written in a clear, popular style. Useful, inexpensive choice.

World Book Encyclopedia, Raleigh, NC: Fields Enterprises Corp. Ranges from young people to general adult in content. Keyed to school curricula.

ENCYCLOPEDIAS, SCIENTIFIC

Scientific encyclopedias are good for general, basic information about a wide variety of subjects. However, they are not detailed technically.

Concise Encyclopedia of the Sciences, New York: Facts on File (1978). Complete guide to the language and history of science and technology. Covers the most commonly used words of science and technology, with background material.

Encyclopedia of the Biological Sciences, New York: Van Nostrand Rheinhold (1970). 1027 pp. Articles on instruments in theory and practice, techniques of microscopy, and preparation disciplines. List of references follows each article.

Encyclopedia of Chemical Technology, New York: John Wiley (1980–84) 24 v. Articles written by specialists, includes bibliographies. Uses SI as well as English units.

Encyclopedia of Computer Science and Engineering, New York: Van Nostrand Rheinhold Co. (1983). 1,664 pages of descriptions of basic computer terms. 550 articles for the nonspecialist. Cross-referenced; excellent index.

Encyclopaedic Dictionary of Physics, Elmsford, NY: Pergammon Press (1961–1975). 9 v. Scholarly work, alphabetically arranged. Articles on general, nuclear, solid-state, molecular, chemical, metal, and vacuum physics.

Encyclopedia of Engineering Signs and Symbols, New York: Odyssey Press (1965). A comprehensive dictionary of signs, symbols, and abbreviations used in engineering.

Harper Encyclopedia of Science, New York: Harper and Row (1967). 1379 pp. One volume. Excellent, readable, covering a wide range of topics. Biographies are included. Entries are brief but informative.

Encyclopedia of Polymer Science and Technology, New York: John Wiley Interscience (1964–72). 14 v. Contains equations, graphs, tabular data, and extensive bibliographic references. Covers all aspects of polymer science (physics, chemistry, biology) and engineering.

McGraw-Hill Encyclopedia of Science and Technology, New York: McGraw-Hill (1982). 15 v. Covers all major scientific subjects. Not too technical, easy to read. Some 7,600 articles by 2,500 scientists and engineers. Many illustrations. Alphabetically arranged with cross references. This is the best of the science and technology encyclopedias, kept up to date by the *McGraw-Hill Yearbook of Science and Technology.*

Van Nostrand Scientific Encyclopedia, New York: Van Nostrand Reinhold (1988). 3264 pp. Defines about 16,500 terms in the physical sciences, computer technology, electrical engineering, electronics, pure and applied math, etc. Over 2,000 illustrations, arranged alphabetically, with excellent cross references.

GOVERNMENT PUBLICATIONS

Guide to Popular U. S. Government Publications, Englewood, CO: Libraries Unlimited (1986). 432 pp. References 2,900 titles of government documents, most of which have been published since 1978, with a brief description of each document's contents.

Monthly Catalog of U. S. Government Publications, Washington, DC: U. S. Govt. Printing Office (1895–). Comprehensive index. Lists all publications of all government departments, bureaucracies, and agencies. Published monthly, with an annual cumulative index. Gives author, title, publication data, price, and availability. Does not include classified documents.

Subject Guide to U. S. Government Reference Sources, Chicago, IL: American Library Association (1985). Subject index, plus a directory. Annotated bibliography of key government documents on general topics, including science and technology. Over 1,300 entries.

U.S. Government Reports Announcement and Index, Springfield, VA: U. S. Dept. of Commerce, National Technical Information Service (NTIS) (1971–). Semimonthly, with annual accumulations. Indexes and abstracts unclassified reports of U. S. government contractors in the public and private sector. Indexed by individual and corporate author, subject, report number, and accession number.

HANDBOOKS

Computer Dictionary and Handbook, Indianapolis, IN: Howard Sams (1980). Dictionary and handbook of 22,000 terms in electronic data processing, information technology, computer sciences, and automation.

CRC Handbook of Chemistry and Physics, Boca Raton, FL: CRC Press (1913–). An indispensable tool for chemists and physicists, this is a ready reference book of chemical and physical data.

Handbook of Mathematical Tables and Formulas, Richard S. Burington. New York: McGraw-Hill (1973). 500 pp. Quick reference to many vital tables of data in math, engineering, physics, chemistry, and physics.

PEOPLE

Authors of articles Why not contact the author of one or more of the reports, books, etc. in your field of interest that you found particularly informative. He or she will usually cooperate and bring you up to date on what has been happening since the article was published.

Magazine editors They're an excellent source for information about what's going on in a certain field, and they can direct you to experts in your specialty.

People at other companies Often even your competitors will consult with you on a problem, particularly if it's on a person-to-person basis, rather than on a company-to-company basis.

Your friendly librarian Librarians are always willing to help and can locate even the most obscure facts, or anything you need, in books you never even dreamed existed.

Your fellow employees Check around your company; you might find an expert or two.

PERIODICALS

When you need to search specific periodicals in your field of interest, the following books will help you locate them.

Catalog of Scientific and Technical Periodicals (2nd ed. 1965), Washington, DC: Smithsonian Institute (1665–1895). Titles of more than 8,600 pure and applied science periodicals published throughout the world, from the rise of literature to the present time.

Business Periodicals Index, Bronx, NY: H. W. Wilson (1958–). Monthly except July. Subject index to about 275 English-language periodicals in all areas of business, computers, accounting, marketing, communications, etc. Arranged alphabetically by subject.

Encyclopedia of Associations, Detroit, MI: Gale Research Company (1956–). 5 v. Excellent reference. Lists some 18,000 national and international organizations. Includes scientific, engineering, and technical associations. Has key word, geographical, and executive index, research activities, and funding programs.

Gale Directory of Publications (Formerly Ayer Directory of Publications), Detroit, MI: Gale Research, Inc. (1989). Index of 25,000 newspapers and magazines published in the United States, Canada, and a few other countries.

Scientific, Engineering, and Medical Societies Publications in Print, New York: R. R. Bowker (1980–81). 626 pp. Lists publications published by 301 technical societies. Arranged alphabetically by society. Author index and periodical title index.

Standard Periodical Directory, New York: Oxbridge (1964/64–). The largest authoritative guide to U. S. and Canadian publications. Information on more than 65,000 publications. Alphabetical arrangement, with index of titles and subjects.

Ulrich's International Periodical Directory, New York: R. R. Bowker (1932–). 2 v. Lists over 70,000 currently published periodicals from more than 120 countries. Grouped by subject, with title and subject indexes. The most comprehensive of scientific (especially scholarly) publications from around the world. Also cites the indexing or abstracting services for each periodical. A very valuable reference.

Appendix B
List of Periodicals

Most of these periodicals have writer's guidelines. To request a copy, include a cover letter and an SASE.

68 Micro Journal
Computer Publishing, Inc.
5900 Cassandra Smith Rd.
Hixson, TN 37343-0794

73 Amateur Radio
70 Route 202, North
Peterborough, NH 03458-1194

A+
Ziff-Davis Pubs. Co.
950 Tower Lane
Foster City, CA 94404

AOPA
Aircraft Owners and Pilots Association
Pilot Aircraft Owners
421 Aviation Way
Frederick, MD 21701

Adhesives Age
6255 Barfield Rd.
Atlanta, GA 30328

Aerospace America
American Institute of Aeronautics and Astronautics
370 L'Enfant Promenade, S.W.
Washington, D.C. 20024

Aerospace Engineer
400 Commonwealth Dr.
Warrendale, PA 15096

Ahoy
Haymarket Group Ltd.
45 W. 34th St., Suite 407
New York, NY 10001

Air Conditioning, Heating & Refrigeration News
Business News Publishing Co.
P.O. Box 2600
Troy, MI 48007

Air & Waste Management Association
P.O. Box 2861
Pittsburgh, PA 15230

Airport Services Management
50 S. Ninth St.
Minneapolis, MN 55402

American Biotechnology Lab
International Scientific
Communications, Inc.
30 Controls Dr.
P.O. Box 870
Shelton, CT 06484-0870

American Ceramic Society Bulletin
757 Brooksedge Plaza Dr.
Westerville, OH 43081-6136

American Machinist Penton Publications
1100 Superior Ave.
Cleveland, OH 44114

AMERICAN MANAGEMENT ASSOCIATION
135 W. 50th St., 15th Floor
New York, NY 10020-1201

AMERICAN SCHOOL AND UNIVERSITY
401 N. Broad St.
Philadelphia, PA 19108

AMERICAN SOCIETY FOR ENGINEERING EDUCATION
11 Dupont Circle, Suite 200
Washington, D.C. 20036

American Tool, Die and Stamping News
Eagle Publications Inc.
31505 Grand River, No. 1
Farmington, MI 48024

America's Textile
2100 Powers Ferry Rd., Suite 125
Atlanta, GA 30357

Analytical Chemistry
American Chemical Society
1155 16th St., N.W.
Washington, DC 20036

Animal Kingdom
New York Zoological Society
185th St. and Southern Blvd.
Bronx, NY 10460

Appliance Manufacturer
Corcoran Communications, Inc.
29100 Aurora Rd.
Solon, OH 44139

Applied Optics
Optical Society of America
1816 Jefferson Place, N.W.
Washington, D.C. 20036

Applied Radiology
30 Vreeland Rd.
Florham Park, NJ 07932

Architectural & Engineering Systems
Softek Communications, Inc.
760 Whalers Way
Suite 100, Bldg. A
Fort Collins, CO 80525

Architectural Record
McGraw-Hill Information Systems
1221 Avenue of the Americas
New York, NY 10020

Architecture
American Institute of Architects
1735 New York Ave., N.W.
Washington, DC 20006

ASHRAE Journal
1791 Tullie Circle N.E.
Atlanta, GA 30329

The Asphalt Contractor
Specialty Publications Corp.
552 S. Brookside
Independence, MO 64053

Assembly Engineering
Hitchcock Publishing Co.
25 W. 550 Geneva Rd.
Wheaton, IL 60188

Astronomy
1027 N. Seventh St.
Milwaukee, WI 53233

Automatic Machining
100 Seneca Ave.
Rochester, NY 14621

Automation
Penton Publishing
1100 Superior Ave.
Cleveland, OH 44114

Automotive Engineering
Society of Automotive Engineers, Inc.
400 Commonwealth Dr.
Warrendale, PA 15096

Automotive Industries
Chilton Company
Chilton Way
Radnor, PA 19089

Avionics
Phillips Publishing, Inc.
7811 Montrose Rd.
Potomac, MD 20854

Barron's National Business and Financial Weekly
200 Liberty St.
New York, NY 10281

Better Roads
P.O. Box 558
Park Ridge, IL 60068

Bio Science
730 11th St., N.W.
Washington, DC 20001-4584

Boxboard Containers
MacLean Hunter Publishing Co.
29 N. Wacker Dr.
Chicago, IL 60606

Broadcast Engineering
Intertec Publishing Co.
P.O. Box 12901
Overland Park, KS 66212

Broadcast Management/Engineering
295 Madison Ave.
New York, NY 10017

Bulletin of the Atomic Scientists
6042 South Kimbark Ave.
Chicago, IL 60637

Business Marketing
Crain Communications
1400 Woodbridge
Detroit, MI 48207

Byte
McGraw-Hill Inc.
One Phoenix Mill Lane
Peterborough, NH 03458

Cable Television Business
Cardiff Publishing
6300 S. Syracuse Way, Suite 650
Englewood, CO 80111

California Engineer
California Engineer Publ. Co.
221 Bechtel Engr. Ctr.
Berkeley, CA 94720

Car Craft
8490 Sunset Blvd.
Los Angeles, CA 90069

Casting World
Continental Communications
P.O. Box 1919
Bridgeport, CT 06601-1919

Certified Engineering Technician
American Society of Certified Engineering Technicians
P.O. Box 371474
El Paso, TX 79937

Chemical and Petroleum Engineering
Plenum Publishing Corp.
233 Spring St.
New York, NY 10013

Chemical Business
Schnell Publishing Co.
80 Broad St.
New York, NY 10004-2203

Chemical Engineering
McGraw-Hill
1221 Avenue of the Americas
New York, NY 10020

Chemical Engineering Progress
American Institute of Chemical Engineers
345 E. 47th St.
New York, NY 10017

Chemical Processing
Penton Publishers
301 E. Erie St.
Chicago, IL 60611

Chief Engineer
Chief Engineers Association of Chicago
11340 W. 159th St.
Overland Park, IL 60462

Civil Engineering
American Society of Civil Engineers
345 East 47th St.
New York, NY 10017-2398

Coal
11 W. 19th St.
New York, NY 10011

Coal
Maclean Hunter Publishing Company
29 N. Wacker Dr.
Chicago, IL 60606

Combustion and Flame
Elsevier Science Publishing Co.
655 Avenue of the Americas
New York, NY 10010

Commodore Magazine
1200 Wilson Dr.
West Chester, PA 19380

Communication News
HBJ Publications, Inc.
124 S. First St.
Geneva, IL 60134

Compressed Air
253 E. Washington Ave.
Washington, NJ 07882-2495

Computer-Aided Engineering
Penton Pub.
1100 Superior Ave.
Cleveland, OH 44114

Computerworld
IDG Communications
375 Cochituate Rd. Box 9171
Framingham, MA 01701

Computer Design
Penwell
One Technology Park
Westford, MA 01886

Computer Graphics Today
Media Horizons Inc.
50 W. 23rd St.
New York, NY 10010

Computers and Industrial Engineering
Pergamon Journals, Inc.
Maxwell House Fairview Park
Elmsford, NY 10523

Computers in Physics
American Institute of Physics
335 E. 45th St.
New York, NY 10017

Concrete Construction Magazine
426 S. Westgate
Addison, IL 60101-9929

Concrete Products
29 N. Wacker Dr.
Chicago, IL 60606

Constructor
1957 E. Street N.W.
Washington, DC 20006

Consulting Engineer
1350 E. Touhy, No. 5080
Des Plaines, IL 60018

Control Engineering
Cahners Publishing Company
1350 E. Touhy, No. 5080
Des Plaines, IL 60018

Cost Engineering
American Assn. of Cost Engineers
308 Monogahela Bldg.
Morgantown, WV 26505-5468

DE/Domestic Engineering
Delta Communications
385 N. York Rd.
Elmhurst, IL 60126

Defense Electronics
1170 E. Meadow Dr.
Palo Alto, CA 94303

Defense Science & Electronics
Rush Franklin Building
300 Orchard City Dr. Suite 234
Campbell, CA 95008

Designfax
International Thomson Indus. Press, Inc.
6521 Davis Industrial Parkway
Solon, OH 44139

Design News
Cahners Publishing Co.
275 Washington St.
Newton, MA 02158-1630

Diesel & Gas Turbine Worldwide
13555 Bishop Court
Brookfield, WI 53005-6286

Dr. Dobbs Journal
M & T Publishing
501 Galveston Dr.
Redwood City, CA 94063

Drilling and Engineering
Society of Petroleum Engineers
P.O. Box 833836
Richardson, TX 75083-3836

EDN
Cahners Publishing Company
275 Washington St.
Newton, MA 02158-1630

EMC Technology
Interference Control Technologies, Inc.
Rte 625, P.O. Box D
Gainesville, VA 22065

ENR Engineering News Record
McGraw-Hill Information Systems
1221 Avenue of the Americas
New York, NY 10020

Electric Light and Power
Pennwell Publishing Co.
1421 S. Sheridan Rd.
Tulsa, OK 74112

Electrical Construction and Maintenance
McGraw-Hill Information Services
1221 Avenue of the Americas
New York, NY 10020

Electrical Wholesaling
1221 Avenue of the Americas
New York, NY 10020

Electrical World
McGraw-Hill
11 W. 19th St.
New York, NY 10011

Electronic Business
Cahners Publishing Co.
275 Washington St.
Newton, MA 02158-1630

Electronic Design
Hayden Publishing/VNU
10 Mulholland Dr.
Hasbrouck Hts., NJ 07604

Electronic/Electrical Product News
707 Westchester Ave.
White Plains, NY 10604

Electronic Engineering Times
CMP Publications
600 Community Dr.
Manhasset, NY 11030

Electronic Manufacturing
Lake Publishing Corp.
17730 W. Peterson Rd. Box 159
Libertyville, IL 60048

Electronic Packaging & Production
1350 E. Touhy Ave.
P.O. Box 5080
Des Plaines, IL 60017-5080

Electronic Products
645 Stewart Ave.
Garden City, NY 11530

Electronic Servicing & Technology
P.O. Box 12901
Overland Park, KS 66212-9981

Electronics
VNU Business Publications, Inc.
Ten Mulholland Dr.
Hasbrouck Heights, NJ 07604

Energy Management Technology
Walter-Davis Publications
2500 Office Center
Willow Grove, PA 19090

Engineered Systems
Business News Publishing Co.
P.O. Box 2600
Troy, MI 48007

Engineering Education
American Society for Engineering Education
Eleven Dupont Circle Suite 200
Washington, D.C. 20036

Engineering and Mining Journal
300 W. Adams St.
Chicago, IL 60606

Engineering Tools
VNU Business Publications
10 Mulholland Dr.
Hasbrouck Heights, NJ 07604

Environmental Science & Technology
American Chemical Society
1155 16th St. N.W.
Washington, D.C. 20036

Equal Opportunity Publications
44 Broadway
Greenlawn, NY 11740

Evaluation Engineering
Nelson Publishing
2504 Tamiami Trail
Nokomis, FL 34275

Executive Computing
Association of Computer Users
P.O. Box 2189
Berkeley, CA 94702

Food Engineering
Chilton Company
Chilton Way
Radnor, PA 19089

Food Processing
301 E. Erie St.
Chicago, IL 60611

Foundry Management and Technology
Penton Publishing Co.
1100 Superior Ave.
Cleveland, OH 44114

Geology
Geological Society of America
3300 Penrose
Boulder, CO 80301

Geotimes
American Geological Inst.
4220 King St.
Alexandria, VA 22302-1507

Graduating Engineer
McGraw-Hill
1221 Avenue of the Americas Ste 4360
New York, NY 10020

Ham Radio
Communications Technology, Inc.
Main St.
Greenville, NH 03048

Harvard Business Review
Soldiers Field
Boston, MA 02163

Heating/Piping/Air Conditioning
Penton Publishing Co.
1100 Superior Ave.
Cleveland, OH 44114

Heating, Air Conditioning & Plumbing Products
P.O. Box 1952
Dover, NJ 07801-0952

High Performance Systems
600 Community Dr.
Manhasset, NY 11030

High Technology Careers
Westech Publishing Co.
4701 Patrick Henry Dr.
Santa Clara, CA 95054

Home Office Computing
730 Broadway
New York, NY 10003

Hybrid Circuit Technology
Lake Publishing Co.
17730 W. Peterson Rd. Box 159
Libertyville, IL 60048

Hydraulics and Pneumatics
Penton Publishing
1100 Superior Ave.
Cleveland, OH 44114

Hydrocarbon Processing
P.O. Box 2608
Houston, TX 77252

IAN-Instrument & Control News
Chilton Company
Chilton Way
Radnor, PA 19089

I&CS-Control Tech. for Engrs. and Engr. Mgmt
Chilton Company
Chilton Way
Radnor, PA 19089

IEEE Spectrum
Inst. of Elec. and Electronic Engrs.
345 E 45th St.
New York, NY 10017

INTECH
Instrument Society of America
P.O. Box 12277
Research Triangle Park, NC 27709

Industrial Chemist
McGraw-Hill
1221 Avenue of the Americas
New York, NY 10020

Industrial Design
Design Publications, Inc.
330 W. 42nd St.
New York, NY 10036

Industrial Distribution
Cahners Publishing Co.
275 Washington St.
Newton, MA 02158

Industrial Engineering
Institute of Industrial Engineers
25 Technology Park/Atlanta P.O. Box 6510
Norcross, GA 30091-6150

Industrial Finishing
Hitchcock Building
Wheaton, IL 60188

Industrial Heating
National Industrial Publishing Co.
1000 Killarney Dr.
Pittsburgh, PA 15234

Industrial Maintenance & Plant Operation
Chilton Company
Chilton Way
Radnor, PA 19089

Industrial Photography
210 Crossways Park Dr.
Woodbury, NY 11797

Industrial Safety & Hygiene News
Chilton Company
Chilton Way
Radnor, PA 19089

Information Display
201 Varick St. Suite 1140
New York, NY 10014

Infoworld
1060 Marsh Rd.
Menlo Park, CA 94025

Iron and Steel Engineer
Three Gateway Center Suite 2350
Pittsburgh, PA 15222

JAPCA
Three Gateway Center, Four West
Pittsburgh, PA 15230

Journal of Accountancy
1211 Avenue of the Americas
New York, NY 10036

Journal of Electronic Defense
Horizon House
685 Canton St.
Norwood, MA 02062

Journal of Petroleum Technology
Society of Petroleum Engineers
P.O. Box 833836
Richardson, TX 75083-3836

Journal of Small Business Management
West Virginia University
P.O. Box 6025
Morgantown, WV 26506-6025

Journal of Systems Management
24587 Bagley Rd.
Cleveland, OH 44138

Laboratory Management
Nature Publishing Co.
65 Bleecker St.
New York, NY 10012-2420

Laser Focus
Penn Well Publishing
One Technology Park Drive
Westford, MA 01886

Lasers and Optronics
High Tech Pubs
P.O. Box 650
Morris Plains, NJ 07950

Lubrication Engineering
Society of Tribologists & Lubrication Engineers
838 Busse Highway
Park Ridge, IL 60068

Manage
2210 Arbor Blvd.
Dayton, OH 45439

Managing Automation
Thomas Publishing
One Penn Plaza
New York, NY 10001

Manufacturing Engineering
One SME Dr.
P.O. Box 930
Dearborn, MI 48121

Material Handling Engineering
Penton Publishing
1100 Superior Ave.
Cleveland, OH 44114

Materials Engineering
Penton Publishing
1100 Superior Ave.
Cleveland, OH 44114

Mechanical Engineering
American Society of Mechanical Engineers
345 E. 47th St.
New York, NY 10017

Medical Electronics and Equipment News
532 Busse Highway
Park Ridge, IL 60068

Metalworking News
Fairchild Publications
Seven E. 12th St.
New York, NY 10003

Metalworking Digest
Gordon Publications Inc.
Box 1952
Dover, NJ 07801

MicroCAD News
Ariel Communications
12710 Research, Suite 250
P.O. Box 203550
Austin, TX 78759

Microtimes
BAM Publications
5951 Canning St.
Oakland, CA 94609

Microwave Journal
685 Canton St.
Norwood, MA 02062

Microwave News
P.O. Box 1799
Grand Central Station
New York, NY 10163-1799

Microwaves and RF
VNU Business Publications
10 Mulholland Dr.
Hasbrouck Heights, NJ 07604

The Military Engineer
Society of American Military Engineers
607 Prince St.
P.O. Box 21289
Alexandria, VA 22320-2289

Mini-Micro Systems
Cahners Publishing Co.
275 Washington St.
Newton, MA 02158-1630

Mining Engineering
Society of Mining Engineers, Inc.
P.O. Box 625002
Littleton, CO 80162-5002

Modern Casting
American Foundrymen's Society
Golf & Wolf Rds.
Des Plaines, IL 60016-2277

Modern Electronics
CQ Commun., Inc.
76 N. Broadway
Hicksville, NY 11801

Modern Machine Shop
6600 Clough Pike
Cincinnati, OH 45244

Modern Materials Handling
Cahners Publishing
275 Washington St.
Newton, MA 02158

Modern Metals
Delta Communications, Inc.
400 N. Michigan Ave.
Chicago, IL 60611

Modern Plastics
1221 Avenue of the Americas
New York, NY 10020

Modern Railroads
Int. Thomson Transport Press
424 33rd St.
New York, NY 10001

National Engineer
National Association of Power Engineers
2350 E. Devon Ave. Suite 115
Des Plaines, IL 60018

National Petroleum News
950 Lee St.
Des Plaines, IL 60016

National Public Accountant
1010 N. Fairfax St.
Alexandria, VA 22314

National Safety and Health News
National Safety Council
444 N. Michigan Ave.
Chicago, IL 60611

Nibble
52 Domino Dr.
Concord, MA 01742

Nuclear News
American Nuclear Society
555 N. Kensington Ave.
La Grange Park, IL 60525

Nuclear Times
1601 Conn. Ave. N.W. No. 300
Washington, DC 20009

Oil and Gas Journal
Penwell Publishing
P.O. Box 1260
Tulsa, OK 74101

Packaging
Cahners Publishing Company
1350 E. Touhy No. 5080
Des Plaines, IL 60017-5080

Paper, Film & Foil Converter
MacLean Hunter Publishing
29 N. Wacker Dr.
Chicago, IL 60606

Paper Trade Journal
Vance Publishing
400 Knightsbridge Pkwy.
Lincolnshire, IL 60069

Personal Computing
VNU Business Pub., Inc.
10 Mullholland Dr.
Hasbrouck Hts., NJ 07604

Petroleum Engineer International
Edgell Communications Inc.
7500 Old Oak Rd.
Cleveland, OH 44130

Photomethods
1090 Executive Way
Des Plaines, IL 60018

Pipeline & Gas Journal
Edgell Communications Inc.
7500 Old Oak Rd.
Cleveland, OH 44130

Pipe Line Industry
3301 Allen Parkway Box 2608
Houston, TX 77001

Plant Engineering
Cahners Publishing Company
249 W. 17th St.
New York, NY 10011

Plant Services
301 E. Erie St.
Chicago, IL 60611

Plastics Engineering
Society of Plastic Engineers
14 Fairfield Dr.
Brookfield, CT 06804

Plastics Technology
110 N. Miller Rd.
Akron, OH 44313

Plastics World
Cahners Publishing Company
275 Washington St.
Newton, MA 02158

Power
McGraw-Hill
11 W. 19th St.
New York, NY 10011

Power Engineering
1421 Sheridan Rd.
Tulsa, OK 74112

Power Transmission Design
Penton Publishers
1100 Superior Ave.
Cleveland, OH 44114

Precision Metal
Penton Publishing Co.
1100 Superior Ave.
Cleveland, OH 44114

Private Pilot
Fancy Publications, Inc.
Box 6050
Mission Viejo, CA 92690

Product Design and Development
Chilton Co.
Chilton Way
Radnor, PA 19089

Programmable Controls
ISA Services, Inc.
67 Alexander Dr.
Box 12277
Research Triangle Park, NC 27707

Public Power
2301 M St., N.W.
Washington, D.C. 20037

Pulp and Paper
500 Howard St.
San Francisco, CA 94105

QST, ARRL
225 Main St.
Newington, CT 06111

Quality
Hitchcock Publishing
25W550 Geneva Rd.
Wheaton, IL 60187

RF Design
Cardiff Publishing Co.
6300 S. Syracuse Way, Suite 650
Englewood, CO 80111

Radio-Electronics
500-B Bi-County Blvd.
Farmingdale, NY 11735

Radiology Today
Slack, Inc.
6900 Grove Rd.
Thorofare, NJ 08086-9447

Railway Age
345 Hudson St.
New York, NY 10014

Research and Development
Cahners Publishing Company
249 W. 17th St.
New York, NY 10011

Rock Products
29 N. Wacker St.
Chicago, IL 60606

Rural Electrification
1800 Massachusetts Ave. N.W.
Washington, D.C. 20036

SAMPE Journal
Society for the Advancement of
Material
P.O. Box 2459
Covina, CA 91722

SMPTE Journal
Society of Motion Picture and Television Engineers
595 W. Hartsdale Ave.
White Plains, NY 10607

Safety & Health
National Safety Council
444 N. Michigan Ave.
Chicago, IL 60611

Science
American Assoc. for the Advancement of Science
1333 H St. N.W.
Washington, D.C. 20005

Science & Technology
48 E. 43rd St.
New York, NY 10017

Science Education
John Wiley & Sons, Inc.
605 Third Ave.
New York, NY 10158

Science News
1719 N Street N.W.
Washington, DC 20036

Scientific American
415 Madison Ave.
New York, NY 10017

Sea Technology
Compass Publications
1117 N. 19th St., Suite 1000
Arlington, VA 22209

Sky and Telescope
P.O. Box 9111
Belmont, MA 02178-9111

Solid State Technology
14 Vanderventer Ave.
Port Washington, NY 11050

Space Technology: Industrial & Commercial Applications
Pergamon Journals, Inc.
Maxwell House Fairview Park
Elmsford, NY 10523

Spectroscopy
Aster Publishing Co.
859 Willamette St.
P.O. Box 10955
Eugene, OR 97440

Supervision
424 N. 3rd St.
Burlington, IA 52601-5224

Telecommunications
685 Canton St.
Norwood, MA 02062

Test & Measurement World
Cahners Publishing Co.
275 Washington St.
Newton, MA 02158

Test Engineering & Management
The Mattingley Pub. Co., Inc.
3756 Grand Ave., Suite 205
Oakland, CA 94610

The Physics Teacher
American Association of Physics Teachers
5112 Berwyn Rd. 2nd Floor
College park, MD 20740

Today's Chemist
American Chemical Society
500 Post Road E.
P.O. Box 231
Westport, CT 06881

Tooling and Production
6521 Davis Industrial Parkway
Solon, OH 44139

Traffic Safety
National Safety Council
444 N. Michigan Ave.
Chicago, IL 60611

Traffic World
1325 G St. N.W. Suite 900
Washington, DC 20005

Transmission & Distribution
Andrews Communications, Inc.
5123 W. Chester Pike
P.O. Box 556
Edgemont, PA 19028

TV Technology
5827 Columbia Pike, Suite 310
Falls Church, VA 22041

Water Engineering and Management
Scranton Gilette Communications, Inc.
380 Northwest Highway
Des Plaines, IL 60016

Welding Journal
American Welding Society
Box 351040
Miami, FL 33125

World Oil
Gulf Publishing Do.
Box 2608
Houston, TX 77252

Appendix C
Addresses of On-Line Vendors

American Open University
New York Institute of Technology
Central Islip, NY 11722
1-800-222-6948

BRS Information Technologies
1200 Route 7
Latham, NY 12110
1-800-468-0908

Chemical Abstracts Service
A Division of the American Chemical Society
2540 Olentangy River Rd.
P.O. Box 3012
Columbus, OH 43210-0012
614-447-3600
1-800-848-6538: Customer Service
1-800-848-6533: STN Search Assistance Desk

CompuServe Information Services
5000 Arlington Centre Blvd.
P.O. Box 20212
Columbus, OH 43220
614-457-8650
1-800-848-8990

Connect.ed
92 Van Cortlandt Park South, #6F
Bronx, NY 10463
212-549-6509

Delphi
3 Blackstone St.
Cambridge, MA 02139
1-800-544-4005
617-491-3393

Dialog Information Services, Inc.
3460 Hillview Ave.
Palo Alto, CA 94304
415-858-3785
1-800-3-DIALOG

Dow Jones
P.O. Box 300
Princeton, NJ 08543-0300
609-452-1511

GEnie
General Electric Information Services
401 N. Washington St.
Rockville, MD 20850
301-340-4000
1-800-638-9636

Knowledge Index
415-858-3785
1-800-334-2564

Mead Data Central
9443 Springboro Pike DM
P.O. Box 933
Dayton, OH 45401
513-865-6800
1-800-227-4908

NEWSNET
945 Haverford Rd.
Bryn Mawr, PA 19010
800-537-0808 in PA
1-800-345-1301

Nova University
3301 College Ave.
Fort Lauderdale, FL 33314
1-305-475-7047

Orbit Search Service
8000 Westpark Dr.
McLean, VA 22102
703-442-0900
1-800-456-7248

Prodigy
445 Hamilton Ave.
White Plains, NY 10601
914-993-8000

Videolog Communications
50 Washington St.
Norwalk, CN 06854
203-838-5100
1-800-843-3656

VU/TEXT Information Services, Inc.
325 Chestnut St. Suite 1300
Philadelphia, PA 19106
215-574-4416
1-800-323-2940

Appendix D
Exchange Sort

A variation to the method of chapter 5, known as an *exchange sort,* is covered here. To demonstrate this method, the identical cards and numbers used in chapter 5 will be used.

Place the cards in a single row as illustrated in FIG. D-1. To begin the procedure, compare the first card, A-1941, with the second card, B-187. B-187 is smaller than A-1941, so exchange their positions, as illustrated in FIG. D-2.

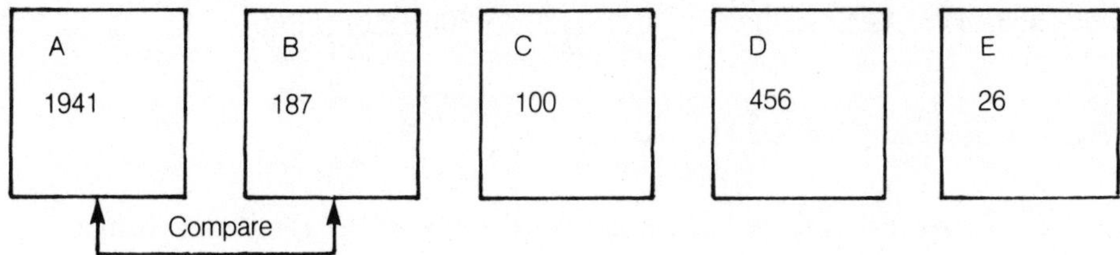

Fig. D-1. Original order.

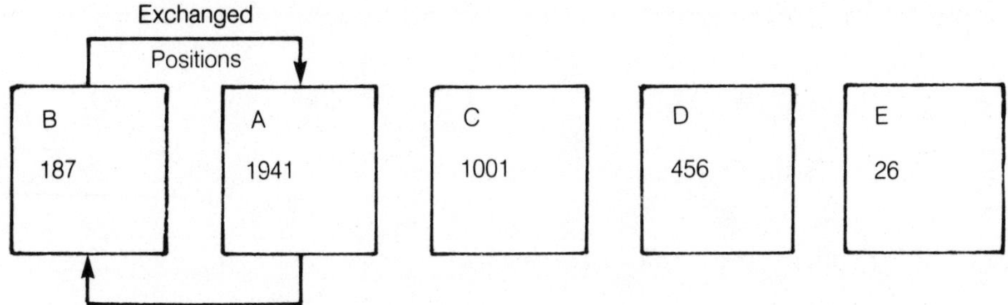

Fig. D-2. Results of the first comparison.

For the second step, compare A-1941 with C-1001. Since C-1001 is smaller, exchange its position with A-1941, as illustrated in FIG. D-3.

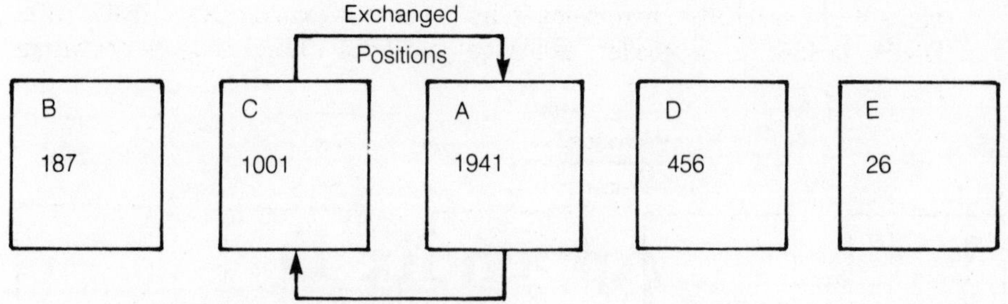

Fig. D-3. Results of the second comparison.

For the next comparison, note that D-456 is smaller than A-1941, so exchange their positions as shown in FIG. D-4.

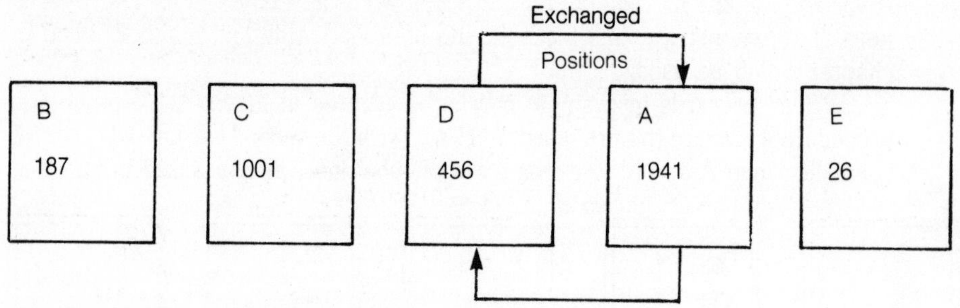

Fig. D-4. Results of the third comparison.

In the next comparison, E-26 is smaller than A-1941 so the two exchange positions as shown in FIG. D-5.

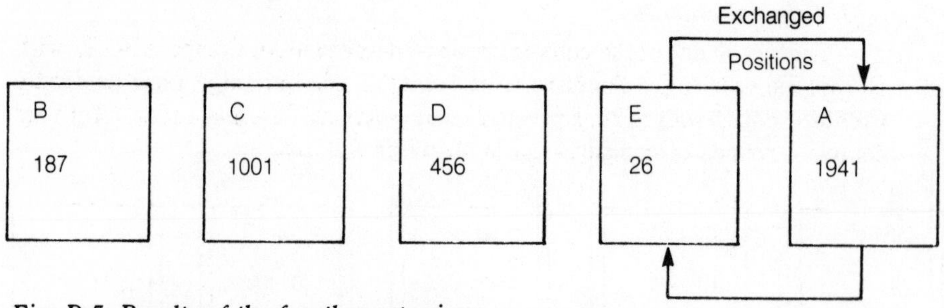

Fig. D-5. Results of the fourth comparison.

So, at the end of the first round of comparisons, the order is as shown in FIG. D-5. Note that the largest number, A-1941, has moved to the rightmost position. Also note that four (*n*—1) comparisons were needed to sort five (*n*) cards.

To begin the second round of comparisons, compare B-187 with C-1001.

Since B-187 is smaller, it remains in its position. Next compare C-1001 with D-456. D-456, being smaller, exchanges positions with C-1001 as shown in FIG. D-6.

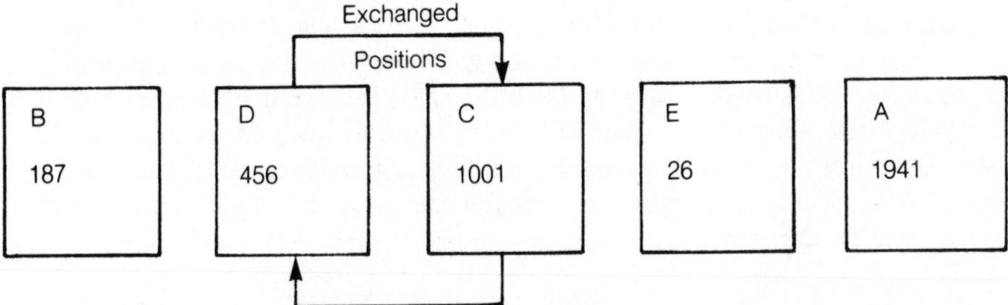

Fig. D-6. Results of the second comparison, Round 2.

For the next comparison, E-26 and C-1001 exchange positions, as shown in FIG. D-7. Note that since the largest number, A-1941, has already moved to the rightmost position, a comparison with C-1001 is not necessary.

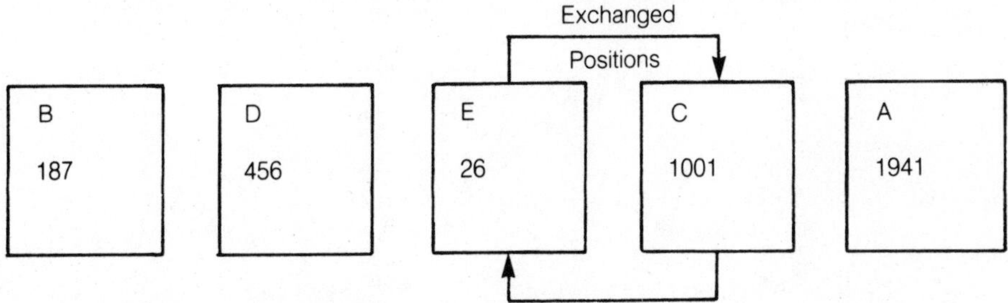

Fig. D-7. Results of the third comparison, Round 2.

The remainder of the comparisons are detailed here: Compare B-187 with D-456. No exchange. Compare D-456 with E-26. Exchange their positions. Compare B-187 with E-26. Exchange their positions. The final order, after four complete rounds of comparisons, is shown in FIG. D-8.

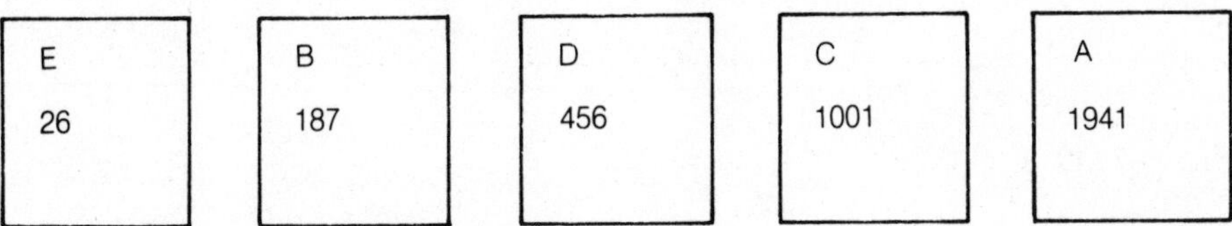

Fig. D-8. The final results—Sorted order.

Practice both this method and the method described in chapter 5 to see which one works the best for you. One disadvantage of the open-card method discussed here is that all the cards are face up and might distract you when you try to make comparisons. When you use this method, it might work best if you pick up the two cards you are comparing so you can concentrate on the two topics without being distracted.

Another possible variation, particularly if you have a large number of topic cards, is to place the cards in one stack and go through the deck, top to bottom repeatedly. Compare one pair of the topics at a time and exchange their positions if one topic is to be covered before the other. Keep exchanging their order as you work through the deck until all of the cards are in the desired order. Whichever method you use, although the mechanics might differ, the results will be the same.

Glossary

abstract A brief summary of an article, book, or report, which also lists author, title, publisher, and date of publishing.

acoustic coupler An early type of modem with rubber cups that cradle over the handset of a telephone.

agate line A unit of measure used in selling newspaper advertising space. Fourteen agate lines equal one column inch.

ampersand The word used to denote "&," the symbol for the word *and.*

analog signals The signals carried over telephone lines.

ASCII code The standard code developed by the American Institute for Information Interchange that is used for all communications and represents letters, numbers, and symbols as bit patterns. ASCII files can be read by practically all software.

asynchronous Of or referring to a method of transmitting signals in which data is sent one bit at a time.

autodial A modem feature that lets the computer automatically dial a telephone number that is stored in the computer's memory.

back matter The printed matter following the end of the text of a book.

basis weight The weight of a ream (500 sheets) of paper. One pound of 20 lb. paper weighs 20 lbs.

baud rate A technical term referring to the rate at which data is transmitted over telephone lines. In most cases, the baud approximately equals bits per second.

bibliography A list of reference literature located in the back matter of a book or at the end of an article or chapter.

boldface Type that is heavier than the text type with which it is used.

BPS Abbreviation for bits per second, the speed at which data is transmitted.

buffer Computer memory set aside to store information obtained from an on-line service.

bullet A solid dot character used ornamentally.

camera-ready Of or referring to a pasteup that can be photographed as is to produce a negative for plate-making, without the need to typeset text or redraw artwork.

contact print A photographic print made with a negative or positive in contact with sensitized paper. No camera is necessary. Images are reversed, as from negative to positive, and prints can only be made the same size as the original.

cropping Marking a drawing or photograph to indicate that only a portion of the artwork is to be used, instead of physically cutting out the desired area. Usually, the entire piece of art is actually reproduced as film and the undesired image area then cut away.

direct connect A design that allows a modular telephone cord to connect directly into a phone line.

downloading The process of receiving information from one computer and storing it on another.

electronic mail A service that lets you send messages to other users on the same on-line system.

elite A type font that has 12 characters per linear inch and 6 lines to the vertical inch.

errata A list of errors, corrections, and deletions discovered after an article or book has been printed.

exploded view An illustration that displays separate parts of a component displaced outward in order of disassembly.

flush Even with margins. For example, for flush right all characters line up vertically at the right-hand margin.

font A complete alphabet of any one typeface in a given point size—uppercase, lowercase, numerals, punctuation marks, etc.

front matter All matter preceding the first page of text.

full duplex Simultaneous transmission of data in both directions.

galley proofs A preliminary reproduction of text composition for the purpose of checking spelling, spacing, etc. before pasteup or makeup.

half duplex The transmission of data in only one direction at a time.

halftone A photo, engraved plate, or printed illustration in which a range of solid tones is reproduced by a pattern of dots of varying sizes.

hardcopy A printed copy of writing.

Hayes-compatible Of or referring to modems that use commands originated by Hayes Microcomputer Products. Hayes compatibility is not an absolute requirement, but is a de facto industry standard.

head, heading A title or caption at the head of a chapter, section, column, list, table, or illustration.

line drawing A drawing made from lines with no gradation of gray tones and no dot pattern.

log-on The process of accessing an on-line service.

masthead The matter printed in each issue of a periodical stating the ownership, title, editorial staff, etc.

modem A contraction of *mo*dulator/*dem*odulator, the interface box or card that converts a signal into two different frequencies so that it can be transmitted over telephone lines and that also performs the reverse function, converting the signal on the phone lines to ones that can be handled by a computer. The modem converts the computer's 1's and 0's into two musical tones that whistle down the telephone line and are turned back into 1's and 0's at the receiving end.

parity A setting for an error-checking bit during transmission of data.

pica A standard type font that has ten characters per inch.

point A printer's measure equal to 1/72 inch.

protocol The agreed upon settings for the transfer of information between systems. Typical protocols are XMODEM, Y-MODEM, Z-MODEM, CompuServe-B, Kermit, and ASCII. The two computers must agree on parity and protocol for them to be able to interchange information.

recto A right-hand, odd-numbered page.

RS-232C The standard established by the Electronic Industries Association for serial transmission of data for telecommunications.

running head A title repeated at the top of each page of a book.

soft copy A copy of written material that is stored in digital form on a diskette or other storage media. A printer converts soft copy to hardcopy.

stet A printer's term meaning "Let it stand," used in editing to retain material that has been crossed off.

stub The list of subjects or entries at the left of a table.

subhead A secondary headline or title.

subtitle A secondary title, often an explanation or expansion of the main title.

synchronous Of or referring to a system of transmission in which data is sent in blocks and the receiver and sender must be in synchronism.

terminal A device that can send and receive information from a computer.

uploading Transmitting information locally and storing it at a remote location.

verso A left-hand, even-numbered page.

widow A single word or partial line of words spilling over from the previous page and appearing at the top of the next page.

Bibliography

Adams, James L. *The Care & Feeding of Ideas.* Reading, MA: Addison-Wesley, 1986.

Alley, Michael. *The Craft of Scientific Writing.* Englewood Cliffs, NJ: Prentice Hall, 1987.

Blicq, R.S. *Technically—Write!* Englewood Cliffs, NJ: Prentice Hall, 1972.

Dodds, Robert H. *Writing for Technical and Business Magazines.* New York: John Wiley & Sons, 1969.

Ehrlich, Eugene, and Daniel Murphy. *The Art of Technical Writing.* New York: Bantam Books, 1964.

Emerson, Connie. *Write on:Target,* Cincinnati, OH: Writer's Digest, 1981.

Ferrarini, Elizabeth. *Infomania.* Boston, MA: Houghton Mifflin, 1985.

Glossbrenner, Alfred. *How to Look it up Online.* New York: St. Martins Press, 1987.

————————. *Personal Computer Communications.* New York: St. Martins Press, 1985.

Glover, John A. *Becoming a More Creative Person.* Englewood Cliffs, NJ: Prentice-Hall, 1980.

Hicks, Tyler G. *Writing for Engineering and Science.* New York: McGraw-Hill, 1961.

Helliwell, John. *Inside Information,* New York: New American Library, 1986.

Bibliography

Hoover, Hardy. *Essentials for the Scientific and Technical Writer.* New York: Dover Publications, 1980.

Li, Tze-chung. *An Introduction to Online Searching.* Westport, CT: Greenwood Press, 1985.

Mills, Gordon H. and John A. Walter. *Technical Writing.* New York: Holt, Rhinehart and Winston, 1962.

Mullins, Carolyn J. *The Complete Writing Guide,* Englewood Cliffs, NJ: Prentice-Hall, 1980.

Sherman, Theodore A. and Simon S. Johnson. *Modern Technical Writing.* New York: Prentice Hall, 1975.

Tichy, H.J. *Effective Writing.* New York: John Wiley & Sons, 1966.

Turner, Barry T. *Effective Technical Writing and Speaking.* London: Business Books, 1978.

Index

Y

Other Bestsellers of Related Interest

ENCYCLOPEDIA OF ELECTRONICS
—2nd Edition—Stan Gibilisco and Neil Sclater, Co-Editors-in-Chief
Praise for the first edition:
". . . a fine one-volume source of detailed informationfor the whole breadth of electronics."

—Modern Electronics

The second edition, newly revised and expanded, brings you more than 950 pages of listings that cover virtually every electronics concept and component imaginable. From basic electronics terms to state-of-the-art applications, this is the most complete and comprehensive reference available for anyone involved in any area of electronics practice! 976 pages, 1400 illustrations, Book No. 3389, $68.95 hardcover only.

AutoCAD PROGRAMMING—
Dennis N. Jump

CAD expert Dennis Jump offers you a straightforward, comprehensive guide to the popular CAD software package that includes version 2.0. Jump explains in detail the data structures and algorithms associated with AutoCAD programming, including numerous sample program listings in both C and BASIC. You'll learn how to write application programs that use AutoCAD as a companion, as well as how to use data within AutoCAD to display the images, drawings, diagrams, and more created with your application programs. 288 pages, 150 illustrations, Book No. 3093, $24.95 paper, $33.95 hardcover.

UNDERSTANDING DIGITAL ELECTRONICS
—2nd Edition—R.H. Warring and Michael J. Sanfilippo

This revised edition of the bestselling guidebook to digital electronics is the perfect tool to help you keep up with the growth and change in technology. It's a quick and complete resource of all the principles and concepts of digital circuits, providing coverage of important areas such as binary numbers, digital logic gates, Boolean algebraic theorems, flip-flops and memories, number systems, and arithmetic logic units (including the 74181 ALU). 196 pages, 172 illustrations, Book No. 3226, $14.95 paperback, $22.95 hardcover.

COMPUTER TECHNICIAN'S HANDBOOK
—3rd Edition—Art Margolis
"This is a clear book, with concise and sensible language and lots of large diagrams . . . use [it] to cure or prevent problems in [your] own system . . . the [section on troubleshooting and repair] is worth the price of the book." ***—Science Software Quarterly***

MORE than just a how-to manual of do-it-yourself fix-it techniques, this book offers complete instructions on interfacing and modification that will help you get the most out of your PC. 580 pages, 97 illustrations, Book No. 3279, $24.95 paperback, $36.95 hardcover only.

Other Bestsellers of Related Interest

ELECTRONIC DATABOOK—
Fourth Edition—Rudolf F. Graf

If it's electronic, it's here—current, detailed, and comprehensive! Use this book to broaden your electronics information base. Revised and expanded to include all up-to-date information, the fourth edition of *Electronic Databook* will make any electronic job easier and less time-consuming. This edition includes information that will aid in the design of local area networks, computer interfacing structure, and more! 528 pages, 131 illustrations, Book No. 2958, $24.95 paperback, $34.95 hardcover.

ELECTRONICS EQUATIONS HANDBOOK—Stephen J. Erst

Here is immediate access to equations for nearly every imaginable application! In this book, Stephen Erst provides an extensive compilation of formulas from his 40 years' experience in electronics. He covers 21 major categories and more than 600 subtopics in offering the over 800 equations. This broadbased volume includes equations in everything from basic voltage to microwave system designs. 280 pages, 219 illustrations, Book No. 3241, $16.95 paperback, $24.95 hardcover.

INVENTING: Creating and Selling Your Ideas—Philip B. Knapp, Ph.D.

If you've ever said, "Somebody ought to invent a (so-and-so) that will do (such-and-such) . . ." you should read this guide. Philip B. Knapp supplies valuable advice and practical guidance on transforming your ideas into working items and selling them! 250 pages, Book No. 3184, $15.95 paperback, $24.95 hardcover.

THE COMPLETE HANDBOOK OF MAGNETIC RECORDING—
3rd Edition—Finn Jorgensen

This book covers virtually every aspect of the magnetic recording science. A recognized classic in its field, this comprehensive reference makes extensive use of illustrations, line drawings, and photographs. Audio recording, instrumentation recording, video recording, FM and PCM recording, as well as the latest digital techniques are thoroughly described. 768 pages, 565 illustrations, Book No. 3029, $44.50 hardcover only.

UNDERSTANDING LASERS—
Stan Gibilisco

If you could have only one book that would tell you everything you need to know about lasers and their applications—this would be the book for you! Covering all types of laser applications—from fiberoptics to supermarket checkout registers—Stan Gibilisco offers a comprehensive overview of this fascinating phenomenon of light. He describes what lasers are and how they work, and examines in detail the different kinds of lasers in use today. 180 pages, 96 illustrations, Book No. 3175, $14.95 paperback, $23.95 hardcover.

EYEWITNESS *TRAVEL GUIDES*

VENICE & THE VENETO

Main contributors:
SUSIE BOULTON
CHRISTOPHER CATLING

DORLING KINDERSLEY
LONDON • NEW YORK • SYDNEY • MOSCOW

A Dorling Kindersley Book

Produced by Pardoe Blacker Publishing Limited,
Lingfield, Surrey
Project Editor Caroline Ball
Art Editor Simon Blacker
Editors Jo Bourne, Molly Perham, Linda Williams
Designers Kelvin Barratt, Dawn Brend, Jon Eland,
Nick Raven, Steve Rowling
Picture Research Jill De Cet
Managing Editor Alan Ross
Project Secretary Cindy Edler

Dorling Kindersley Limited
Deputy Editorial Director Douglas Amrine
Deputy Art Director Gaye Allen
Map Co-ordinators Simon Farbrother, David Pugh
Production Hilary Stephens

Contributor (Travellers' Needs) Sally Roy

Maps
Phil Rose, Jennifer Skelley, Jane Hanson
(Lovell Johns Ltd, Oxford UK)
Street Finder maps based upon digital data, adapted
with permission from L.A.C. (Italy)

Photographers
John Heseltine (Venice), Roger Moss (Veneto)

Illustrators
Arcana Studios, Donati Giudici Associati srl,
Robbie Polley, Simon Roulstone

•

Film outputting by The Best Bureau Centre Limited
(East Grinstead)
Reproduced by Colourscan (Singapore)
Printed and bound by G. Canale & C. (Italy)

First published in Great Britain in 1995
by Dorling Kindersley Limited
9 Henrietta Street, London WC2E 8PS
Reprinted with revisions 1995, 1997 (twice)

Visit us on the World Wide Web at http://www.dk.com

A CIP catalogue record is available from the British Library.

ISBN 0-7513-0103-5

•

Every effort has been made to ensure that the information in this book is as up-to-date as possible at the time of going to press. However, details such as telephone numbers, opening hours, prices, gallery hanging arrangements and travel information are

Contents

The Venetian explorer Marco Polo

Introducing Venice and the Veneto

Venice Area by Area

The Rialto Bridge, on the Grand Canal

Veronese's *Passion and Virtue* in the Villa Barbaro at Masèr

The Veneto Area by Area

The medieval Palio dei Dieci Comuni at Montagnana

Travellers' Needs

Survival Guide

Anguilla in umido

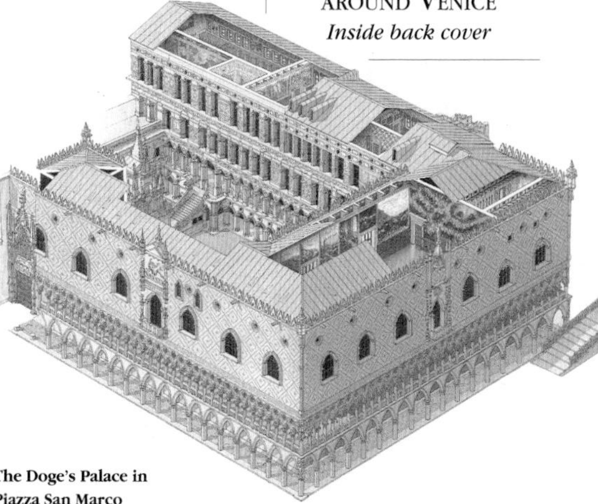
The Doge's Palace in Piazza San Marco

How to Use this Guide

This guide helps you get the most from your stay in Venice and the Veneto. It provides both expert recommendations and detailed practical information. *Introducing Venice and the Veneto* maps the region and sets it in its historical and cultural context. *Venice Area by Area* and *The Veneto Area by Area* describe the important sights, with maps, pictures and detailed illustrations. Suggestions for food, drink, accommodation, shopping and entertainment are in *Travellers' Needs*, and the *Survival Guide* has tips on everything from the Italian telephone system to travelling around Venice by *vaporetto*.

Venice Area by Area

The city has been divided into five sightseeing areas. The lagoon islands make up a sixth area. Each area has its own chapter, which opens with a list of the sights described. All the sights are numbered and plotted on an *Area Map*. The detailed information for each sight is presented in numerical order, making it easy to locate within the chapter.

Each area of Venice can be quickly identified by its colour coding.

A locator map shows where you are in relation to other areas of the city.

Sights at a Glance lists the chapter's sights by category: Churches; Museums and Galleries; Historic Buildings; Palaces; Streets, Bridges and Squares.

1 Area Map
For easy reference, the sights are numbered and located on a map. The sights are also shown on the Venice Street Finder *on pages 280–89.*

2 Street-by-Street Map
This gives a bird's eye view of the heart of each sightseeing area.

Stars indicate the sights that no visitor should miss.

A suggested route for a walk covers the more interesting streets in the area.

3 Detailed information on each sight
All the sights in Venice are described individually. Addresses, telephone numbers, nearest vaporetto *stop, opening hours and information on admission charges are also provided.*

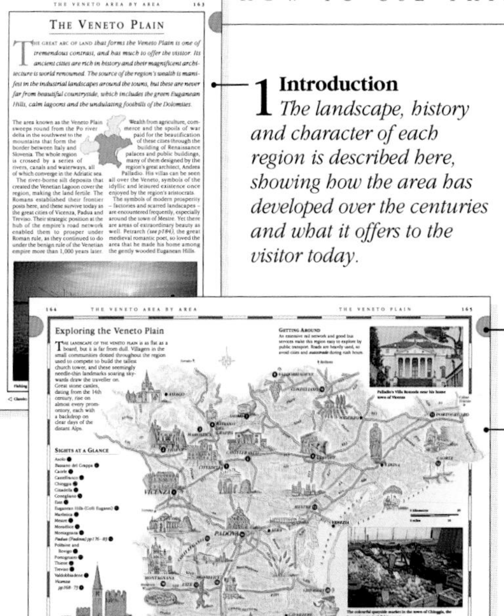

1 Introduction
The landscape, history and character of each region is described here, showing how the area has developed over the centuries and what it offers to the visitor today.

THE VENETO AREA BY AREA
In this book, the Veneto has been divided into three regions, each of which has a separate chapter. The most interesting sights to visit have been numbered on a *Pictorial Map*.

Each area of the Veneto can be quickly identified by its colour coding.

2 Pictorial Map
This shows the road network and gives an illustrated overview of the whole region. All the sights are numbered and there are also useful tips on getting around the region by car, bus and train.

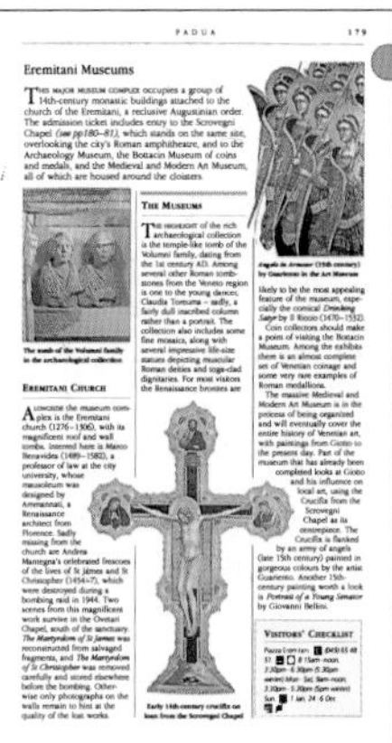

3 Detailed information on each sight
All the important towns and other places to visit are described individually. They are listed in order, following the numbering on the Pictorial Map. *Within each town or city, there is detailed information on important buildings and other sights.*

Stars indicate the best features and works of art.

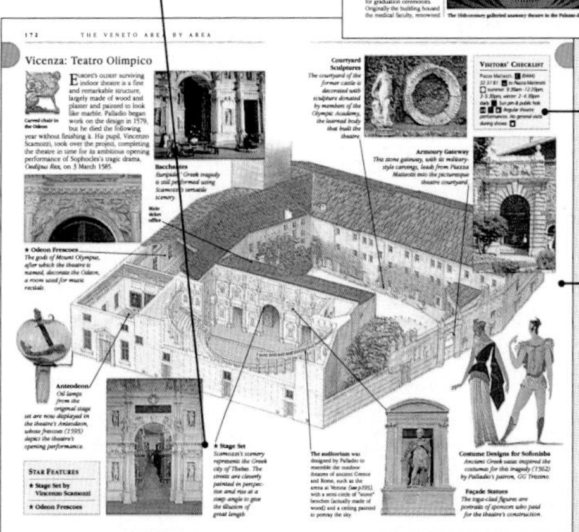

For all the top sights, a Visitors' Checklist provides the practical information you will need to plan your visit.

4 The top sights
These are given two or more full pages. Historic buildings are dissected to reveal their interiors; museums and galleries have colour-coded floorplans to help you locate the most interesting exhibits.

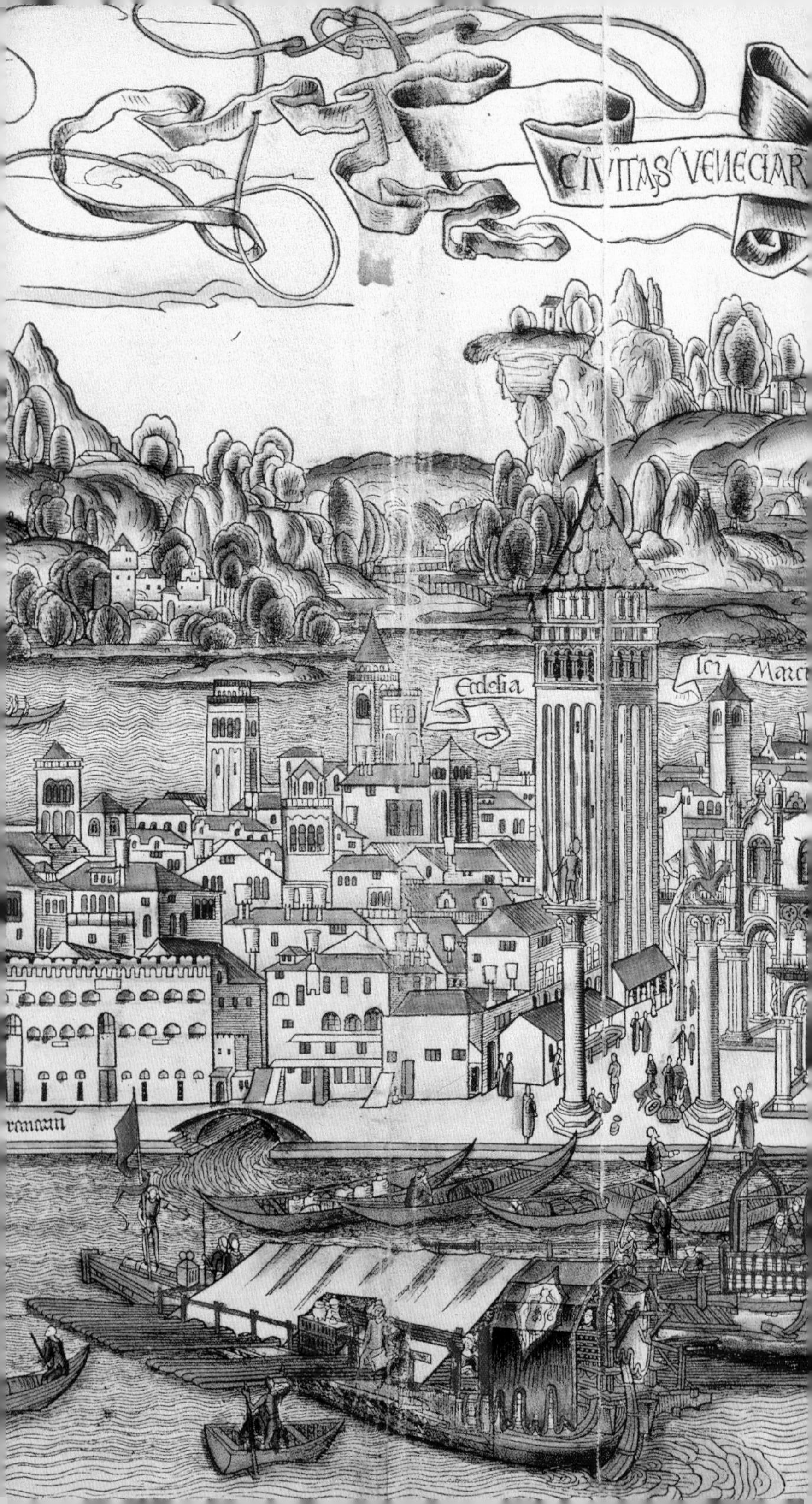

CIVITAS VENECIAR
Ecclesia
Sci Marci

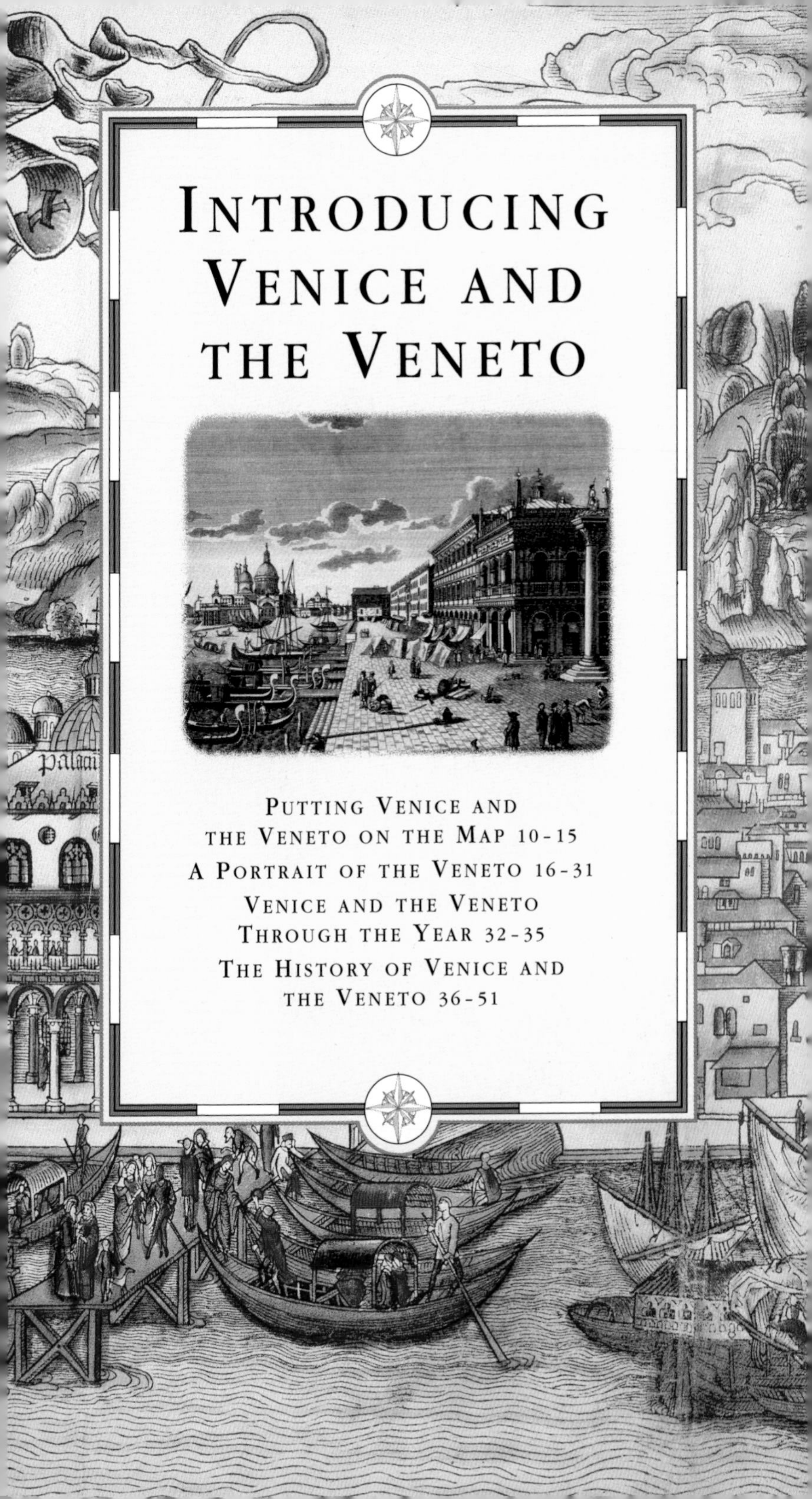

Introducing Venice and the Veneto

Putting Venice and the Veneto on the Map

THE VENETO LIES in the northernmost sector of Italy, and stretches from the Dolomite mountains in the north to the flatlands of the Venetian lagoon in the south. One of the most prosperous regions of Italy, the Veneto covers an area of 47,562 sq km (18,364 sq miles), and has a population of 4.5 million. Rail and road links with the rest of Europe are excellent, and three international airports serve the region: Valerio Catullo in Verona, Marco Polo on the edge of the lagoon, and Treviso.

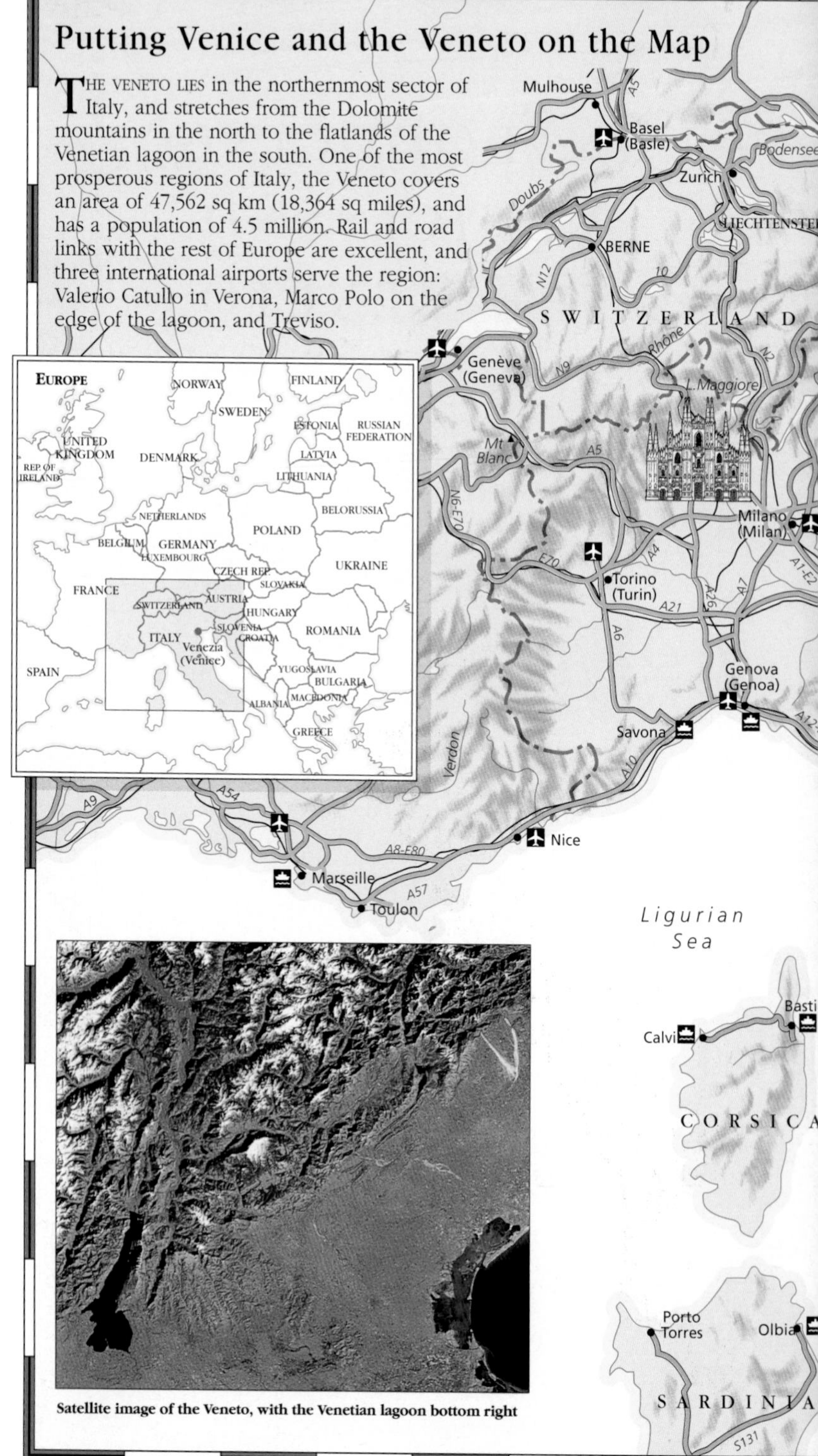

Satellite image of the Veneto, with the Venetian lagoon bottom right

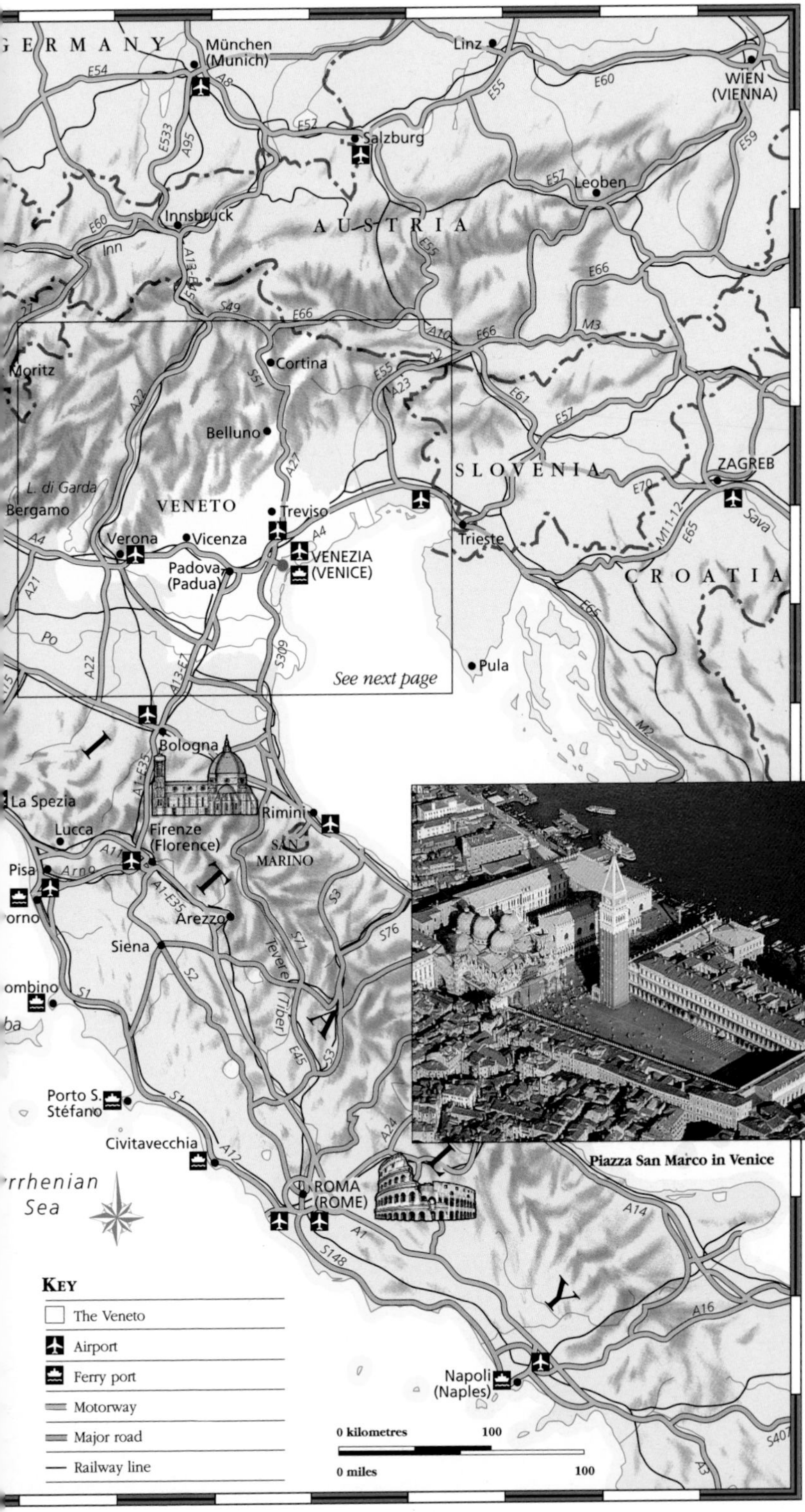

See next page

Piazza San Marco in Venice

KEY

- The Veneto
- Airport
- Ferry port
- Motorway
- Major road
- Railway line

0 kilometres 100

0 miles 100

Road Map of the Veneto

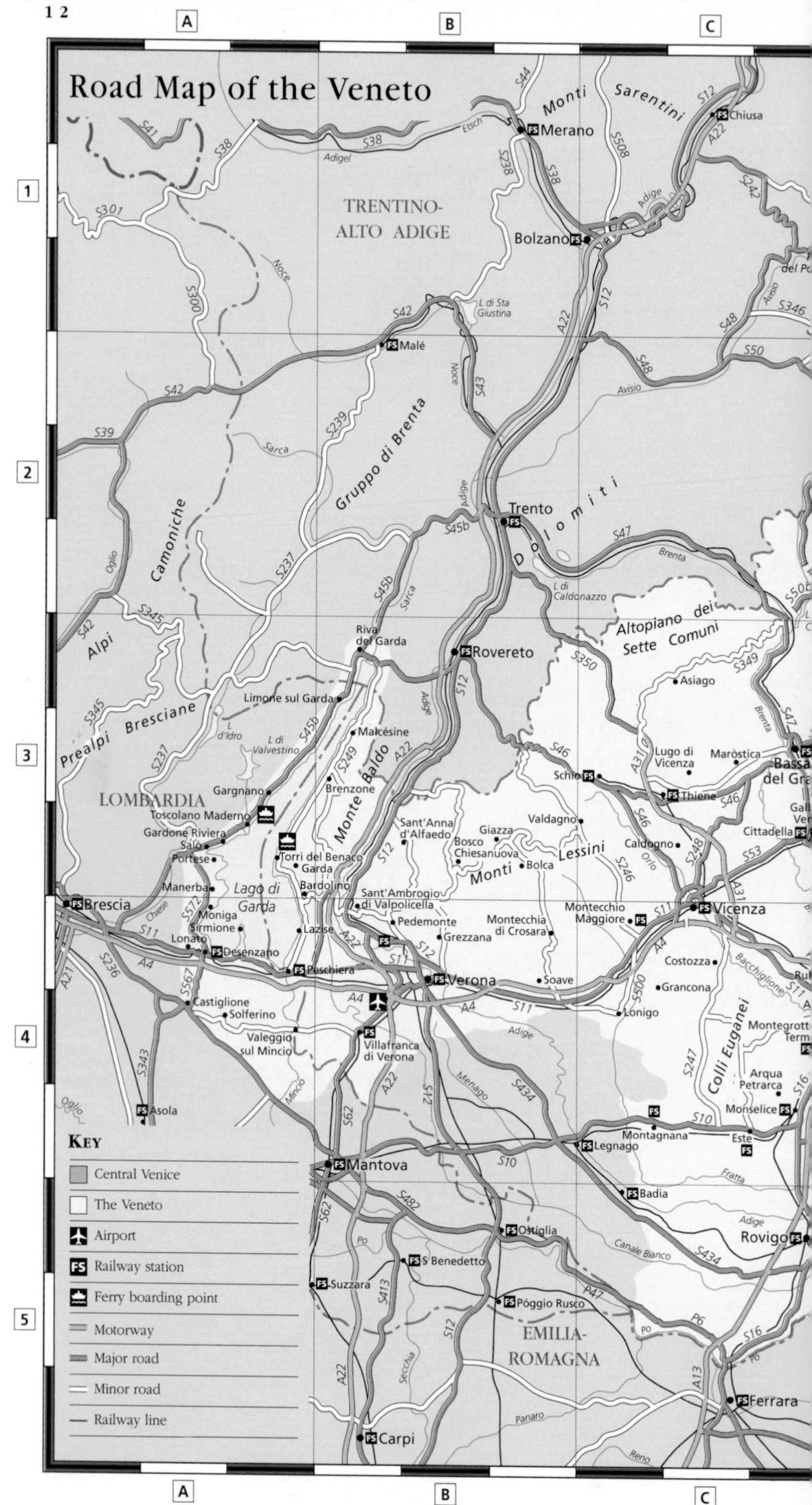

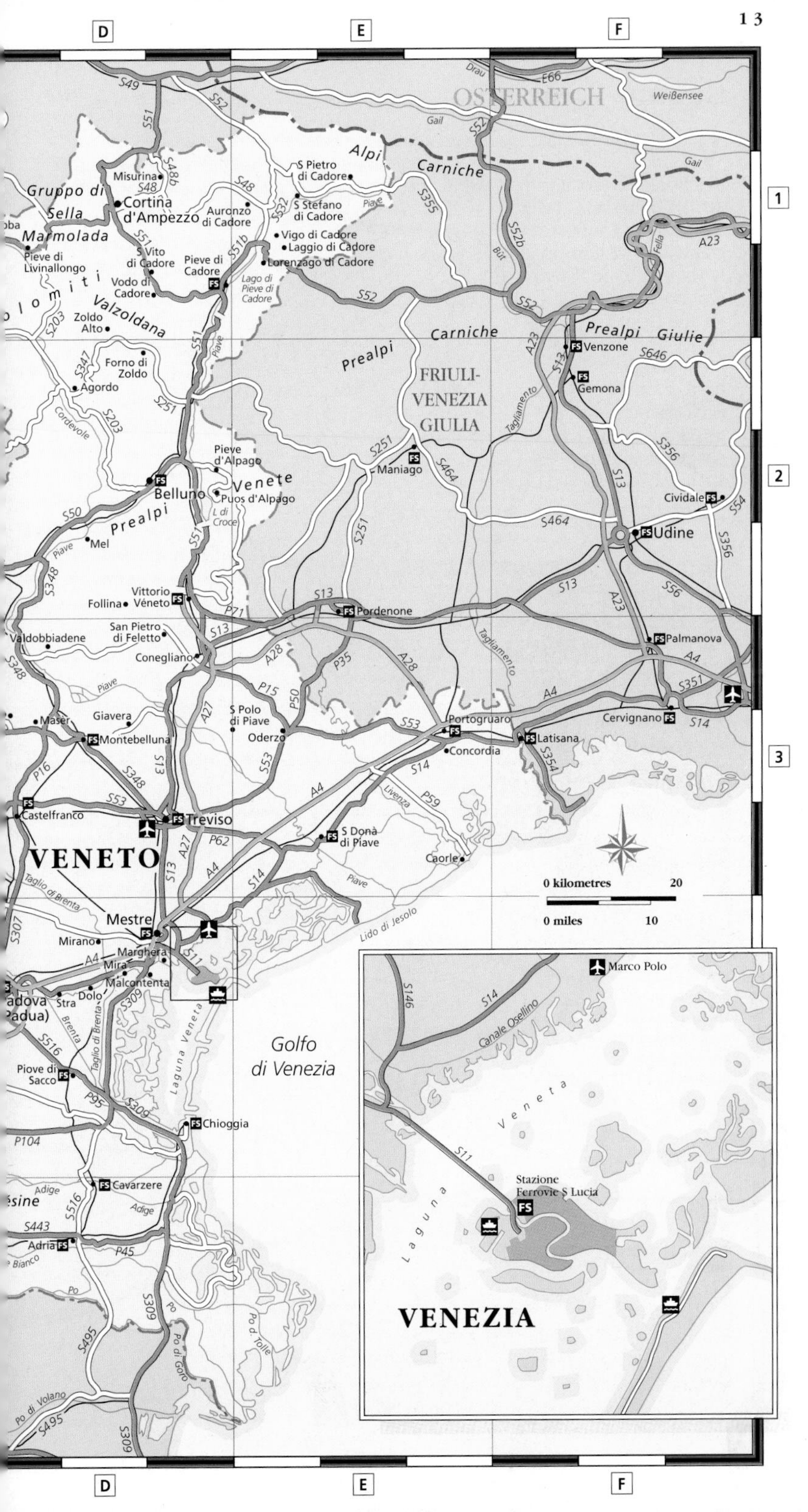

D
E
F
1
2
3
OSTERREICH
Alpi Carniche
Prealpi Carniche
Prealpi Giulie
FRIULI-VENEZIA GIULIA
VENETO
Gruppo di Sella
Marmolada
Dolomiti
Valzoldana
Prealpi Venete
Misurina
Cortina d'Ampezzo
Auronzo di Cadore
S Pietro di Cadore
S Stefano di Cadore
Vigo di Cadore
Laggio di Cadore
Lorenzago di Cadore
S Vito di Cadore
Pieve di Cadore
Lago di Pieve di Cadore
Pieve di Livinallongo
Vodo di Cadore
Zoldo Alto
Forno di Zoldo
Agordo
Venzone
Gemona
Maniago
Pieve d'Alpago
Puos d'Alpago
L di Croce
Belluno
Mel
Cividale
Udine
Vittorio Véneto
Follina
Pordenone
Valdobbiadene
San Pietro di Feletto
Conegliano
Palmanova
Maser
Giavera
Montebelluna
S Polo di Piave
Oderzo
Portogruaro
Concordia
Latisana
Cervignano
Castelfranco
Treviso
S Donà di Piave
Caorle
Mestre
Mirano
Marghera
Mira
Malcontenta
Dolo
Stra
Padova (Padua)
Piove di Sacco
Chioggia
Cavarzere
Adria
Golfo di Venezia
Laguna Veneta
Lido di Jesolo
Taglio di Brenta
Brenta
Piave
Livenza
Tagliamento
Adige
Po
Po di Goro
Po d. Tolle
Po di Volano
Cordevole
Drau
Gail
Fella
Büt
Weißensee
0 kilometres 20
0 miles 10
VENEZIA
Marco Polo
Canale Osellino
Laguna Veneta
Stazione Ferrovie S Lucia

Central Venice

VENICE IS DIVIDED into six ancient administrative districts or *sestieri*. The areas described in this book for the most part follow the *sestieri* boundaries, with San Polo and Santa Croce combined. Visitors usually start with the Piazza San Marco, heading for the Doge's Palace and the breath-taking basilica, but each district has its own distinct character, and time spent exploring each will be fully rewarded.

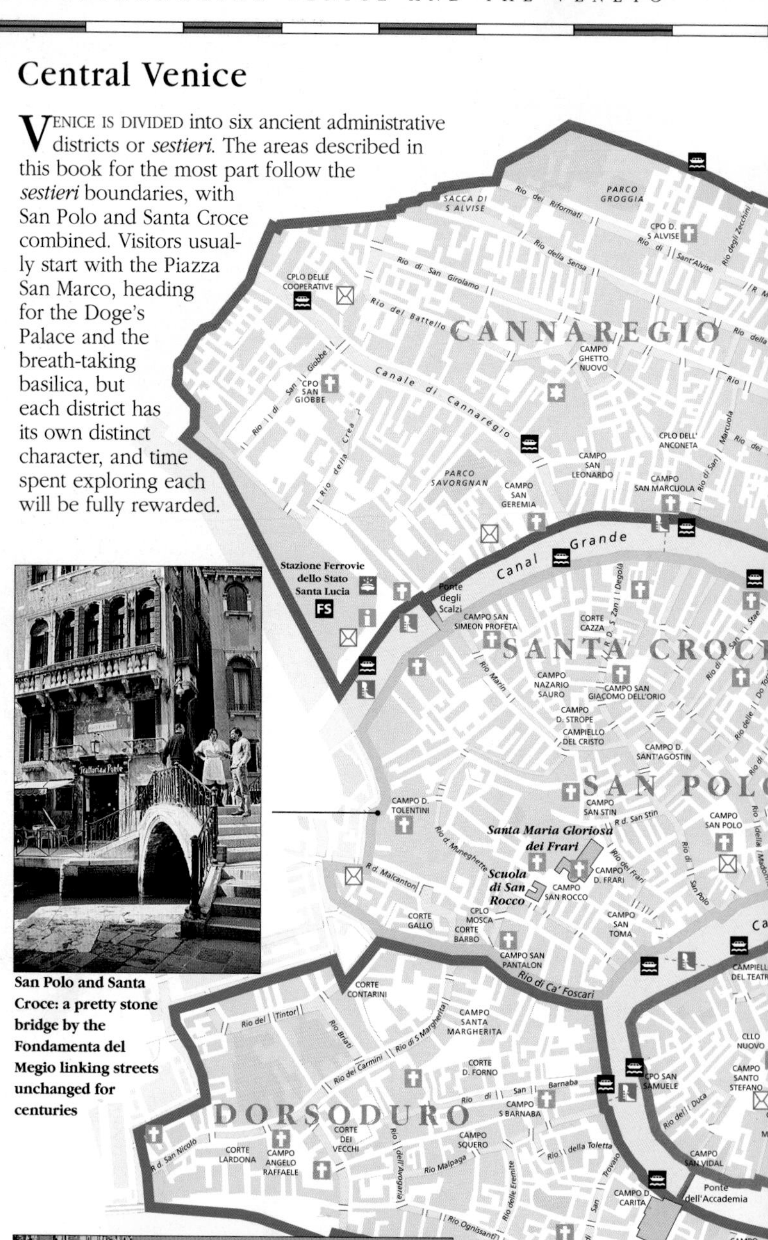

San Polo and Santa Croce: a pretty stone bridge by the Fondamenta del Megio linking streets unchanged for centuries

Dorsoduro: the mouth of the Grand Canal

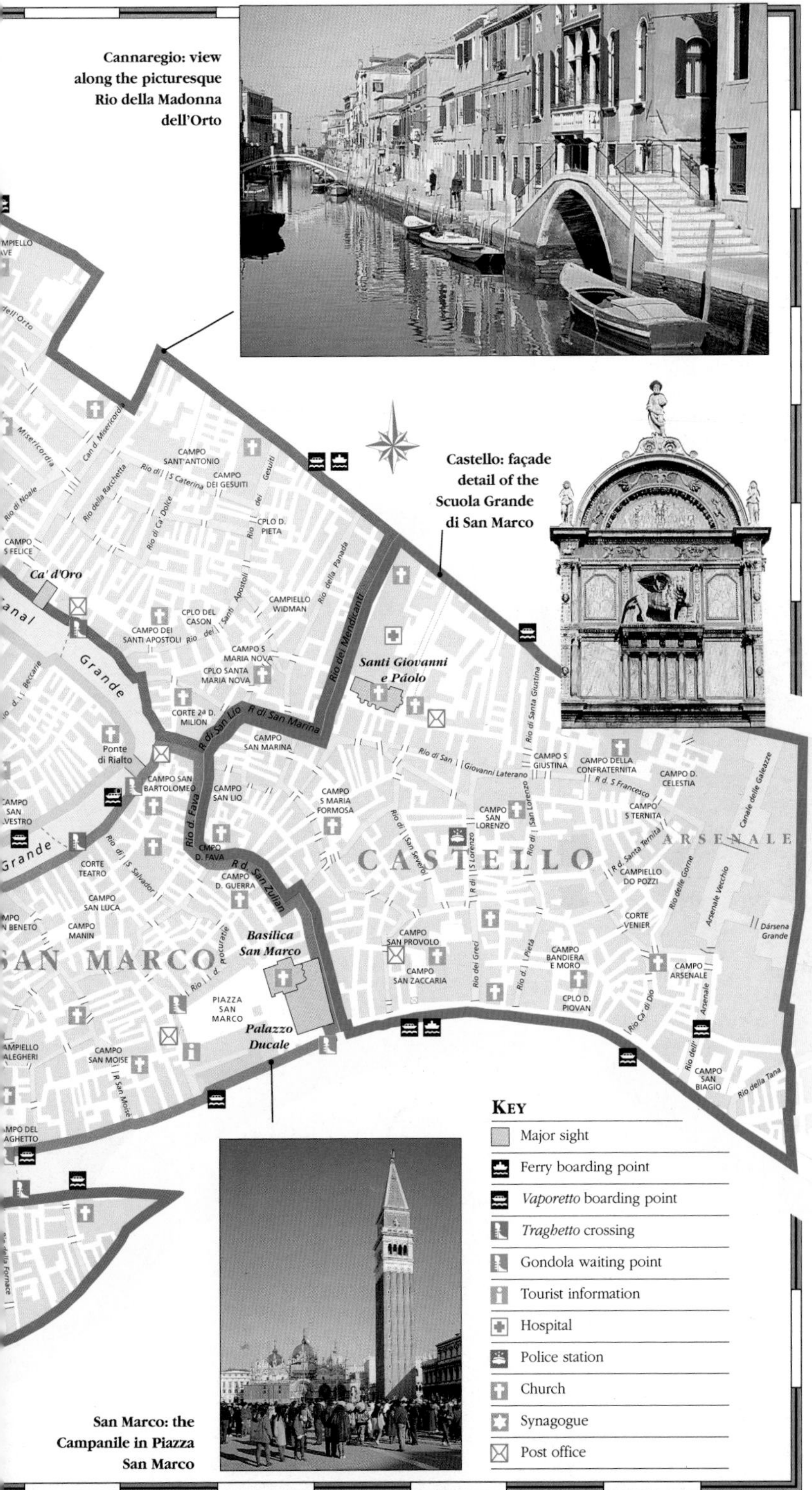

Cannaregio: view along the picturesque Rio della Madonna dell'Orto

Castello: façade detail of the Scuola Grande di San Marco

San Marco: the Campanile in Piazza San Marco

KEY

- Major sight
- Ferry boarding point
- *Vaporetto* boarding point
- *Traghetto* crossing
- Gondola waiting point
- Tourist information
- Hospital
- Police station
- Church
- Synagogue
- Post office

A PORTRAIT OF THE VENETO

VENICE AND THE VENETO *form, on the face of it, an unlikely parnership. Venice is a romantic tourist city frozen in time, the Veneto a forward-thinking and cosmopolitan part of the new Europe. Yet the commercial dynamism of the mainland cities is a direct legacy of the Old Lady of the Lagoon who, in her prime, ruled much of the Mediterranean.*

Venice is one of the few cities in the world that can truly be described as unique. It survives against all the odds, built on a series of low mud banks amid the tidal waters of the Adriatic and regularly subject to floods. Once a powerful commercial and naval force in the Mediterranean, Venice has found a new role. Her *palazzi* have become shops, hotels and apartments, her warehouses have been transformed into museums and her convents have been turned into centres for art restoration. Yet little of the essential fabric of Venice has altered in 200 years. A prewar guide to the city is just as useful today as when it was published, a rare occurrence on a continent scarred by the aerial bombing of World War II and the demands of postwar development. More than 12 million visitors a year succumb to the magic of this improbable city whose streets are full of water and where the past has more meaning than the present.

The lion of St Mark, symbol of imperial Venice

For all this Venice has had a price to pay. So desirable is a Venetian apartment that rents are beyond the means of the Venetians themselves. Many of the city's apartments are owned by wealthy foreigners who use them perhaps for two or three weeks a year – unlit windows at night are indicative of absent owners.

Children attending their first communion at Monte Berico, outside Vicenza

◁ **Venice's Carnival, an historic celebration revived in 1979**

An elderly Venetian in an ageing Venice

In 1994 the population of the city was 70,000 (compared with 150,000 in 1950) and death carries off another 2,000 a year. The average age of the Venetian population is nearly 50 and the city's schools and maternity hospitals are closing for lack of use – pampered children, ubiquitous everywhere else in Italy, are markedly absent from the streets of Venice. One reason why the city shuts down so early at night is that the waiters, cooks and shop assistants all have to catch the last train home across the causeway to Mestre.

Mestre, by contrast, is a bustling city of 340,000 inhabitants, with a busy oil terminal and an expanding industrial base, as well as some of the liveliest discos in Italy. Governed by the same mayor and city council, Mestre and Venice have been described as the ugliest city in the world married to the most beautiful. Yet Mestre, founded by Venetians who foresaw a day when development land would run out in the lagoon, is simply an extension of that same entrepreneurial spirit that characterized mercantile Venice in her heyday, a spirit that is now typical of the region as a whole.

Fruit seller in Sirmione, on Lake Garda

The Industrious North

The creativity and industry demonstrated by the people of the Veneto contradict all the clichés about the irrationality and indolence of the Italian character. For a tiny area, with a population of 4.5 million, the Veneto is remarkably productive. Many world-renowned companies have manufacturing bases in the area, from Jacuzzi Europe, manufacturers of whirlpool baths, to Benetton, Zanussi, Olivetti and Iveco Ford. As a result, poverty is rare, and the region has successfully progressed from its prewar agricultural base to a modern manufacturing and distribution economy.

Benetton shop in Treviso

Unencumbered by the rest of Italy, the three northern provinces of Piedmont, Lombardy and Venetia alone would qualify for membership of the G10 group of the world's richest nations, a fact exploited by the region's politicians in separatist calls for independence from Rome. Cold-shouldering the rest of the Italian peninsula, the Veneto looks east to Slovenia for an example of a small state which has recently achieved independence, and north to Germany as a model of political federalism and sound economic management.

Valle di Cadore in the Dolomites, close to the Austrian border

Despite the ferocity of battles fought against them down the ages, the people in the north of the Veneto have a close relationship with their Germanic neighbours. Today, German signs, language and food dominate the towns around Lake Garda and the Dolomites. Here, the pretty Tyrolean farmsteads and onion-domed churches are a marked contrast to the isolated fishing communities of the lagoon, where Venice's maritime heritage is still evident. Between these two extremes, however, the urbane and likeable cities of the Veneto plain, with their wealth of culture, provide a more typical view of Italian life.

Traditional Venetian rowing

Italian Tradition

Padua is a perfect example of the *salotta città*, a city built like a salon on a human scale, where the streets are an extension of the home and where the doorless Caffè Pedrocchi is treated like the city's main square. Here Paduans come to drink coffee or write a letter, read a newspaper or talk to friends. Just like the salons of old, the café provides a meeting place for intellectual discourse and entertainment.

It is not just the Paduans who treat their streets and squares like so many corridors and rooms in one vast communal palace. After 5pm crowds throng Verona's Via Mazzini, taking part in the evening stroll, the *passeggiata*. Against the backdrop of the Roman arena or medieval *palazzi* they argue, swap gossip, forge alliances and strike deals. Younger strollers dress to impress, while young mothers bring their babies out to be admired. For all their modernity, the people of the Veneto still understand the powerful part played by ancient rituals such as this in cementing a strong sense of community.

Wedding Ferrari decorated with typical Italian style

The Building of Venice

VENICE IS BUILT on a patchwork of more than 100 low-lying islands in the middle of a swampy lagoon. To overcome these extremely challenging conditions, early Venetian builders evolved construction techniques unique to the city, building with impermeable stone supported by larchwood rafts and timber piles. This method proved effective and most Venetian buildings are remarkably robust, many having stood for at least 400 years. By 1500 the city had taken on much of its present shape and only in the 20th century has further building begun to alter the outline.

Campo Santa Maria Mater Domini *is a typical medieval square, with its central wellhead and its business-like landward façades – decoration on buildings was usually reserved for the canal façades.*

Pinewood piles *were driven 7.5 m (25 ft) into the ground before building work could begin. They rest on the solid* caranto *(compressed clay) layer at the bottom of the lagoon.*

Campaniles often lean because of compaction of the underlying subsoil.

Istrian stone, a type of marble, was used to create damp-proof foundations.

Bricks

Closely packed piles do not rot in the waterlogged subsoil because there is no free oxygen, vital for microbes that cause decay.

Water grilles

Sand acting as a filter

The well was the source of the fresh water supply. Rainwater was channelled through pavement grilles into a clay-lined cistern filled with sand to act as a filter.

Ornate wellheads*, such as this one in the Doge's Palace courtyard photographed in the late 19th century, indicate the importance of a reliable water supply for the survival of the community. Strict laws protected the purity of the source, prohibiting "beasts, unwashed pots and unclean hands".*

The Campanile Foundations

When the Campanile in the Piazza San Marco *(see p76)* collapsed in 1902, the ancient pilings, underpinning the 98.5-m high (323-ft) landmark, were found to be in excellent condition, after 1,000 years in the ground. Like the Campanile, all buildings in Venice are supported on slender oak and pine piles, harvested in the forests of the northern Veneto and floated downriver to the Venetian lagoon. Once driven through the lagoon subsoil, they create an immensely strong and flexible foundation. Even so, there is a limit to how much weight the piles can carry – the Campanile, its height having been increased several times, simply grew too tall and collapsed. When the tower was rebuilt, timber foundations were again used, but this time more than double the size.

Strengthening the Campanile foundations

***Palazzo* roofs**, built of light, glazed tiles, had gutters to channel rainwater to the well.

Façades were built of light-weight rose-coloured bricks, sometimes left bare, sometimes weatherproofed with plaster.

Bridges *were often privately owned and tolls were charged for their use. Originally, none had railings, creating a night-time hazard for the unwary in the dark streets.*

High water level

Low water level

Accumulated rubbish is regularly removed by dredging to prevent the canal silting up.

Sand and clay

Caranto is compacted clay and sand in alternate layers, which provides a stable base for building.

The Campo (Santa Maria Mater Domini)

The fabric of Venice is made up of scores of self-contained island communities, linked by bridges to neighbouring islands. Each has its own water supply, church and bell-tower, centred on a *campo* (square), once the focus of commercial life. *Palazzi*, with shops and warehouses at ground floor level, border the *campo* which is connected to workshops and humbler houses by a maze of side alleys.

The Venetian Palazzo

Baroque statue

Venetian houses evolved to meet the needs of a city without roads. Visitors usually arrived by boat, so the façade facing the canal was given lavish architectural treatment, while the landward side, which was accessible from a square or alley, was rarely so ornate. Most Venetian houses were built with three storeys, with kitchens located on the ground floor for ready access to water, or in the attic to enable cooking smells to escape. Typically, a *palazzo* served as a warehouse and business premises, as well as a family home, reflecting the city's mercantile character.

Renaissance doorcase with lion

Byzantine (12th and 13th Centuries)

The earliest surviving private *palazzi* in Venice date from the 13th century and reflect the architectural influence of the Byzantine world. Façades are recognizable by their ground-floor arcades and arched open galleries which run the entire length of the first floor. Simple motifs feature leaves or palm trees.

Byzantine roundel, Fondaco dei Turchi

The Byzantine arcades of the Fondaco dei Turchi (built 1225)

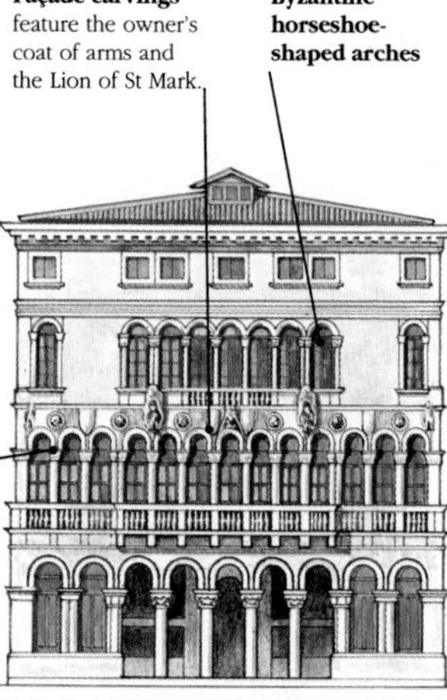
Façade carvings feature the owner's coat of arms and the Lion of St Mark.

Byzantine horseshoe-shaped arches

Cushion capitals have only simple motifs.

Palazzo Loredan (see p64) *has an elegant ground floor arcade and first floor gallery typical of a 13th-century Byzantine palace.*

Gothic (13th to mid 15th Centuries)

Elaborate Gothic *palazzi* are more numerous than any other style in Venice. Most famous of all is the Doge's Palace *(see pp82–3)*, with elegant arches in Istrian stone and fine tracery which give the façade a delicate, lace-like appearance. This style, emulated throughout the city, can be identified through its use of pointed arches and carved window heads.

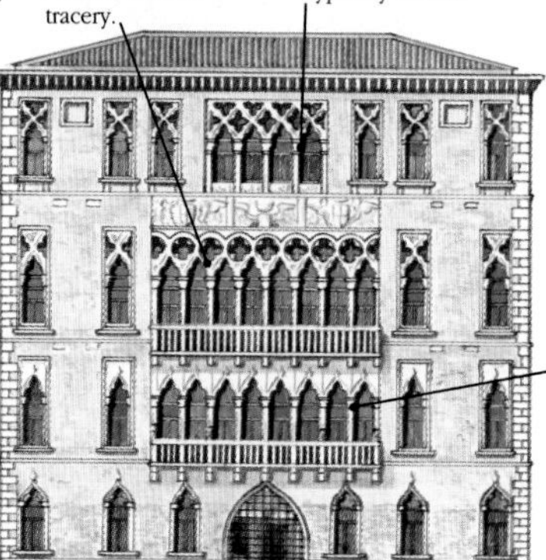
The interlacing ribs of pointed ogee arches create a delicate tracery.

Trefoil "three leaved" window heads are typically Gothic.

Quatrefoil patterns on elegant gallery windows

Gothic capitals are adorned with foliage, animals and faces.

Palazzo Foscari (see p66) *is a fine example of the 15th-century Venetian Gothic style, with its finely carved white Istrian stone façade and pointed arches.*

Gothic capitals (Doge's Palace)

Renaissance (15th and 16th Centuries)

Houses of the Renaissance period were often built in sandstone rather than traditional Venetian brick. The new style was based on Classical architecture, with emphasis on harmonious proportions and symmetry. The new decorative language, borrowing motifs from ancient Rome and Greece, typically incorporated fluted columns, Corinthian capitals and semi-circular arches.

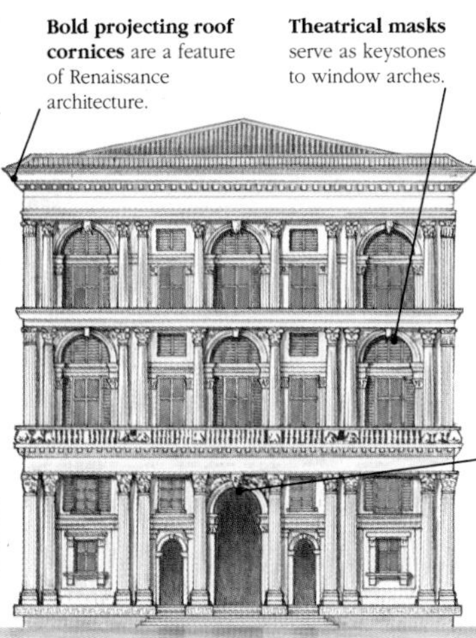

Bold projecting roof cornices are a feature of Renaissance architecture.

Theatrical masks serve as keystones to window arches.

The Venetian door, a very popular Renaissance motif, has a rounded central arch flanked by narrower side openings. This combination was also used for windows.

Palazzo Grimani (see p64) *has lavish stone carving which none but the wealthy could afford; massive foundations were constructed to bear the incredible weight.*

Corinthian pilasters on the portal to San Giovanni Evangelista

Baroque (17th Century)

Venetian Baroque has its roots in the Renaissance Classical style but is far more exuberant. Revelling in bold ornamentation that leaves no surface uncarved, garlands, swags, cherubs, grotesque masks and rosettes animate the main façades of buildings such as the 17th-century Ca' Pesaro.

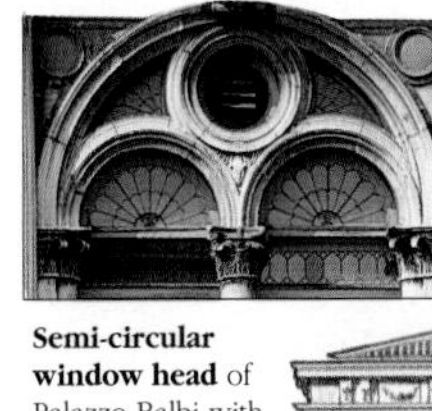

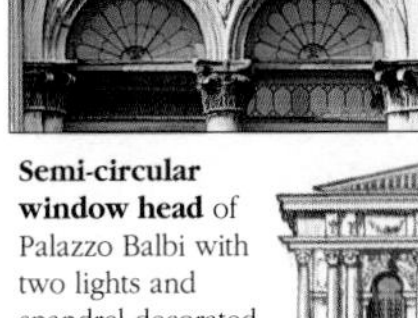

Semi-circular window head of Palazzo Balbi with two lights and spandrel decorated with a circle.

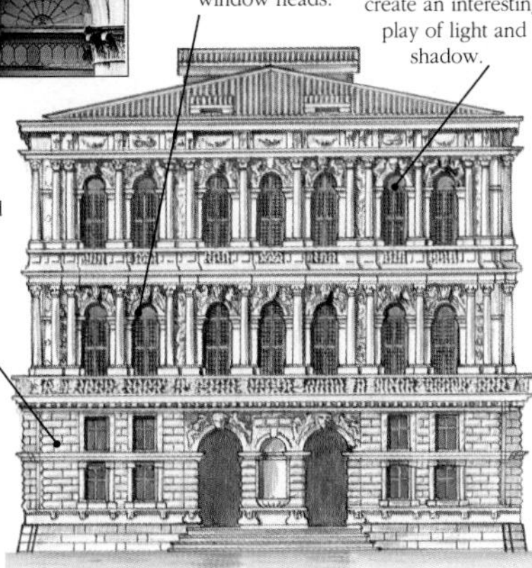

Cherubs and plumed heads are carved into Baroque stone window heads.

Recessed windows and column clusters create an interesting play of light and shadow.

Massive blocks with deep ridges give solidity to the lower walls.

Ca' Pesaro (see p62) *is an example of Baroque experimentation, with its flat façade broken into a three-dimensional stone pattern of deep recesses and strong projections.*

Baroque cartouche

The Venetian House

The layout of a typical *palazzo* (often called Ca', short for *casa*, or house) has changed little over the centuries, despite the very different styles of external decoration.

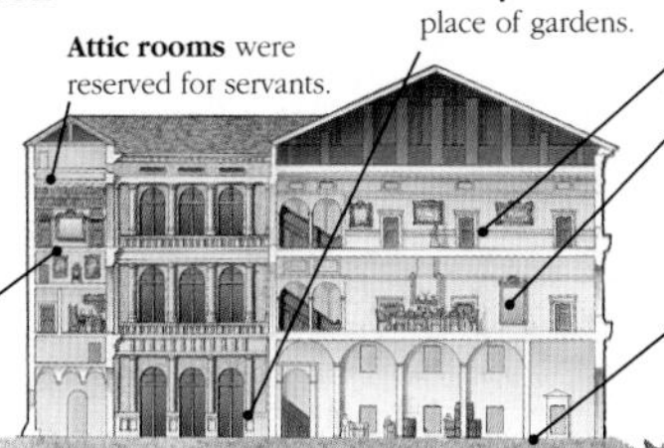

Attic rooms were reserved for servants.

Courtyards took the place of gardens.

The upper floor housed the family.

The *piano nobile* (grand floor), often lavishly decorated, was used to entertain visitors.

Offices, used for storing business records, evolved into libraries.

The ground floor storerooms and offices were used for the transaction of business.

The Villas of Palladio

Andrea Palladio

When it became fashionable in the 16th century for wealthy Venetians to acquire rural estates on the mainland, many turned to the prolific architect, Andrea Palladio (1508–80) for the design of their villas. Inspired by ancient Roman prototypes, described by authors such as Vitruvius and Virgil, Palladio provided his clients with elegant buildings in which the pursuit of pleasure could be combined with the functions of a working farm. Palladio's designs were widely imitated and continue to inspire architects to this day.

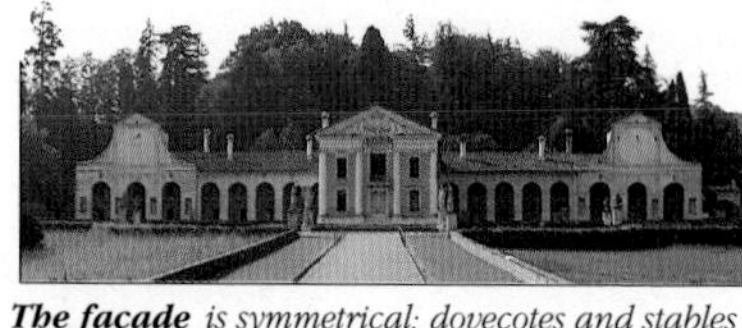

The façade *is symmetrical; dovecotes and stables in the wings balance the central block.*

The Room of the Little Dog *is ornate and lavishly decorated with frescoes by Veronese. Look closely to see the detail of a spaniel in one of the panels.*

The Nymphaeum *combines utility with art; the same spring that feeds the statue-lined pool also supplies water to the villa.*

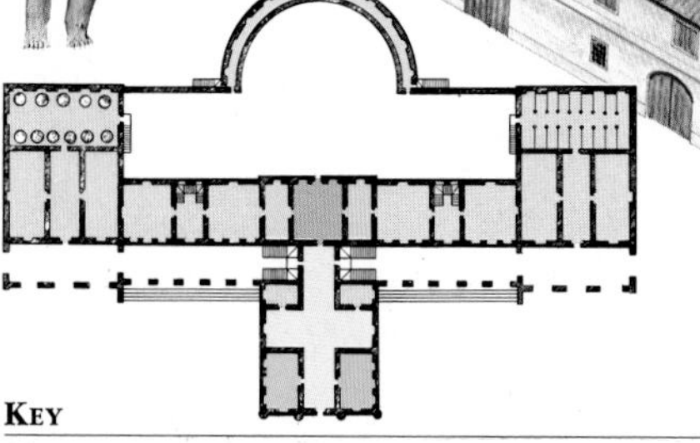

Key

Crociera	Room of the Little Dog
Bacchus Room	Room of the Oil Lamp
Room of the Tribunal of Love	Nymphaeum
Hall of Olympus	Non-exhibition space

The Villa Barbaro

Palladio and Veronese worked closely to create this splendid villa (commissioned in 1555, *see p167*). Lively frescoes of false balconies, doors, windows and rural views create the illusion of greater space, perfectly complementing Palladio's light, airy rooms.

Development of the Villa

Palladio experimented with many different designs which he published in his influential *Quattro Libri (Four Books)* in 1570, illustrating the astonishing fertility of his mind and his ability to create endless variations on the Classical Roman style.

The portico statues reflect Palladio's study of ancient Roman buildings.

The pedimented pavilion is all that survives of Palladio's ambitious design; the main residence was never built.

Stables and storerooms

Villa Thiene (1546), now the town hall, Quinto Vicentino

The Hall of Olympus *shows Giustiniana, mistress of the house and wife of Venetian ambassador Marcantonio Barbaro, with her youngest son, wetnurse and family pets.*

In the Crociera*, the cross-shaped central hall, servants peer round false doors, while imaginary landscapes blur the boundary between the house interior and the garden.*

The Room of the Oil Lamp *symbolizes virtuous behaviour; here Strength, with the club, leans on Truth, with the mirror.*

The Bacchus Room*, with its winemaking scenes and chimneypiece carved with the figure of Abundance, reflects the bucolic ideal of the villa as a place of good living and plenty.*

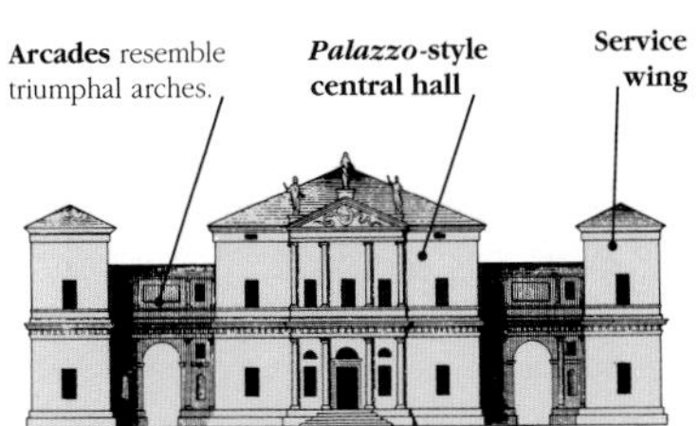
Arcades resemble triumphal arches.
***Palazzo*-style central hall**
Service wing

Villa Pisani (1555), Montagnana *(see p184)*

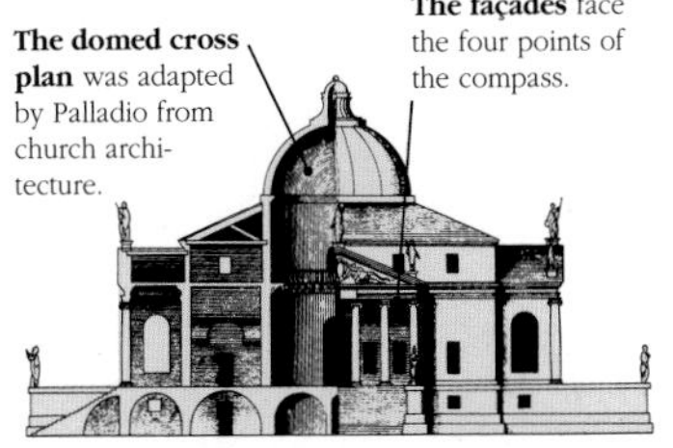
The domed cross plan was adapted by Palladio from church architecture.
The façades face the four points of the compass.

Villa Capra "La Rotonda" (1569), Vicenza *(see p171)*

Styles in Venetian Art

VENETIAN ART grew out of the Byzantine tradition of iconographic art, designed to inspire religious awe. Because of the trade links between Venice and Constantinople, capital of Byzantium, the Eastern influence lasted longer here than elsewhere in Italy. Andrea Mantegna introduced the Renaissance style to the Veneto in the 1460s, and his brother-in-law Giovanni Bellini became Venice's leading painter. In the early 16th century Venetian artists began to develop their own style, in which soft shading and dramatic use of light distinguishes the works of Venetian masters Titian, Giorgione, Tintoretto and Veronese. The development of this characteristic Venetian style, which the prolific but lesser known artists of the Baroque and Rococo periods continued, can be seen in the chronological arrangement of the Accademia *(see p130–33)*.

Detail from Veneziano's *Coronation of the Virgin*

The Last Judgment *(12th century) from Torcello: in the damp climate, mosaics, not frescoes, were used to decorate Venetian churches.*

Byzantine Gothic

Paolo Veneziano is credited with the move from grand-scale mosaics to more intimate altarpieces. His painting mixes idealized figures with the hairstyles, costumes and textiles familiar to 14th-century Venetians. The typically lavish use of jewel colours and gold, symbol of purity, can also be seen in the work of Veneziano's pupil (and namesake) Lorenzo, and in the gilded warrior angels of Guariento *(see p179)*.

Veneziano's entire dazzling polyptych *(1325) of which this is the centrepiece, is in the Accademia* (see p132).

Paolo Veneziano's *Coronation of the Virgin*

The Madonna's gentle face *reinforces the courtly refinement of Veneziano's work.*

The composition and colours reflect the style of the early Byzantine icons which influenced the artist.

Arabesque patterns on the tunics reflect Moorish influence.

Musicians like these played at grand ceremonies in San Marco.

Timeline of Venetian Artists

1356–72 (active)	Lorenzo Veneziano
1338–c.1368	Guariento
1430–1516	Giovanni Bellini
1431–1506	Andrea Mantegna
1415–84	Antonio Vivarini
1483–1539	Giovanni Pordenone
1450–1526	Vittore Carpaccio
1480–1528	Palma il Vecchio
1480–1556	Lorenzo Lotto
1300	
1350	
1400	
1450	
1395–1455	Antonio Pisanello
1400–71	Jacopo Bellini
1321–62 (active)	Paolo Veneziano
1429–1507	Gentile Bellini
1432–99	Bartolomeo Vivarini
1441–1507	Alvise Vivarini
1467–1510	"Il Morto da Feltre"
1477–1510	Giorgione
1487–1576	Titian

Byzantine Gothic

Early Renaissance

Illusionistic details fool the eye: the real moulding copies the painted one.

Early Renaissance

Renaissance artists were fascinated by Classical sculpture and developed new techniques of perspective and shading to give their figures a three-dimensional look. Using egg-based tempera gave crisp lines and bold blocks of colour, but with little tonal gradation. The Bellini family dominated art in Renaissance Venice, and Giovanni, who studied anatomy for greater accuracy in his work, portrays the feelings of his subjects through their facial expressions.

In Bellini's 1488 Frari altarpiece, *the Madonna is flanked by Saints Peter, Nicholas, Benedict and Mark* (see p102).

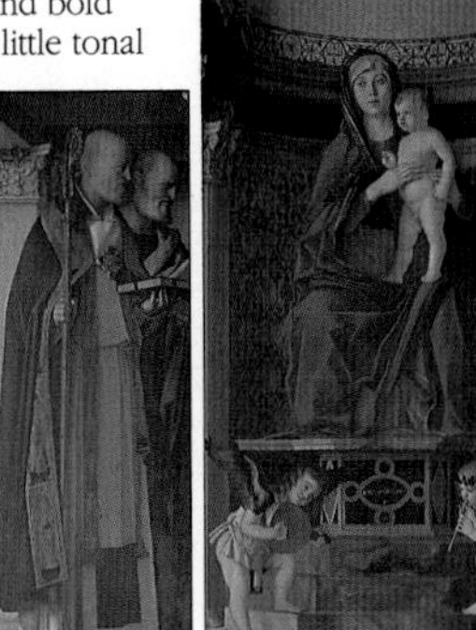

Giovanni Bellini's *Madonna and Child with Saints*

St Benedict carries the Benedictine book of monastic rule.

Musical cherubs playing at the feet of the Virgin are a Bellini trademark; music was a symbol of order and harmony.

High Renaissance

Oil-based paints, developed in the late 15th century, liberated artists. This new medium enabled them to create more fluid effects, an advantage Titian exploited fully. The increasingly expressive use of light by Titian and contemporaries resulted in a distinctive Venetian style, leading to Tintoretto's masterly combination of light and shade *(see p106–7).*

Titian began this Madonna *in 1509 for the Pesaro family altar in the great Frari church* (see p102), *after his* Assumption *was hung above the high altar.*

Titian's *Madonna di Ca' Pesaro*

The Virgin is placed off centre, contrary to a centuries-old rule, but Titian's theatrical use of light ensures that she remains the focus of attention.

Saint Peter looks down at Venetian nobleman Jacopo Pesaro, who kneels to give thanks to the Virgin.

Members of the Pesaro family, *Titian's patrons, attend the Virgin; Lunardo Pesaro, gazing outwards, was heir to the family fortune.*

1500–71 Paris Bordone
1518–94 Tintoretto
1600–38 Francesco Maffei
1712–93 Francesco Guardi
1707–88 Francesco Zuccarelli
1708–85 Pietro Longhi
1696–1770 Giambattista Tiepolo

1500 | **1550** | **1600** | **1650** | **1700**

1548–1628 Palma il Giovane
1528–88 Paolo Veronese
1517–92 Jacopo Bassano
1581–1644 Bernardo Strozzi
1675–1758 Rosalba Carriera
1676–1729 Marco Ricci
1697–1768 Canaletto
1727–1804 Giandomenico Tiepolo

High Renaissance **Baroque, Rococo & Later Artists**

Gondolas and Gondoliers

Hippocampus (sea horse) ornament

GONDOLIERS ARE PART of the symbolism and mythology of Venice. Local legend has it that they are born with webbed feet to help them walk on water. Their intimate knowledge of the city's waterways is passed down from father to son (this is still very much a male preserve). The gondola, with its slim hull and flat underside, is perfectly adapted to negotiating narrow, shallow canals. Once essential for the transport of goods from the markets to the *palazzi*, gondolas today are largely pleasure craft and a trip on one is an essential part of the Venetian experience *(see p276)*. It gives an entirely different perspective on the city, gliding past grand palatial homes, using a form of transport that dates back over 1,000 years.

Squero San Trovaso (see p129) *is the oldest of Venice's three surviving* squeri *(boatyards). Here, new wood is seasoned, while skilled craftsmen build new gondolas and repair some of the 400 craft in use.*

Traditional dress for a gondolier is a beribboned straw hat, striped vest and black trousers.

The gondolier, unusually for an oarsman, stands upright and pushes on the oar to row the boat in the direction he is facing.

Passengers sit on upholstered cushions and low stools.

The rowlock *(forcola)* can hold the oar in eight different positions for steering the craft.

The oar has a ribbed blade.

The asymmetrical shape *of the gondola counteracts the force of the oar. Without the leftward curve to the prow, 24 cm (9.5 inches) wider on the left than the right, the boat would go round in circles.*

CONTINUING A TRADITION

Gondolas are hand-crafted from nine woods – beech, cherry, elm, fir, larch, lime, mahogany, oak and walnut – using techniques established in the 1880s. A new gondola takes three months to build and costs £10,000.

GONDOLA DECORATION

Black pitch, or tar, was originally used to make gondolas watertight. In time this sombre colour gave way to bright paintwork and rich carpets, but such displays of wealth were banned in 1562. Today all except ceremonial gondolas are black, ornamented only with their *ferro*, and a golden hippocampus on either side. For special occasions such as weddings, the *felze* (the traditional black canopy) and garlands of flowers appear, while funeral craft, now seldom seen, have gilded angels.

Ceremonial gondolas

Upper Reaches of the Grand Canal *(1483) is one of many paintings by Canaletto to capture the everyday life of gondoliers and their craft. Since they were first recorded in 1094, gondolas have been a Venetian institution, inspiring writers, artists and musicians.*

Races and parades *are part of the fun during Venice regattas. Professional gondoliers race in pairs or in teams of six, using boats specially designed for competition. Many amateur gondoliers also participate in the events.*

The *ferro* with its metal teeth symbolizes the six *sestieri* of Venice, beneath a doge's cap.

Seven layers of black lacquer give the gondola its gloss.

The main frame is built of oak.

More than 280 separate pieces of wood are used in constructing a gondola.

Mooring posts *and channel markers feature prominently in the crowded waterways of Venice. The posts may be topped with a family crest, to indicate a private mooring.*

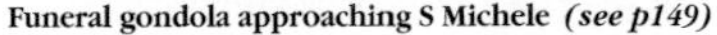

Funeral gondola approaching S Michele ***(see p149)***

Wedding gondola

Venetian Masks and the Carnival

Flamboyant Carnival costume

THE VENETIAN GIFT for intrigue comes into its own during the Carnival, a vibrant, playful festival preceding the abstinence of Lent *(see p32)*. Masks and costume play a key role in this anonymous world; social divisions are dissolved, participants delight in playing practical jokes, and anything goes. The tradition of Carnival in Venice began in the 11th century and reached its peak of popularity and outrageousness in the 18th century. Industrialization left little leisure time and Carnival fell into decline, but was successfully revived in 1979.

Modern Carnival Revellers
Since 1979, each year sees more lavish costumes and impromptu celebrations.

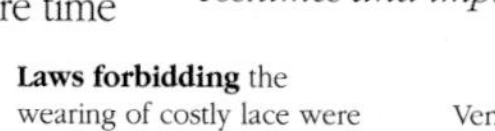

Laws forbidding the wearing of costly lace were suspended at Carnival.

The high spirits of Venetian women scandalized many foreign observers.

The Plague Doctor
This sinister Carnival garb is based on the medieval doctor's beaked face-protector and black gown, worn as a precaution against plague.

TRADITIONAL MASK CELEBRATION

Carnival in the 18th century began with a series of balls in the Piazza San Marco, as in this fresco on the walls of Quadri's famous café in the square *(see p74)*.

Gambling at the Ridotto
Fortunes were squandered every night of Carnival at the state-run casino depicted in Guardi's painting (c.1768).

Street Entertainers
Musicians and comedians attract the crowds in the piazza San Marco.

The satyr-like profile of this dancer hints that he is the devil in disguise.

Columbine
A classic Carnival figure, Columbine wears lace and an apron, but no mask.

Making a Mask

Many masks, and the characters they represent, are deeply rooted in Venetian history. Though instantly recognizable by such features as the beaked nose of the Plague Doctor, each character can be interpreted in a style that is unique to its maker, making each piece a true work of art.

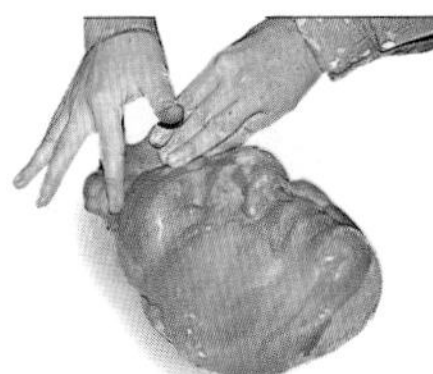

① ***The form of the mask*** *is first modelled out of clay. Then a plaster of Paris mould is made using the fired clay sculpture as a pattern.*

② ***Papier mâché paste,*** *made from a pulpy fibrous mixture of rags and paper dipped in glue, is used to make the mask itself.*

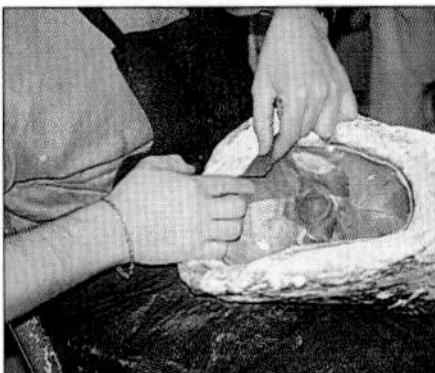

③ ***To shape the mask****, papier mâché paste is pushed into the plaster mould, then put aside to set. It becomes hard yet flexible as it dries.*

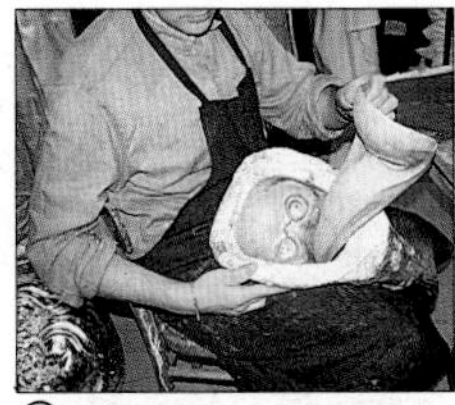

④ ***The size, or glue,*** *used to make the papier mâché gives the mask a smooth, shiny surface, similar to porcelain, when it is extracted.*

⑤ ***An abrasive polish*** *is used to buff the surface of the mask, which is then ready to receive the white base coat.*

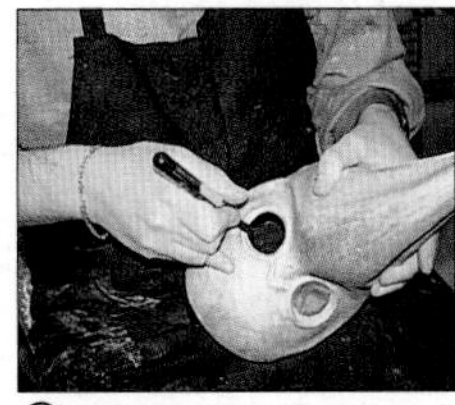

⑥ ***Cutting the eye holes*** *and other features requires the mask maker to have a steady hand.*

⑦ ***The features*** *are painted on the mask and the final touches are added with a few clever brushstrokes.*

⑧ ***The finished mask*** *is ready to wear at the Carnival or to hang on a wall – the perfect Venetian souvenir.*

VENICE AND THE VENETO THROUGH THE YEAR

VENICE IS A CITY that can be enjoyed at all times of the year. Even winter's mists add to the city's romantic appeal, though clear blue skies and balmy weather make spring and autumn the best times to go. This is especially true if you combine a visit to Venice with a tour of the Veneto, where villa gardens and alpine meadows put on a colourful display from the beginning of April. Autumn sees the beech, birch and chestnut trees of the region turn every shade of red and gold. In summer the waters of Lake Garda, fed by melted snow from the Alps, serve to moderate the heat. Winters are mild, allowing some of the crops typical of the southern Mediterranean, like lemons and oranges, to grow.

Festive flag throwers in Feltre

Winter in the delta of the River Po

WINTER

ONCE A QUIET time of year, winter now brings an increasing number of visitors to the city of Venice, especially over Christmas, New Year and Carnival. Many a day that begins wet and overcast ends in a blaze of colour – the kind of sunset reflected off rain-washed buildings that Canaletto liked to paint. In the resorts of the Venetian Dolomites, popular for winter sports, the conditions are perfect for skiing from early December throughout the winter months.

DECEMBER

Nativity. Churches all over Venice and the Veneto mount elaborate Nativity scenes in the days leading up to Christmas. Attending mass is a moving experience at this time, even for non-Christians.

Canto della Stella. In Desenzano, on Lake Garda *(see p204)*, Christmas is marked by open-air processions called *Canto della Stella*, literally "singing to the stars".

JANUARY

Epiphany *(6 Jan)*. Children of the Veneto get another stocking full of presents at Epiphany, supposedly brought by the old witch Befania (also known as Befana, Refana or Berolon). She forgot about Christmas, according to the story, because she was too busy cleaning her house. Good children traditionally get sweets, but naughty children get cinders from her hearth. Images of the witch appear in cake-shop windows, along with evil-looking biscuits made to resemble charcoal.

FEBRUARY

Carnival *(ten days up to Shrove Tuesday)*. The pre-Lent festival of Carnevale *(see p30)*, which means "farewell to meat", is celebrated throughout the Veneto. First held in Venice in the 11th century, it consisted of two months of revelry every year. Carnival fell into decline during the 18th century, but was revived in 1979 with such success that the causeway has to be closed at times to prevent overcrowding in the city.

Today the ten-day festival is mainly an excuse for donning a mask and costume and parading around the city. Various events are organized for which the Tourist Board will have details, but anyone can buy a mask and participate while watching the gorgeous costumes on show in the Piazza San Marco *(see pp74–5)*.

Bacanal del Gnoco *(11 Feb)*. Traditional masked procession in Verona, with groups from foreign countries and allegorical floats from the Verona area. Masked balls are held in the town's squares.

Masked revellers at the Carnival

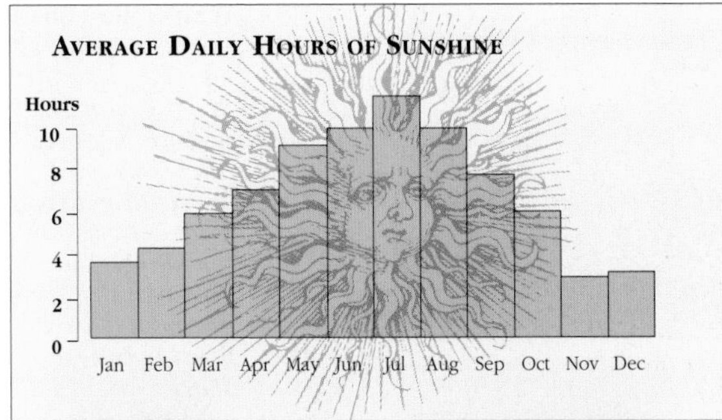

Sunshine Chart
Few days are entirely without sunshine in Venice and the Veneto. The amount of sunshine progressively builds up to mid-summer, when it is dangerous to venture out without adequate skin protection.

Spring wisteria in Verona's Giardini Giusti ***(see p203)***

SPRING

THIS IS THE SEASON when many fine gardens all over the Veneto and round Lake Garda come into their own. As the snow melts, there is time to catch the brief glory of the alpine meadows and the region's nature reserves, renowned for rare orchids and gentians. Verona holds its annual cherry market and many other towns celebrate the arrival of early crops.

MARCH

La Vecia *(mid Lent).* Gardone and Gargnano, villages on Lake Garda *(see p204)*, play host to festivals of great antiquity, when the effigy of an old woman is burnt on a bonfire. The so-called Hag's Trials are an echo of the darker side of medieval life.
Su e zo per i ponti *(second Sun in Mar).* A gruelling marathon-style race in Venice. Participants run through the city's streets *su e zo per i ponti* (up and down the bridges).

APRIL

Festa di San Marco *(25 Apr).* The feast of St Mark, patron saint of Venice, is marked by a gondola race across St Mark's Basin between Sant' Elena *(see p121)* and Punta della Dogana *(see p135)*, and by the consumption of the traditional dish of *risi e bisi* *(see p236).* Men give their wives or lovers a red rose.

MAY

Festa della Sparesca *(1 May).* Festival and regatta for the new season's asparagus held on Cavallino, in the lagoon, where the crop is grown.

Spring produce in the Rialto's vegetable market

La Sensa *(Sun after Ascension Day).* The ceremony of Venice's Marriage with the Sea draws huge crowds, as it has every year since Doge Pietro Orseolo established the custom in AD 1000. Once the ceremony was marked with all the pomp that the doge and his courtiers could muster. Today the words: "We wed thee, O Sea, in token of true and lasting dominion" are spoken by a local dignitary who then casts a laurel crown and ring into the sea.

Celebrating La Sensa, Venice's annual Marriage with the Sea

Vogalonga *(Sun following La Sensa).* Hundreds of boats take part in the Vogalonga (the "Long Row") from the Piazza San Marco to Burano *(see p150)* and back – a distance of 32 km (20 miles).
Festa Medioevale del vino Soave Bianco Soave *(5 May).* Sumptuous medieval-style celebration of the investiture of the Castillian of Suavia. There is a procession with a historical theme, music in the town square, theatrical performances and displays of various sports.
Valpollicellore *(8 May).* Festival of local wine in Cellore with exhibitions and displays.

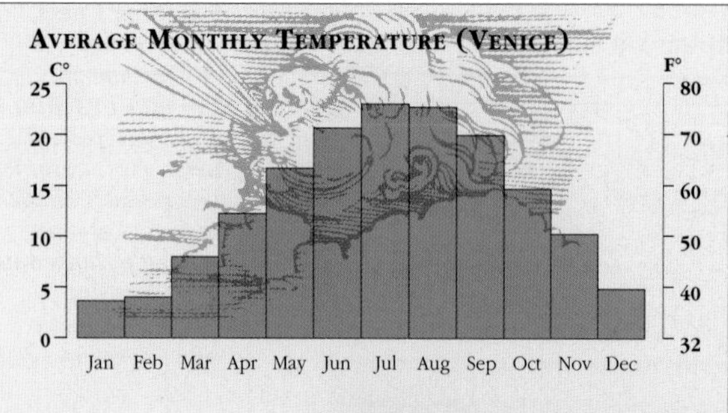

Temperature Chart
Venice, being by the sea, rarely suffers from frost in winter, but summers in the city can be unbearably humid. Temperatures in the Dolomites are considerably lower, with snow and freezing conditions from November to March.

SUMMER

SUMMER BRINGS the crowds to Venice. Queues for museums and popular sites are long, and hotels are frequently fully booked. Avoid visiting the city during the school holidays (mid Jul–end Aug). Verona, too, will be full of opera lovers attending the famous festival, but elsewhere in the Veneto it is possible to escape the crowds and enjoy the spectacular countryside.

Boats for hire at Sirmione on Lake Garda

JUNE

Sagra di Sant'Antonio *(13 Jun).* The Feast of St Anthony has been celebrated in Padua for centuries. The day is marked by a lively fair in Prato della Valle *(see p183).*
Biennale *(Jun–Sep).* The world's biggest contemporary art exhibition takes place in Venice in odd-numbered years *(see p256).*
Festa di Santi Pietro e Paolo *(end Jun).* The feast day of Saints Peter and Paul is celebrated in many towns with fairs and musical festivals.
Regata di Santi Giovanni e Paolo *(third Sun in Jun).* Adriatic Classic sailing regatta in Caorle *(see p175).*

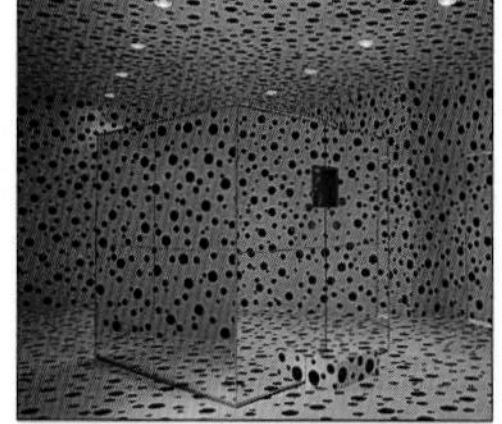
Exhibit by Japanese artist Yayoi Kusama at the Biennale

JULY

Opera Festival *(Jul–Sep).* Verona's renowned opera festival overlaps with the equally famous **Shakespeare Festival**, providing culture lovers with a feast of music, drama, opera and dance in the stimulating setting of the Roman Arena and the city's churches *(see pp256–7).*
Festa del Redentore *(third Sun in Jul).* The city of Venice commemorates its deliverance from the plague of 1576. An impressive bridge of boats stretches across the Giudecca Canal so that people can walk to the Redentore church to attend mass. At night, crowds line the Zattere or row their boats into the lagoon to watch a spectacular firework display *(see p154).*
Sardellata al Pal del Vo *(22 Jul).* Moonlit sardine fishing displays on Lake Garda at Pal del Vo. Boats are illuminated and decorated, and the catch is cooked and distributed to guests and participants.

AUGUST

Village Festivals. The official holiday month is marked by local festivals throughout the Veneto, giving visitors the chance to sample food and wines and see local costume and dance. Around Lake Garda these are often accompanied by firework displays and races in boats like large gondolas.
Palio di Feltre *(first week in Aug).* Medieval games, horse-racing and feasts commemorate Feltre's inclusion in the Venetian empire *(see p219).*
Festa dell'Assunta *(6–14 Aug).* Spectacular nine-day celebration in Treviso *(see p174).* The colourful festivities feature dance, poetry, cabaret and music competitions.

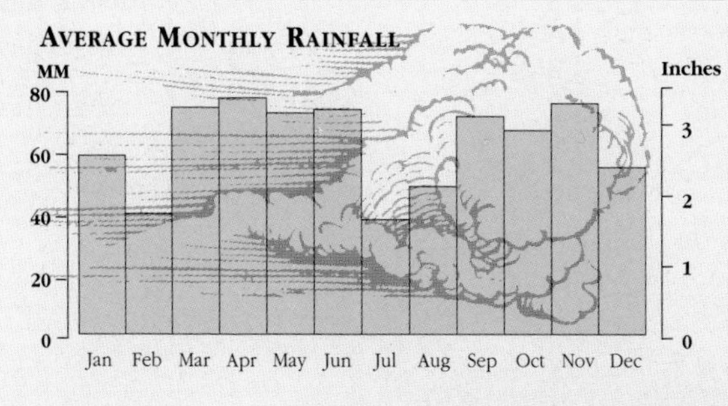

Rainfall Chart
The mountains and sea combine to give Venice and the Veneto higher rainfall than is normal in the rest of Italy, with the possibility of rain on just about any day of the year. The driest months are February and July.

AUTUMN

EXPECT TO SEE a profusion of market stalls selling a huge range of wild fungi as soon as the climatic conditions are right for them to grow. Local people go on expeditions to harvest them, and mushroom dishes will also feature high on the restaurant menus along with game. Another feature of autumn is the grape harvest, a busy time of year in the wine-producing regions of Soave, Bardolino and Valpolicella *(see pp208–9).*

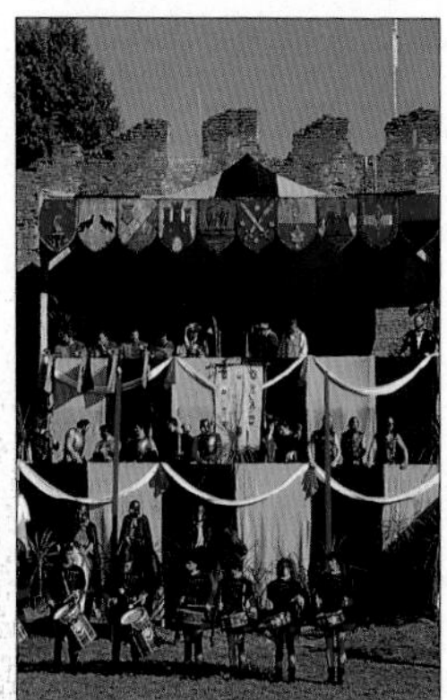

Medieval costume at Montagnana's Palio dei Dieci Comuni

SEPTEMBER

Venice Film Festival *(early Sep).* The International Film Festival attracts an array of filmstars and paparazzi to the Lido *(see p157).*
Regata Storica *(first Sun in Sep).* Gondoliers and other boatsmen compete in a regatta which starts with an historic pageant down the Grand Canal.

Grapes ripening in the Bardolino area

Partita a Scacchi *(second weekend in Sep).* In Maròstica's chequerboard main square, a human chess game is re-enacted in medieval costume *(see p166).*
Palio dei Dieci Comuni *(first Sun in Sep).* The liberation of the town of Montagnana is celebrated with a pageant and horse race *(see p184).*

OCTOBER

Bardolino Grape Festival *(first Sun in Oct).* A festival for wine lovers, held to celebrate the completion of the harvest.
Festa di Mosto *(first weekend in Oct).* The Feast of the Must on Sant'Erasmo, the market-garden island in the lagoon *(see p149).* Here, grapes are still pressed underfoot by bare-legged dancers.

NOVEMBER

Festa della Salute *(21 Nov).* Deliverance from the plague is celebrated with the erection of a pontoon bridge across the Grand Canal to La Salute *(see p135).* Venetians light candles in the church to give thanks for a year's good health.

PUBLIC HOLIDAYS
New Year (1 Jan)
Epiphany (6 Jan)
Good Friday (variable)
Easter Monday (variable)
Liberation Day (25 Apr)
Labour Day (1 May)
Assumption (15 Aug)
All Saints (1 Nov)
Immaculate Conception (8 Dec)
Christmas Day (25 Dec)
St Stephen (26 Dec)

Rowers practising for the Regata Storica

THE HISTORY OF VENICE AND THE VENETO

THE WINGED LION of St Mark is a familiar sight to anyone travelling in the Veneto. Mounted on top of tall columns in the central square of Vicenza, Verona, Chioggia and elsewhere, it is a sign that these cities were once part of the proud Venetian empire. The fact that the lion was never torn down as a hated symbol of oppression is a credit to the benign nature of Venetian authority.

Doge Giovanni Mocenigo (1478–85)

In the 6th century AD, Venice had been no more than a collection of small villages in a swampy lagoon. By the 13th century she ruled Byzantium and, in 1508, the pope, the kings of France and Spain and the Holy Roman Emperor felt compelled to join forces to stop the advances of this powerful empire. As the League of Cambrai, their combined armies sacked the cities of the Veneto, including those such as Vicenza which had initially sided with the League. Venetian territorial expansion was halted, but she continued to dominate the Eastern Mediterranean for another 200 years.

The Venetian system of government came as close to democracy as anyone was to devise until the 19th century, and it stood the city and its empire in good stead until the bumptious figure of Napoleon Bonaparte dared to intrude in 1797. But by then Venice had become a byword for decadence and decline, the essential mercantile instinct that had created and sustained the Serene Republic for so long having been extinguished. As though exhausted by 1,376 years of independent existence, the ruling doge and his Grand Council simply resigned, but their legacy lives on, to fascinate visitors with its extraordinary beauty and remarkable history.

A map dated 1550, showing how little Venice has changed in nearly 500 years

◁ **Tintoretto's *Triumph of Doge Nicolò da Ponte* (1580–84), Sala del Maggior Consiglio, Doge's Palace**

Roman Veneto

THE VENETO TAKES ITS NAME from the Veneti, the pre-Roman inhabitants of the region, whose territory fell to the superior military might of the Romans in the 3rd century BC. Verona was then built as a base for the thrusting and ambitious Roman army which swept northwards over the Alps to conquer much of modern France and Germany. While the Roman empire remained intact the Veneto prospered, but the region bore the brunt of fierce and destructive barbarian attacks that began in the 4th century AD. Riddled by in-fighting and the split between Rome and Constantinople, the imperial administration began to crumble.

A Roman bust in Vicenza

Horsemen in Roman Army
Goths, Huns and Vandals served as mercenaries in the Roman cavalry but later turned to plunder.

Horse-Drawn Carriage
Finds from the region show the technological skills and luxurious lifestyles of the inhabitants.

The Forum (market square)

The Arena was completed in AD 30 to entertain the troops stationed in Verona. It could hold 30,000 spectators.

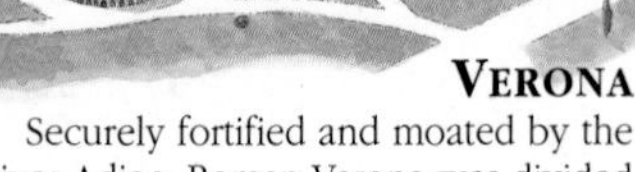

Chariot Racing
A pre-Roman chariot in Adria's museum (see p185) *suggests the Romans adopted the sport from their predecessors.*

VERONA
Securely fortified and moated by the River Adige, Roman Verona was divided into square blocks (*insulae* or "islands"). The Forum has since been filled in by medieval palaces, but several landmarks are still discernible today *(see p192).*

TIMELINE

6th century BC Veneto region occupied by the Euganei and the Veneti

87 BC Catullus, Roman love poet, born in Verona

89 BC The citizens of Verona, Padua, Vicenza, Este and Treviso granted full rights of Roman citizenship

600 BC | 500 | 400 | 300 | 200 | 10

3rd century BC Veneto conquered by the Romans. The Veneti and Euganei adopt Roman culture and lose their separate identities

Catullus (87–c.54 BC)

Hunting in the Lagoon
The wild lagoon, future site of Venice, attracted fishermen and huntsmen in pursuit of game and wildfowl. It also became a place of refuge during raids by Huns and Goths.

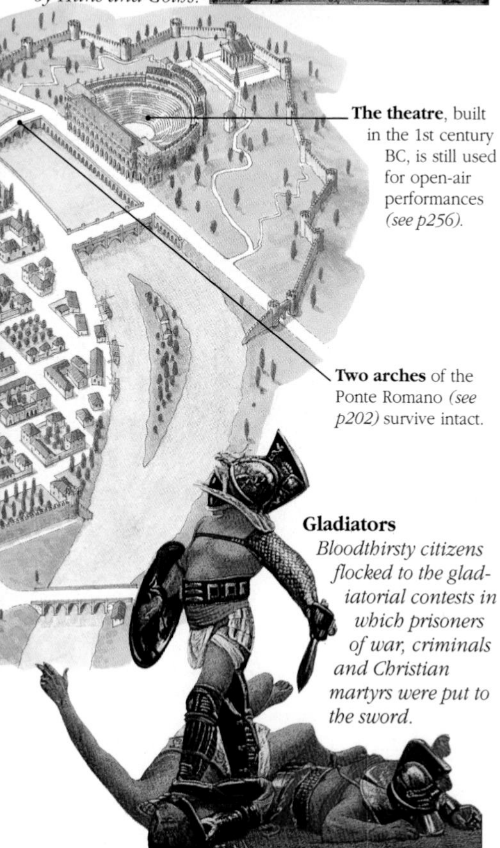

The theatre, built in the 1st century BC, is still used for open-air performances *(see p256).*

Two arches of the Ponte Romano *(see p202)* survive intact.

Gladiators
Bloodthirsty citizens flocked to the gladiatorial contests in which prisoners of war, criminals and Christian martyrs were put to the sword.

WHERE TO SEE ROMAN VENETO

Verona *(p192)* has the highest concentration of Roman sites in the region; the archaeological museum *(p202)* is full of fine mosaics and sculptures, and Castelvecchio *(p193)* has some very rare early Christian glass and silver. Good museums can also be found at Este *(p184)*, Adria, Treviso *(p174)* and Portogruaro, situated near Concordia *(p175).*

This fine mosaic *of a nightingale in Treviso Museum is from Trevisium, the town's Roman predecessor.*

Verona's Arena *is an awe-inspiring home for the city's opera festival, despite the loss of its outer wall to earthquakes.*

AD 100 The Arena, Verona's amphitheatre, is built. Near Eastern merchants bring Christianity to the region

360 The Roman Empire's northern borders under attack from Slavic and Teutonic tribes

401 Led by Alaric, the Goths invade northern Italy; the Veneto bears the brunt of the attack

Fierce Visigoth

AD 1 | 100 | 200 | 300 | 400

59 BC Livy, Roman historian, born in Padua

313 Constantine the Great grants official status to Christianity

331 Constantinople takes over from Rome as capital of the Roman Empire

395 Roman Empire splits into eastern and western halves

410 Alaric succeeds in sacking Rome itself, but dies the same year

The Birth of Venice

9th-century Venetian coin

FLEEING THE GOTHS, who were systematically looting and burning their way southwards to Rome, the people of the Veneto sought refuge among the wild and uninhabited islands of their marshy coast. There they formed villages, and from the ashes of the Roman past rose the city of Venice (founded, as tradition has it, in AD 421). Exploiting its easily defended maritime position, important trade links with Byzantium were created. Venice proclaimed its brash self-confidence by brazenly stealing the relics of St Mark the Evangelist from Alexandria, in Egypt.

Early Venetian Settlements
The Rialto Bridge (from Rivo Alto, or "high bank") marks the spot of one of many early settlements.

The First Crusade *(1095–9)*
Venice cunningly used the Crusades to her advantage, gaining valuable trading rights in captured cities such as Antioch and Tripoli.

San Marco as it was before 14th-century rebuilding.

The Bishop of Altino
The cathedral at Torcello was founded in AD 639, when Altino's bishop led a mass exodus to the lagoon island, fleeing Lombardic invaders.

THE ARRIVAL OF THE RELICS

This 13th-century mosaic from the façade of San Marco depicts the body of St Mark being carried into the newly built basilica for reburial in AD 832. By securing the relics of such an important saint, Venice signalled its ambition to be considered one of the foremost cities in Christendom, on a par with Rome.

TIMELINE

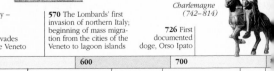

421 Venice founded, traditionally – and conveniently – on St Mark's Day, 25 March

452 Attila the Hun invades Italy and plunders the Veneto

So-called "Attila's throne" in Torcello

552 Totila the Goth invades Italy and destroys many towns in the Veneto

570 The Lombards' first invasion of northern Italy; beginning of mass migration from the cities of the Veneto to lagoon islands

639 Torcello cathedral founded

697 According to legend, Paoluccio Anafesta is elected first doge

726 First documented doge, Orso Ipato

Charlemagne (742–814)

774 Charlemagne invited to drive Lombards from Italy

800 Charlemagne is crowned first Holy Roman Emperor by Pope Leo III

400 | 500 | 600 | 700

Diplomacy
Strategically placed between the powers of Rome and Byzantium, Venice was continually exerting her powers of diplomacy. Here, Doge Ziani receives Holy Roman Emperor Frederick I, whom he reconciled with Pope Alexander III in 1177.

Looting the remains of St Mark from Alexandria was seen as an act of anti-Moslem piety.

The doge and his entourage are wearing Byzantine-style caps and robes.

St Theodore
The Byzantine emperor nominated Theodore as the patron saint of Venice. Venice chose St Mark instead, an act of defiance against Byzantine rule.

WHERE TO SEE EARLY VENICE

The cathedral at Torcello *(pp152–3)* is the oldest surviving building in Venice, and the Basilica San Marco *(pp78–83)* has many period treasures. Early Venetian coins are in the Correr Museum *(p77)*. The original statue of St Theodore is in the Doge's Palace courtyard.

Torcello cathedral's *jewel-like mosaics (11th century) are masterpieces of Byzantine art, probably the work of craftsmen from Constantinople.*

The Pala d'Oro*, St Mark's 10th-century altarpiece, shows merchants bringing St Mark's plundered relics to Venice.*

14 First Venetian coins minted; ork begins on first Doge's Palace

832 First Basilica San Marco completed

828 Venetian merchants steal body of St Mark from Alexandria

900

888 King Berengar I of Italy chooses Verona as his seat

1000

1000 Doge Pietro Orseolo rids the Adriatic of pirates, commemorated by the first Marriage of Venice to the Sea ceremony

1095 First Crusade; Venice provides ships and supplies

1104 Arsenale founded

1100

1120 Verona's San Zeno church begun

1128 First street lighting in Venice

1171 Six districts *(sestieri)* of Venice established

1173 First Rialto Bridge built

1177 Emperor Frederick I Barbarossa agrees to peace terms with Pope Alexander III

1200

1202 Venice diverts the Fourth Crusade to its own ends, the conquest of Byzantium

The Growth of the Empire

The doge's hat, the *zogia*

DURING THE MIDDLE AGES, Venice expanded in power and influence throughout the eastern Mediterranean, culminating in the conquest of Byzantium in 1204. At home, in contrast to the fractional strife of most of the area, Venice enjoyed a uniquely ordered administration headed by the doge, an elected leader whose powers were carefully defined by the Venetian constitution. Real power lay with the Council of Ten and the 2,000 or so members of the Grand Council, from whose number the doge and his advisers were elected.

Bocca di Leone
Such letterboxes were used to report crimes anonymously and were often abused (p89).

Cangrande I
Founder of the Veronese Scaligeri dynasty (see p207), *Cangrande I ("Big Dog") typified the totalitarian rule of most Italian cities.*

Doge Enrico Dandolo boldly led the attack on Constantinople, despite being over 90 and completely blind.

Marco Polo in China
Renowned Venetian merchant, Marco Polo (see p143) *spent over 20 years at the court of Kublai Khan.*

SIEGE OF CONSTANTINOPLE
Facing financial difficulties, the leaders of the Fourth Crusade agreed to attack the capital of Byzantium, as payment for warships supplied by Venice. The city fell in 1204, leaving Venice ruler of Byzantium.

TIMELINE

1204 Conquest of Constantinople; Venice's plunder includes four bronze horses

1222 University of Padua founded

1260 Scaligeri family rules Verona

1271–95 Marco Polo's journey to China

1309 Present Doge's Palace begun

1325 The names of Venice's ruling families are fixed and inscribed in the Golden Book

1200 | 1250 | 1300 | 1350

The Four Horses of San Marco

1284 Gold ducats first minted in Venice

1301 Dante, exiled from his native Florence, is welcomed to Verona by the Scaligeri rulers

1310 The Venetian Constitution is passed; Council of Ten formed

1304–13 Giotto paints the Scrovegni Chapel frescoes *(pp180–81)* in Padua

1348–9 Black Death plague kills half Venice's population

Decapitation
Doge Marin Falier was beheaded in 1355 for plotting to become absolute ruler of Venice. His execution was a warning to future doges.

Imperial treasures and ancient buildings were lost when the 900-year-old city was looted and burned.

Electing the Doge
This pointer was used for counting votes during dogal elections, using a convoluted system designed to prevent candidates bribing their way to power.

Troops scaled the fortifications from galleys moored against the city walls.

Queen of Cyprus
Venice shamelessly gained Cyprus in 1489 by arranging for Caterina Cornaro, from one of Venice's noblest families, to marry the island's king, then poisoning him.

Where to See Imperial Venice

The Doge's Palace combines ceremonial splendour and the grimmer business of imprisonment and torture *(pp84–9)*. Aspects of the constitution are on display in the Correr Museum *(p77)*. A *bocca di leone* survives on the Zattere *(p129)*.

***Many doges** are commemorated by Renaissance-style monuments in the church of Santi Giovanni e Paolo* (pp116–17).

***Meetings of the Grand Council**, dominated by the merchant class, were held in the Sala del Maggior Consiglio* (p87) *in the Doge's Palace.*

Battle of Chioggia

1489 Cyprus ceded to Venice by Queen Caterina Cornaro

1508 Titian's *Assumption* hung in Frari *(p102)*

1400 | 1450 | 1500

1380 Battle of Chioggia: Venice defeats Genoa to win undisputed maritime supremacy in the Adriatic and Mediterranean

1435 Giovanni Bellini born, greatest of the artistic family

1453 Constantinople falls to the Turks; Venice's empire reaches its zenith

1508 Andrea Palladio, architect, born in Vicenza

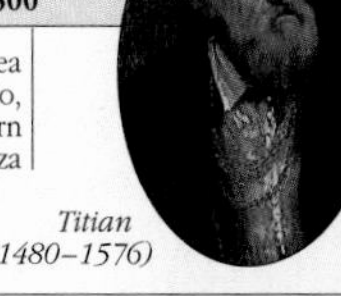

Titian (1480–1576)

The Queen of the Adriatic

BY THE 16TH CENTURY, Venice held a monopoly on Mediterranean trade and had colonized the whole of northeastern Italy, from the Adriatic to the Alps. Keeping hold of such a vast empire meant being in a constant state of war. The League of Cambrai, dedicated to destroying Venice, was formed in 1508 by the most powerful men in Europe, Pope Julius II and the Holy Roman Emperor Maximilian. Their troops sacked the cities of the Veneto, but the region remained loyal to Venice's relatively benign rule. Far more of a threat were the Turks. They carved out the Ottoman Empire from 1522, driving Venice from the eastern Mediterranean and eventually taking Cyprus in 1570.

16th-century armour from the Doge's Palace

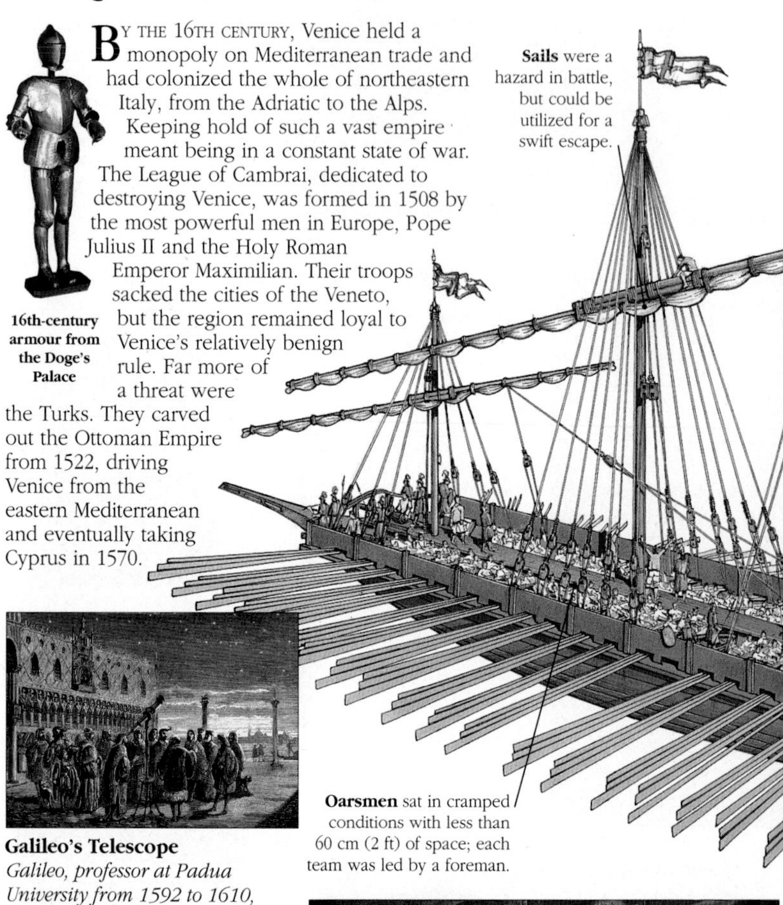
Sails were a hazard in battle, but could be utilized for a swift escape.
Oarsmen sat in cramped conditions with less than 60 cm (2 ft) of space; each team was led by a foreman.

Galileo's Telescope
Galileo, professor at Padua University from 1592 to 1610, demonstrated his telescope to Doge Leonardo Donà in 1609.

Battle of Lepanto
Venice led the combined forces of the Christian world in this bloody victory over the Turks, fought in 1571.

TIMELINE

1500 — **1550** — **1600**

- **1501** Doge Leonardo Loredan, great diplomat, begins 20-year rule
- **1514** Fire destroys the original timber Rialto Bridge
- **1516** Jews confined to the Venetian Ghetto. End of League of Cambrai wars
- **1519** Tintoretto born
- **1528** Paolo Veronese born
- **1529** Death of Luigi da Porto of Vicenza, author of the story of Romeo and Juliet
- **1570** Cyprus lost to the Turks
- **1571** Battle of Lepanto: decisive victory for the western fleet, led by Venice, over the Turks
- **1577** Palladio designs the Redentore church *(p154)* to mark the end of the plague that took 51,000 lives
- **1585** First performance at Vicenza's Teatro Olimpico *(p172)*
- **1592** Galileo appointed professor of mathematics at Padua University
- **1595** Shakespeare's *Romeo and Juliet*

Celebrating the End of the Plague
More deadly than any opposing army, plague hit Venice in 1575 and again in 1630, carrying off Titian among its 100,000 victims.

Where to See Maritime Venice

The triumph of Venice over the sea is celebrated in the Museo Storico Navale *(p118)*. For a glimpse of the extensive and disused Arsenale shipyard in Castello, take a trip on *vaporetto* route No. 52 or 23 *(p275)*.

Arsenale lions*, plundered from Piraeus in 1687, guard the forbidding gates of the Arsenale shipyard* (p119).

Santa Maria della Salute *was built in thanksgiving for deliverance from the 1630 plague* (p135).

The Venice Arsenale
Venice was at the forefront of maritime construction. Her heavily defended shipyards were capable of turning out warships at the rate of one a day.

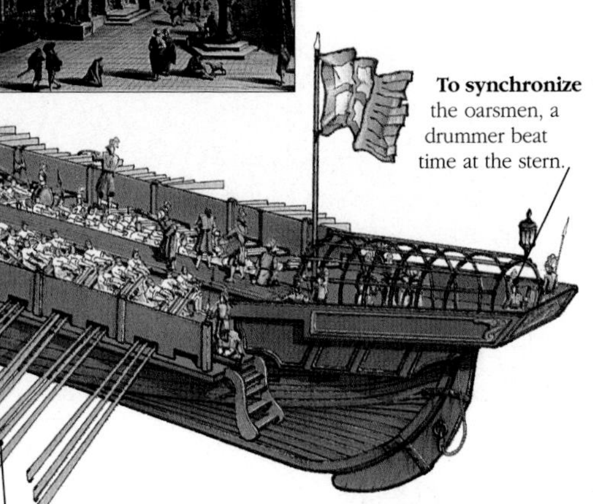

To synchronize the oarsmen, a drummer beat time at the stern.

The trireme was so named because the oars were grouped in threes. Each trireme had up to 150 oars.

Venetian Trireme

Venetian naval supremacy was based on the swift and highly manoeuvrable trireme, used to sink enemy ships by means of its pointed battering ram and its bow-mounted cannon.

Monteverdi (1567–1643)

1613 Monteverdi appointed choirmaster at Basilica San Marco

1630 Plague strikes Venice again, reducing the city's population to 102,243, its smallest for 250 years

1650

1669 Venice loses Crete to the Turks

1678 Elena Piscopia receives doctorate from Padua University, the first woman in the world ever to be awarded a degree *(p178)*

Elena Piscopia (1646–84)

1700

1703 Vivaldi joins La Pietà as musical director

1708 In a bitter winter, the lagoon freezes over and Venetians can walk to the mainland

1718 Venetian maritime empire ends with the surrender of Morea to the Turks

Glorious Decadence

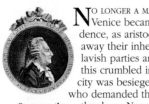

Casanova, the Venetian libertine

No longer a major power, 18th-century Venice became a byword for decadence, as aristocratic Venetians frittered away their inherited wealth in lavish parties and gambling. All this crumbled in 1797 when the city was besieged by Napoleon, who demanded the abdication of the doge. Napoleon granted the city to his allies, the Austrians, whose often authoritarian rule drove many people of the Veneto to join the vanguard of the revolutionary Risorgimento. This movement, led in Venice by Daniele Manin, was dedicated to creating a free and united Italy, a dream not fully realized until 1870, four years after Venice was freed from Austrian rule.

The State-Run Casino
The notorious Ridotto, open to anyone wearing a mask, closed in 1774 as many Venetians had bankrupted themselves.

Gambling fever so gripped the city that gaming tables were set up between the columns in the Piazza.

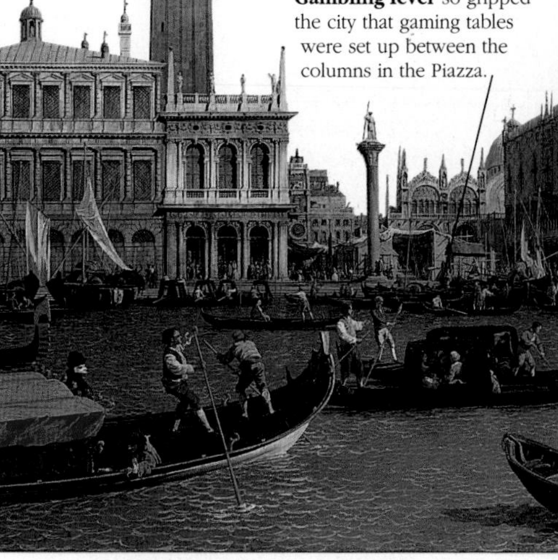

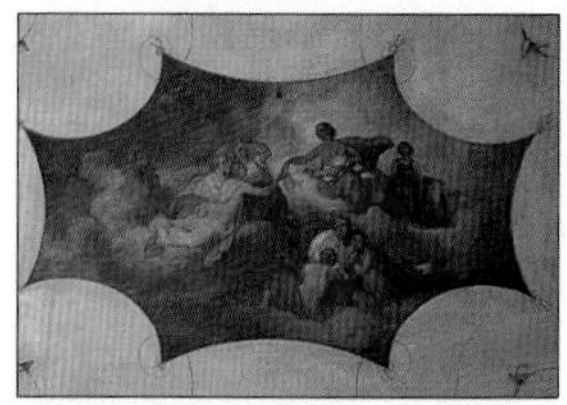
Caffè Pedrocchi
Several intellectuals who had used this lavishly decorated café (see p178) *in Padua as their base, were executed for leading a revolt against Austrian rule in 1831.*

Imperial Ritual

Canaletto's *St Mark's Basin on Ascension Day* (c.1733) captures the empty splendour of Venice on the eve of her demise. The doge's gold and scarlet barge has been launched for the annual ceremony of Venice's Marriage to the Sea.

The Horses of St Mark
Among the art treasures looted by Napoleon were the Four Horses of St Mark, symbols of Venetian liberty. The horses were returned in 1815.

Timeline

1720 Florian's café opens in Venice *(p246)*

1725 Casanova born in Venice

1752 Completion of sea walls protecting the lagoon entrances

1755 Casanova imprisoned in Doge's Palace

1757 Canova, Neo-Classical sculptor, born in Venice

1775 Quadri's café *(p246)* opens in Venice

1789 The Dolomites named after Déodat Dolomieu (1750–1801)

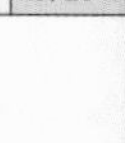

1720 | 1770

Florian's café

1790 Venetian opera house, La Fenice, opens

1797 Napoleon invades the Veneto; Doge Lodovico Manin abdicates; Venetian Republic ends

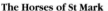

1798 Napoleon grants Venice and its territories to his Austrian allies in return for Lombardy

Antonio Vivaldi
(1678–1741) Fashionable Venetians flocked to hear the red-haired priest's latest compositions, performed by the orphan girls of La Pietà. Vivaldi's most famous work, The Four Seasons *(1725), was a great success throughout Europe.*

The Bucintoro, the doge's ceremonial barge

Sumptuary laws, passed in 1562, decreed that all Venetian gondolas must be black to prevent lavish displays of wealth.

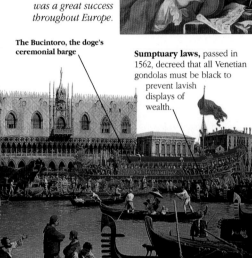

No Longer an Island
Venice lost its isolation in 1846 when a causeway joined the city to the mainland and the Italian rail network.

WHERE TO SEE 18TH-CENTURY VENICE

The Museo Storico Navale *(p118)* displays a beautifully crafted model of the Bucintoro and its original banner. Vivaldi concerts are a regular feature at La Pietà church *(p112)*. Paintings by Guardi, Canaletto and Longhi capture the spirit of the age and are found in the Accademia *(pp130–3)*, Correr Museum and Ca' Rezzonico *(p126)*.

Fortunes were spent *on opulent wigs, jewels and clothing for costume balls and the theatre. This high-heeled shoe is in the Correr Museum* (p77).

The comic antics *of Harlequin and Pantaloon at La Fenice* (p93) *ensured the popularity of the theatre with opera-loving Venetians.*

1804 Napoleon crowned King of Italy and takes back Venice

1814–15 Austrians drive French from Venice; Congress of Vienna returns the Veneto to Austria

Daniele Manin (1804–57)

1859 Second War of Italian Independence; after Battle of Solferino, Red Cross founded

1861 Vittorio Emanuele crowned King of Italy

1820 | 1870

1818 Byron swims up the Grand Canal

1846 Venetian rail causeway links the city to the mainland for the first time

1848 First Italian War of Independence. Venice revolts against Austrian rule

1853 Ruskin publishes *The Stones of Venice*

1849 Hunger and disease force Venetian rebels, led by Daniele Manin, to surrender

1866 Venice and Veneto freed from Austrian rule

Venice in Vogue

From being an introverted and unchanging city, Venice developed with remarkable speed. The opening of the Suez Canal in 1869 brought new prosperity; a new harbour was built for ocean-going ships and Venice became a favourite embarkation point for colonial administrators and rich Europeans travelling east. The fashion for sea-bathing and patronage by wealthy socialites reawakened interest in the city, and the founding of the Biennale attracted Europe's leading artists, who expressed their enthusiasm for the city in novels, paintings and music.

Peggy Guggenheim *(1898–1979)*
Patron of the avant garde, Peggy Guggenheim brought her outstanding art collection (see p134) *to Venice in 1949.*

The Hotel Excelsior's Moorish exterior is distinctive.

Bathing huts, designed for modesty in the 1920s, are still a feature of the Lido.

Igor Stravinsky *(1882–1971)*
Along with Turgenev, Diaghilev and Ezra Pound, Stravinsky was one of many émigrés enchanted by the magic of Venice.

Hotel Excelsior
When it was built in 1907, the Hotel Excelsior (see p231) *was the world's largest hotel.*

The Lido

From the turn of the century, grand hotel developments along the sandy Adriatic shore turned the Lido into Europe's most stylish seaside resort. The island has since given its name to bathing establishments the world over.

Timeline

1870 · 1880 · 1890 · 1900 · 1910

1881 Venice becomes second largest port in Italy after Genoa

1883 Wagner dies in Palazzo Vendramin-Calergi

Richard Wagner (1813–83)

1885 First Biennale art exhibition

1889 Poet Robert Browning dies in Ca' Rezzonico

1902 Collapse of campanile in Piazza San Marco

1903 Patriarch Sarto of Venice becomes Pope Pius X

1912 Opening of rebuilt campanile; Thomas Mann writes *Death in Venice*

The International Exhibition of Modern Art
Venice became a showcase for all that was new in world art and architecture when the Biennale was launched. The first exhibition, in 1895, showed work by Renoir and Monet.

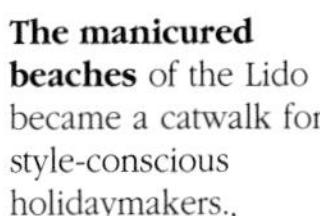

The manicured beaches of the Lido became a catwalk for style-conscious holidaymakers.

The Campanile
After the appearance of ominous warning cracks, the 1,000-year-old bell tower crashed to the ground in 1902. It was rebuilt within a decade (see p76).

Where to See Turn-of-the-Century Venice

Regular *vaporetto* services link Venice to the Lido *(p156)*, with its deluxe hotels, sports facilities and beaches. The pavilions of the Biennale *(p121)* are usually only open during the exhibition. A lift carries visitors to the top of the rebuilt Campanile *(p76)* for panoramic views of Venice.

San Michele*, the cemetery isle* (p151), *is the last resting place of eminent foreigners, such as Serge Diaghilev, Igor Stravinsky and Ezra Pound.*

The exclusive *Grand Hôtel des Bains* (p231) *on the Lido has retained its Art Deco style and private section of beach.*

1917 Work starts on constructing the port of Marghera

1918 Fierce fighting in mountain passes of the Veneto in the last weeks of World War I

1920

1926 Mestre is formally granted town status

1930

1931 Venice is linked to the mainland by a road causeway

1932 First Venice Film Festival

German travel poster from 1936

1940

1943–5 Mussolini rules a puppet state, the Salò Republic

1950

1951 Stravinsky's *The Rake's Progress* premièred in Venice

1954 Britten's *Turn of the Screw* premièred in Venice

1956 Cortina d'Ampezzo hosts Winter Olympics

1959 Patriarch Roncalli elected Pope John XXIII

1960

1960 Venice airport opens

Venice Preserved

IN NOVEMBER 1966 Venice was hit by the worst floods in its history, sparking worldwide concern for the future of the city's delicate and decaying fabric. Major steps have since been taken to protect Venice and its unique heritage, though some difficult issues remain, including the erosion caused by large numbers of visitors to the city and pollution from the economically buoyant mainland. However, the allure of Venice, set in its watery lagoon, is as compelling as ever.

Pink Floyd in Venice
Pink Floyd's 1989 rock concert threatened the city's equilibrium.

Venice as Film Set
Venice has served as the backdrop to countless films, including Fellini's Casanova *(1976) and* Indiana Jones and the Last Crusade *(1989).*

The Regata Storica, held in September, is an annual trial of strength and skill for gondoliers.

After the Flood
During the 1966 floods, the waters rose nearly 2 m (6 ft). Great damage was done by fuel oil, washed out of broken tanks. It is now banned from the city in favour of gas.

TOURISM

Venetian regattas are part of a rich tradition that enhances the city's attraction to tourists, providing employment for many on the mainland as well as in Venice itself. Even so, some complain that tourism has turned Venice from a living city into one vast museum.

TIMELINE

1960

1966 Floods cause devastation in Venice. UNESCO launches its Save Venice appeal

1968 Protestors prevent part of the lagoon being drained to extend Marghera's industrial zone

1970

1970 Luchino Visconti's film, *Death in Venice*

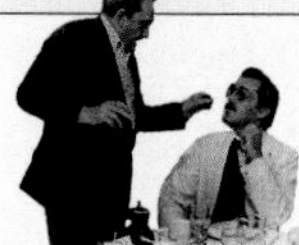

Visconti and Dirk Bogarde on the set of Death in Venice

1973 Laws passed to reduce pollution, subsidence and flooding.

1978 Patriarch Luciani of Venice elected Pope John Paul I, but dies 33 days later

Benetton
The famous clothing firm, originating in Treviso, represents the modern face of Veneto industry.

Restoration in Venice

One positive result of the 1966 floods was a major international appeal for funds to pay for the cleaning of historic buildings, statues and paintings. Funds raised are coordinated under the auspices of UNESCO, with offices in Venice.

Restorers *learn how to repair and conserve fragile works of art at the School of Craftsmanship on San Servolo* (p154).

Madonna dell'Orto (p140) *was restored by the Italian Art and Archives Rescue Fund (later renamed Venice in Peril).*

Venice plays host to over 12 million visitors every year.

Glass Blowing
This age-old tradition still contributes to the economy.

Water is now piped into Venice to combat subsidence caused by water extraction from the subsoil.

The Acqua Alta
High tides cause frequent flooding in San Marco. Plans to complete a flood barrier across the lagoon are subject to controversy.

1979 Venetian Carnival is revived

1983 Venice officially stops sinking after extraction of underground water prohibited

1988 First experimental stage of MOSE, the lagoon flood barrier, is completed

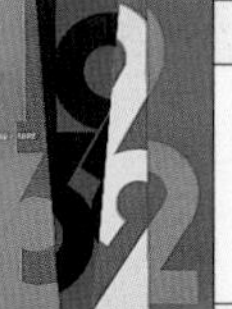

1932–1992 Venice Film Festival poster

1992 Venice Film Festival celebrates 60 years

1995 Centenary of Biennale Exhibition

1980

1990

Carnival reveller

1990 Plans to drain Venetian lagoon to create site for Expo 2000 defeated by Italian parliament

1992 Venice rocked by corruption scandals. Approval given for metro network beneath lagoon

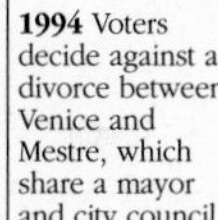

1994 Voters decide against a divorce between Venice and Mestre, which share a mayor and city council

Venice Area by Area

Venice at a Glance

VENICE IS SMALL and most of the sights can be comfortably visited on foot. The heart of the city is the Piazza San Marco, which is overlooked by the great Basilica and the Doge's Palace. For many, these are attractions enough, but there are delights worth exploring beyond the Piazza, such as the galleries of the Accademia, Ca' Rezzonico and the imposing Frari church. Unique to Venice are the naval Arsenale to the east and the Ghetto in the north.

Ghetto
Established in the early 16th century, this fascinating quarter was the world's first ghetto (see p145).

Santa Maria Gloriosa dei Frari
This soaring Gothic edifice, founded by the Franciscans in 1340, is a rich repository of Venetian painting and sculpture (see pp102–3).

CANNAREGIO
Pages 136–45

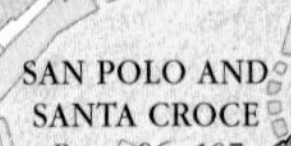

SAN POLO AND SANTA CROCE
Pages 96–107

DORSODURO
Pages 122–35

SAN MAR
Pages 72-

0 metres 500

0 yards 500

Ca' Rezzonico
The splendid rooms of this palace, overlooking the Grand Canal, are decorated with 18th-century furniture and paintings (see p126).

Accademia
Carpaccio's St Ursula *cycle (1490–5) is one of the treasures of the Accademia, which has a comprehensive collection of Venetian art* (see pp130–3).

Rialto Bridge
The bustling Rialto Bridge (see p100) *was named after the ancient commercial seat of Venice, where the first inhabitants settled.*

Ca' d'Oro
This ornate palace is the finest example of Venetian Gothic style (see p142).

Basilica San Marco
Magnificent mosaics sheathe the domes, walls and floor of the Byzantine Basilica (see pp78–83).

Arsenale
The great dockyard, first of its kind in Europe, was the naval nerve centre of the Venetian Empire (see p119).

CASTELLO
Pages 108–21

Doge's Palace
The colonnaded Gothic palace was the seat of government as well as home to the doge and his family (see pp84–89).

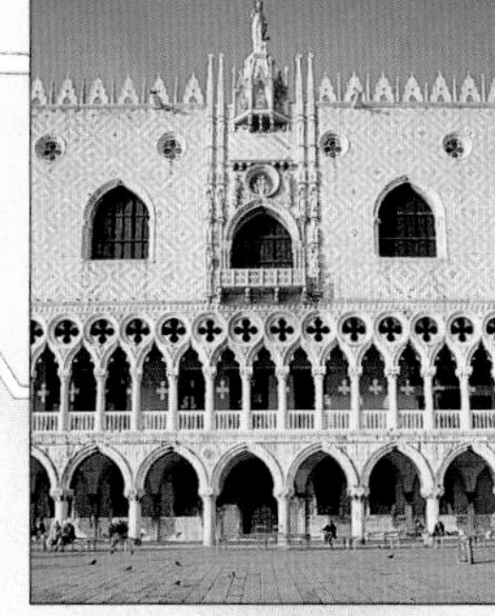

Santa Maria della Salute
Marking the southern end of the Grand Canal, this great Baroque church is one of the city's landmarks (see p135).

202

A View of the Grand Canal

Known to the Venetians as the *Canalazzo*, the Grand Canal sweeps through the heart of Venice, following the course of an ancient river bed. Since the founding days of the empire it has served as the city's main thoroughfare. Once used by great galleys or trading vessels making their stately way to the Rialto, it is nowadays teeming with *vaporetti*, launches, barges and gondolas. Glimpses of its glorious past, however, are never far away. The annual re-enactment of historic pageants, preserving the traditions of the Venetian Republic, brings a blaze of colour to the canal. The most spectacular is the Regata Storica held in September *(see p35)*, a huge procession of historic craft packed with crews in traditional costumes, followed by boat and gondola races down the Grand Canal.

Venetian gondolier

The parade of palaces bordering the winding waterway, built over a span of around 500 years, presents some of the finest architecture of the Republic. Historically it is like a roll-call of the old Venetian aristocracy, with almost every *palazzo* bearing the name of a once-grand family. Bright frescoes may have faded, precious marbles worn, and foundations frayed with the tides, but the Grand Canal is still, to quote Charles VIII of France's ambassador in 1495, "the most beautiful street in the world".

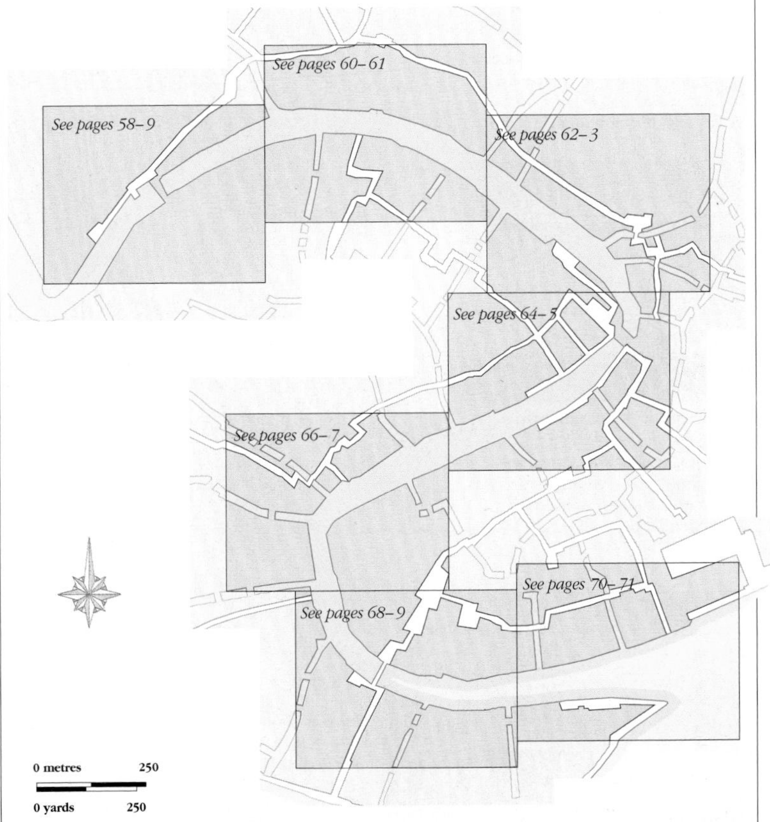

◁ **The Grand Canal at its most colourful, during the Regata Storica**

Santa Lucia to Palazzo Flangini

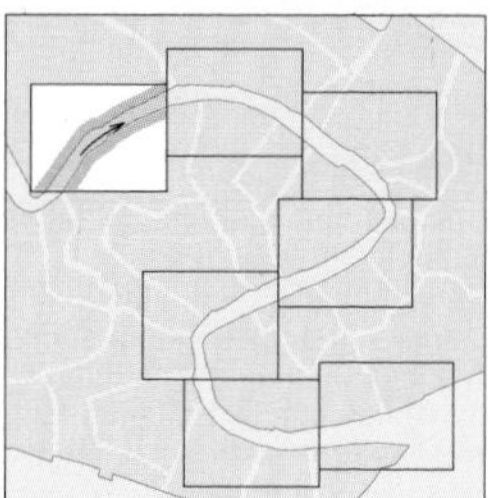

Locator Map

Vaporetto ticket office, Grand Canal

THE GRAND CANAL is best admired from a gondola or, more cheaply, from a *vaporetto*. Several lines travel the length of the canal *(see p275)* but only the No. 1 goes sufficiently slowly for you to take in any of the individual palaces. The journey from the station to San Zaccaria takes about 40 minutes. Ideally you should take a return trip, absorbing one bank at a time. Nearly 4 km (2½ miles) long, the canal varies in width from 30 to 70 m (98 to 230 ft) and is spanned by three bridges, the Scalzi, the Rialto and the Accademia.

***Santa Maria di Nazareth** is known today as the Scalzi, after the supposedly "shoeless" Carmelites who founded it* (see p145). *Within is the tomb of Ludovico Manin, last of the doges.*

Santa Lucia railway station (see p272), *built in the mid 19th century and remodelled in the 1950s, links the city with the mainland.*

***La Direzione Compartimentale**, the administration offices for the railway, was built at the same time as the station, on the site of the church of Santa Lucia and other ancient buildings.*

***Palazzo Diedo**, also known as Palazzo Emo, is a Neo-Classical palace of the late 18th century. It is believed to be the birthplace of Angelo Emo (1731–92), the last admiral of the Venetian fleet. The palace was built by Andrea Tirali, an engineer who worked on the restoration of San Marco.*

Palazzo Calbo Crotta *is now the 4-star Hotel Principe. Fine antiques and fabrics which once decorated the palace are now in Ca' Rezzonico* (see p126).

Palazzo Flangini *was designed by Giuseppe Sardi, a leading 17th-century architect.*

Palazzo Gritti *was built in the 16th century. The Grittis were a wealthy family who produced one of the most intelligent doges, Andrea Gritti (reigned 1523–38).*

The Scalzi Bridge was built in 1934, replacing the original wrought iron bridge.

Campo San Simeone Grande*, named after the nearby church (otherwise called San Simeone Profeta), is one of the few* campi *overlooking the canal.*

Casa Adoldo and Palazzo Foscari-Contarini *were both rebuilt in the 16th century. According to local tradition, the great Doge Francesco Foscari (ruled 1423–57) was born in the original Foscari-Contarini palace.*

San Simeone Piccolo *is a large church, despite its name (*piccolo *means small). Built in 1738, its design was based partly on the Pantheon in Rome. It is only open for concerts.*

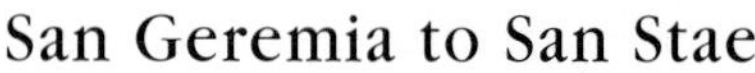

San Geremia to San Stae

THIS STRETCH sees the start of the great palaces. The most remarkable is the Vendramin Calergi, which became a model for other Venetian palaces.

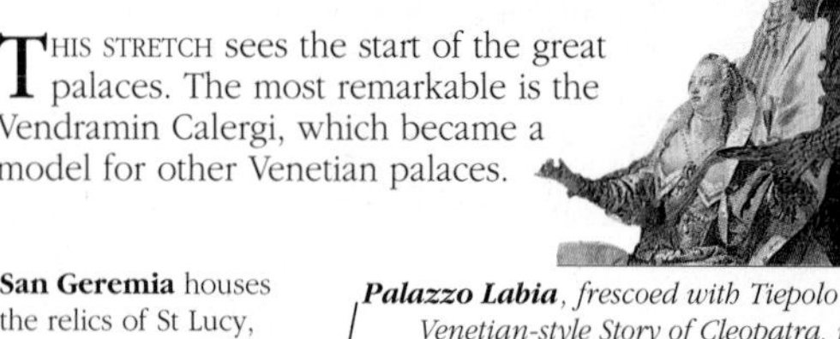

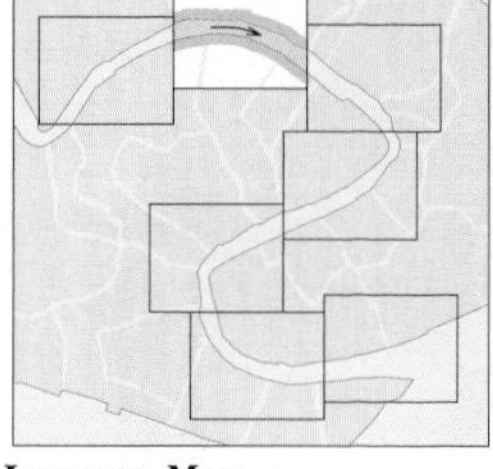

Locator Map

San Geremia houses the relics of St Lucy, formerly preserved in Santa Lucia where the station now stands.

Palazzo Labia, *frescoed with Tiepolo's Venetian-style Story of Cleopatra, is open to the public* (see p143).

Palazzo Querini has the family coat of arms on the façade.

Palazzo Corner-Contarini is also called Ca' dei Cuori after the hearts in the family coat of arms.

Palazzo Giovanelli, *a restored Gothic palace, was acquired by the Giovanellis in 1755. This titled non-Venetian family had been admitted into the Great Council in 1668 for a fee of 100,000 ducats.*

Fondaco dei Turchi *was a splendid Veneto-Byzantine building before last century's brutal restoration. Today it houses the Natural History Museum* (see p105).

Palazzo Donà Balbi, *built in the 17th century, is named after two great Venetian families who intermarried. The Donà family produced four doges.*

Deposito del Megio, *a crenellated building with a reconstructed Lion of St Mark, was a granary in the 15th century.*

***San Marcuola**, dedicated to St Ermagora and St Fortunatus, was built in 1728–36 by Giorgio Massari, but the façade was never completed.*

***Palazzo Vendramin Calergi**, an early Renaissance palace, was designed by Mauro Coducci. The composer Richard Wagner died here in 1883. The palace houses the municipal casino in winter.*

Palazzo Marcello, rebuilt in the early 18th century, was the birthplace of composer Benedetto Marcello in 1686.

Palazzo Erizzo has two huge paintings depicting the feats of Paolo Erizzo, who died heroically fighting the Turks in 1469.

Palazzo Emo belonged to the family of a famous Venetian admiral *(see p58)*.

San Marcuola

San Stae

***Palazzo Belloni Battagia**, with its distinctive pinnacles, was built by Longhena in the mid 17th century for the Belloni family, who had bought their way into Venetian aristocracy.*

***Palazzo Tron**, built in the late 16th century, hosted a famous ball in 1775 in honour of Emperor Joseph II of Austria.*

***San Stae** is striking for its Baroque façade, graced by marble statues. It was funded by a legacy left by Doge Alvise Mocenigo in 1709* (see p143).

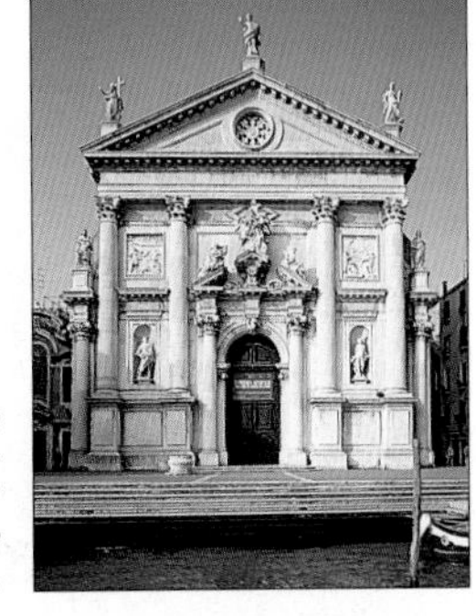

Palazzo Barbarigo to the Markets

HERE THE CANAL is flanked by stately palaces, built over a period of five centuries. The most spectacular is the Gothic Ca' d'Oro, whose façade once glittered with gold.

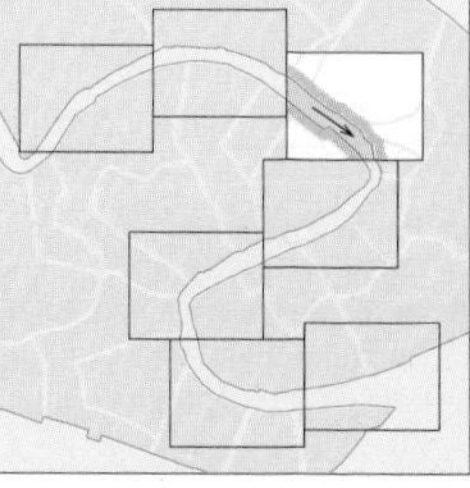
LOCATOR MAP

Palazzo Barbarigo retains the vestiges of its 16th-century frescoed façade paintings.

Palazzo Gussoni-Grimani's façade once had frescoes by Tintoretto. It was home to the English ambassador in 1614–18.

San Stae

Palazzo Fontana Rezzonico was the birthplace of Count Rezzonico (1693), the fifth Venetian pope.

Ca' Foscarini, a Gothic building of the 15th century, belonged to the Foscari family before it became the residence of the Duke of Mantua in 1520.

Ca' Pesaro, a huge and stately Baroque palace designed by Longhena (see p23), *today houses the Gallery of Modern Art and the Oriental Museum* (see p105). *It was built for Leonardo Pesaro, a Procurator of San Marco.*

Casa Favretto (Hotel San Cassiano) was the home of the painter Giacomo Favretto (1849–87).

***Palazzo Morosini Brandolin** belonged to the Morosini family, one of the* Case Vecchie *families, deemed to be noble before the 9th century.*

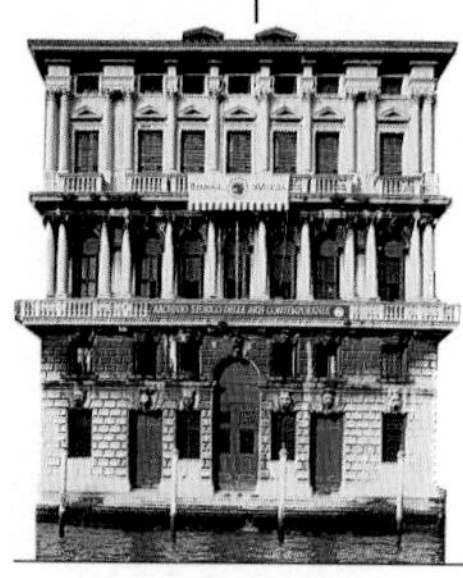
***Ca' Corner della Regina** is named after Caterina Cornaro, Queen of Cyprus, who was born here in 1454. The present building (1724–7) was designed by Domenico Rossi.*

***The Pescheria** has been the site of a busy fish market for six centuries. Today it takes place in the striking mock-Gothic market hall, built in 1907.*

***Ca' d'Oro**, the most famous of Venetian Gothic palaces* (see p144)*, houses paintings, frescoes and sculpture from the collection of Baron Giorgio Franchetti, who bequeathed the palace and all its contents to the State.*

CANALETTO

Antonio Canale (Canaletto) (1697–1768) is best known for his *vedute* or views of Venice. He studied in Rome, but lived here for most of his life. One of his patrons was Joseph Smith *(see below)*. Sadly there are very few of his paintings left on view in the city.

***Palazzo Sagredo** passed from the Morosini to the Sagredo family in the early 18th century. The façade shows characteristics of both Veneto-Byzantine and Gothic styles.*

***Palazzo Foscarini** was the home of Marco Foscarini, a diplomat, orator and scholar who rose to the position of doge in 1762.*

Palazzo Michiel dalle Colonne was named after its distinctive colonnade.

***Palazzo Mangili Valmarana** was designed by Antonio Visentini (above) in Classical style for Joseph Smith, who became the English consul in Venice. Smith (1682–1770) was a patron of both Visentini and Canaletto.*

Palazzo Michiel del Brusà was rebuilt and named after the great fire *(brusà)* that swept the city in 1774.

***Ca' da Mosto** is a good example of 13th-century Veneto-Byzantine style. Alvise da Mosto, the 15th-century navigator, was born here in 1432.*

Tribunale Fabbriche Nuove, Sansovino's market building (1555), is now the seat of the Assize Court.

The Rialto Quarter

THE AREA AROUND THE RIALTO BRIDGE is the oldest and busiest quarter of the city. Traditionally a centre of trade, crowded quaysides and colourful food markets still border the canal south of the bridge.

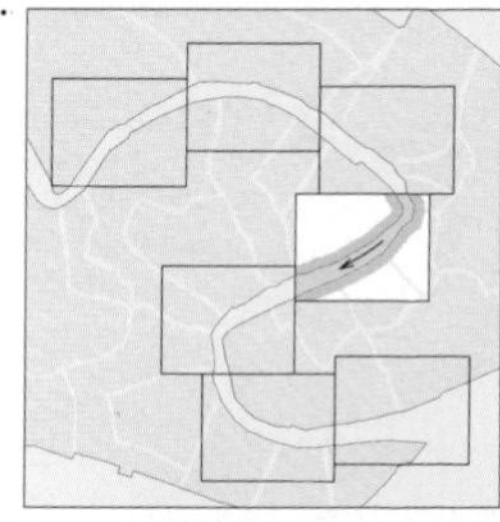

Locator Map

Palazzo Papadopoli, *formerly known as Coccina-Tiepolo, was built in 1560. Its splendid hall of mirrors has been preserved.*

Riva del Vin *is one of the few spots where you can sit and relax on the banks of the Grand Canal* (see p98).

Ca' Corner-Martinengo-Ravà *became the Leon Bianco Hotel in the 19th century. The American writer, James Fenimore Cooper, stayed here in 1838.*

Palazzo Barzizza, rebuilt in the 17th century, still preserves its early 13th-century façade.

San Silvestro

Palazzo Grimani, *a fine, if somewhat austere looking, Renaissance palace* (see p23), *was built in 1556 by Michele Sanmicheli for the Procurator, Girolamo Grimani. The State purchased the palace in 1807 and it is now occupied by the city's Court of Appeal.*

Palazzo Farsetti and Palazzo Loredan, *both occupied by the City Council, were built in the late 12th century and merged in 1868. Palazzo Farsetti became an academy for young artists, one of whom was Canova.*

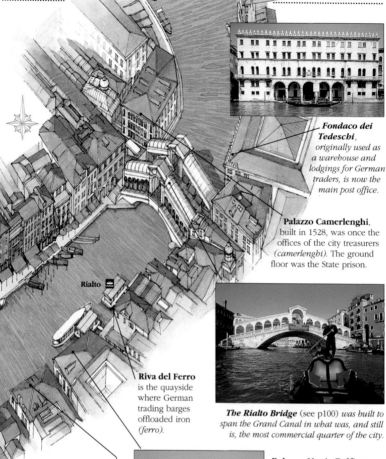

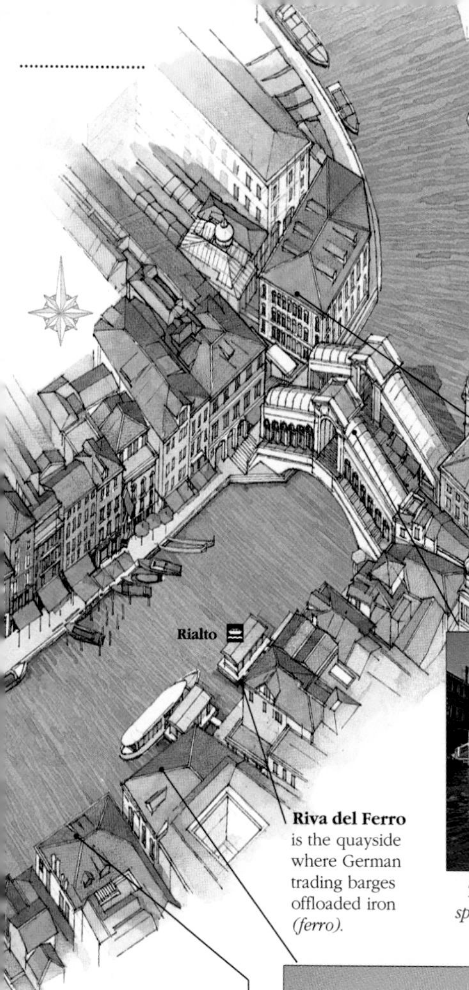

Fondaco dei Tedeschi, *originally used as a warehouse and lodgings for German traders, is now the main post office.*

Palazzo Camerlenghi, built in 1528, was once the offices of the city treasurers *(camerlenghi)*. The ground floor was the State prison.

Rialto

Riva del Ferro is the quayside where German trading barges offloaded iron *(ferro)*.

The Rialto Bridge (see p100) *was built to span the Grand Canal in what was, and still is, the most commercial quarter of the city.*

Casetta Dandolo's predecessor is said to have been the birthplace of Doge Enrico Dandolo (ruled 1192–1205).

Palazzo Manin-Dolfin *was built by Sansovino in 1538–40 but only his Classical stone façade survives. The interior was completely transformed for Ludovico Manin, last doge of Venice (died 1797). He intended to turn the house into a magnificent palace extending as far as Campo San Salvatore.*

Palazzo Bembo*, a 15th-century Gothic palace, was the birthplace of the Renaissance cardinal and scholar, Pietro Bembo, who wrote one of the earliest Italian grammars.*

The Dandolo Family

Doge Enrico Dandolo

The illustrious Dandolo family produced four doges, 12 procurators of San Marco, a patriarch of Grado and a queen of Serbia. The first of the doges was Enrico who, despite being old and blind, was the principal driving force in the Crusaders' plan to take Constantinople in 1204 *(see p42)*. The other remarkable doge in the family was the humanist and historian, Andrea Dandolo (died 1354).

La Volta del Canal

THE POINT WHERE THE CANAL doubles back sharply on itself is known as La Volta – the bend. This splendid curve was long ago established as the finishing stretch for the annual Regata Storica *(see p35)*.

LOCATOR MAP

***Palazzo Marcello**, which belonged to an old Venetian family, is also called "dei Leoni" because of the lions either side of the doorway.*

Palazzo Persico, on the corner of Rio San Polo, is a 16th-century house in Lombardesque style.

Palazzo Civran-Grimani is a Classical building of the early 17th century.

***Palazzo Balbi**, seat of the regional government, was built for Nicolò Balbi, who is said to have died of a chill surveying its construction. From here, Napoleon viewed the 1807 regatta, held in his honour.*

San Tomà

***Ca' Foscari** was built for Doge Francesco Foscari in 1437*(see p22). *It is now part of the University of Venice.*

Palazzo Giustinian was the residence of Wagner in 1858–9, when he was composing the second act of *Tristan and Isolde*.

San Samuele

Ca' Rezzonico

***Ca' Rezzonico**, now the museum of 18th-century Venice* (see p126), *became the home of the poet Robert Browning and his son, Pen, in 1888.*

Palazzo Barbarigo, built in the 1560s, was celebrated for its roof terrace. It is now home to the German Institute.

Palazzo Capello-Layard was the home of Sir Austen Henry Layard, excavator of Nineveh.

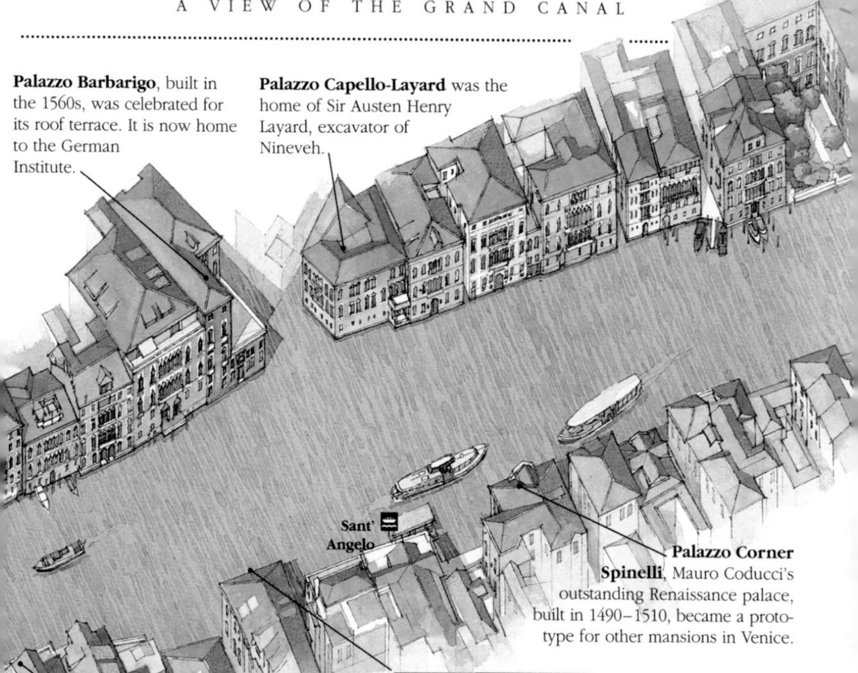

Sant' Angelo

Palazzo Corner Spinelli, Mauro Coducci's outstanding Renaissance palace, built in 1490–1510, became a prototype for other mansions in Venice.

***Palazzo Mocenigo**, formed by four palaces linked together, has a plaque to the poet Byron who stayed here in 1818.*

***Palazzo Garzoni**, a renovated Gothic palace, is now part of the university. The traghetto service, which links the neighbouring Calle Garzoni to San Tomà on the other side of the canal, is one of the oldest in Venice.*

***Palazzo Moro Lin**, also known as the "palace of the 13 windows", was created in the 17th century for the painter Pietro Liberi by merging two Gothic houses.*

***Palazzo Grassi**, built in the 1730s, was bought by Fiat in 1984 and turned into a venue for art exhibitions.*

***Palazzo Capello Malipiero**, a Gothic palace, was reconstructed in 1622. Beside it, in Campo di San Samuele, stands the church of San Samuele which has a 12th-century Veneto-Byzantine campanile.*

Ca' Rezzonico to the Guggenheim

THIS SOUTHERN STRETCH of the canal, widening after the Accademia, is lined by a rich and varied parade of palaces.

Palazzo del Duca, planned in the 15th century as a sumptuous palace but never finished, houses a collection of porcelain.

Palazzo Falier *was said to have been home to Doge Marin Falier, who was beheaded for treason in 1355* (see p43).

Palazzo degli Scrigni, built in 1609, acquired its name from the coffers *(scrigni)* inherited by the Contarini in 1418.

Palazzo Loredan (see p22), *home of Doge Francesco Loredan (1752–62), became the Austrian embassy.*

The wooden Accademia Bridge *was built in 1932 as a temporary structure to replace a 19th-century iron bridge. By popular demand it has been retained.*

The Accademia *galleries, within the former church, monastery and Scuola della Carità, house the world's greatest collection of Venetian paintings* (see pp130–33).

Palazzo Contarini del Zaffo, *a magnificent Renaissance palace of the late 1400s, was built for a branch of the ubiquitous Contarini family. Early this century it was acquired by the Polignac family.*

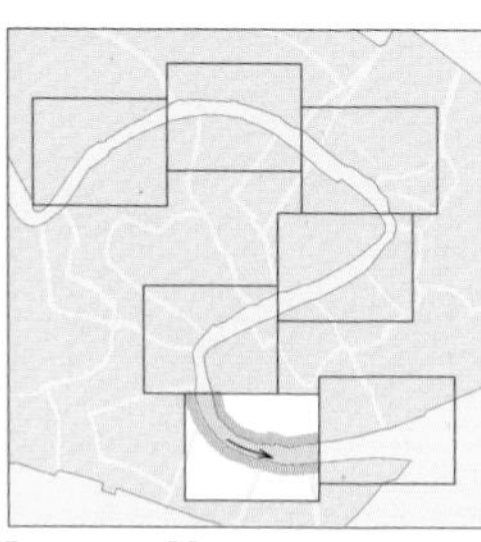

LOCATOR MAP

Ca' Grande, *a huge Classical palace, was designed in 1545 by Sansovino for Giacomo Cornaro, nephew of the Queen of Cyprus. The family was one of the richest in Venice and spared no expense in the palace's decoration. This family tree illustrates the extent of the Cornaro's wealth and influence in Venice.*

Palazzo Franchetti Cavalli belonged to Archduke Frederick of Austria, who died here in 1836.

Palazzo Barbaro *comprises two palaces, one of which was bought by the Curtis family in 1885. Monet and Whistler painted here and Henry James (right) wrote* The Aspern Papers.

Casetta delle Rose, *one of the smallest houses on the canal, was the home of Italian poet Gabriele d'Annunzio during World War I. Canova (above) had his studio here in 1770.*

Palazzo Barbarigo, *beside the Campo San Vio, stands out for the harsh mosaics, added in 1887.*

Peggy Guggenheim *established her collection of modern art in Venice in 1951* (see p134). *She chose as her venue the Palazzo Venier dei Leoni, which had been built in 1749 and never finished.*

Palazzo Dario, *built in 1487, is a charming but strangely ill-fated palace* (see p135).

To La Salute and San Marco

THE VIEW ALONG THE FINAL STRETCH of the canal is one of the finest – and most familiar – in Venice. Near the mouth rises the magnificent church of La Salute with busy St Mark's Basin beyond.

***The Palazzo Gritti-Pisani**, where Ruskin stayed in 1851, is better known today as the luxurious 5-star Hotel Gritti Palace* (see p229).

LOCATOR MAP

***Palazzo Contarini Fasan**, a tiny 15th-century palace with an elegant façade, is popularly known as the House of Desdemona from Shakespeare's play.*

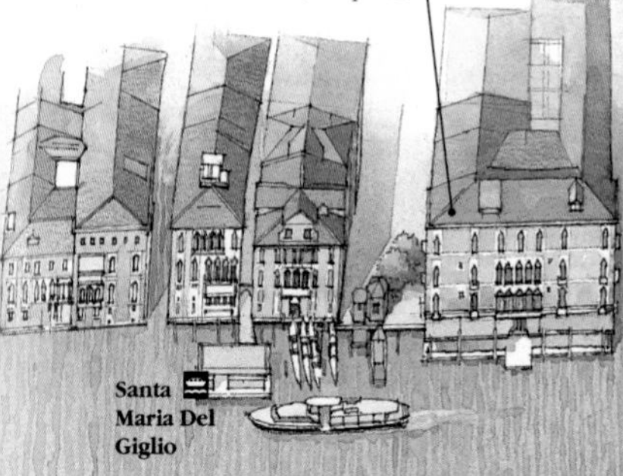

***Palazzo Salviati** is the head-quarters of the Salviati glass-producing company, hence the glass mosaics on the façade.*

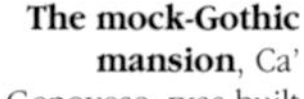

The mock-Gothic mansion, Ca' Genovese, was built in 1892 in the place of the second Gothic cloister of the San Gregorio monastery.

***The deconsecrated Gothic brick church** of Abbazia San Gregorio and a little cloister are all that survive of what was for centuries a powerful monastic centre. The church is now used as a laboratory for the renovation of large-scale paintings.*

***Palazzo Tiepolo**, the Hotel Europa and Regina, was formerly owned by the Tiepolo family, associated with an unsuccessful uprising in 1310.*

Harry's Bar (see p92) *was popular with Hemingway and other writers. This was the very first Harry's Bar in the world.*

Palazzo Giustinian, headquarters of the Biennale, used to be a hotel, where Turner, Verdi and Proust stayed.

San Marco Vallaressa

Giardinetti Reali, the Royal Gardens, were created by Napoleon to improve his view from the Procuratie Nuove.

Palazzo Treves Bonfili, a Classical building of the 17th century, is decorated with Neo-Classical frescoes, paintings and statuary.

***The view from the Dogana**, taking in the Doge's Palace, the Campanile of San Marco and the Zecca, is one of the most memorable in Venice.*

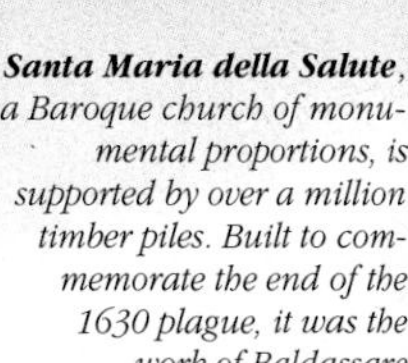

***Santa Maria della Salute**, a Baroque church of monumental proportions, is supported by over a million timber piles. Built to commemorate the end of the 1630 plague, it was the work of Baldassare Longhena* (see p135).

***Dogana di Mare**, the customs house, is topped by a weathervane figure of Fortune* (see p135).

SAN MARCO

HOME OF THE POLITICAL and judicial nerve centres of Venice, the *sestiere* of San Marco has been the heart of Venetian life since the early days of the Republic. The great show-piece of the Serenissima was the Piazza San Marco, conceived as a vista for the Doge's Palace and the Basilica. The square, described by Napoleon as "the most elegant drawing room in Europe", was the only one deemed fit to be called a piazza – the others were merely *campi*, or fields.

Adam and Eve on the corner of the Doge's Palace

The San Marco area has the bulk of luxury hotels, restaurants and shops. It is also home to several imposing churches, three theatres, including the famous Fenice, and a wealth of handsome *palazzi*. Many of these line the sweeping southern curve of the Grand Canal which borders the *sestiere*.

SIGHTS AT A GLANCE

Churches

Basilica San Marco pp78–83 3
Santa Maria Zobenigo 13
San Moisè 11
San Salvatore 18
Santo Stefano 16
San Zulian 21

Museums and Galleries

Libreria Sansoviniana 5
Museo Archeologico 6
Museo Correr 8
Museo Fortuny 17

Palaces

Doge's Palace pp84–9 4
Palazzo Contarini del Bovolo 12

Historic Buildings and Monuments

Campanile 1
Columns of San Marco and San Teodoro 7
San Giorgio Maggiore 22
Torre dell'Orologio 2

Streets and Squares

Campo San Bartolomeo 19
Campo Santo Stefano 15
Mercerie 20

Bars

Harry's Bar 9

Theatres

La Fenice 14
Ridotto 10

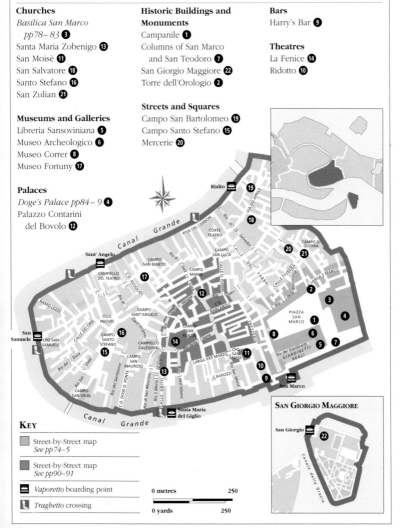

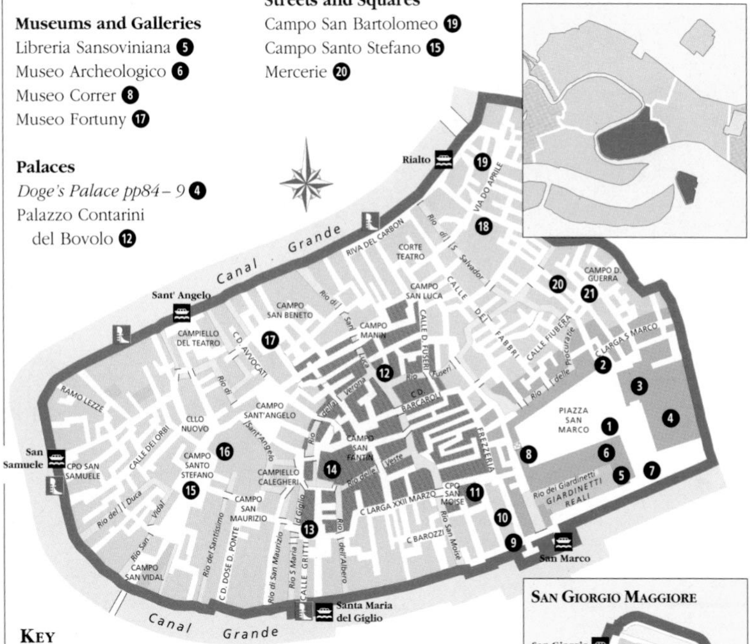

SAN GIORGIO MAGGIORE

KEY

Street-by-Street map *See pp74–5*

Street-by-Street map *See pp90–91*

Vaporetto boarding point

Traghetto crossing

0 metres 250

0 yards 250

◁ **Central dome of the Basilica San Marco**

Street-by-Street: Piazza San Marco

Lion of St Mark

THROUGHOUT ITS long history the Piazza San Marco has witnessed pageants, processions, political activities and countless Carnival festivities. Tourists flock here in their thousands, for the Piazza's eastern end is dominated by two of the city's most important historical sights – the Basilica and the Doge's Palace. In addition to these magnificent buildings there is plenty to entertain, with elegant cafés, open-air orchestras and smart boutiques beneath the arcades of the Procuratie. So close to the waters of the lagoon, the Piazza is one of the first points in the city to suffer at *acqua alta* (high tide). Tourists and Venetians alike can then be seen picking their way across the duckboards which are set up to crisscross the flooded square.

Gondolas customarily moor in the Bacino Orseolo, named after Doge Pietro Orseolo who established a hospice for pilgrims here in 977.

Quadri's café was the favourite haunt of Austrian troops during the Occupation *(see p48)*.

PROCURATIE VECCHIE

PIAZZA SAN MAR

PROCURATIE NUOV

Museo Correr
Giovanni Bellini's Pietà *(1455–60) is one of many Renaissance masterpieces hanging in the picture galleries of the Correr* 8

The Ala Napoleonica is the most recent wing enclosing the square, built by Napoleon to create a new ballroom.

0 metres 75

0 yards 75

Florian's café *(see p246)* was the favourite haunt of 19th-century literary figures such as Byron, Dickens and Proust.

The Giardinetti Reali (royal gardens) were laid out in the early 19th century.

San Marco Vallaresso

STAR SIGHTS

- ★ **Basilica San Marco**
- ★ **Doge's Palace**
- ★ **Campanile**

Torre dell'Orologio
The Madonna on the clock tower is greeted each Epiphany and Ascension by clockwork figures of the Magi ❷

Piazzetta dei Leoncini was named after the pair of porphyry lions which stand in the square.

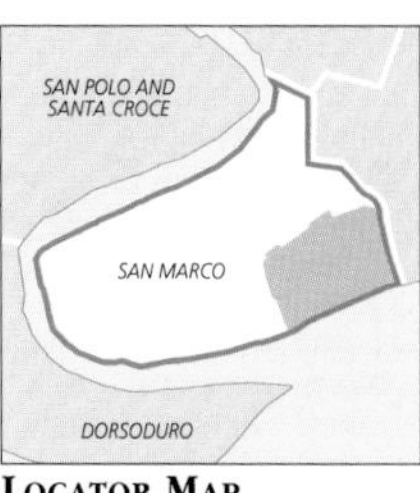

LOCATOR MAP
See Street Finder, map 7

★ Basilica San Marco
The remarkable Basilica of St Mark is a glorious reflection of the city's Byzantine connection ❸

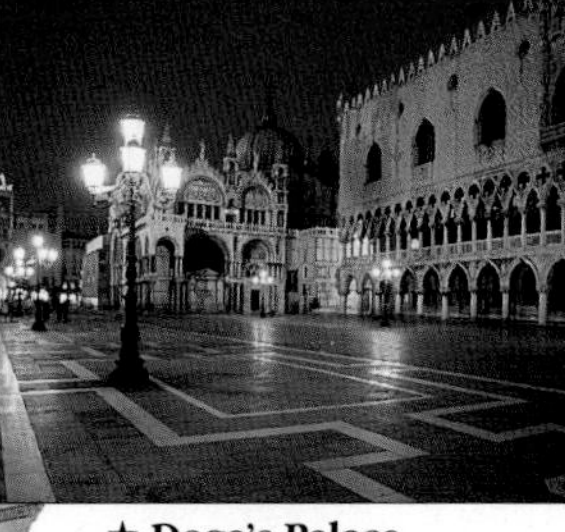

★ Doge's Palace
Once the Republic's seat of power and home to its rulers, the Doge's Palace, beside the Basilica, is a triumph of Gothic architecture ❹

★ Campanile
Today's tower replaced the one that collapsed in 1902. The top provides spectacular views of the city ❶

Museo Archeologico
The museum sculptures had a marked influence on Venetian Renaissance artists ❻

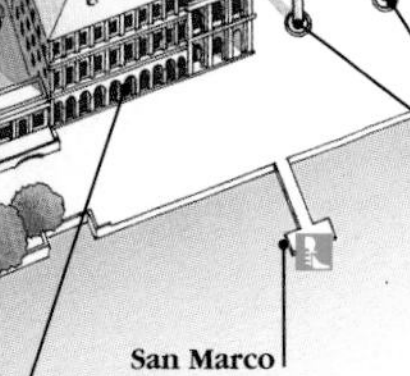

Columns of San Marco and San Teodoro
The columns marked the main entrance to Venice when the city could be reached only by sea ❼

San Marco Giardinetti

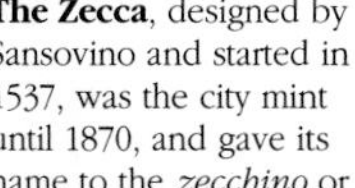

The Zecca, designed by Sansovino and started in 1537, was the city mint until 1870, and gave its name to the *zecchino* or Venetian ducat.

Libreria Sansoviniana
The ornate vaulting of the magnificent library stairway is decorated with frescoes and gilded stucco ❺

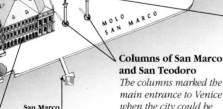

Campanile ❶

Piazza San Marco. **Map** 7 B2.
(041) 522 40 64. *San Marco.*
9:30am–3:45pm daily. *Jan.*

FROM THE TOP of St Mark's campanile, high above the Piazza, visitors can enjoy sublime views of the city, the lagoon and, visibility permitting, the peaks of the Alps. It was from this viewpoint that Galileo demonstrated his telescope to Doge Leonardo Donà in 1609. To do so he would have climbed the internal ramp. Access these days is achieved far less strenuously via a lift which can carry 14 people. Nevertheless there is almost always a queue. If you are at the top of the tower on the hour, beware the resonant ringing of the five bells.

The first tower, completed in 1173, was built as a lighthouse to assist navigators in the lagoon. It took on a less benevolent role in the Middle Ages as the support for a torture cage where offenders were imprisoned and in some cases left to die. The tower's present appearance dates from the early 16th century, when it was restored by Bartolomeo Bon after an earthquake.

The tower survived the vicissitudes of time until 14 July 1902 when, with little warning, its foundations gave way and it suddenly collapsed. The only casualties were the Loggetta at the foot of the tower and the custodian's cat. Donations for reconstruction came flooding in and the following year the foundation stone was laid for a campanile *"dov'era e com'era"* ("where it was and how it was"). The new tower was finally opened on 25 April (St Mark's Day) 1912.

The spire, 98.5 m (323 ft) high, is topped with a golden weathervane which was designed by Bartolomeo Bon.

The five bells in the tower each had their role during the Republic. The *marangona* tolled the start and end of the working day; the *maleficio* warned of an execution; the *nona* rang at noon; the *mezza terza* summoned senators to the Doge's Palace; and the *trottiera* announced a session of the Great Council.

An internal lift, installed in 1962, provides visitors with access to one of the most spectacular views across Venice.

The Loggetta was built in the 16th century by Jacopo Sansovino. Its Classical sculptures celebrate the glory of the Republic.

The allegorical reliefs in red marble from Verona depict Justice representing Venice, Jupiter as Crete and Venus as Cyprus. All were carefully rebuilt after the campanile's collapse in 1902.

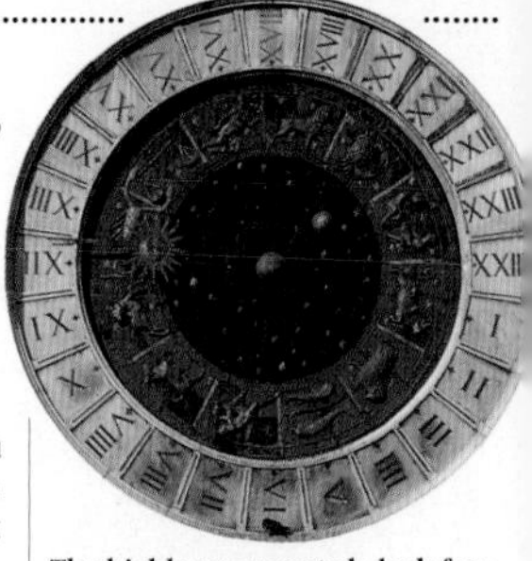

The highly ornamented clock face of the Torre dell'Orologio

Torre dell'Orologio ❷

Piazza San Marco. **Map** 7 B2. *San Marco.* *for restoration.*

THE RICHLY decorated Renaissance clock tower stands on the north side of the Piazza, over the archway leading to the Mercerie *(see p95)*. It was built in the late 15th century, and the central section is thought to have been designed by Mauro Coducci. Displaying the phases of the moon and the zodiac, the gilt and blue enamel clock was originally designed with seafarers in mind. A story was spread by scandalmongers that once the clock was complete, the two inventors of the complex clock mechanism had their eyes gouged out to prevent them creating a replica.

During Ascension Week, crowds gather on the hour to watch the figures of the Magi emerge from side doors to pay their respects to the Virgin and Child, whose figures are set in the niche above the clock. On the upper level, the winged lion of St Mark stands against a star-spangled blue backdrop. At the very top the two huge bronze figures, known as the *Mori*, or Moors, because of their dark patina, strike the bell on the hour.

Basilica San Marco ❸

See pp78–83.

Doge's Palace ❹

See pp84–9.

Libreria Sansoviniana ❺

Piazzetta. **Map** 7 B3. *(041) 520 87 88.* *San Marco.* *9am–7pm Mon–Sat by appointment only.* *public hols and 2 weeks Aug.*

PRAISED BY Andrea Palladio as the finest building since antiquity, the library was designed in the Classical style by the Tuscan architect, Jacopo Sansovino. A graceful building, it is surmounted by a procession of statues of mythological gods. During construction (1537–88) the vaulting collapsed: Sansovino was blamed and imprisoned. He was freed only after appeals from eminent acquaintances.

Today the national library of St Mark, the Biblioteca Marciana, is housed here. Its greatest treasure is the Grimani Breviary, a manuscript illuminated by 15th- and 16th-century Flemish artists. The salon is sumptuously decorated, and two fine ceiling paintings by Paolo Veronese, *Arithmetic and Geometry* and *Music*, won for the artist the prize of a golden necklace.

Columns of San Marco and San Teodoro

Museo Archeologico ❻

Piazzetta. **Map** 7 B3. *(041) 522 59 78.* *San Marco.* *9am–2pm daily.* *1 Jan, 1 May, 25 Dec.*

HOUSED IN ROOMS in both the Libreria Sansoviniana and the Procuratie Nuove, the museum provides a quiet retreat from the bustle of San Marco. The collection owes its existence to the generosity of Domenico Grimani, son of Doge Antonio Grimani, who bequeathed all of his Greek, Roman and earlier sculpture, together with his library, to the State in 1523.

Columns of San Marco and San Teodoro ❼

Piazzetta. **Map** 7 C3. *San Marco.*

ALONG WITH ALL the bounty from Constantinople came the two huge granite columns which now tower above the Piazzetta. These were said to have been erected in 1172 by the engineer Nicolò Barattieri, architect of the very first Rialto Bridge. For his efforts he was granted the right to set up gambling tables between the columns. A more gruesome spectacle on the same spot was the execution of criminals, which took place here until the mid 18th century. Even today, superstitious Venetians will not be seen walking between the columns.

The western column is crowned by a marble statue of St Theodore, who was the patron saint of Venice before St Mark's relics were smuggled from Alexandria in AD 828. The statue is a modern copy – the original is kept for safety in the Doge's Palace *(see p88)*.

The second column is surmounted by a huge bronze of the Lion of St Mark. Its origin remains a mystery, though it is thought to be a Chinese chimera with wings added to make it look like a Venetian lion. In September 1990 the 3,000-kg (3-ton) beast went to the British Museum in London for extensive restoration, and was returned with great ceremony and skill to the top of the column.

Fragment from a monumental statue, in the Museo Archeologico

***A Portrait of a Young Man in a Red Hat* by Carpaccio (c.1490)**

Museo Correr ❽

Procuratie Nuove. Entrance in Ala Napoleonica. **Map** 7 B2. *(041) 522 56 25.* *San Marco.* *10am–5pm daily.* *1 Jan, 1 May, 25 Dec.*

THE WEALTHY Abbot Teodoro Correr bequeathed his extensive collection of works of art and documents to the city in 1830. This forms the nucleus of the civic museum.

The first rooms form a suitably Neo-Classical backdrop for early statues by Antonio Canova (1757–1822). The rest of the floor covers the history of the Venetian Republic, with maps, coins, armour and a host of doge-related exhibits.

On the second floor, the Museo del Risorgimento is devoted to the history of the city, until Venice became part of unified Italy in 1866. Also here is the Quadreria, or picture gallery. The paintings are hung chronologically and the rooms have the bonus of explanations in English. The collection enables you to trace the evolution of Venetian painting, and to see the influence that Ferrarese, Paduan and Flemish artists had on the Venetian school. The most famous works in the gallery are the Carpaccios: *A Portrait of a Young Man in a Red Hat* (c.1490), and *Two Venetian Ladies* (c.1507). The latter is traditionally, but probably incorrectly, known as *The Courtesans* because of the ladies' décolleté dresses.

Basilica San Marco ❸

THIS AWESOME BASILICA, built on a Greek cross plan and crowned with five huge domes, is the third church to stand on this site. The first, built to enshrine the body of St Mark in the 9th century, was destroyed by fire. The second was pulled down in the 11th century in order to make way for a more spectacular edifice designed by an unknown architect (1063–94), reflecting the escalating power of the Republic. The basilica continued to be remodelled over the following centuries, and in 1807 it succeeded San Pietro in the *sestiere* of Castello *(see p120)* as the cathedral of Venice; it had until then served as the doge's private chapel for State ceremonies.

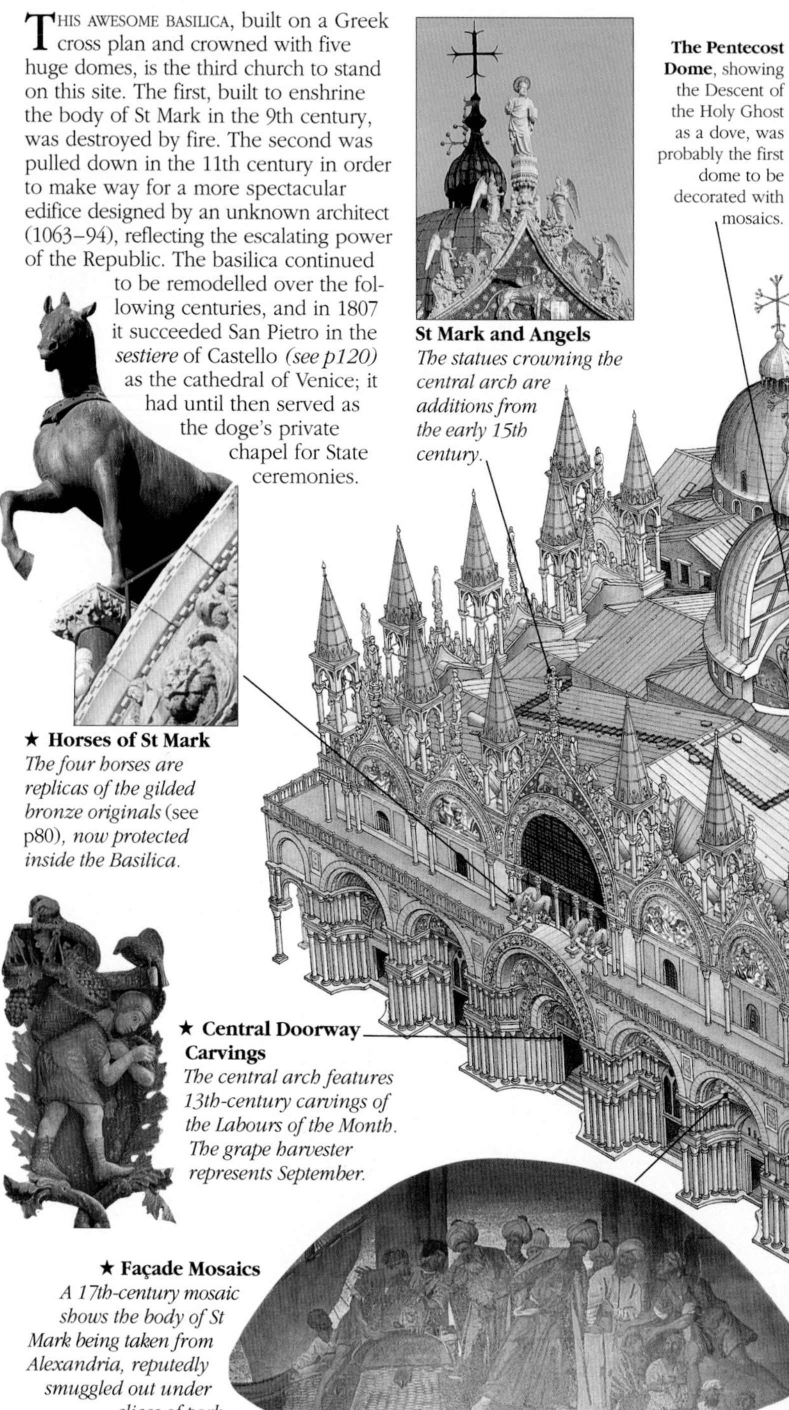

St Mark and Angels
The statues crowning the central arch are additions from the early 15th century.

The Pentecost Dome, showing the Descent of the Holy Ghost as a dove, was probably the first dome to be decorated with mosaics.

★ Horses of St Mark
The four horses are replicas of the gilded bronze originals (see p80), *now protected inside the Basilica.*

★ Central Doorway Carvings
The central arch features 13th-century carvings of the Labours of the Month. The grape harvester represents September.

★ Façade Mosaics
A 17th-century mosaic shows the body of St Mark being taken from Alexandria, reputedly smuggled out under slices of pork.

Baldacchino
The fine alabaster columns of the altar canopy, or baldacchino, are adorned with scenes from the New Testament.

Visitors' Checklist

Piazza San Marco. **Map** 7 B2. *(041) 522 52 05.* San Marco. **Basilica** *10am–5pm Mon–Sat, 2–5pm Sun.* **Museum, Treasury and Pala d'Oro** *10am–5pm Mon–Sat, 2–5pm Sun (Museum 9:45am–2pm Sun).* *Museum, Treasury and Pala d'Oro only.* *9 times a day. Sightseeing is limited during services.* *in English twice a week in season.*

The Ascension Dome features a magnificent 13th-century mosaic of Christ surrounded by angels, the 12 Apostles and the Virgin Mary.

St Mark's body, believed lost in the fire of AD 976, supposedly reappeared in this spot in 1094, when the new church was consecrated.

★ The Tetrarchs
This charming sculptured group in porphyry (4th-century Egyptian) is thought to represent Diocletian, Maximian, Valerian and Constance. Collectively they were the tetrarchs, appointed by Diocletian to help rule the Roman Empire.

Allegorical mosaics

St Mark's Treasury

The so-called Pilasters of Acre in fact came from a 6th-century church in Constantinople.

Baptistry

Baptistry Mosaics
Herod's Banquet *(1343–54) is part of the cycle of scenes from the life of St John the Baptist.*

Star Features

- ★ **Façade Mosaics**
- ★ **Horses of St Mark**
- ★ **The Tetrarchs**
- ★ **Central Doorway Carvings**

Inside the Basilica

DARK, MYSTERIOUS and enriched with the spoils of conquest, the Basilica is a unique blend of Eastern and Western influences. This oriental extravaganza, embellished over a period of six centuries with fabulous mosaics, marble and carvings, made a fitting location for the ceremonies of the Serene Republic. It was here that the doge was presented to the city following his election, that heads of State, popes, princes and ambassadors were received, and where sea captains came to pray for protection before embarking on epic voyages.

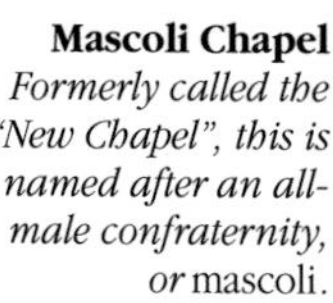

Mascoli Chapel
Formerly called the "New Chapel", this is named after an all-male confraternity, or mascoli.

North Aisle
The gallery leading off the museum affords visitors a splendid overall view of the mosaics.

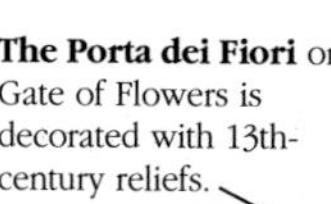

The Porta dei Fiori or Gate of Flowers is decorated with 13th-century reliefs.

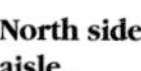

North side aisle

★ Pentecost Dome
Showing the Apostles touched by tongues of flame, the Pentecost Dome was decorated in the 12th century.

The columns of the inner façade are thought to be fragments of the first basilica.

Main entrance

★ Atrium Mosaics
In the glittering Genesis Cupola the Creation of the World is described in concentric circles. Here, God creates the fish and birds.

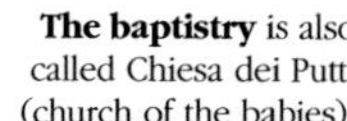

The baptistry is also called Chiesa dei Putti (church of the babies).

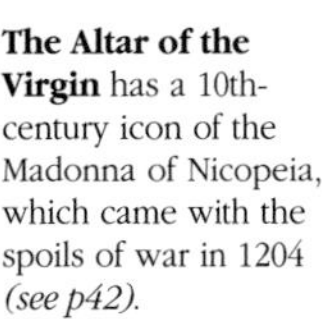

The Altar of the Virgin has a 10th-century icon of the Madonna of Nicopeia, which came with the spoils of war in 1204 *(see p42)*.

The Chapel of St Peter has a 14th-century altar screen relief of St Peter worshipped by two Procurators.

★ Pala d'Oro
The magnificent altarpiece, created in the 10th century by medieval goldsmiths, is made up of 250 panels such as this one, each adorned with enamels and precious stones.

The sacristy door (often locked) has fine bronze panels by Sansovino, which include portraits of himself with Titian and Aretino.

★ Ascension Dome
A mosaic of Christ in Glory decorates the enormous central dome. This masterpiece was created by 13th-century Venetian craftsmen, who were strongly influenced by the art and architecture of Byzantium.

The Altar of the Sacrament is decorated with mosaics of the parables and miracles of Christ dating from the late 12th or early 13th century.

South side aisle

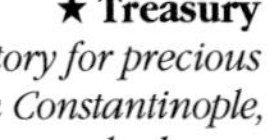

★ Treasury
A repository for precious booty from Constantinople, the Treasury also houses ancient works of art from Italy, such as this 11th-century silver-gilt coffer.

STAR FEATURES

- **★ Pala d'Oro**
- **★ Atrium Mosaics**
- **★ Treasury**
- **★ Ascension and Pentecost Domes**

Exploring the Basilica

THE BASILICA cannot comfortably be covered in one visit. The glittering mosaics, the rich store of bounty from the east, the dim, mysterious lighting and the sheer size of the place create a feeling of confusion for first-time visitors. To get the most out of it, make several visits, ideally at different times of the day. The mosaics look especially splendid during the hour 11:30am–12:30pm when the church is fully illuminated. It is also a rewarding experience to tour the basilica while a service is in progress. Visitors are often led towards the (fee-paying) Pala d'Oro and Treasury and miss out on other sections of the church. To avoid the crowds, visit early in the morning or in the evening.

The Genesis Cupola of the atrium

Mosaics

CLOTHING THE DOMES, walls and floor of the basilica are over 4,000 sq m (40,000 sq ft) of gleaming golden mosaics. The earliest, dating from the 12th century, were the work of mosaicists from the east. Their techniques were adopted by Venetian craftsmen who gradually took over the decoration, combining Byzantine inspiration with western influences. During the 16th century, sketches and cartoons by Tintoretto, Titian, Veronese, and other leading artists were reproduced in mosaic. The original iconographical scheme, depicting stories from the Testaments, has more or less been preserved by careful restoration.

Among the finest mosaics in the basilica are those decorating the 13th-century central Dome of the Ascension and the 12th-century Dome of the Pentecost over the nave.

The *pavimento*, or basilica floor, spreads out like an undulating Turkish carpet. Mosaics, made of marble, porphyry and glass are used to create complex and colourful geometric patterns and beautiful scenes of beasts and birds. Some of these scenes are allegorical. The one in the left transept of two cocks carrying a fox on a stick was designed to symbolize cunning vanquished by vigilance.

Atrium (Vestibule)

THE 13TH-CENTURY mosaics decorating the cupolas, vaults and lunettes of the atrium are among the finest in the basilica. The scenes depict Old Testament stories, starting at the southern end with the Genesis Cupola (showing 26 detailed episodes of the Creation), to the Stories of Joseph and of Moses in the domes at the north end. The figures of saints on either side of the main doorway date from the 11th century and are among the earliest mosaics in the church. Just in front of the central doorway there is a lozenge of porphyry to mark the spot where the Emperor Frederick Barbarossa was obliged to make peace with Pope Alexander III in 1177 *(see p41)*.

Museo Marciano

A PRECARIOUS STAIRWAY from the atrium, marked *Loggia dei Cavalli*, takes you up to the church museum. The gallery gives a splendid view into the basilica, while from the exterior loggia you can survey the Piazza San Marco and take a close look at the replica horses on the church façade. It was from this panoramic balcony that doges and dignitaries once looked down on ceremonies taking place in the square. The original gilded bronze horses, housed in a room at the far end of the museum, were stolen from the top of the Hippodrome (ancient racecourse) in Constantinople in 1204 but their origin, either Roman or Hellenistic, remains a mystery. In the same room is Paolo Veneziano's 14th-century *pala feriale*, painted with stories of St Mark, which once covered the Pala d'Oro. Also on show are medieval

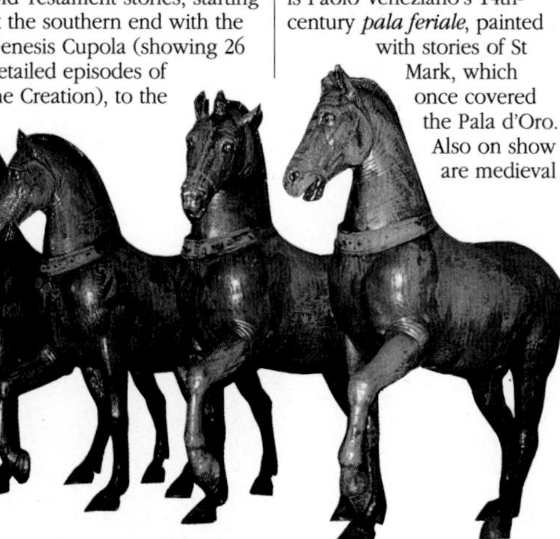

The Quadriga, the original gilded bronze horses in the museum

Noah and the Flood – atrium mosaics from the 13th century

illuminated manuscripts, fragments of ancient mosaics and antique tapestries.

Sanctuary and Pala d'Oro

Beyond the chapel of St Clement, tickets are sold to view the most valuable treasure of San Marco: the Pala d'Oro. This jewel-spangled altarpiece situated behind the high altar consists of 250 enamel paintings on gold foil, enclosed within a gilded silver Gothic frame. Originally commissioned in Byzantium in AD 976, the altarpiece was embellished over the centuries. Following the fall of the Republic, Napoleon helped himself to some of the precious stones, but the screen still gleams with pearls, rubies, sapphires and amethysts.

Statue of St Mark on the iconostasis

The iconostasis, the screen dividing nave from chancel, is adorned with marble Gothic statues of the Virgin and Apostles, and was carved in 1394 by the Dalle Masegne brothers. Above the high altar the imposing green marble baldacchino is supported by finely carved alabaster columns featuring scenes from the New Testament.

Baptistry and Chapels

The baptistry (closed for restoration) was added in the 14th century by Doge Andrea Dandolo (1343–54) who is buried here. Under his direction the baptistry was decorated with outstanding mosaics depicting scenes from the lives of Christ and John the Baptist. Sansovino, who designed the font, is buried by the altar.

The adjoining Zen Chapel (also closed for restoration) originally formed part of the atrium. It became a funeral chapel for Cardinal Zen in 1504 in return for his bequest to the State.

In the left transept of the basilica the Chapel of St Isidore, normally accessible only for worship, was also built by Dandolo. Mosaics in the barrel vault ceiling tell the tale of the saint, whose body was stolen from the island of Chios and transported to Venice in 1125. To its left the Mascoli Chapel, used in the early 17th century by the confraternity of Mascoli (men), is decorated with scenes from the life of the Virgin Mary. The altarpiece has statues depicting the Virgin and Child between St Mark and St John.

The third chapel in the left transept is home to the icon of the Madonna of Nicopeia. Looted in 1204, she was formerly carried into battle at the head of the Byzantine army.

The revered icon of the Nicopeia Madonna, once a war insignia

Treasury

Although plundered after the fall of the Republic and much depleted by the fund-raising sale of jewels in the early 19th century, the treasury nevertheless has a precious collection of Byzantine silver, gold and glasswork. Today, most of the treasures are housed in a room whose remarkably thick walls are believed to have been a 9th-century tower of the Doge's Palace. Exhibits include chalices, goblets, reliquaries, two intricate icons of the archangel Michael and an 11th-century silver-gilt reliquary made in the form of a five-domed basilica *(see p81)*. The sanctuary, with over 100 reliquaries, is normally closed to the public.

The archangel Michael, a Byzantine icon from the 11th century in the Treasury

Doge's Palace ❹

THE PALAZZO DUCALE started life in the 9th century as a fortified castle, but this and several subsequent buildings were destroyed by a series of fires. The existing palace owes its external appearance to the building work of the 14th and early 15th centuries. The designers broke with tradition by perching the bulk of the pink Verona marble palace on lace-like Istrian stone arcades, with a portico supported by columns below. The result is a light and airy masterpiece of Gothic architecture.

Arco Foscari
The Adam and Eve figures on this triumphal arch in the courtyard are copies of the 15th-century originals by Antonio Rizzo.

★ Porta della Carta
This 15th-century Gothic gate is the principal entrance to the palace. From it, a vaulted passageway leads to the Arco Foscari and the internal courtyard.

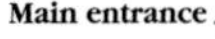

STAR FEATURES

- ★ Giants' Staircase
- ★ Porta della Carta

The balcony on the west façade was added in 1536 to mirror the early 15th-century balcony looking on to the quay.

★ Giants' Staircase
This late 15th-century staircase by Antonio Rizzo was used for ceremonial purposes. It was on the landing at the top that the doges were crowned with the glittering zogia.

Torture Chamber
"The court of the room of the Cord" recalls the practice of interrogating suspects as they hung by their wrists.

Visitors' Checklist

Piazzetta. **Map** 7 C2. (041) 522 49 51. San Marco. Mar–end Sep: 8:30am–6pm daily; Oct–end Feb: 8:30am–4pm daily. **Last adm**: one hour before closing. ***Tickets for Secret Itineraries*** *(twice a day, Thu–Tue) must be reserved in advance:* (041) 520 42 87. 1 Jan, 1 May, 25 Dec.

Bridge of Sighs
The famous bridge once crossed by offenders on their way to the State interrogators.

Sala dei Tre Capi
(Chamber of the Three Heads of the Council of Ten)

Sala della Bussola
("Compass" Room)

Drunkenness of Noah
This early 15th-century sculpture, symbolic of the frailty of man, is set on the corner of the palace.

Ponte della Paglia
(see p113)

Adam and Eve
with the serpent are depicted in stone on the corner of the Piazzetta.

Sala del Maggior Consiglio
An entire wall of the Great Council Hall is taken up by Domenico and Jacopo Tintoretto's Paradise *(1588–92).*

Inside the Doge's Palace

Intricate carved Gothic capital

FROM THE early days of the Republic, the Doge's Palace was the seat of the government, the Palace of Justice and the home of the doge. For centuries this was the only building in Venice entitled to the name *palazzo* (the others were merely called *Ca'*, short for *Casa*). The power of the Serenissima is ever present in the large and allegorical historical paintings which embellish the walls and ceilings of the splendid halls and chambers. These ornate rooms are testament to the glory of the Venetian Republic, and were designed to impress and overawe visiting ambassadors and dignitaries.

Colonnade
Sunlight streams through the arches of the Loggia on the first floor of the palace.

Star Features

- ★ **Sala del Maggior Consiglio**
- ★ **Collegiate Rooms**
- ★ **Prisons**

Mars
The Giants' Staircase is named after Sansovino's monumental figures, statues of Mars and Neptune, sculpted in 1567.

Scala d'Oro
Visitors enter the palace via Sansovino's lavish staircase, built between 1554 and 1558. The arched ceiling is embellished with gilded stucco by Alessandro Vittoria.

Ground floor

Main ticket office

Main entrance through Porta della Carta

Key to Floorplan

- State Apartments
- Collegium and Senate Rooms
- Council of Ten and Armoury
- Great Council Rooms
- Prisons
- Non-exhibition space

Wellhead
The two 16th-century bronze wellheads in the courtyard are considered to be the finest in Venice.

★ Collegiate Rooms
Bacchus and Ariadne Crowned by Venus *is the finest of four mythological scenes by Tintoretto in the Anticollegio.*

Third floor

The Sala del Consiglio dei Dieci has a ceiling decorated with paintings by Veronese (1553–54).

Sala dello Scudo
The walls of this room are covered with maps of the world. In the centre are two huge 18th-century globes.

First floor

Second floor

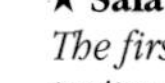

★ Sala del Maggior Consiglio
The first 76 doges, with the exception of the traitor Marin Falier, are portrayed on a frieze round the upper walls of the room.

★ Prisons
These 16th-century cells were mainly used for petty offenders. Serious criminals were lodged in the dank pozzi *(wells).*

The Secret Itinerary

The fascinating, though poorly publicized, Secret Itinerary *(Itinerari Segreti)* tour *(see Visitors' Checklist p85)*, takes you behind the scenes in the palace to the offices and Hall of the Chancellery, the State Inquisitors' room, the Torture Chamber and the prisons. It was from these cells that Casanova made his spectacular escape in 1755. The tour is limited to 20 people and lasts for 90 minutes. Unfortunately it is conducted in Italian only.

Casanova's cell door

Exploring the Doge's Palace

A TOUR OF THE PALACE takes you through a succession of richly decorated chambers and halls. The rooms are on three floors and, unless you are visiting the state apartments (only accessible during exhibitions), you start at the top and work your way down.

The rooms within the palace are neither named nor numbered, and without a guide, the place can be very confusing. The latest equipment available is the Light and Man infra-red Walkman which you can hire for a commentary on the whole palace or just the areas which interest you.

St Theodore in the palace courtyard

COURTYARD

THE COURTYARD is reached via a vaulted passage from the Porta della Carta. At the top of the Giants' Staircase, which rises to the first floor, new doges were crowned with the *zogia* or dogal cap.

SCALA D'ORO AND STATE APARTMENTS

THE SUMPTUOUS Scala d'Oro (golden staircase) takes its name from the elaborate gilt stucco vault by Alessandro Vittoria (1554–58). The doge's private apartments on the first floor can be seen only when they are being used for temporary exhibitions (a separate ticket is usually required). The apartments, built after the fire of 1483 and later looted on the orders of Napoleon, are bare of furnishings, but the lavish ceilings and colossal carved chimney-pieces in some of the rooms give you an idea of the doges' lifestyle. The most ornate is the Sala degli Scarlatti, with a richly carved gilt ceiling, a fireplace (c.1501) designed by Antonio and Tullio Lombardo and a relief (1501–21) by Pietro Lombardo of Doge Leonardo Loredan at the feet of the Virgin.

The Sala dello Scudo, or map room, contains maps and charts. The picture gallery further on features works by Vittore Carpaccio and Giovanni Bellini, and some incongruous wooden demoniac panels by Hieronymous Bosch.

A *bocca di leone* used for denouncing tax evaders

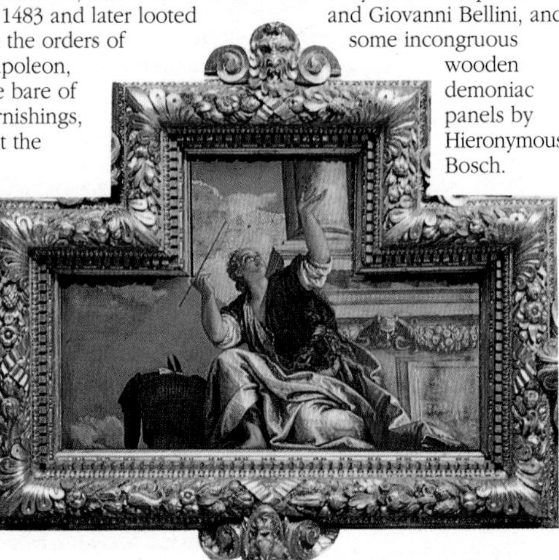

Veronese's *Dialectic* (c.1577), Sala del Collegio

SALA DELLE QUATTRO PORTE TO SALA DEL SENATO

THE SECOND FLIGHT of the Scala d'Oro leads to the third floor and its council chambers. The first room, the Sala delle Quattro Porte, was completely rebuilt after the 1574 fire, its ceiling designed by Andrea Palladio and frescoed by Tintoretto.

The next room, the Anticollegio, was the waiting room. The end walls are decorated with mythological scenes by Tintoretto: *Vulcan's Forge*; *Mercury and the Graces*; *Bacchus and Ariadne* and *Minerva Dismissing Mars*, all painted in 1578. Veronese's masterly *Rape of Europa* (1580), opposite the window, is one of the most eyecatching works in the palace.

Off the Anticollegio, the Sala del Collegio was the hall where the doge and his counsellors met to receive ambassadors and discuss matters of State. Embellishing the magnificent ceiling are 11 paintings by Veronese (c.1577), of which the most notable – in the centre, far end – is *Justice and Peace Offering Sword, Scales and Olive Branch to Venice.*

It was in the next room, the Sala del Senato, that the doge would sit with some 200 senators to discuss matters such as foreign affairs or nominations of ambassadors. The wall and ceiling paintings, by pupils of Tintoretto or the master himself, are further propaganda for the Republic.

SALA DEL CONSIGLIO DEI DIECI TO THE ARMERIA

THE ROUTE RETURNS through the Sala delle Quattro Porte to the Sala del Consiglio dei Dieci. This was the meeting room of the awesomely powerful Council of Ten, founded in 1310 to investigate and prosecute crimes concerning the security of the State. Napoleon pilfered some of the Veroneses from the ceiling but two of the finest found their way back here in 1920: *Age and Youth* and *Juno Offering the Ducal Crown to Venice* (both 1553–54).

In the next room, the Sala della Bussola, offenders awaited their fate in front of the Council of Ten. The room's *bocca di leone* (lion's mouth), used to post secret denunciations, was just one of several within the palace. The wooden door here leads to the rooms of the Heads of the Ten, the State Inquisitors' Room and thence to the torture chamber and prisons. This is the route taken by those on the Secret Itinerary.

Age and Youth (1553–54) by Veronese

Others follow the flow to the Armoury – one of the finest collections in Europe, thanks in part to bequests by European monarchs.

Sala del Maggior Consiglio

Another staircase, the Scala dei Censori, takes you down again to the second floor, along the hallway and past the Sala del Guariento with fresco fragments of *The Coronation of the Virgin* by Guariento (1365–67). From the *liagò*, or veranda, where Antonio Rizzo's marble statues of Adam and Eve (1480s) are displayed, you pass into the magnificent Sala del Maggior Consiglio or Hall of the Great Council. A chamber of monumental proportions, it was here that the Great Council convened to vote on constitutional questions, to pass laws and elect the top officials of the Serene Republic. The hall was also used for State banquets. When Henry III of France paid a royal visit, 3,000 guests were entertained in this spectacular room.

By the mid 16th century the Great Council had around 2,000 members. Any Venetian of high birth over 25 was entitled to a seat – with the exception of those married to a commoner. From 1646, by which time the Turkish wars had depleted state coffers, nobility from the *terra firma* or those from merchant or professional classes with 100,000 ducats to spare could purchase their way in.

Tintoretto's huge, highly restored work called *Paradise* (1587–90) occupies the eastern wall. Measuring 7.45 by 24.65 m (25 by 81 ft) it is one of the largest paintings in the world. For a man in his late seventies, albeit assisted by his son, it is a remarkably vigorous composition.

The ceiling of the hall is decorated with panels glorifying the Republic. One of the finest is Veronese's *Apotheosis of Venice* (1583). A frieze along the walls illustrates 76 doges by Tintoretto's pupils. The portrait covered by a curtain is Marin Falier, beheaded for treason in 1355. The other 42 doges are portrayed in the Sala dello Scrutinio, where new doges were nominated.

Prisons

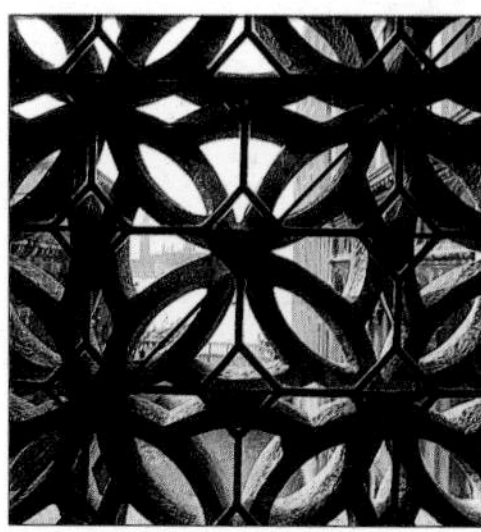

View of the lagoon through a grille on the Bridge of Sighs

From the Sala del Maggior Consiglio a series of passageways and stairways leads to the Bridge of Sighs *(see p113)* which links the palace to what were known as the New Prisons, built between 1556 and 1595.

Situated at the top of the palace, just below the leaded roof, are the *piombi* cells (*piombo* means lead). These cells are hardly inviting but prisoners here were far more comfortable than the criminals who were left to fester in the *pozzi* – the dark dank dungeons at ground level. The windowless cells of these ancient prisons are still covered with the graffiti of the convicts. Visitors on the Secret Itinerary tour are shown Casanova's cell in the *piombi* and told of how he made his daring escape from the palace through a hole in the roof.

Visits end with the offices of the Avogaria, where the state prosecutors *(avogadori)* prepared the trials.

The splendid Sala del Maggior Consiglio, the hall of the Great Council

Street-by-Street: Around La Fenice

West of the huge expanse of the ever-crowded Piazza San Marco there is a labyrinth of alleys to explore. At the centre of this part of the *sestiere* is Campo San Fantin, flanked by the Renaissance church of San Fantin. Nearby is the Ateneo Veneto, formerly a *scuola* whose members had the unenviable role of escorting prisoners to the scaffold. The narrow streets around these sights have some wonderfully exotic little shops, while the more recent Calle Larga XXII Marzo further south boasts big names in Italian fashion. The quarter in general has some excellent restaurants but, being San Marco, you will find that prices in the majority of establishments are steep.

Campo San Fantin has a late Renaissance church, San Fantin, with a particularly beautiful apse designed by Jacopo Sansovino.

★ La Fenice
The opera house gained its name (the phoenix) after a fire in 1836. Sadly, it was again destroyed by fire in 1996 ⓮

The Rio delle Veste leads past the rear of the theatre. This is the route taken by those fortunate enough to arrive for their night out by gondola.

Key

— — — Suggested route

0 metres 75

0 yards 75

Santa Maria Zobenigo
The carvings feature the Barbaro family who paid for the church façade. Ground-level reliefs show towns where the family held high ranking posts ⓭

Star Sights

★ **La Fenice**

★ **San Moisè**

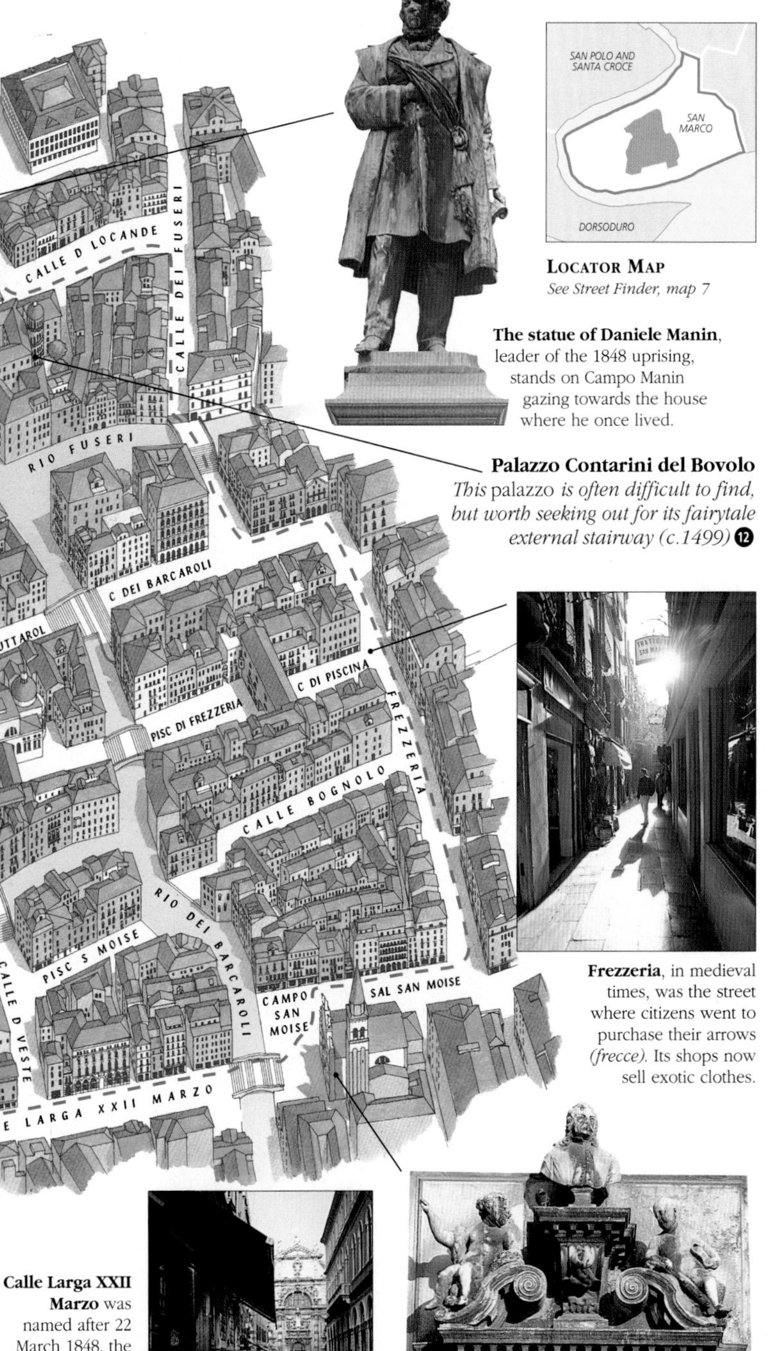

Locator Map
See Street Finder, map 7

The statue of Daniele Manin, leader of the 1848 uprising, stands on Campo Manin gazing towards the house where he once lived.

Palazzo Contarini del Bovolo
This palazzo is often difficult to find, but worth seeking out for its fairytale external stairway (c.1499) ⓬

Frezzeria, in medieval times, was the street where citizens went to purchase their arrows *(frecce)*. Its shops now sell exotic clothes.

Calle Larga XXII Marzo was named after 22 March 1848, the day of Manin's rebellion. Today the street is best known for its trendy designer boutiques.

★ San Moisè
The exuberant Baroque façade of San Moisè (c.1668) was funded by a legacy from the patrician Vincenzo Fini, whose bust features above a side door ⓫

Harry's Bar ❾

Calle Vallaresso 1323. **Map** 7 B3.
San Marco. *See also* **Restaurants, Cafés and Bars** *pp246–7.*

CELEBRATED FOR cocktails, *carpaccio* and American clientèle, Harry's Bar is famous throughout Venice. Founded in 1931 by the late Giuseppe Cipriani, it was financed by a Bostonian called Harry who thought Venice had a dearth of decent bars. They chose a storeroom at the Grand Canal end of the Calle Vallaresso as their location, conveniently close to the Piazza San Marco. Since then, the bar has seen a steady stream of American visitors, among them Ernest Hemingway who used to come here after shooting in the lagoon. The bar became the most popular venue in Venice, patronized by royalty, film stars and heads of state.

These days there are far more American tourists than famous figures, often there to sample the Bellini cocktail that Cipriani invented *(see p239)*. Aesthetically, the place is unremarkable and there is no terrace for meals *al fresco*.

Ernest Hemingway, a regular at Harry's Bar

Ridotto ❿

Calle del Ridotto, 1332 San Marco.
Map 7 B3. *(041) 522 29 39.*
San Marco. ***No public access*** *except for theatre performances.*
See **Entertainment** *p255.*

IN AN EFFORT to control the gambling mania that swept the city in the 17th century, the State gave Marco Dandolo permission to use his palace as the first public gaming house in Europe. In 1638 the Ridotto was opened, with the proviso that players came disguised in a mask. In 1774 the Great Council closed its doors on account of the number of Venetians ruined at its tables.

In 1947 the old Palazzo Dandolo became a theatre, and today makes a charming setting for Italian plays.

San Moisè ⓫

Campo San Moisè. **Map** 7 A3.
(041) 528 58 40. *San Marco.*
3:30–7pm daily.

ONE OF the churches in Venice that people love to hate, San Moisè displays a ponderous Baroque façade. Completed in 1668, it is covered in grimy statues, swags and busts. John Ruskin, in a characteristic anti-Baroque outrage, described it as the clumsiest church in Venice. The interior has a mixed collection of paintings and sculpture from the 17th and 18th centuries. In the nave is the tombstone of John Law, a financier from Scotland who founded the Compagnie d'Occident to develop the Mississippi Valley. His shares collapsed in 1770 in the notorious South Sea Bubble, and he fled to Venice, surviving on his winnings at the Ridotto.

Façade of San Moisè, encrusted with Baroque ornamentation

Palazzo Contarini del Bovolo ⓬

Corte Contarini del Bovolo.
Map 7 A2. *Rialto or Sant'Angelo.*
Apr–Oct: 10:30am–6:30pm daily.

TUCKED AWAY in a maze of alleys (follow signs from Calle della Vida), the Palazzo Contarini del Bovolo is best known for its graceful external stairway. The word *bovolo* in Venetian dialect means snail shell, appropriate to the spiral shape of the Lombardesque stairway. The Contarini, who had the 15th-century palace built, were a learned Venetian family, known as "the philosophers". The collection of wellheads within the enclosure belongs to the present owners.

The external stairway of the Palazzo Contarini del Bovolo

Santa Maria Zobenigo ⓭

Campo Santa Maria del Giglio.
Map 6 F3. *(041) 522 57 39.*
Santa Maria del Giglio.
9am–noon, 3:30–6pm daily.

NAMED AFTER the Jubanico family who are said to have founded it in the 9th century, this church is also referred to as "del Giglio" ("of the lily"). The exuberant Baroque façade was financed by the affluent Barbaro family and was used to glorify their naval and diplomatic achievements.

Inside is a tiny museum of church ornaments and paintings including *The Sacred Family* attributed to Rubens. To see Tintoretto's Evangelist paintings (both 1552) you will need coins for the light meter.

La Fenice ⓮

Campo San Fantin. **Map** 7 A3. San Marco. (041) 520 40 10 *(Palafenice booking office).* ***Theatre*** *until further notice.*

THEATRE HOUSES were enormously popular in the 18th century and La Fenice, the city's oldest theatre, was no exception. Built in 1792 in Classical style, it was one of several privately owned theatres showing plays and operas to audiences from all strata of society. In December 1836 a fire destroyed the interior but a year later it was resurrected, just like the mythical bird, the phoenix *(fenice)* which is said to have arisen from its ashes.

Another fire in early 1996 again completely destroyed the theatre. La Fenice's season is currently being held at Palafenice, a temporary structure near the parking area just outside the city.

Throughout the 19th century the name of La Fenice was linked with great Italian composers. Some of the many operatic premières that took place here were Rossini's *Tancredi* (1813), *Semiramide* (1823), and Verdi's *La Traviata* (1853). During the Austrian Occupation *(see p48)* red, white and green flowers, symbolizing the Italian flag, were thrown on stage, amid shouts of "Viva Verdi" – the letters of the composer's name standing for Vittorio Emanuele Re d'Italia. More recently, the theatre saw premières of Stravinsky's *The Rake's Progress* (1951) and Britten's *Turn of the Screw* (1954).

La Fenice, destroyed by fire in 1996

Shop in Campo Santo Stefano selling antiques and masks

Campo Santo Stefano ⓯

Map 6 F3. *Accademia or Sant'Angelo.*

ALSO KNOWN as Campo Francesco Morosini after the 17th-century doge who once lived here, this *campo* is one of the most spacious in the city. Bullfights were staged until 1802, when a stand fell and killed some of the spectators. It was also a venue for balls and Carnival festivities. Today it is a pleasantly informal square where children play and visitors drink coffee in open-air cafés.

The central statue is Nicolò Tommaseo (1802–74), a Dalmatian scholar who was a central figure in the 1848 rebellion against the Austrians.

At the southern end of the square the austere-looking Palazzo Pisani, overlooking the Campiello Pisani, has been the Conservatory of Music since the end of the 19th century. In summer months, music wafts from its open windows. On the opposite side of the square No. 2945, Palazzo Loredan, is the home of the Venetian Institute of Sciences, Letters and Arts.

The ceiling of Santo Stefano, in the form of a ship's keel

Santo Stefano ⓰

Campo Santo Stefano. **Map** 6 F2. *(041) 522 50 61. Accademia or Sant'Angelo. 8am–noon, 4–7pm Mon–Sat, 7:30am–12:30pm, 6–8pm Sun (phone to check).*

DECONSECRATED six times on account of the murder and violence that took place within its walls, Santo Stefano today is remarkably serene. Built in the 14th century and radically altered in the 15th, the church has a notable carved portal by Bartolomeo Bon and a campanile which has a typical Venetian tilt. The interior has a splendid ship's keel ceiling, carved tie-beams and tall pillars of Veronese marble. The most notable works of art, including some paintings by Tintoretto, are housed in the damp sacristy.

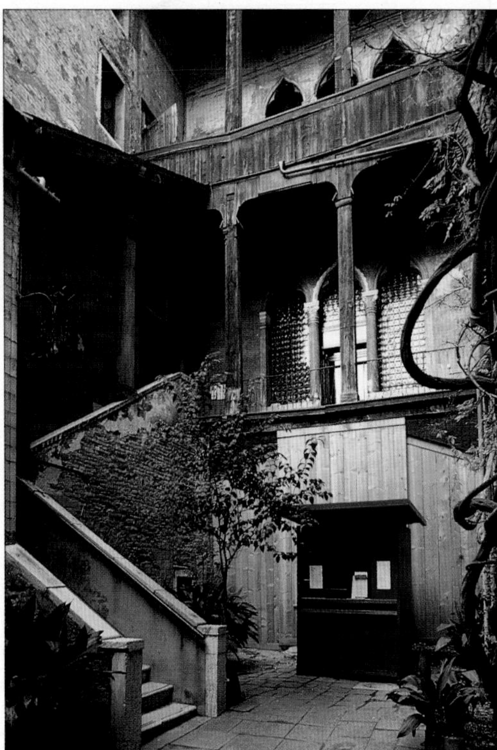

Courtyard of the Palazzo Pesaro, where Fortuny lived

Museo Fortuny 17

Palazzo Pesaro degli Orfei, Campo San Benedetto, San Marco 3958. **Map** 6 F2. (041) 520 09 95. *Sant'Angelo. Temporarily for restoration, but occasionally open for exhibitions.*

KNOWN PRINCIPALLY for his fantastic pleated silk dresses, Fortuny was also a painter, sculptor, set designer, photographer, and scientist. One of his inventions was the Fortuny Dome which is used in theatre performances to create the illusion of sky.

Mariano Fortuny y Madrazo, or Don Mariano as he liked to be called, was born in 1871 in Granada and moved to Venice in 1889. In the early 20th century he purchased the Palazzo Pesaro, a late Gothic *palazzo* which had originally been owned by the fabulously rich and influential Pesaro family. Fortuny spent the remainder of his life here and the house and its contents were bequeathed to the city by his wife in 1956.

The large rooms and *portego* make a splendid and appropriate setting for the precious Fortuny fabrics. Woven with gold and silver threads, these were created by Fortuny's reintroduction of Renaissance techniques and use of ancient dyes. The collection also includes paintings by Fortuny (less impressive than the fabrics), decorative panels and a few of the finely pleated, clinging silk dresses regarded as a milestone in early 20th-century women's fashion.

San Salvatore 18

Campo San Salvatore. **Map** 7 B1. *(041) 523 67 17. Rialto. 10am–noon, 5–7pm daily.*

THE INTERIOR of the church of San Salvatore is a fine example of Venetian Renaissance architecture. If the main door is closed you can enter by the side entrance, which is squeezed between shops along the Mercerie. The present church was designed by Giorgio Spavento in the early 16th century, and continued by Tullio Lombardo and Jacopo Sansovino. The pictorial highlight is Titian's *Annunciation* (1566) over the third altar on the right. Nearby, Sansovino's monument to Doge Francesco Venier (1556–61), is one of several Mannerist tombs in the church.

On the high altar is Titian's *Transfiguration of Christ* (1560). The end of the right transept is dominated by a vast monument to Caterina Cornaro, Queen of Cyprus *(see p43)*. Executed by the sculptor Bernardino Contino in 1580–84, the tomb shows the queen handing over her kingdom to the doge.

Campo San Bartolomeo 19

Map 7 B1. *Rialto.*

CLOSE TO the Rialto, the square of San Bartolomeo bustles with life, particularly in the early evening when young Venetians make this their rendezvous. They meet at cafés, bars or by the statue of Carlo Goldoni (1707–93), Venice's most celebrated playwright *(see p101)*. The statue, sculpted in 1883 by Antonio del Zotto, echoes Goldoni's humorous yet somewhat haughty personality.

The beautiful Renaissance interior of the church of San Salvatore

St George and Dragon bas-relief on a corner of the Mercerie

Mercerie ⓴

Map 7 B2. *San Marco or Rialto.*

DIVIDED INTO the Merceria dell'Orologio, Merceria di San Zulian and Merceria di San Salvatore, this is, and always has been, a principal shopping thoroughfare. Linking Piazza San Marco with the Rialto, it is made from a string of narrow, bustling alleys, lined by small shops and boutiques. The 17th-century English author, John Evelyn, described it as "the most delicious streete in the World for the sweetnesse of it . . . tapisstry'd as it were, with Cloth of Gold, rich Damasks & other silk." He wrote of perfumers, apothecary shops and nightingales in cages. Today all this has been replaced with fashions, footwear and glass.

At the southern end, the relief over the first archway on the left portrays the woman who in 1310 accidentally stopped a revolt. She dropped her pestle out of the window, killing the standard-bearer of a rebel army. They retreated, and the woman was given a guarantee that her rent would never be raised.

Bronze statue of Tommaso Rangone

Eng

O N... ch... Giulian... ge from the ... hopping alleys. Its ...rior is rich with gilded woodwork, 16th- and 17th-century paintings, and sculpture. The central panel of the frescoed ceiling portrays *The Apotheosis of St Julian*, painted by Palma il Giovane in 1585.

The 16th-century church façade was designed by Sansovino and financed by the rich and immodest physician, Tommaso Rangone. His bronze statue, surrounded by books, stands out against the white Istrian stone walls.

San Giorgio Maggiore ㉒

Map 8 D4. *(041) 528 99 00.* *San Giorgio.* *9am–12:30pm, 3–5:30pm. Phone in advance for visits to the Foundation.* ***Campanile*** *9am–12:30pm, 3–5pm (phone to check).*

APPEARING LIKE a stage set across the water from the Piazzetta, the little island of San Giorgio Maggiore has been captured on canvas countless times.

The church and monastery, built between 1559–80, are among Andrea Palladio's greatest architectural achievements. The church's temple front and the spacious, serene interior with its perfect proportions and cool beauty are typically Palladian in that they are modelled on the Classical style of ancient Rome. Within the church, the major works of art are the two late Tintorettos on the chancel walls: *The Last Supper* and *Gathering of the Manna* (both 1594). In the Chapel of the Dead is his last work, *The Deposition* (1592–94), finished by his son Domenico.

The top of the tall campanile, reached by a lift, affords a superb panorama of the city and lagoon.

Centuries ago, when Benedictines occupied the original monastery, it was known as the Isle of Cypresses. The monastery was rebuilt in the 13th century following an earthquake and later became a centre of learning and a residence for eminent foreign visitors. Following the Fall of the Republic in 1797 (*see p48*) the monastery was suppressed and its treasures plundered.

In 1829 the island became a free port, and in 1851 the headquarters of the artillery. By this time it had changed out of recognition. The complex regained its role as an active cultural centre when the monastery, embracing Palladio's cloisters, refectory and library, was purchased in 1951 by Count Vittorio Cini (*see p134*). Today it is a thriving centre of Venetian culture, with international events and exhibitions. There is also a small open-air theatre. Visits are arranged by appointment.

Cloisters designed by Palladio in the monastery of San Giorgio Maggiore

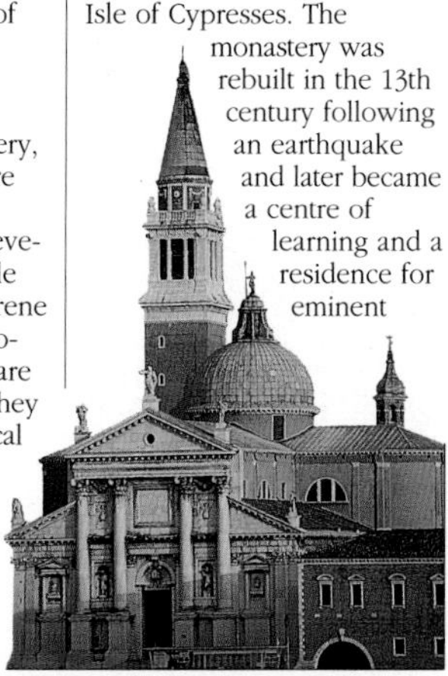

Palladio's church of San Giorgio Maggiore on the island of the same name

San Polo and Santa Croce

The sestieri of San Polo and Santa Croce, bordered by the upper sweep of the Grand Canal, were both named after churches which stood within their boundaries. The first inhabitants are said to have settled on the cluster of small islands called *Rivus Altus* (high bank) or Rialto. When markets were established in the 11th century, the quarter became the commercial hub of Venice. San Polo is still one of the liveliest *sestieri* of the city, with its market stalls, small shops and local bars. The bustle of the market gives way to a maze of narrow alleys opening on to squares. Focal points are the spacious Campo San Polo, the Frari church and the neighbouring Scuola di San Rocco. Santa Croce for the most part is a *sestiere* of very narrow, tightly packed streets and squares where you will see the humbler side of Venetian life. Its grandest *palazzi* line the Grand Canal. Less alluring is the Piazzale Roma, the city's giant car park, lying to the west.

Shuttered window in Campo Sant'Aponal

Sights at a Glance

Churches
San Cassiano 4
San Giacomo dell'Orio 14
San Giacomo di Rialto 2
San Giovanni Evangelista 13
Santa Maria Gloriosa dei Frari pp102–3 8
San Nicolò da Tolentino 12
San Pantalon 11
San Polo 6
San Rocco 10
San Stae 16

Museums and Galleries
Ca' Mocenigo 17
Ca' Pesaro 18
Fondaco dei Turchi (Natural History Museum) 15
Museo Goldoni 7
Scuola Grande di San Rocco pp106–7 9

Streets and Squares
Campo San Polo 5

Bridges
Rialto Bridge 1

Markets
Rialto Markets 3

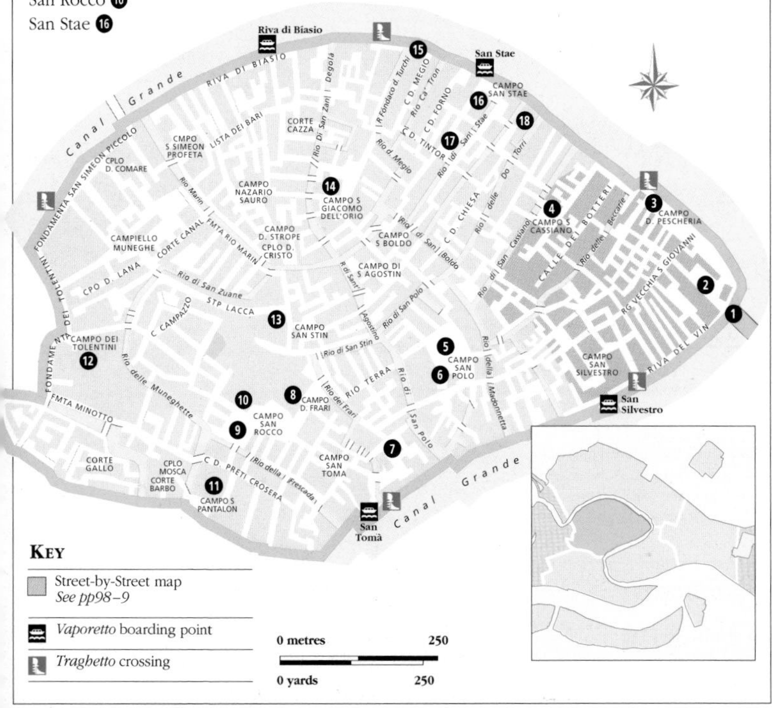

Key

Street-by-Street map
See pp98–9

Vaporetto boarding point

Traghetto crossing

0 metres 250

0 yards 250

◁ **Ponte del Megio, in a quiet corner of Santa Croce**

Street-by-Street: San Polo

The Rialto Bridge and markets make this a magnet for tourists. Traditionally the city's commercial quarter, it was here that bankers, brokers and merchants conducted their affairs. Streets are no longer lined with stalls selling spices and fine fabrics, but the food markets and pasta shops are a colourful sight. The old-fashioned standing-only bars called *bacari* are packed with locals. In contrast, Riva del Vin to the south, by the Grand Canal, is strictly tourist territory.

San Cassiano
Inside this church is a carved altar (1696) and a Crucifixion *by Tintoretto (1568)* ❹

SAN CASSIANO
CAMPO SAN CASSIANO
CALLE DEL CAMPANIEL
RIO DI
C D CRISTI
CALLE DEI BOTTERI
BECCARIE
RT D CARAMPANE
CAMPIELLO ALBRIZZI
C SALVIATI
RIO DELLE
CAMPO SANT'APONAL
RUGA R
C DELL' OLIO
C DEI LUGANEGHER
CAMPO SAN SILVESTRO

Ponte Storto is crooked, like many bridges in the city. It leads under a portico to Calle Stretta, a narrow alley that is only 1 m (3 ft) wide in places.

Sant'Aponal, founded in the 11th century, rebuilt in the 15th, is now deconsecrated. Gothic reliefs decorate the façade.

Riva del Vin, where wine was offloaded from boats, is one of the few accessible quaysides along the Grand Canal.

San Silvestro

Star Sights

★ **Rialto Bridge**

★ **Rialto Markets**

Locator Map
See Street Finder, maps 2, 3, 7

★ Rialto Markets
The Rialto markets have been in operation for centuries. The Pescheria (above) sells fresh fish and seafood, and the Erberia sells fruit and vegetables ❸

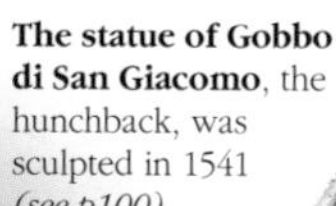

The statue of Gobbo di San Giacomo, the hunchback, was sculpted in 1541 *(see p100).*

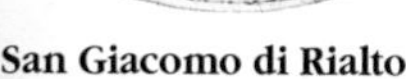

San Giacomo di Rialto
Since its installation in 1410, the clock on this church has been a notoriously poor time-keeper ❷

Key

– – – Suggested route

0 metres 75

0 yards 75

Trattoria alla Madonna is a popular seafood restaurant *(see p240).*

★ Rialto Bridge
A beloved landmark of the Grand Canal, the bridge marks the geographical centre of the city. The balustrades afford fine views of the canal ❶

Rialto Bridge ❶

Ponte di Rialto. **Map** 7 A1. *Rialto.*

THE RIALTO BRIDGE has been a busy part of the city for centuries. At any time of day you will find swarms of crowds jostling on the bridge, browsing among the souvenirs or taking a break to watch the constant swirl of activity on the Grand Canal from the bridge's balustrades.

Stone bridges were built in Venice as early as the 12th century, but it was not until 1588, after the collapse, decay or sabotage of earlier wooden structures, that a solid stone bridge was designed for the Rialto. One of the early wood crossings collapsed in 1444 under the weight of spectators at the wedding ceremony of the Marchese di Ferrara.

Vittore Carpaccio's painting *The Healing of the Madman* (1496, *see p133*) in the Accademia shows the fourth bridge – a rickety looking structure with a drawbridge for the tall-masted galleys. By the 16th century this was in a sad state of decay and a competition was held for the design of a new bridge to be built in stone. Michelangelo, Andrea Palladio and Jacopo Sansovino were among the eminent contenders, but after months of deliberation it was the aptly named Antonio da Ponte who won the commission. The bridge was built between 1588 and 1591 and, until 1854, when the Accademia Bridge was constructed, this remained the only means of crossing the Grand Canal on foot.

Busy canalside restaurant near the Rialto Bridge

San Giacomo di Rialto ❷

Campo San Giacomo, San Polo. **Map** 3 A5. *(041) 522 47 45.* *Rialto.* *9am–noon Mon–Fri.*

THE FIRST CHURCH to stand on this site was allegedly founded in the 5th century, making it the oldest church in Venice. The present building dates from the 11th–12th centuries, with major restoration in 1601. The original Gothic portico and the huge 24-hour clock are the most distinctive features.

The crouching stone figure on the far side of the square is the so-called Gobbo (hunchback) of the Rialto. In the 16th century this was a welcome sight for minor offenders who were forced to run the gauntlet from Piazza San Marco to this square at the Rialto.

***Traghetto* ferrying passengers across to the Erberia**

Rialto Markets ❸

San Polo. **Map** 3 A5. *Rialto.* **Erberia** *(fruit and vegetable market) until noon Mon–Sat,* **Pescheria** *(fish market) until noon Tue–Sat.*

VENETIANS HAVE come to the Erberia to buy fresh produce for hundreds of years. Heavily laden barges arrive at dawn and offload their crates on to the quayside by the Grand Canal. Local produce includes red radicchio from Treviso, and succulent asparagus and baby artichokes from the islands of Sant'Erasmo and Vignole *(see p149)*. In the adjoining fish market are sole, sardines, skate, squid, crabs, clams and other species of seafood and fish. To see it all in full swing you must arrive early in the morning – by noon the vendors are packing up.

RIALTO BRIDGE

Until the 19th century this was the only link between the two sides of the Grand Canal.

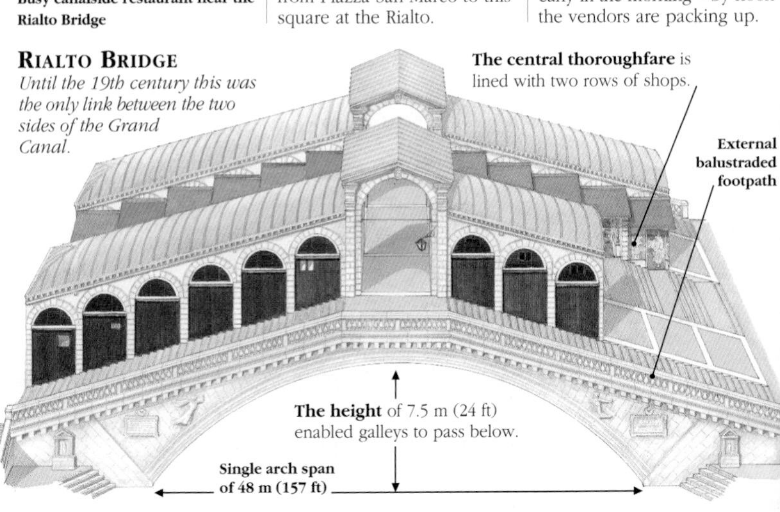
The central thoroughfare is lined with two rows of shops.
External balustraded footpath
The height of 7.5 m (24 ft) enabled galleys to pass below.
Single arch span of 48 m (157 ft)

San Cassiano ❹

Campo San Cassiano, San Polo. **Map** 2 F5. *(041) 72 14 08.* *San Stae.* *7:30am–noon, 4:30–7pm daily.*

THE MEDIEVAL CHURCH of San Cassiano suffered heavily at the hands of 19th-century restorers and the building is a bizarre mix of architectural styles. Of the original church, only the campanile survives. The highlight of the interior is Jacopo Tintoretto's immensely powerful *Crucifixion* (1568), which is in the sacristy.

The campo in which the church stands was notorious for prostitutes in the 1500s – as was the Rialto in general.

A detail of the Gothic façade of Palazzo Soranzo, Campo San Polo

Campo San Polo ❺

Map 6 F1. *San Silvestro.*

THE SPACIOUS SQUARE of San Polo has traditionally been host to spectacular events. As far back as the 15th century it was the venue for festivities, masquerades, ceremonies, balls and bullbaiting.

The most dramatic event was the assassination of Lorenzo de' Medici in 1548. He took refuge in Venice after brutally killing his cousin Alessandro, Duke of Florence. Lorenzo was stabbed in the square by two assassins who were in the service of Cosimo de' Medici, and both were handsomely rewarded by the Florentine duke.

On the eastern side of the square is the beautiful Gothic Palazzo Soranzo. This was originally two palaces – the one on the left is the older. The building is still owned by the Soranzo family.

Palazzo Corner Mocenigo, which is situated in the north-west corner (No. 2128), was once the residence of the eccentric English writer Frederick Rolfe (1860–1913), alias Baron Corvo. He was thrown out of his lodgings when his English hostess read his manuscript of *The Desire and Pursuit of the Whole* – a cruel satirization of English society in Venice.

Since 1979 the square has enjoyed a revival of Carnival festivities. This wide open space is also a haven for local youngsters, who ride bikes, rollerskate or play football. Such activities would not have gone down well in the 17th century – a plaque on the apse of the church, dated 1611, forbids all games (or selling merchandise) on pain of prison, galley service or exile.

San Polo ❻

Campo San Polo. **Map** 6 F1. *(041) 523 76 31.* *San Silvestro.* *8am–noon, 3:30–6:30pm Mon–Sat, 8am–noon Sun.*

FOUNDED IN THE 9th century, rebuilt in the 15th and revamped in the early 19th in Neo-Classical style, the church of San Polo lacks any sense of homogeneity. Yet it is worth visiting for individual features such as the lovely Gothic portal and the Romanesque lions at the foot of the 14th-century campanile – one holds a serpent between its paws, the other a human head.

Inside, follow the signs for the *Via Crucis del Tiepolo* – fourteen canvases of the Stations of the Cross by Giandomenico Tiepolo. The church also has paintings by Veronese, Palma il Giovane (the Younger) and a dark and dramatic *Last Supper* by Tintoretto.

A lion at the foot of the campanile, Church of San Polo

Carlo Goldoni 1707–93

Museo Goldoni ❼

Palazzo Centani, Calle dei Nomboli, San Polo 2794. **Map** 6 E1. *(041) 523 63 53.* *San Tomà.* *8:30am–1:30pm Mon–Sat.* *public hols.*

CARLO GOLDONI, one of the city's favourite sons, wrote over 250 comedies, many based on Commedia dell'Arte figures. Goldoni was born in the beautiful Gothic Palazzo Centrani (or Zantani) in 1707. The house was left to the city in 1931 and is now a centre for theatrical studies and has a collection of theatrical memorabilia. If the house is closed, peep into the enchanting courtyard with its 15th-century open stairway and see a magnificent wellhead, which features carved lions and a coat of arms bearing a hedgehog.

Santa Maria Gloriosa dei Frari 8

Known by all simply as the Frari (a corruption of *Frati*, meaning brothers), this huge, plain Gothic church dwarfs the eastern section of San Polo. The first church was built by Franciscan friars in 1250–1338, but was replaced by a larger building which was completed by the mid 15th century. The interior is striking for its sheer size and for the quality of its works of art. These include masterpieces by Titian and Giovanni Bellini *(see pp26–7)*, a statue by Donatello and a number of imposing monuments to famous Venetians.

The campanile is 262 ft (83 m) high, the tallest in the city after that of San Marco.

Foscari Monument
Doge Foscari set a record by reigning for 34 years (1423–57).

★ Assumption of the Virgin
Titian's glowing and spectacular work (1518) inevitably draws the eye through the monk's choir towards the altar.

★ Monks' Choir
This consists of three-tiered stalls (1468), carved with bas-reliefs of saints and Venetian city scenes.

Rood Screen *(1475)*
Pietro Lombardo and Bartolomeo Bon carved this and decorated it with marble figures.

Entrance

Madonna di Ca' Pesaro (1526) shows Titian's mastery of light and colour.

Floorplan

Exploration of the huge interior can be daunting. The floorplan pinpoints 12 highlights that should not be missed.

Key To Floorplan

1 Canova's tomb
2 Monument to Titian
3 Titian's *Madonna di Ca' Pesaro*
4 Choir stalls
5 Corner Chapel
6 Tomb of Monteverdi
7 Tomb of Doge Nicolò Tron
8 High altar with Titian's *Assumption of the Virgin*
9 Tomb of Doge Francesco Foscari
10 Donatello's *John the Baptist* (c.1450)
11 B Vivarini's altar painting (1474), Bernardo Chapel
12 Giovanni Bellini's *Madonna Enthroned with Saints* (1488)

Visitors' Checklist

Campo dei Frari. **Map** 6 D1. *(041) 522 26 37.* *San Tomà.* *9–11:45am, 2:30–5:30pm Mon–Sat, 3–5:30pm Sun & public hols.* *except for those attending mass and on public hols.* *frequent.*

Monument to Titian *(1853)*
Canova's pupils, Luigi and Pietro Zandomeneghi, built this monument to Titian in place of the one conceived by Canova himself.

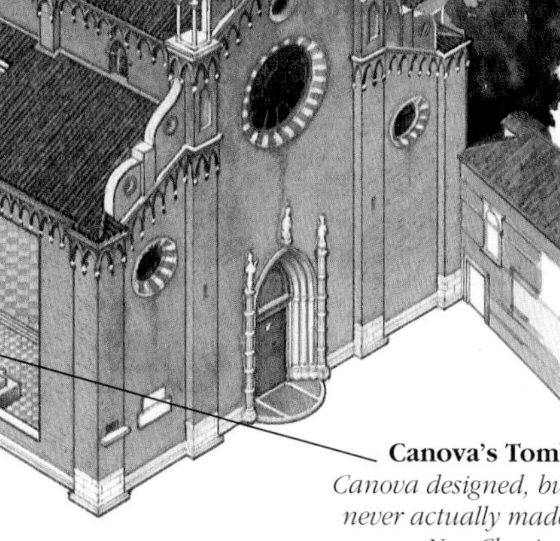

The former monastery, which houses the State Archives, has two cloisters, one in the style of Sansovino, another designed by Palladio.

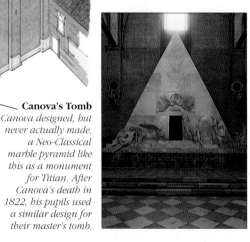

Canova's Tomb
Canova designed, but never actually made, a Neo-Classical marble pyramid like this as a monument for Titian. After Canova's death in 1822, his pupils used a similar design for their master's tomb.

Star Features

★ **Assumption of the Virgin by Titian**

★ **Monks' Choir**

Scuola Grande di San Rocco ❾

See pp106–7.

San Rocco ❿

Campo San Rocco, San Polo. **Map** 6 D1. *(041) 523 48 64.* *San Tomà.* *7:30am–12:30pm daily (also 3–5pm Sun).*

SHARING THE LITTLE square with the celebrated Scuola Grande di San Rocco is the church of the same name. Designed by Bartolomeo Bon in 1489 and largely rebuilt in 1725, the exterior has suffered from a mixture of architectural styles. The façade, similar in concept to the Scuola, was added in 1765–71.

Inside, the main interest lies in Tintoretto's paintings in the chancel, which depict scenes from the life of St Roch, patron saint of contagious diseases. Of these the most notable is *St Roch Curing the Plague Victims* (1549).

San Pantalon ⓫

Campo San Pantalon, Dorsoduro. **Map** 6 D2. *San Tomà.* *8–11am, 4–6:30pm Mon–Sat.*

Fumiani's ceiling painting (1680–1704), San Pantalon

THE OVERWHELMING feature of this late 17th-century church is the painted ceiling, dark, awe-inspiring and remarkable for its illusionistic effects. The ceiling comprises a total of 40 scenes (admirers claim this makes it the world's largest work of art on canvas), depicting the martyrdom and apotheosis of the physician St Pantalon. The artist, Gian Antonio Fumiani, took 24 years (1680–1704) to achieve this masterpiece, but then allegedly fell to his death from the scaffolding.

Paolo Veronese's emotive painting *St Pantalon Healing a Boy* (second chapel on the right) was his final work of art (1587). If you would like to see Antonio Vivarini and Giovanni d'Alemagna's *Coronation of the Virgin* (1444) and *The Annunciation* (1350) attributed to Paolo Veneziano, ask the custodian for access to the Chapel of the Holy Nail *(Cappella del Sacro Chiodo).*

San Nicolò da Tolentino ⓬

Campo dei Tolentini, Santa Croce. **Map** 5 C1. *(041) 522 58 06.* *Piazzale Roma.* *8–11am, 4:30–7pm daily.*

CLOSE TO Piazzale Roma *(see p271)*, San Nicolò da Tolentino is an imposing 17th-century church with a Classical portico. The interior, decorated with 17th-century paintings, is the resting place of Francesco Morosini (d.1694), the great Venetian admiral. A cannonball embedded in the façade is a memento of an Austrian bombardment during the siege of 1849.

San Giovanni Evangelista ⓭

Campiello de la Scuola, San Polo. **Map** 6 D1. *(041) 523 39 97.* *San Tomà. No official entry for the public. Phone for an appointment or ring the bell.*

A CONFRATERNITY of flagellants founded the Scuola of St John the Evangelist in 1261. The complex, just north of the Frari *(see pp102–3)*, has a church, *scuola* and courtyard. Separating the square from the street is Pietro Lombardo's elegant white and grey screen and portal (1480) and in the arch crowning the portal, the carved eagle is the symbol of St John the Evangelist.

Lombardo's marble screen and portal, San Giovanni Evangelista

The main hall of the Scuola is reached by climbing a splendid 15th-century double stairway by Mauro Coducci (1498). Large, dark canvases decorate the ceiling and walls of the 18th-century hall. The Scuola's greatest art treasure, the cycle of paintings which depicts *The Stories of the Cross*, is now on display in the Accademia gallery *(see p133)*. The cycle formerly embellished the oratory (off the main hall) where the Reliquary of the True Cross is still carefully preserved.

San Giacomo dell'Orio ⓮

Campo San Giacomo dell'Orio, Santa Croce. **Map** 2 E5. *(041) 524 06 72.* *Riva di Biasio or San Stae.* *8am–noon, 5–7pm Mon–Sat am, Sun.*

THIS CHURCH is a focal point of a quiet quarter of Santa Croce. The name "dell'Orio" (locally dall'Orio) may derive from a laurel tree *(alloro)* that once stood near the church.

Founded in the 9th century, rebuilt in 1225 and repeatedly modified, the church is a mix of architectural styles. The campanile, basilica ground plan and Byzantine columns survive from the 13th century. The ship's keel roof and the columns are from the Gothic period, and the apses are Renaissance. For access to the Veronese ceiling and altar paintings in the new sacristy, apply to the custodian.

Fondaco dei Turchi ⓯

Canal Grande, Santa Croce 1730.
Map 2 E4. *(041) 524 08 85.*
San Stae. *for renovation.*

THE BUILDING that now contains Venice's natural history museum has a chequered history. In the 13th century it was one of the largest *palazzi* on the Grand Canal. In 1381 it was bought by the state for the Dukes of Ferrara and its lavishly decorated rooms were used for banquets and state functions. In 1621 the Turks set up a warehouse *(fondaco)*, and the spacious portico was used for loading merchandise. As commerce with the Orient declined further, the structure fell into disrepair until, roused by Ruskin's passionate interest, the Austrians began restoration work in the 1850s.

Since 1924 the Fondaco has housed the natural history museum (Museo di Storia Naturale). There is a collection of stuffed animals, crustacea and dinosaur fossils and a section on lagoon life. Prize exhibits include a skeleton of an *Ouranosaurus nigeriensis*, 7 m (23 ft) long and 3.6 m (12 ft) tall, and a fossil of an *Sarcosuchus imperator* – an ancestor of the crocodile.

Ouranosaurus skeleton in the Fondaco dei Turchi

San Stae ⓰

Campo San Stae, Santa Croce.
Map 2 F4. *San Stae.* *9am–noon, 4–6pm daily.*

RESTORED IN 1977–8 by the Pro Venezia Foundation, San Stae (or Sant'Eustachio) has a spick-and-span sculpted façade. It was built in 1709 by Domenico Rossi. Works by Piazzetta, Tiepolo and other 18th-century artists decorate the chancel. Near the second altar on the left is the bust of Antonio Foscarini, executed for treason in 1622 but pardoned the following year.

One of the finely furnished rooms of Ca' Mocenigo

Ca' Mocenigo ⓱

Salizzada San Stae, Santa Croce 1992.
Map 2 F5. *(041) 72 17 98.*
San Stae. *10am–4pm Tue–Sun.* *public hols.*

ONE OF the oldest and greatest of all Venetian families, the Mocenigos produced no fewer than seven doges. There were various branches of the family, one of which resided in this handsome 17th-century mansion. Count Alvise Nicolò Mocenigo, the last of this particular branch, died in 1954, bequeathing the palace to the Comune di Venezia (city authorities).

The entrance façade is unremarkable, but the interior is elegantly furnished and gives you a rare opportunity of seeing inside a *palazzo* preserved more or less as it was in the 18th century. The frescoed ceilings and other works of art are celebrations of the family's achievements. The illustrious Mocenigos are portrayed in a frieze around the portego on the first floor.

The Museo del Tessuto e del Costume inside the house contains antique fabrics and exquisitely made costumes.

Ca' Pesaro ⓲

Canal Grande, Santa Croce 2076.
Map 2 F5. *San Stae.* **Galleria d'Arte Moderna** *(041) 72 11 27.*
for restoration until 1999.
Museo Orientale *(041) 524 11 73.*
9am–2pm Tue–Sun. *1 Jan, 1 May, 25 Dec.*

IT TOOK 58 YEARS to complete this magnificent Baroque palace. Built for the Pesaro family, it was the masterpiece of Baldassare Longhena, who worked on it until his death in 1682. Antonio Gaspari then took over Longhena's design, eventually completing the structure in 1710.

In the 19th century the Duchess of Bevilacqua La Masa bequeathed the palace to the city for exhibiting the works of unestablished Venetian artists. The Galleria d'Arte Moderna was founded in 1897. Today this features a permanent exhibition of work by artists such as Bonnard, Matisse, Miró, Klee, Klimt and Kandinsky, in addition to works by Italian artists of the 19th and 20th centuries.

The Museo Orientale has an idiosyncratic collection of Chinese and Japanese artifacts collected by the Count of Barbi during his 19th-century travels in the Far East.

Gustav Klimt's *Salome*, Gallery of Modern Art, Ca' Pesaro

Scuola Grande di San Rocco ❾

Pianta's caricature of Tintoretto

FOUNDED IN HONOUR of St Roch (San Rocco), the Scuola was set up as a charitable institution for the sick. Construction began in 1515 under Bartolomeo Bon and was completed in 1549 by Scarpagnino, financed largely by donations from Venetians who believed that St Roch, the patron saint of contagious diseases, would save them from the plague. In 1564 Tintoretto *(see p140)* was commissioned to decorate the walls and ceilings of the Scuola. His remarkable cycle of paintings starts in the Sala dell'Albergo *(see* Gallery Guide*)*.

Restored main entrance to the Scuola di San Rocco

SALA DELL'ALBERGO

A self-portrait was often a feature of Tintoretto's paintings.

The subsidiary figures are full of life but do not lessen the central drama.

THE CRUCIFIXION

In this panorama of Calvary, Tintoretto reached a pitch of religious feeling never hitherto achieved in Venetian art.

Figure of Christ
The crucified figure of the Redeemer is raised and leaning, accentuating His divinity and saving grace.

A COMPETITION was held in 1564 to select an artist for the central ceiling panel of the Sala dell'Albergo. To the fury of his rivals, Tintoretto pre-empted his fellow competitors by installing his painting *in situ* prior to judging. He won the commission and was later made a member of the Scuola. Over the next 23 years, Tintoretto decorated the entire building.

The series of paintings, completed in 1587, reveals Tintoretto's revolutionary use of light, mastery of foreshortening and visionary use of colour. The winning painting, *St Roch in Glory* ①, can be seen on the ceiling of the Sala dell'Albergo. The most moving work in the cycle is the *Crucifixion* (1565) ②. Henry James wrote: "Surely no single picture contains more of human life; there is everything in it, including the most exquisite beauty." Of the paintings on the entrance wall, portraying the Passion of Christ, the most notable is *Christ Before Pilate* (1566–7) ③. The easel painting, *Christ Carrying the Cross,* once attributed to Giorgione, is now believed to be a Titian.

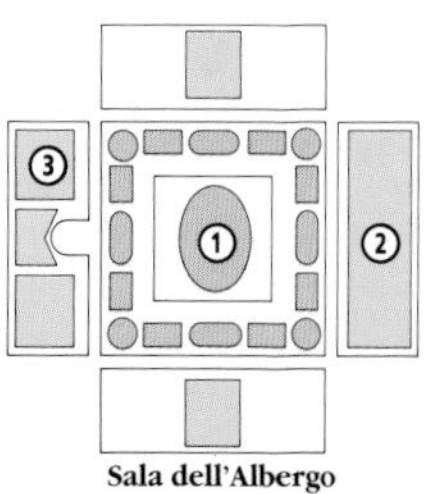

Sala dell'Albergo

Upper Hall

SCARPAGNINO'S great staircase (1544–6), decorated with two vast paintings commemorating the plague of 1630, leads to the Upper Hall. The biblical subjects, decorating the ceiling and walls, were painted in 1575–81. The ceiling paintings (which can be viewed most comfortably with a hired mirror) portray scenes from the Old Testament. The three large and dynamic central square paintings represent: *Moses Striking Water from the Rock* ④, *The Miracle of the Bronze Serpent* ⑤ and *The Gathering of the Manna* ⑥, all alluding to the charitable aims of the Scuola in alleviating thirst, sickness and hunger respectively. All three paintings are crowded compositions with much violent movement.

The vast wall paintings in the hall feature episodes from the New Testament, linking with the ceiling paintings. Two of the most striking paintings are *The Temptation of Christ* ⑦, which shows a handsome young Satan offering Christ two loaves of bread, and *Adoration of the Shepherds* ⑧. Like *The Temptation of Christ*, the *Adoration* is composed in two halves, with a female figure, shepherds and ox below, and the Holy Family and onlookers above.

The beautiful carvings below the paintings were added in the 17th century by sculptor Francesco Pianta. The figures are allegorical and include (near the altar) a caricature of Tintoretto with his palette and brushes, which is meant to represent Painting. Near the entrance to the Sala dell'Albergo you can see a portrait of Tintoretto.

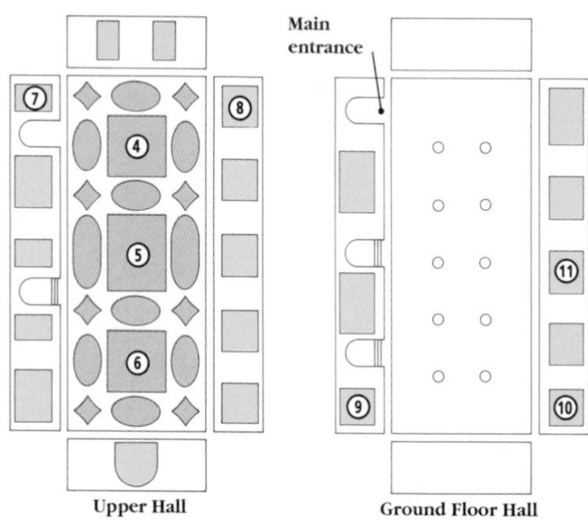

***The Temptation of Christ*, 1578–81 (detail)**

Main entrance

⑦ ⑧ ④ ⑤ ⑥ ⑨ ⑩ ⑪

Upper Hall

Ground Floor Hall

Visitors' Checklist

Campo San Rocco. **Map** 6 D1. *(041) 523 48 64.* *San Tomà.* *Apr–Oct: 9am–5:30pm daily; Nov–Mar: 10am–4pm daily.* *1 Jan, Easter, 1 May, 23–29 Dec.*

Ground Floor Hall

***The Flight into Egypt* (1582–7) (detail)**

THIS FINAL CYCLE, executed in 1583–7, consists of eight paintings illustrating the life of Mary. The series starts with an *Annunciation*, and ends with an *Assumption*, which has been poorly restored. The tranquil scenes of *St Mary of Egypt* ⑨, *St Mary Magdalene* ⑩ and *The Flight into Egypt* ⑪, painted when Tintoretto was in his late sixties, are remarkable for their serenity. This is portrayed most lucidly by the Virgin's isolated spiritual contemplation in the *St Mary of Egypt*. In all three paintings, the landscapes, rendered with rapid strokes, play a major role.

Gallery Guide

The paintings, which unfortunately are not well lit, have no labels, but a useful plan of the Scuola is available (in several languages) free of charge at the entrance.

To see the paintings in chronological order, start in the Sala dell'Albergo (off the Upper Hall), followed by the Upper Hall and finally the Ground Floor Hall.

Key

Wall paintings

Ceiling paintings

CASTELLO

THE LARGEST *sestiere* of the city, Castello stretches from San Marco and Cannaregio in the west to the modern blocks of Sant'Elena in the east. The area takes its name from the 8th-century fortress that once stood on what is now San Pietro, the island which for centuries was the religious focus of the city. The church here was the episcopal see from the 9th century and the city's cathedral from 1451 to 1807. The industrial hub of Castello was the Arsenale, where the great shipyards produced Venice's indomitable fleet of warships. Castello's most popular and solidly commercial area is the Riva degli Schiavoni promenade. Behind the waterfront it is comparatively quiet, characterized by narrow alleys, elegantly faded *palazzi* and fine churches, including the great Santi Giovanni e Paolo *(see pp116–17)*.

Water stoup, Santa Maria Formosa

SIGHTS AT A GLANCE

Churches

La Pietà 3
San Francesco della Vigna 14
San Giorgio dei Greci 2
San Giovanni in Bragora 17
Santi Giovanni e Paolo pp116–17 12
San Lorenzo 15
San Zaccaria 1

Historic Buildings and Monuments

Arsenale 19
Hotel Danieli 4
Ospedaletto 13
Statue of Colleoni 10

Streets, Bridges and Squares

Campo Santa Maria Formosa 9
Ponte della Paglia and Bridge of Sighs 6
Riva degli Schiavoni 5

Walk

Exploring Eastern Castello 20

Museums, Galleries and Scuole

Fondazione Querini Stampalia 8
Museo Diocesano d'Arte Sacra 7
Museo Storico Navale 18
Scuola di San Giorgio degli Schiavoni 16
Scuola Grande di San Marco 11

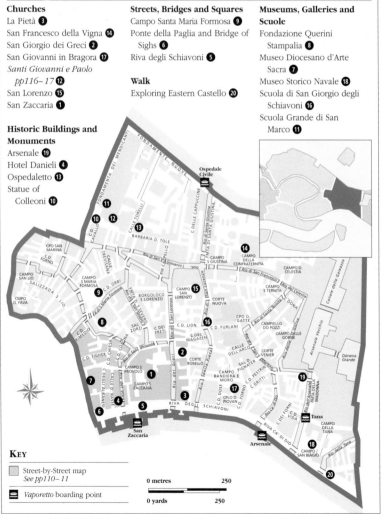

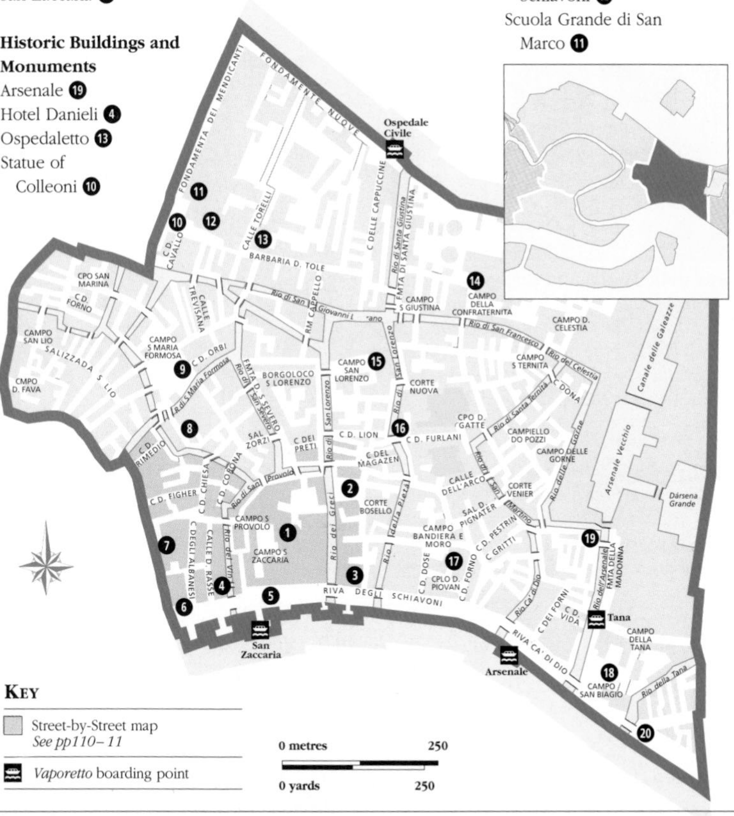

KEY

Street-by-Street map *See pp110–11*

Vaporetto boarding point

0 metres 250

0 yards 250

◁ **Bas relief on Rio Terrà Garibaldi, eastern Castello**

Street-by-Street: Castello

13th-century Madonna in the Museo Diocesano

A STROLL ALONG the Riva degli Schiavoni is an integral part of a visit to Venice. Glorious views of San Giorgio Maggiore compensate for the commercialized aspects of the quayside: souvenir stalls, excursion touts and an overabundance of tourists. Associations with literary figures are legion. Petrarch lived at No. 4145, Henry James was offered "dirty" lodgings at No. 4161, and Ruskin stayed at the Hotel Danieli. Inland, the quiet, unassuming streets and squares of Castello provide a contrast to the bustling waterfront.

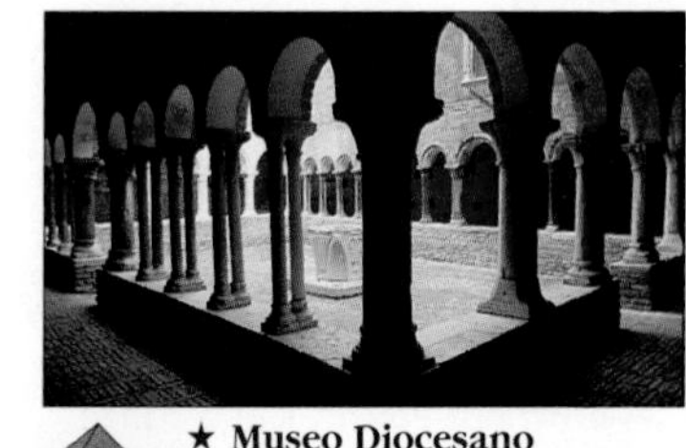

★ Museo Diocesano
The cloisters of the ancient Benedictine monastery of Sant'Apollonia herald the museum ❼

Palazzo Trevisan-Cappello, used as a showroom for Murano glass, was the home of Bianca Cappello, wife of Francesco de' Medici.

CAMPO SS FILIPPO E GIACOMO

CALLE DRIO LA CHIESA

SALIZZADA SAN P...

RIO DEL PALAZZO

CALLE DEGLI ALBANESI

CALLE DELLE RASSE

RIO DEL VIN

RIVA DEGLI SCHIAVONI

Ponte della Paglia and Bridge of Sighs
Crowds throng the Istrian stone Ponte della Paglia – the "straw bridge" – for the best views of the neighbouring Bridge of Sighs, the covered bridge that links the Doge's Palace to the old prisons ❻

Riva degli Schiavoni
This paved quayside was established over 600 years ago, and widened in 1782 ❺

San Zaccaria Jolanda

San Zaccaria Danieli

STAR SIGHTS

- ★ La Pietà
- ★ San Zaccaria
- ★ Museo Diocesano

Hotel Danieli
Joseph da Niel, after whom this hotel was named, turned the Palazzo Dandolo into a haunt for 19th-century writers and artists ❹

Palazzo Priuli, overlooking the quiet Fondamenta Osmarin, is a fine Venetian Gothic palace. The corner window is particularly beautiful, but the early 16th-century façade frescoes have long since disappeared.

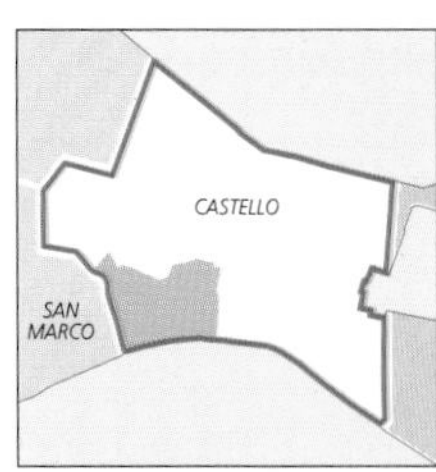

Locator Map
See Street Finder, maps 7, 8

San Giorgio dei Greci
Subsidence is the cause of the city's tilting bell-towers: San Giorgio dei Greci's looks particularly perilous ❷

★ San Zaccaria
Coducci added Renaissance details such as this panel to the Gothic façade ❶

Key

— — — Suggested route

0 metres 75
0 yards 75

San Zaccaria

MVE

Henry James stayed here and completed *Portrait of a Lady* (1881).

★ La Pietà
In Vivaldi's day, the church became famous for the superb quality of its musical performances ❸

The Statue of Vittorio Emanuele II, the first king of a united Italy, was sculpted by Ettore Ferrari in 1887.

San Zaccaria ❶

Campo San Zaccaria. **Map** 8 D2. *(041) 522 12 57.* *San Zaccaria.* *10am–noon, 4–6pm daily.* *to chapels only.*

SET IN A QUIET square just a stone's throw from the Riva degli Schiavoni, the church of San Zaccaria is a successful blend of Flamboyant Gothic and Classical Renaissance styles. Founded in the 9th century, it was completely rebuilt between 1444 and 1515. Antonio Gambello began the façade in Gothic style and, when Gambello died in 1481, Mauro Coducci completed the upper section, adding all the Classical detail.

The adjoining Benedictine convent, which had close links with the church, became quite notorious for the riotous behaviour of its nuns. The majority were from families of Venetian nobility, many of them sent to the convent to avoid the expense of a dowry.

Every Easter the doge came with his entourage to San Zaccaria – a custom which originated as an expression of gratitude to the nuns, who had relinquished part of their garden so that Piazza San Marco could be enlarged.

The artistic highlight of the interior (illuminate with coins in the meter) is Giovanni Bellini's sumptuously coloured and superbly serene *Madonna and Child with Saints* (1505) in the north aisle.

On the right of the church is a door to the Chapel of St Athanasius which leads to the Chapel of San Tarasio. The chapel is decorated with vault frescoes (1442) by Andrea del Castagno of Florence, and Gothic polyptychs painted in 1443–4 by Antonio Vivarini and Giovanni d'Alemagna. The relics of eight doges lie buried in the waterlogged crypt.

Distant view of San Giorgio dei Greci's tilting campanile

San Giorgio dei Greci ❷

Map 8 D2. *San Zaccaria.* **Church** *(041) 523 95 69.* *10am–5pm daily.* **Museo dei Icone** *(041) 522 65 81.* *9am–12:30pm, 2–5pm.* *Sun & public hols.*

THE MOST remarkable feature of this 16th-century Greek church is the listing campanile which looks as if it is about to topple into the Rio dei Greci. A characteristic feature of the interior is the *matroneo* – the gallery where, in keeping with Greek Orthodox custom, the women sat apart from men. Note also the iconostasis, separating the sanctuary from the nave. The nearby Scuola di San Nicolò dei Greci, redesigned by Baldassare Longhena in 1678, is now the museum of icons of the Hellenic Institute.

Detail from *The Nun's Parlour at San Zaccaria* by Francesco Guardi

La Pietà ❸

Riva degli Schiavoni. **Map** 8 D2. *(041) 520 44 31.* *San Zaccaria.* *9:30–12:30am daily.*

THE CHURCH of La Pietà (or Santa Maria della Visitazione) originally dates from the 15th century. It was rebuilt between 1745 and 1760 by Giorgio Massari, though the Classical façade was added only in 1906. The church has a cool, elegant interior, with an oval plan. The resplendent ceiling fresco, *Triumph of Faith* (1755), was painted by Giambattista Tiepolo.

The Pietà started its life as a foundling home for orphans. It proved so popular that a warning plaque was set up (still to be seen on the side wall), threatening damnation to parents who tried to pass off their children as orphans.

From 1703 until 1740 Antonio Vivaldi directed the musical groups and wrote numerous oratorios, cantatas and vocal pieces for the Pietà choir, and the church became famous for its performances.

Today the church is a popular venue for concerts – with a strong emphasis on Vivaldi. These are held throughout the year, usually on Mondays and Thursdays.

Bas relief on La Pietà's early 20th-century façade

Hotel Danieli ❹

Riva degli Schiavoni 4196. **Map** 7 C2. *San Zaccaria. See also* ***Where to Stay*** *p230.*

ONE OF THE MOST celebrated hotels in Europe, the Danieli's deep pink façade is a landmark on the Riva degli Schiavoni. Built in the 14th century, it became famous as the venue for the first opera performed in Venice, Monteverdi's *Proserpina Rapita*

(1630). The palace became a hotel in 1822 and soon gained popularity with the literary and artistic set. Its famous guests included Balzac, Proust, Dickens, Cocteau, Ruskin, Debussy and Wagner. In the 1830s Room 10 witnessed an episode in the love affair between the French poet and dramatist Alfred de Musset, and novelist George Sand: when de Musset fell ill after a surfeit of orgies, Sand ran off with his Venetian doctor.

Riva degli Schiavoni – the city's most famous promenade

Riva degli Schiavoni 5

Map 8 D2. *San Zaccaria.*

THE SWEEPING promenade that forms the southern quayside of Castello was named after the traders from Dalmatia (Schiavonia) who used to moor their boats and barges here. For those who arrive in Venice by water, this long curving quayside is a spectacular introduction to the charms of the city.

At its western end, close to Piazza San Marco, the broad promenade teems during the day with tourists thronging around the souvenir stalls and people hurrying to and from the *vaporetto* stops. Nothing can detract, however, from the glorious views across the lagoon to the island of San Giorgio Maggiore *(see p95)*.

The Riva degli Schiavoni has always been busy with boats. Canaletto's drawings in the 1740s and 1750s show the Riva bustling with gondolas, sailing boats and barges. The gondolas are still here, but it is also chock-a-block with water taxis, *vaporetti*, excursion boats, tugs, and – beyond the Arsenale – naval ships and ocean-going liners.

The modern annexe of the Hotel Danieli caused a great furore when it was built in 1948. Intruding on a waterfront graced by fine Venetian palaces and mansions, its stark outline is still something of an eyesore. The annexe marks the spot where Doge Vitale Michiel II was stabbed to death in 1172. Three centuries earlier, in 864, Doge Pietro Tradonico had suffered the same fate in nearby Campo San Zaccaria.

Ponte della Paglia and Bridge of Sighs 6

Map 7 C2. *San Zaccaria.*

THE PONTE della Paglia, whose name may derive from the boats made of straw *(paglia)* that once moored here, was originally built in 1360. The existing structure dates from 1847.

According to legend the Bridge of Sighs, built in 1600 to link the Doge's Palace with the new prisons, takes its name from the lamentations of the prisoners as they made their way to the offices of the State Inquisitors. Access is available to the public via the Secret Itinerary tour in the Doge's Palace *(see p87)*.

Ponte della Paglia behind the Bridge of Sighs

Museo Diocesano d'Arte Sacra 7

Sant'Apollonia, Ponte della Canonica. **Map** 7 C2. *(041) 522 91 66.* *San Zaccaria.* *10:30am–12:30pm Mon–Sat.* *public hols.* ***Donations** appreciated.*

ONE OF the architectural gems of Venice, the cloister of Sant'Apollonia is the only Romanesque building in the city. Only a few steps from St Mark's, the cloister provides a quiet retreat from the hubbub of the Piazza.

The monastery was once the home of Benedictine monks, but its non-ecclesiastical uses have been manifold. In 1976 its cloisters became the home of the diocesan museum of sacred art, founded in order to provide a haven for works of art from closed or deconsecrated churches. The collection includes paintings, statues, crucifixes and many pieces of valuable silver. The museum has two workshops, staffed by volunteers who restore the paintings and statues. The collection is ever-changing, but among the major permanent exhibits are works by Luca Giordano (1634–1705), which came from the Church of Sant'Aponal, and a 16th-century wood and crystal tabernacle.

Fondazione Querini Stampalia ❽

Campiello Querini, Castello 4778. **Map** 7 C1. (041) 522 52 35. *San Zaccaria.* **Museum** *10am–1pm, 3–6pm Tue–Thu & Sun, 10am–1pm, 3–10pm Fri & Sat.* *Mon.* **Library** *4–11:30pm Mon–Fri, 2:30–11:30pm Sat, 3–7pm Sun.*

THE LARGE PALAZZO Querini Stampalia was commissioned in the 16th century by the descendants of the old Venetian Querini family. Great art lovers, they filled the palace with fine paintings.

In 1868 the last member of the dynasty bequeathed the palace and the family collection of art to the foundation that bears his name. The paintings include works by Giovanni Bellini, Giambattista Tiepolo, and some vignettes by Pietro and Alessandro Longhi. The library on the first floor, which is open to the public, contains over 200,000 books.

Campo Santa Maria Formosa ❾

Campo Santa Maria Formosa. **Map** 7 C1. *Rialto.*

LARGE, RAMBLING and flanked by handsome palaces, the market square of Santa Maria Formosa is one of the most characteristic *campi* of Venice. On the southern side, distinctive for its swelling apses, stands the church of Santa Maria Formosa. Built on ancient foundations, it was designed by Mauro Coducci in 1492 but took over a century to assume its current form. Unusually, it has two main façades – one overlooking the *campo*, the other the canal. The campanile was added in 1688. Its most notable feature is the truly grotesque stone face that decorates its foot.

Inside, Palma il Vecchio's polyptych, *St Barbara and Saints* (c.1510), ranks among the great Venetian masterpieces and looks particularly splendid since its restoration by the American Save Venice organization. Palma il Vecchio's portrayal of the handsome and dignified figure of St Barbara glorifies Venice's ideal female beauty. She is surrounded by saints, with a central lunette of the *pietà* above. St Barbara was the patron saint of soldiers: in wartime they prayed to her for protection, in victory they came for thanksgiving.

Palma il Vecchio's *St Barbara* in Santa Maria Formosa

Statue of Colleoni ❿

Campo Santi Giovanni e Paolo. **Map** 3 C5. *Ospedale Civile.*

BARTOLOMEO COLLEONI, the famous *condottiere* or commander of mercenaries, left his fortune to the Republic on condition that his statue was placed in front of San Marco. A prominent statue in the Piazza would have broken with precedent, so the Senate cunningly had Colleoni raised before the Scuola di San Marco instead of the basilica. A touchstone of early Renaissance sculpture, the equestrian statue of the proud warrior (1481–8) is by the Florentine Andrea Verrocchio and, after his death, was cast in bronze by Alessandro Leopardi. The statue has a strong sense of power and movement which arguably ranks it alongside works of Donatello.

Statue of Bartolomeo Colleoni

Scuola Grande di San Marco ⓫

Campo Santi Giovanni e Paolo. **Map** 3 C5. *(Library) (041) 529 43 23.* *Ospedale Civile.* **Church and Library** *9am–noon daily.*

FEW HOSPITALS can boast as rich and unusual a façade as that of Venice's Ospedale Civile. It was built originally as one of the six great confraternities of the city *(see p127)*. Their first headquarters were destroyed by fire in 1485, but the Scuola was rebuilt at the end of the 15th century.

The delightful asymmetrical façade, with its arcades, marble panels and *trompe l'oeil* designs, was the work of Pietro Lombardo working in conjunction with his sons and Giovanni Buora. The upper order was finished by Mauro Coducci in 1495. The interior was revamped in the last century and, since then, most of the artistic masterpieces have been dispersed.

Occasionally, visitors are allowed in to see the library with its finely carved 16th-century ceiling, and the hospital chapel, the Church of San Lazzaro dei Mendicanti, which has an early Tintoretto and a work by Veronese.

Santi Giovanni e Paolo ⓬

See pp116–17.

Ospedaletto ⓭

Calle Barbaria delle Tole. **Map** 4 D5. *(041) 520 06 33.* *Ospedale Civile.* *4–7pm Thu–Sat.*

Beyond the south flank of Santi Giovanni e Paolo *(see pp116–17)* is the façade of the Ospedaletto or, more correctly, Santa Maria dei Derelitti. The Ospedaletto was set up by the Republic in 1527 as a charitable institution to care for the sick and aged, and to educate orphans and abandoned girls. Such an education consisted largely of the study of music. The girls became leading figures in choirs and orchestras, with concerts bringing in funds for the construction in 1630 of a *sala della musica,* which became the main performance venue. This elegant room, frescoed by Jacopo Guarana, can be visited on request.

The church, which formed part of the Ospedaletto, was built by Baldassare Longhena in 1662–74. John Ruskin regarded the huge, hideous heads on the façade as anti-Classical abominations, likening them to "masses of diseased figures and swollen fruit". The interior of the church is decorated with less provocative works of art and notable paintings from the 18th century, including *The Sacrifice of Isaac* (1715–16) by Giambattista Tiepolo.

Fresco by Guarana in the *sala della musica* of the Ospedaletto

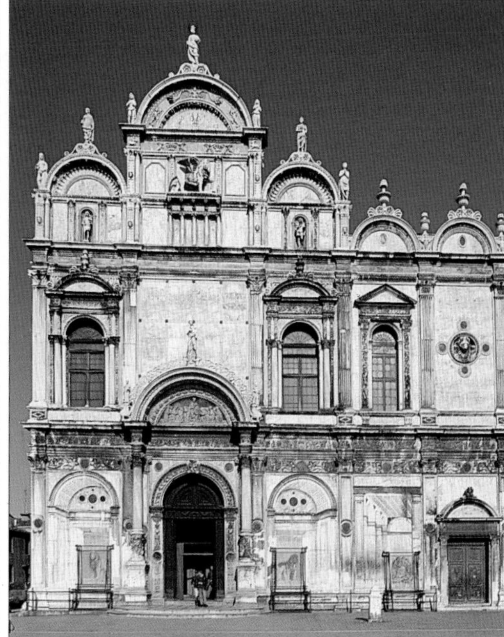

The decorative façade of the Scuola Grande di San Marco

San Francesco della Vigna ⓮

Ramo San Francesco. **Map** 8 E1. *(041) 520 61 02.* *Celestia.* *8am–12:30pm, 3–7pm daily.*

The name "della Vigna" derives from a vineyard which was bequeathed to the Franciscans in 1253. The church which the order built here in the 13th century was rebuilt under Jacopo Sansovino in 1534, with a façade added in 1562–72 by Palladio.

The interior has a rich collection of works of art, including sculpture by Alessandro Vittoria, Paolo Veronese's *The Holy Family with Saints* (1562) and Antonio da Negroponte's *Virgin and Child* (c.1450). The *Madonna and Child with Saints* (1507) by Giovanni Bellini hangs near the cloister.

San Lorenzo ⓯

Campo San Lorenzo. **Map** 8 D1. *(041) 520 61 02.* *San Zaccaria.* *for restoration.*

Deconsecrated and closed for restoration, the church of San Lorenzo's only claim to fame is as the alleged burial place of Marco Polo *(see p143).* Unfortunately there is nothing to show for it because his sarcophagus disappeared during rebuilding in 1592. A collection of paintings was dispersed, and for many years the church was abandoned.

In 1987 restorers discovered the foundations of two earlier churches, dating from 850 AD and the late 12th century. The foundations of the present medieval structure, as well as substantial remains of the marble floor, have suffered damage by water seeping in from the adjacent canal. Restoration work by the British Venice in Peril Fund continues indefinitely.

Marco Polo

Santi Giovanni e Paolo ⓬

Figure in left transept

MORE FAMILIARLY known as San Zanipolo, Santi Giovanni e Paolo vies with the Frari *(see pp102–3)* as the city's greatest Gothic church. It was built in the late 13th to early 14th centuries by the Dominican friars, and is striking for its huge dimensions and architectural austerity. Known as the Pantheon of Venice, it houses monuments to no less than 25 doges. Many of these are outstanding works of art, executed by the Lombardi family and other leading sculptors of the day.

★ Cappella del Rosario
The Adoration of the Shepherds *is one of many works by Paolo Veronese which decorate the Rosary Chapel.*

The sacristy has paintings that celebrate the Dominican Order.

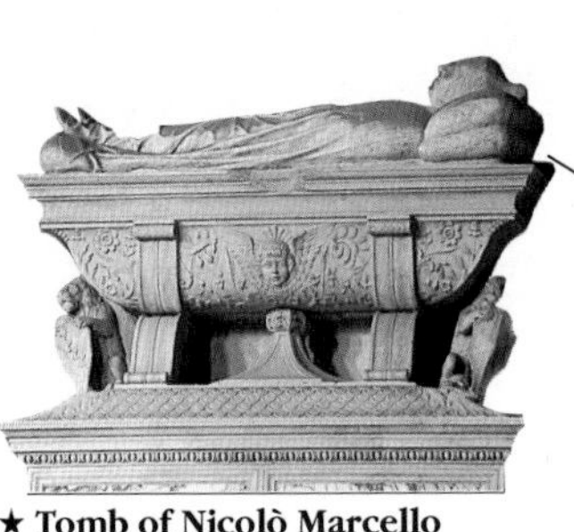

★ Tomb of Nicolò Marcello
This magnificent Renaissance monument to Doge Nicolò Marcello (d.1474) was sculpted by Pietro Lombardo.

The doorway, which is decorated with Byzantine reliefs, is one of the earliest Renaissance architectural features in Venice. The portico carvings are attributed to Bartolomeo Bon.

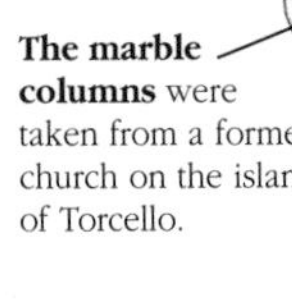

The marble columns were taken from a former church on the island of Torcello.

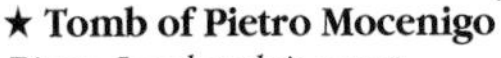

★ Tomb of Pietro Mocenigo
Pietro Lombardo's great masterpiece (1481) commemorates the doge's military pursuits when he was Grand Captain of the Venetian forces. This west side wall is largely devoted to Mocenigo monuments.

STAR FEATURES

- **★ Doges' Tombs**
- **★ Cappella del Rosario**
- **★ Cappella di San Domenico**

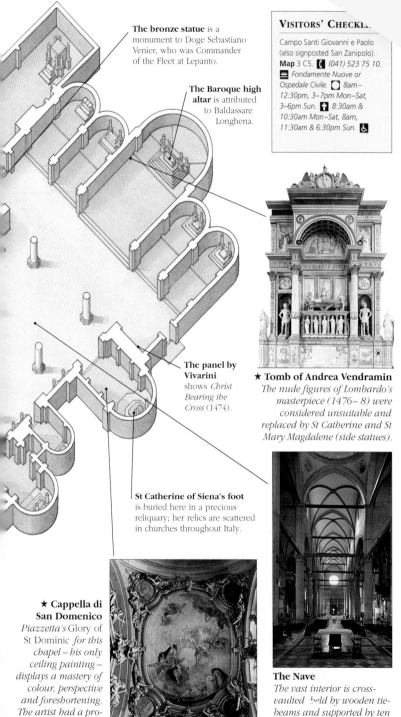

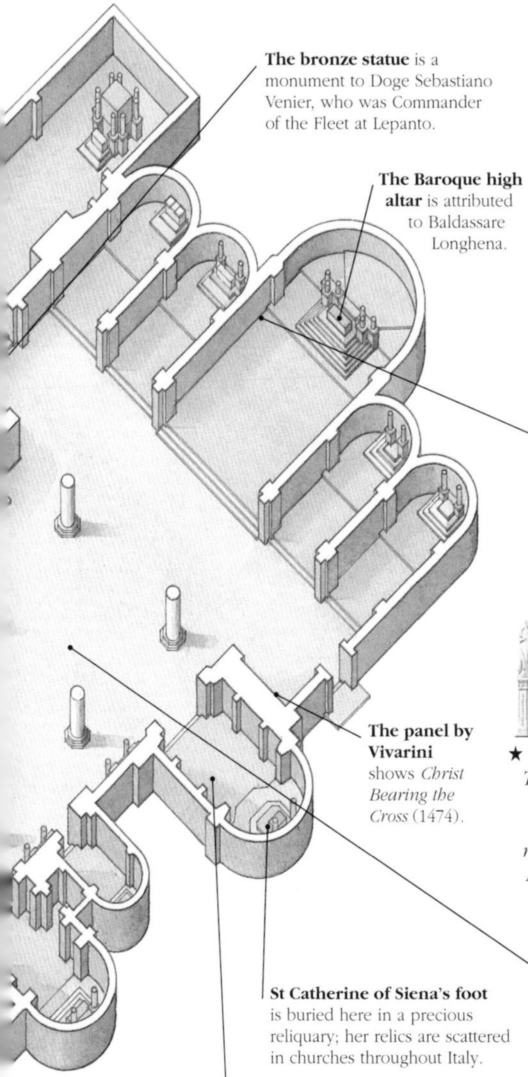

The bronze statue is a monument to Doge Sebastiano Venier, who was Commander of the Fleet at Lepanto.

The Baroque high altar is attributed to Baldassare Longhena.

Visitors' Checkl

Campo Santi Giovanni e Paolo (also signposted San Zanipolo). **Map** 3 C5. *(041) 523 75 10.* *Fondamente Nuove or Ospedale Civile.* *8am–12:30pm, 3–7pm Mon–Sat, 3–6pm Sun.* *8:30am & 10:30am Mon–Sat, 8am, 11:30am & 6:30pm Sun.*

★ Tomb of Andrea Vendramin
The nude figures of Lombardo's masterpiece (1476–8) were considered unsuitable and replaced by St Catherine and St Mary Magdalene (side statues).

The panel by Vivarini shows *Christ Bearing the Cross* (1474).

The Nave
The vast interior is cross-vaulted, held by wooden tie-beams and supported by ten huge columns of Istrian stone blocks.

St Catherine of Siena's foot is buried here in a precious reliquary; her relics are scattered in churches throughout Italy.

★ Cappella di San Domenico
Piazzetta's Glory of St Dominic *for this chapel – his only ceiling painting – displays a mastery of colour, perspective and foreshortening. The artist had a profound influence on the young Tiepolo.*

***St George slaying the Dragon* by Carpaccio, in the Scuola di San Giorgio degli Schiavoni**

Scuola di San Giorgio degli Schiavoni ⑯

Calle Furlani. **Map** 8 E1.
(041) 522 88 28. *San Zaccaria.*
Apr–Sep: 9:30am–12:30pm Tue–Sun; 3–6:30pm Tue–Sat; Oct–Mar: 10am–6pm daily. *1 Jan, 1 May, 25 Dec.*

WITHIN THIS surprisingly simple Scuola are some of the finest paintings of Vittore Carpaccio, commissioned by the Schiavoni community in Venice.

From the earliest days of the Republic, Venice forged trade links with the coastal region of Schiavonia (Dalmatia) across the Adriatic. By 1420 permanent Venetian rule was established there, and many of the Schiavoni came to live in Venice. By the mid 15th century the Slav colony in the city had grown considerably and the State gave permission for them to found their own confraternity *(see p127)*.

The Scuola was established in 1451. It is a delightful spot to admire Carpaccio's really exceptional works of art, and has changed very little since the rebuilding of the Scuola in 1551. The exquisite frieze, executed between 1502 and 1508, shows scenes from the lives of favourite saints: St George, St Tryphon and St Jerome. Each episode of the narrative cycle is remarkable for its vivid colouring, minutely observed detail and historic record of Venetian life. Outstanding among them are *St George Slaying the Dragon*, *St Jerome Leading the Tamed Lion to the Monastery*, and *The Vision of St Jerome.*

San Giovanni in Bragora ⑰

Campo Bandiera e Moro. **Map** 8 E2.
(041) 520 59 06. *Arsenale.*
8–11am, 3–5pm daily.

THE FOUNDATIONS of this simple church date back to ancient times but the existing building is essentially Gothic (1475–9). The intimate interior has major works of art which demonstrate the transition from Gothic to early Renaissance. Bartolomeo Vivarini's altarpiece, *Madonna and Child with Saints* (1478) is unmistakably Gothic. Contrasting with this is Cima da Conegliano's *Baptism of Christ* (1492–5), which adorns the main altar. This large-scale narrative scene, in a realistic landscape, set a precedent for later Renaissance painters.

Model of the *Bucintoro* in the Museo Storico Navale

Museo Storico Navale ⑱

Campo San Biagio, Arsenale.
Map 8 F3. *(041) 520 02 76.*
Arsenale. *8:45am–1:30pm Mon–Fri, 8:45am–1pm Sat.*
1 Jan, 1 May, 25 Dec and other religious hols.

IT WAS THE Austrians who, in 1815, first had the idea of assembling the remnants of the Venetian navy and creating a historical naval museum. They began with a series of models of vessels produced in the 17th century by the Arsenale, and to these added all the naval paraphernalia they could obtain. Exhibits include friezes preserved from famous galleys of the past, maritime firearms and a replica of the Doge's ceremonial barge, the *Bucintoro.*

The collection has been housed in an ex-warehouse on the waterfront since 1958, and now traces Venetian and Italian naval history to the present day.

The first exhibits you see on entering are the World War II human torpedoes or "pigs". Torpedoes such as these helped sink HMS *Valiant* and HMS *Queen Elizabeth*: they were guided to their target by naval divers who jumped off just before impact.

The rest of the museum is divided into the Venetian navy, the Italian navy from 1860 to today, Adriatic vessels and the Swedish room. The museum is well laid out and there are very informative explanations in English.

Arsenale ⓳

Map 8 F1. *Arsenale or Tana.*
Limited public access.

HEART OF the city's maritime power, the Arsenale was founded in the 12th century and enlarged in the 14th to 16th centuries to become the greatest naval shipyard in the world. The word "arsenal" derives from the Arabic *darsina'a*, house of industry – which indeed it was.

At its height in the 16th century, a workforce of 16,000, the *arsenalotti*, were employed to construct, equip and repair the great Venetian galleys *(see pp44–5)*. One of the first production lines in Europe, it was like a city within a city, with its own workshops, warehouses, factories, foundries and docks.

THE ASSEMBLY-LINE SYSTEM

The *arsenalotti*, master shipbuilders of the 16th century

During the Arsenale's heyday, a Venetian galley could be constructed and fully equipped with remarkable speed and efficiency. From the early 16th century the hulls, which were built in the New Arsenal, were towed past a series of buildings in the Old Arsenal to be equipped in turn with rigging, ammunition and food supplies. By 1570, when Venice was faced with the Turkish threat to take Cyprus, the Arsenale proved capable of turning out an entire galley in 24 hours. Henry III of France witnessed the system's efficiency in 1574 when the *arsenalotti* completed a galley in the time it took for him to partake in a state feast.

Entrance to the Arsenale, guarded by 16th-century towers

Surrounded by crenellated walls, the site today is largely abandoned. The huge gateway and vast site are the only evidence of its former splendour. The gateway, in the form of a triumphal arch, was built in 1460 by Antonio Gambello and is often cited as Venice's first Renaissance construction.

The two lions guarding the entrance were pillaged from Piraeus (near Athens) by Admiral Francesco Morosini in 1687. A third lion, bald and sitting upright, bears runic inscriptions on his haunches, thought to have been carved by Scandinavian mercenaries who in 1040 fought for the Byzantine emperor against some Greek rebels.

By the 17th century, when the seeds of Venetian decline were well and truly sown, the number of *arsenalotti* plummeted to 1,000. Following the Fall of the Republic in 1797, Napoleon destroyed the docks and stripped the *Bucintoro* (the Doge's ceremonial ship) of its precious ornament. Cannons and bronzes were melted down to contribute to victory monuments celebrating the French Revolution.

Today the area is under military administration and for the most part closed to the public. The bridge by the arched gateway affords partial views of the shipyard, but for a better view, take a trip on a *vaporetto* (No. 23 or 52) which takes you through the heart of the Arsenale Vecchio.

The future of the Arsenale hangs in the air. Ideas such as a new cultural centre or a port are constantly being put forward. So far the only area put to good use is the Corderia, the old rope factory. This serves as a secondary exhibiton centre for the Biennale *(see p256)*.

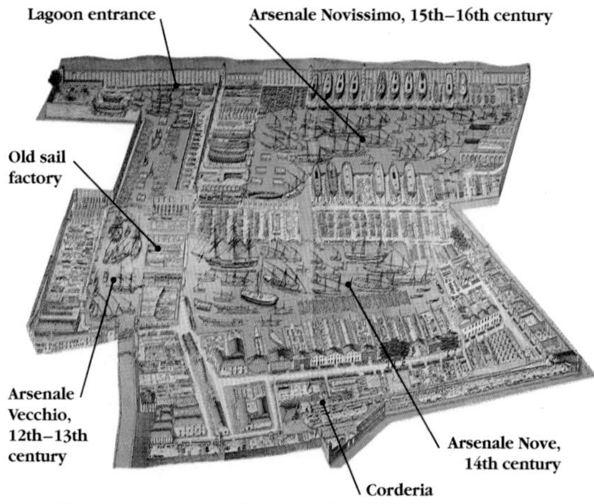

Late 18th-century engraving of the Arsenale

Exploring Eastern Castello ⑳

THIS PEACEFUL STROLL takes you from the animated Castello quayside to the quieter eastern limits of the city. The focal point of the tour is the solitary island of San Pietro di Castello, site of the former cathedral of Venice. From here you head south to the island of Sant'Elena with its historic church and Venice's football stadium, and return via the public gardens along the scenic waterfront.

The calm and leafy Giardini Pubblici ⑯

A tribute to the women fallen in World War II ⑱

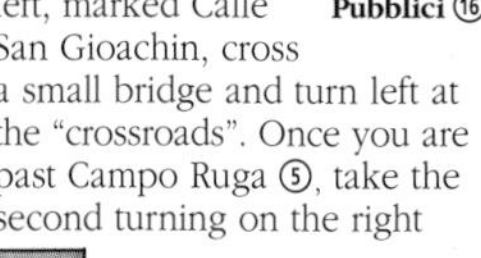

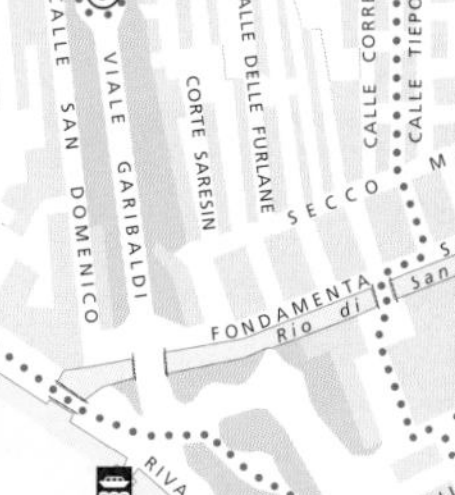

Via Garibaldi

This broad, busy street ① was created by Napoleon in 1808 by filling in a canal. The first house on the right ② was the home of John Cabot and his son Sebastian, the Italian navigators who in 1497 found what they thought to be the coast of China (but in reality was the Labrador coast of Newfoundland). Near the end of the street, through a gate on the right, a bronze monument of Garibaldi ③ by Augusto Benvenuti (1885) marks the northern end of the Viale Garibaldi, which leads to the public gardens.

Returning to Via Garibaldi, take the left-hand embankment at the end of the street, pausing on the bridge ④ for distant views of the Arsenale *(see p119)*. Then take the first on the left, marked Calle San Gioachin, cross a small bridge and turn left at the "crossroads". Once you are past Campo Ruga ⑤, take the second turning on the right and cross the bridge over the broad Canale di San Pietro.

The island of San Pietro di Castello

The old church of San Pietro di Castello ⑥ and its free-standing, tilting campanile ⑦ overlook a grassy square. The island, once occupied by a fortress *(castello)*, was one of Venice's earliest settlements. The church, which was probably founded in the 7th century, became the cathedral of Venice and remained so until 1807 when San Marco took its place *(see p80)*. The existing church, built to a Palladian design in the mid 16th century, has several notable features. These include the Lando Chapel, the Vendramin Chapel and the marble throne from an Arabic tombstone, originally said to have been the Seat of St Peter.

In the south of the square, Mauro Coducci's elegant stone campanile was built in 1482–8, and the cupola was added in 1670. Beside the church, the Palazzo Patriarcale (Bishop's

The busy Via Garibaldi, with John Cabot's house on the far right ②

KEY

••• Walk route

View point

 *Vaporetto* boarding point

Palace) ⑧ was turned into barracks by Napoleon. The old cloisters are overgrown and strung with washing and fishing nets.

From the Bishop's Palace take the Calle drio il Campanile south from the square and turn left when you come to the canal. The first turning right takes you across the Ponte di Quintavalle ⑨, a wooden bridge with good views of brightly coloured boats anchored on either side of the waterway.

npanile
Pietro

The island of San Pietro, with its curious leaning campanile ⑦

San Pietro to Sant'Elena

The large and semi-derelict building at the foot of the bridge is the ex-church and monastery of Sant'Anna ⑩. Take the first left off the *fondamenta*, cross Campiello Correr and then take Calle GB Tiepolo and cross the Secco Marina. Continue straight ahead and over the bridge for the Church of San Giuseppe ⑪. On the rare occasions it is open you can see Vincenzo Scamozzi's monument to Doge Marino Grimani (1595–1605). Cross the square beyond the church and zigzag left, right and left again for Paludo San Antonio, an uninspiring modern street that has been reclaimed from marshland *(palude)*. At the far end cross the bridge over the Rio dei Giardini ⑫ and take the street ahead. A right turn along Viale 4 Novembre brings you down to the spacious gardens of Parco delle Rimembranze ⑬. At the southern end of the park, cut left at Calle Buccari ⑭, then right for the bridge over Rio di Sant'Elena. In front, the Church of Sant'Elena ⑮ is a pretty Gothic church founded in the 13th century. Retrace your steps over the bridge and turn left, following the waterfront back through the park.

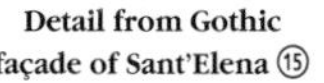

Detail from Gothic façade of Sant'Elena ⑮

Giardini Pubblici and the Biennale Pavilions

At the far side of the park, the bridge across the Rio dei Giardini brings you to the public gardens and to the Biennale gate entrance ⑯. If it happens to be summer in an odd-numbered year, the gardens will be set up with the Biennale pavilions ⑰ at which 40 to 50 nations exhibit many examples of contemporary art *(see p256)*.

Riva dei Sette Martiri

Beyond the Giardini landing stage, cross the bridge for the Riva dei Sette Martiri. Just here, a large bronze statue can be seen at low tide lying on the steps of the embankment. Known as La Donna Partigiana, this is a memorial to the women who were killed fighting in World War II ⑱.

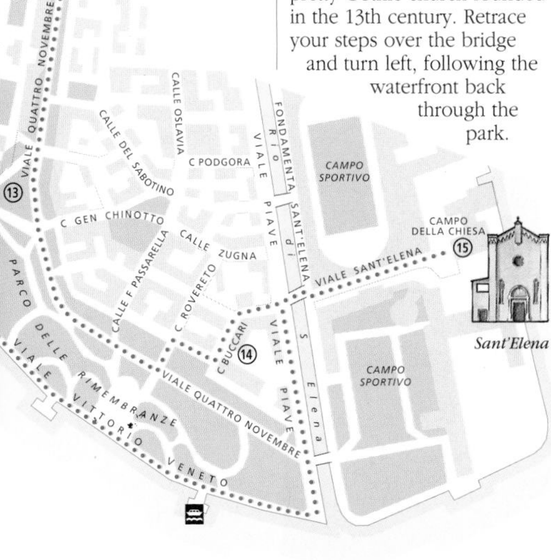

Tips for Walkers

Starting point: *The western end of Via Garibaldi.*
Length: *Just under 5 km (3 miles).*
Getting there: *Vaporetto No. 1 to Arsenale or No. 52 to Tana.*
Stopping-off points: *There are a handful of simple cafés and trattorias along the route, most of them on Via Garibaldi. The waterside Caffè Paradiso at the entrance to the Giardini Pubblici has excellent views. For good seafood, try the Hostaria Da Franz (see p241) along Fondamenta San Giuseppe (No. 754). The green shady parks are a welcome retreat from the bustle of the city.*

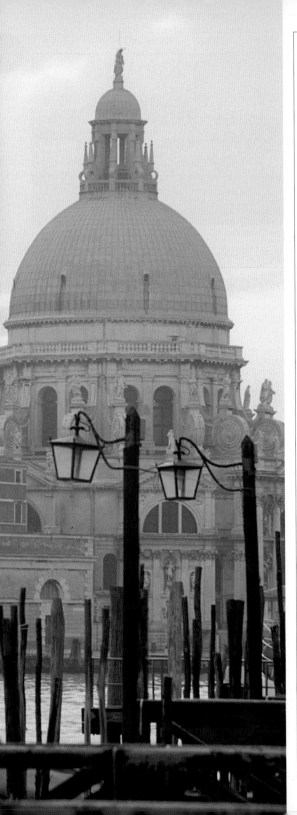

Sights at a Glance

Churches
Angelo Raffaele ❽
Gesuati ⓮
Santa Maria dei Carmini ❻
Santa Maria della Salute ⓳
Santa Maria della Visitazione ⓭
San Nicolò dei Mendicoli ❼
San Sebastiano ❾
San Trovaso ⓬

Museums and Galleries
Accademia see pp130–33 ⓯
Ca' Rezzonico ❸
Cini Collection ⓰
Peggy Guggenheim Collection ⓱

Historic Buildings
Dogana di Mare ⓴
Scuola Grande dei Carmini ❺
Squero di San Trovaso ⓫

Streets, Bridges and Squares
Campiello Barbaro ⓲
Campo San Barnaba ❶
Campo Santa Margherita ❹
Ponte dei Pugni ❷
Zattere ❿

Key

Street-by-Street map *See pp124–5*

Vaporetto boarding point

Traghetto crossing

◁ **View across the Grand Canal to Santa Maria della Salute**

Dorsoduro

Dorsoduro is named after the solid subsoil on which this area has been built up (the name means "hard backbone"). The western part, the island of Mendigola, was colonized centuries before the Rialto was established in AD 828 as the permanent seat of Venice. The settlement then spread eastwards, covering another six islands.

East of the Accademia, the Dorsoduro is a quiet and pretty neighbourhood with shaded squares, quiet canals and picturesque residences belonging to wealthy Venetians and foreigners. In the early 1900s the area was favoured by British expatriates who used to attend the Anglican church of St George in Campo San Vio. Among the area's attractions are the wide-embracing lagoon views, both from the eastern tip near the Salute and from the Zattere across to the island of Giudecca. West of the Accademia, the *sestiere* is more vibrant, with the busy Campo Santa Margherita as its attractive focal point. Further west, the shabbier area around the beautiful church of San Nicolò dei Mendicoli was originally the home of fishermen and sailors. The Dorsoduro plays host to several major collections of art, notably the Accademia gallery and the Peggy Guggenheim Collection of 20th-century art. The churches are also rich repositories of paintings and sculpture: San Sebastiano has fine paintings by Paolo Veronese; the Scuola Grande dei Carmini and the church of the Gesuati have ceilings painted by Giambattista Tiepolo.

Squero di San Trovaso, the gondola boatyard

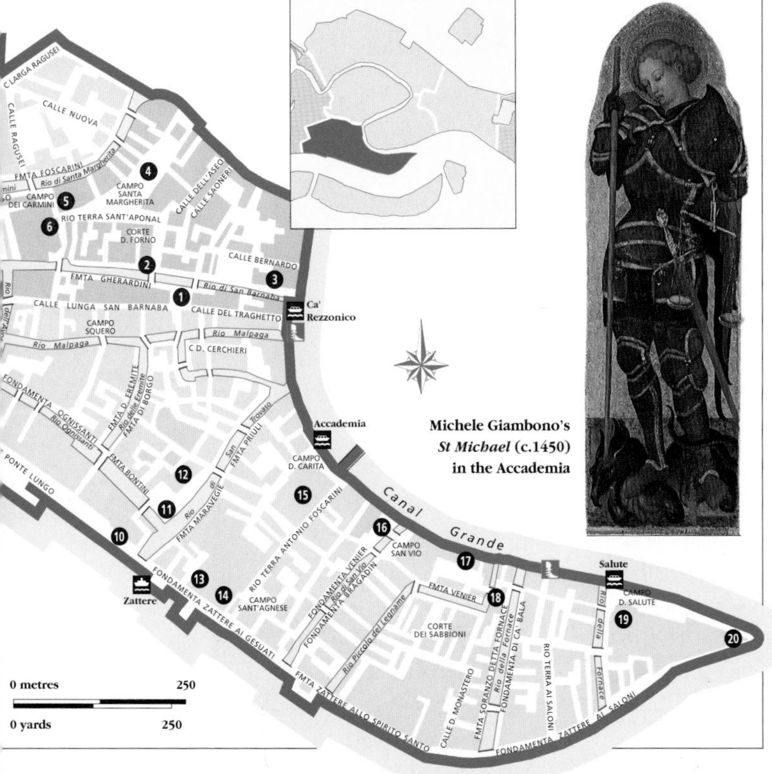

Michele Giambono's *St Michael* (c.1450) in the Accademia

Street-by-Street: Dorsoduro

Reliefs on a house at Ponte Trovaso

BETWEEN THE IMPOSING palaces on the Grand Canal and the Campo Santa Margherita lies an almost silent neighbourhood of small squares and narrow alleys. The delightful Rio San Barnaba is best appreciated from the Ponte dei Pugni, near the barge selling fruit and vegetables. The Rio Terrà canal, though architecturally uninspiring, has a fascinating mask shop and some cafés that are lively at night-time. All roads seem to lead to Campo Santa Margherita, the heart of Dorsoduro. The square bustles with activity, particularly in the morning when the market stalls are functioning.

★ Scuola Grande dei Carmini
Tiepolo painted nine ceiling panels for the Scuola in 1739–44. The central panel features the Virgin and St Simeon Stock 5

Palazzo Zenobio has been an Armenian college since 1850. Occasionally visitors can see the sumptuous 18th-century ballrooom.

Santa Maria dei Carmini
The church's oldest feature is the Gothic side porch with fragments of decorative Byzantine reliefs 6

KEY

— — — Suggested route

0 metres 50

0 yards 50

STAR SIGHTS

- **★ Scuola Grande dei Carmini**
- **★ Ca' Rezzonico**
- **★ San Barnaba**

Fondamenta Gherardini runs beside the Rio San Barnaba, one of the prettiest canals in the *sestiere.*

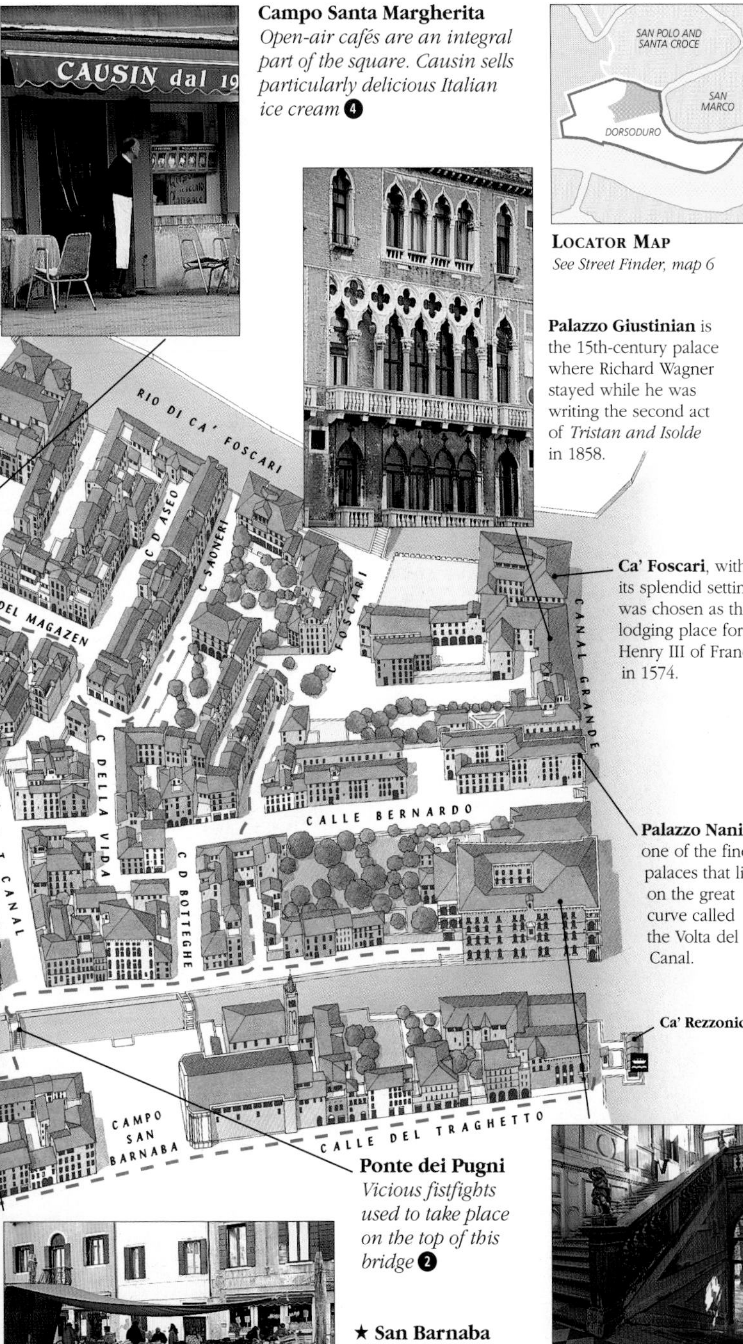

Campo Santa Margherita
Open-air cafés are an integral part of the square. Causin sells particularly delicious Italian ice cream ❹

Locator Map
See Street Finder, map 6

Palazzo Giustinian is the 15th-century palace where Richard Wagner stayed while he was writing the second act of *Tristan and Isolde* in 1858.

Ca' Foscari, with its splendid setting, was chosen as the lodging place for Henry III of France in 1574.

Palazzo Nani is one of the fine palaces that lie on the great curve called the Volta del Canal.

Ponte dei Pugni
Vicious fistfights used to take place on the top of this bridge ❷

★ San Barnaba
A floating barge crammed with crates of fruit and vegetables lends a colourful note to the area ❶

★ Ca' Rezzonico
The grand stairway has two putti, symbolizing winter and autumn ❸

Campo San Barnaba ❶

Map 6 D3. *Ca' Rezzonico.*

THE PARISH OF San Barnaba, with its canalside square at the centre, was known in the 18th century as the home of impoverished Venetian patricians. These *barnabotti* were attracted by the cheap rents, and while some relied on state support or begging, others worked as bankers in the State gambling house.

Today the square and canal, with its laden vegetable barge, are quietly appealing. The church is fairly unremarkable, apart from a Tiepolesque ceiling and a *Holy Family* attributed to Paolo Veronese.

Ponte dei Pugni ❷

Fondamenta Gherardini. **Map** 6 D3. *Ca' Rezzonico.*

VENICE HAS SEVERAL Ponti dei Pugni ("bridges of fists"), but this is the most famous. Spanning the peaceful Rio San Barnaba, the small bridge is distinguished by two pairs of footprints set in white stone on top of the bridge. These mark the starting positions for the fights which traditionally took place between rival factions. Formerly there were no balustrades and contenders hurled each other straight into the water. The battles became so bloodthirsty that they were banned in 1705.

Boats and barges moored along the Rio San Barnaba

Tiepolo's *New World* fresco, part of a series in Ca' Rezzonico

Ca' Rezzonico ❸

Fondamenta Rezzonico 3136.
Map 6 E3. *(041) 241 85 06.*
Ca' Rezzonico. Apr–Sep: 10am–5pm; Oct–Mar: 10am–4pm daily except Fri. 1 Jan, 1 May, 25 Dec.

THIS RICHLY furnished Baroque palace is one of the most splendid in Venice. It is also one of the few palaces on the Grand Canal, or indeed anywhere in the city, which opens its doors to the public. Since 1934 it has housed the museum of 18th-century Venice, its rooms furnished with frescoes, paintings and period pieces taken from other local palaces or museums.

The building was begun by Baldassare Longhena (architect of La Salute, *see p135*) in 1667, but the funds of the Bon family, who commissioned it, ran dry before the second floor was started. In 1712, long after Longhena's death, the unfinished palace was bought by the Rezzonicos, a family of merchants-turned-bankers from Genoa. A large portion of the Rezzonico fortune was spent on the purchase, construction and decoration of the palace. By 1758 it was in a fit state for the Rezzonicos to throw the first of the huge banquets and celebratory parties for which they later became renowned.

In 1888 the palace was bought by the poet Robert Browning and his son, Pen, who was married to an American heiress. Browning spoke of the "gaiety and comfort of the enormous rooms" but had little time to enjoy them. In 1889 he died of bronchitis.

The outstanding attraction in the palace today is Giorgio Massari's ballroom, which occupies the entire breadth of the building. It has been beautifully restored and is embellished with gilded chandeliers, carved furniture by Andrea Brustolon and a ceiling with *trompe l'oeil* frescoes. Three rooms between the ballroom and Grand Canal side of the palace have ceilings with frescoes by Giambattista Tiepolo including, in the Sala della Allegoria Nuziale, his lively *Nuptial Allegory* (1758).

Allegory of Strength, Andrea Brustolon

Eighteenth-century paintings occupy the *piano nobile* (second floor). A whole room is devoted to Pietro Longhi's portrayals of everyday Venetian life. Other paintings worthy of note are Francesco Guardi's *Ridotto* (1748) and *Nuns' Parlour* (1768), and one of the few Canalettos in Venice, his *View of the Rio dei Mendicanti* (1725). Giandomenico Tiepolo's fascinating series of frescoes painted for his villa at Zianigo (1770–1800) are also to be found here. On the floor above is a reconstructed 18th-century apothecary's shop and a puppet theatre.

Campo Santa Margherita ❹

Map 6 D2. *Ca' Rezzonico.*

THE SPRAWLING square of Santa Margherita, lined with houses from the 14th and 15th centuries, is the hub of western Dorsoduro. Market stalls, off-beat shops, and cafés attract a cross-section of the community. The fish stalls sell live eels and lobster, the *erborista* alternative medicine, and the bakers some of the tastiest loaves in Venice.

Santa Margherita, now no longer a church and under restoration, lies off the square at the northern end. Visitors can see sculptural fragments from the original 18th-century church, including dragons and gargoyles, on the truncated campanile in the square and on the house beside it. The Scuola dei Varotari (of the tanners), the isolated building in the centre of the square, has a faded relief of the Madonna della Misericordia protecting the tanners.

A 15th-century carving of Santa Margherita and the dragon

Scuola Grande dei Carmini ❺

Campo Carmini. **Map** 5 C2. *(041) 528 94 20. Ca' Rezzonico.* *9am–noon, 3–6pm daily except Sun. 1 May, 25 Dec.*

THE HEADQUARTERS of the Carmelite confraternity was built beside their church in 1663. In the 1740s Giambattista Tiepolo was commissioned to decorate the ceiling of the *salone* (hall) on the upper floor. These nine ceiling paintings so impressed the Carmelites that Tiepolo was promptly made an honorary member of the brotherhood.

Scuola Grande dei Carmini

The ceiling (which is best viewed with the mirrors that are supplied) shows *St Simeon Stock Receiving the Scapular of the Carmelite Order from the Virgin.* The scapular, two strips of cloth hung over the shoulders and tied by string, was widely believed at the time to protect the wearer from the pains of purgatory after death. The Carmelites honoured St Simeon Stock because he re-established the order in Europe after its expulsion from the Holy Land in the 13th century.

Santa Maria dei Carmini ❻

Campo Carmini. **Map** 5 C3. *(041) 522 65 53. Ca' Rezzonico or San Basilio. 7:30am–noon, 3–7pm Mon–Sat, 8:30am–noon, 4:30–7pm Sun.*

KNOWN ALSO as Santa Maria del Carmelo, this church was built in the 14th century but has since undergone extensive alterations.

The most prominent external feature is the lofty campanile, whose perilous tilt was skilfully rectified in 1688. The interior is large, sombre and richly decorated. The arches of the nave are adorned with gilded wooden statues, and a series of paintings illustrating the history of the Carmelite Order.

There are two interesting paintings in the church's side altars. Cima da Conegliano's *Adoration of the Shepherds* (c.1509) is in the second altar on the right (coins in the light meter are essential). In the second altar on the left is Lorenzo Lotto's *St Nicholas of Bari with Saints Lucy and John the Baptist* (c.1529). This painting demonstrates the artist's religious devotion, personal sensitivity and his love of nature. On the right-hand side of this highly detailed, almost Dutch-style landscape, there is a tiny depiction of St George killing the dragon.

SCUOLE

The *scuole* were peculiarly Venetian institutions. Founded mainly in the 13th century, they were lay confraternities existing for the charitable benefit of the neediest groups of society, the professions or resident ethnic minorities (such as the Scuola dei Schiavoni, *see p118*). Some became extremely rich, spending large sums on buildings and paintings, often to the disadvantage of their declared beneficiaries.

Upper Hall of the Scuola Grande dei Carmini

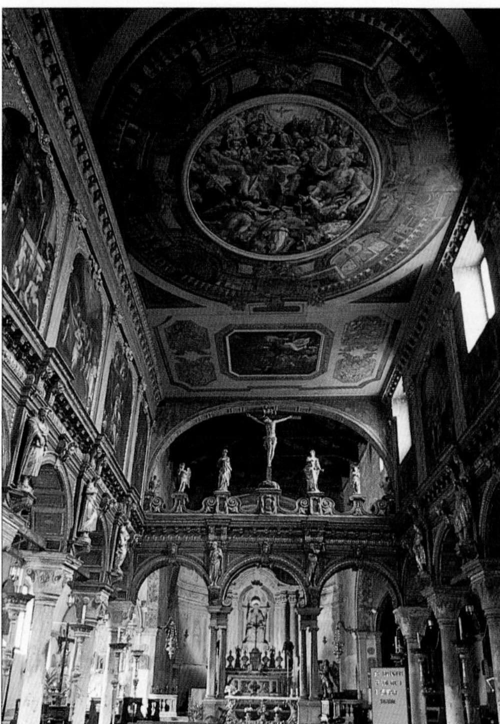

Nave of San Nicolò dei Mendicoli, one of the oldest churches in Venice

San Nicolò dei Mendicoli ❼

Campo San Nicolò. **Map** 5 A3. *(041) 528 59 52.* *San Basilio.* *10am–noon, 4–6pm Mon–Sun.*

Contrasting with the remote and rundown area which surrounds it, this church remains one of the most charming and delightful in Venice. Originally constructed in the 12th century, it has been rebuilt extensively over the years; the little porch on the north flank is 15th century.

Thanks to the Venice in Peril Fund, in the 1970s the church underwent one of the most comprehensive restoration programmes since the floods of 1966 *(see p50)*. The floor, which was 30 cm (1 ft) below the level of the canals, was rebuilt and raised slightly to prevent further damage, the roofs and lower walls were reconstructed, and paintings and statues restored. The interior is richly embellished, particularly the nave with its 16th-century gilded wood statues. On the upper walls is a series of paintings of the life of Christ by Alvise dal Friso and other pupils of Veronese.

Angelo Raffaele ❽

Campo Angelo Raffaele. **Map** 5 B3. *(041) 522 85 48.* *San Basilio.* *7:30am–noon, 4–6pm Mon–Sat, 7:30am–12:30pm, 5–6pm Sun & public hols.*

The saving grace of this dilapidated 17th-century church is the series of panel paintings on the organ balustrade. These were executed in 1749 by Antonio Guardi, brother of the more famous Francesco. They tell the tale of Tobias, the blind prophet cured by the archangel Raphael, after whom the church is named.

San Sebastiano ❾

Campo San Sebastiano. **Map** 5 C3. *San Basilio.* *2:30–5:30pm daily.*

Should you be lucky enough to find this 16th-century church open, you will see one of the most colourful and homogeneous interiors of Venice. This is thanks to the artist Veronese who, from 1555 to 1560 and again in the 1570s, was commissioned to decorate the sacristy ceiling, the nave ceiling, the frieze, the east end of the choir, the high altar, the doors of the organ panels and the chancel – in that order. The paintings, which are typical of Veronese, are rich and radiant, with sumptuous costumes and colours. Among the finest of his works are the three ceiling paintings which tell the story of Esther, Queen of Xerxes I of Persia, who brought about the deliverance of the Jewish people. Appropriately, the artist is buried in San Sebastiano, alongside the organ.

Zattere ❿

Map 5 C4. *Zattere or San Basilio.*

Stretching along the southern part of the *sestiere*, the Zattere is the long quayside looking across to the island of Giudecca. From the 11th to 15th centuries, the city held the salt monopoly for the region, and the quays' name may derive from the

San Sebastiano, viewed from the bridge of the same name

Café tables laid out along the Zattere

floating rafts *(zattere)* where cargo was offloaded. On a sunny day it is a pleasure to sit at a waterside café here, looking across to the Church of the Redentore *(see p154)* or watching the waterbuses as they cross back and forth between the shores.

Squero di San Trovaso ⓫

Rio San Trovaso. **Map** 6 D4.
Zattere. ***No public access.***

THIS IS ONE of the few surviving gondola workshops in Venice *(see pp28–9)* and the most picturesque. Its Tyrolean look dates from the days when craftsmen came down from the Cadore area of the Dolomites *(see p215)*.

It is not open to the public, but from the far side of the Rio San Trovaso you can often watch the upturned gondolas being given their scraping and tarring treatment. You may see a new one under construction, but nowadays only around ten are made each year.

San Trovaso ⓬

Campo San Trovaso. **Map** 6 D4.
(041) 522 21 33. Zattere or Accademia. 8–11am Mon–Sat, 3–6pm Mon–Fri (7pm Sat); 8:30am–1pm Sun.

THE CHURCH OF Santi Gervasio e Protasio, which in the eccentric Venetian dialect is slurred to San Trovaso, was built in 1590. Unusually it has two identical façades, one overlooking a canal, the other a quiet square. The church stood on neutral ground between the parishes of the rival factions of the Castellani and Nicolotti families, and tradition has it that this necessitated a separate entrance for each party.

The interior houses some late paintings by Jacopo Tintoretto, and there are two notable works of art worth seeking out. Michele Giambono's 15th-century Gothic painting, *St Chrysogonus on Horseback*, is situated in the chapel on the right of the chancel, and exquisite marble reliefs of angels with instruments decorate the altar of the Clary chapel opposite.

Santa Maria della Visitazione ⓭

Fondamenta delle Zattere. **Map** 6 E4.
Zattere. for restoration.

SITUATED BESIDE the Gesuati, this deconsecrated Renaissance church was built between 1494 and 1524 by the Order of the Gesuati. Part of it is used as a school and there is not much to see apart from a fine wooden ceiling painted by 16th-century Umbrian and Tuscan artists. The exterior *bocca di leone* to the right of the façade is one of several "lion's mouth" denunciation boxes surviving from the rule of the Council of Ten *(see p42)*; this one was used to complain about the state of the streets.

Gesuati ⓮

Fondamenta delle Zattere. **Map** 6 E4.
(041) 523 06 25. Zattere.
8am–noon, 5–7pm daily.

NOT TO BE CONFUSED with the Gesuiti in northern Venice *(see p142)*, this church was built by the Dominicans, who took possession of the site in the 17th century when the Gesuati Order was suppressed. Work began in 1726 and the stately façade reflects that of Palladio's great Redentore church across the Giudecca. It is the most conspicuous landmark of the long Zattere quayside. Unlike the Redentore, the interior of the church is richly decorated. Tiepolo's frescoed ceiling, *The Life of St Dominic* (1737–39) demonstrates the artist's mastery of light and colour. Equally impressive (and far easier to see) is his *Virgin with Saints* (1740), situated in the first chapel on the right. The church also boasts two altar paintings by Sebastiano Ricci and Giambattista Piazzetta.

Gesuati façade statue

Squero di San Trovaso, where gondolas are given a facelift

Accademia ⓯

Exterior detail of the Accademia

THE LARGEST COLLECTION of Venetian art in existence, the Gallerie dell'Accademia, is housed in three former religious buildings. The basis of the collection was the Accademia di Belle Arti, founded in 1750 by the painter Giovanni Battista Piazzetta. In 1807 Napoleon moved the academy to these premises, and the collection was greatly enlarged by works of art from churches and monasteries he suppressed.

Ceiling Sketch
Tiepolo's The Translation of the Holy House to Loreto *(c.1742) was a sketch for the ceiling of the Scalzi church* (see p145).

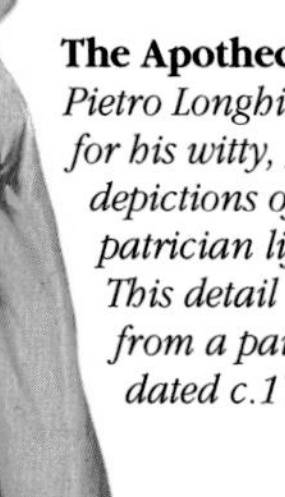

The Apothecary's Shop
Pietro Longhi is best known for his witty, gently satirical depictions of domestic patrician life in Venice. This detail comes from a painting dated c.1752.

KEY TO FLOORPLAN

- Byzantine and International Gothic
- Early Renaissance
- High Renaissance
- Baroque, genre and landscapes
- Ceremonial paintings
- Temporary exhibitions
- Non-exhibition space

★ **Cycle of St Ursula** *(1495–1500) (detail)*
The Arrival of the English Ambassadors *is one of Vittore Carpaccio's eight paintings chronicling the tragic story of St Ursula.*

Entrance

The former Church of Santa Maria della Carità was rebuilt by Bartolomeo Bon in the mid 15th century.

The inner courtyard was designed by Andrea Palladio.

Audio-visual display room

The Stealing of St Mark
Jacopo Tintoretto's painting of 1562 shows the Christians of Alexandria abducting the body of St Mark, which was about to be burnt by the pagans.

Visitors' Checklist

Campo della Carità. **Map** 6 E3. (041) 522 22 47. Accademia. 9am–7pm Mon–Sat, 9am–2pm Sun & public hols. **Last admission** 30 min before closing. 1 Jan, 1 May, 25 Dec.

★ **The Tempest** *(c.1507)*
In his enigmatic landscape, Giorgione was probably indulging his imagination rather than portraying a specific subject.

★ **Coronation of the Virgin**
Paolo Veneziano's polyptych (1325) has a central image of the Virgin surrounded by a panoply of religious scenes. This detail shows episodes from the Life of St Francis.

Gallery Guide

The paintings are housed on one floor divided into 24 rooms or sections. Where the rooms are numbered (some of them on the inside walls) they are given Roman numerals. Restoration work is ongoing, so be prepared for absent paintings or whole sections closed off. The paintings are dependent on natural light, so in order to see them at their best you should try to visit the Accademia on a bright morning.

Star Paintings

- ★ **Cycle of St Ursula by Carpaccio**
- ★ **Coronation of the Virgin by Veneziano**
- ★ **The Tempest by Giorgione**

Exploring the Accademia's Collection

SPANNING FIVE CENTURIES, the fascinating collection of paintings in the Accademia provides a complete spectrum of the Venetian school, from the medieval Byzantine period through the Renaissance to the Baroque and Rococo *(see pp26–7)*. The order is more or less chronological, with the exception of the final rooms, which take you back to the Renaissance.

***Portrait of a Gentleman* (c.1525) by Lorenzo Lotto (detail)**

BYZANTINE AND INTERNATIONAL GOTHIC

ROOM 1 SHOWS the influence of Byzantine art on the early Venetian painters. Paolo Veneziano, the true founder of the Venetian school, displays a blend of both western and eastern influences in his sumptuous *Coronation of the Virgin* (1325). The linear rhythms are quite unmistakably Gothic, yet the overall effect and the glowing gold background are distinctly Byzantine.

In the same room, *Coronation of the Virgin* (1448) by Michele Giambono shows the influence of International Gothic style, which was brought to Venice by Gentile da Fabriano and Pisanello. This particular style was characterized by delicate naturalistic detail, as typified by the birds and animals in the foreground of Giambono's painting.

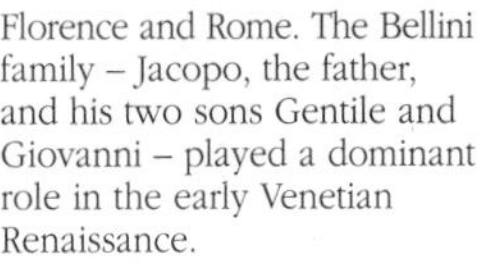

***Coronation of the Virigin* (c.1448) by Michele Giambono**

EARLY RENAISSANCE

THE RENAISSANCE came late to Venice, but by the second quarter of the 15th century it had transformed the city into an art centre rivalling those of Florence and Rome. The Bellini family – Jacopo, the father, and his two sons Gentile and Giovanni – played a dominant role in the early Venetian Renaissance.

Central to Venetian art in the 15th century was the *Sacra Conversazione*, where the Madonna is portrayed in a unified composition with saints. Giovanni Bellini's altarpiece for San Giobbe (c.1487) in Room 2 is one of the finest examples. Giovanni, the younger Bellini, was profoundly influenced by the controlled rational style and mastery of perspective in the works of his brother-in-law, Andrea Mantegna, whose work *St George* (c.1460) is in Room 4. To Mantegna's rationality and harsh realism Giovanni added humanity. This is seen in his Madonna paintings (Rooms 4 and 5), which are masterpieces of warmth and harmony. Outstanding examples are *The Madonna and Child between St Catherine and St Mary Magdalene* (c.1490) in Room 4; *Madonna of the Little Trees* (c.1487) and *Madonna and Child with John the Baptist and a Saint* (c.1505) in Room 5.

The inventive young artist Giorgione was influenced by Bellini, but went way beyond his master in his development of the landscape to create mood. In the famous, atmospheric *Tempest* (c.1507) in Room 5, this treatment of the landscape and the use of the figures to intensify that mood was an innovation adopted in Venetian painting of the 16th century and beyond.

Out on a limb from the main 16th-century Venetian tradition was the enigmatic Lorenzo Lotto, best known for portraits conveying moods of psychological unrest. His melancholic *Portrait of a Gentleman* (c.1525) in Room 7 is a superb example. More in the Venetian tradition, Palma il Vecchio's sumptuously coloured *Sacra Conversazione* in Room 8, painted around the same time, shows the unmistakable influence of the early work of Titian.

HIGH RENAISSANCE

OCCUPYING an entire wall of Room 10, the monumental *Feast in the House of Levi* by Paolo Veronese (1573) was originally commissioned

***Feast in the House of Levi* (detail)**

as *The Last Supper*. However, the hedonistic detail in the painting, such as the drunkard and the dwarfs, was not well received and Veronese found himself before the Inquisition. Ordered to eliminate the profane content of the picture, he simply changed the title.

Jacopo Tintoretto made his reputation with *The Miracle of the Slave* (1548), which is also in Room 10. The painting shows his mastery of the dramatic effects of light and movement. This was the first of a series of works painted for the Scuola Grande di San Marco *(see p114)*. In the next room, Veronese's use of rich colour is best admired in the *Mystical Marriage of St Catherine* (c.1575).

***Healing of the Madman* (c.1496) by Vittore Carpaccio**

Baroque, Genre and Landscapes

***The Rape of Europa* (1740–50) by Francesco Zuccarelli (detail)**

Venice suffered from a lack of native Baroque painters, but a few non-Venetians kept the Venetian school alive in the 17th century. The most notable among these was the Genoese Bernardo Strozzi (1581–1644). The artist was a great admirer of the work of Veronese, as can be seen in his *Feast at the House of Simon* (1629) in Room 11. Also represented in this room is Giambattista Tiepolo, the greatest Venetian painter of the 18th century.

The long corridor (12) and the rooms which lead from it are largely devoted to light-hearted landscape and genre paintings from the 18th century. Among them are pastoral scenes by Francesco Zuccarelli, works by Marco Ricci, scenes of Venetian society by Pietro Longhi and a view of Venice by Canaletto (1763). This was the painter's entry for admission to the Accademia, and is a fine example of his sense of perspective.

Ceremonial Paintings

Rooms 20 and 21 return to the Renaissance, featuring two great cycles of paintings from the late 16th century. The detail in these large-scale anecdotal canvases provide a fascinating glimpse of the life, customs and appearance of Venice at the time. Room 20 houses *The Stories of the Cross* by Venice's leading artists, commissioned by the Scuola of San Giovanni Evangelista *(see p104)*. Each one depicts an episode of the relic of the Holy Cross, which the kingdom of Cyprus donated to the Scuola. In *The Procession in St Mark's Square* (1496) by Gentile Bellini, you can compare the square with how it looks today. Another, Vittore Carpaccio's *Healing of the Madman* (1496), shows the Rialto bridge which collapsed in 1524.

The second series, minutely detailed *Scenes from the Legend of St Ursula* (1490s) by Carpaccio in Room 21, provides a brilliant kaleidoscope of life. Mixing reality and imagination, Carpaccio relates the episodes from the life of St Ursula using settings and costumes of 15th-century Venice.

Sala dell'Albergo

When the Scuola della Carità became the site of the Academy of Art in the early 19th century, the Scuola's *albergo* (where students lodged) retained its original panelling and 15th-century ceiling. The huge *Presentation of the Virgin* (1538) is one of the surprisingly few Titians in the gallery, and was painted for this very room. The walls are also adorned with a grandiose triptych (1446) by Antonio Vivarini and Giovanni d'Alemagna.

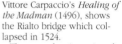

Detail from Titian's *Presentation of the Virgin* (1538)

Cini Collection 16

Palazzo Cini, San Vio 864. **Map** 6 E4. *521 07 55. Accademia. 10am–6pm Tue–Sun.*

THE PALAZZO CINI belonged to Count Vittorio Cini (1884–1977), a collector and patron of the arts. In 1951–56 he restored San Giorgio Maggiore *(see p95)* and created the Cini Foundation as a memorial to his son, who was killed in an air crash in 1949.

The collection displayed here includes china, ivories, books, illuminated manuscripts, miniatures, porcelain and furniture, but the outstanding works of art are the Tuscan Renaissance paintings that Cini collected. These include works by or attributed to Botticelli, Piero di Cosimo, Piero della Francesca, Filippo Lippi and Pontormo.

The Cini Collection has been open to the public since 1984 and is supposedly accessible most afternoons during the summer. However, in many seasons it does not open its doors at all.

***Madonna col Bambino* (c.1437) by Filippo Lippi, the Cini Collection**

Peggy Guggenheim Collection 17

Palazzo Venier dei Leoni, San Gregorio 701. **Map** 6 F4. *(041) 520 62 88. Accademia. 11am–6pm Wed–Mon. 25 Dec.*

INTENDED AS A four-storey palace, the 18th-century Palazzo Venier dei Leoni in fact never rose beyond the ground floor – hence its nickname, *Il Palazzo Nonfinito* (The Unfinished Palace). In 1949 the building was bought as a home by the American millionairess Peggy Guggenheim (1898–1979), a collector, dealer and patron of the arts. A perspicacious and high-spirited woman, she befriended and furthered the careers of many innovative abstract and surrealist artists. One was Max Ernst, who was the second of her husbands.

The collection consists of 200 paintings and sculptures, representing almost every modern art movement. The dining room has notable Cubist works of art including *The Poet* by Pablo Picasso. An entire room is devoted to Jackson Pollock, who was a Guggenheim discovery. Other artists represented are Miró, de Chirico, Magritte, Kandinsky, Mondrian and Malevich.

***Interno Olandese II* (c.1928) by Joan Miró**

Sculpture is laid out in the house and garden. One of the most elegant works is Constantin Brancusi's *Maiastra* (1912). The most provocative piece is Marino Marini's *Angelo della Città* (Angel of the Citadel, 1948), a prominently displayed man sitting on a horse, erect in all respects. Embarrassed onlookers avert their gaze to enjoy views of the Grand Canal.

The Guggenheim is one of the most visited sights of the city. The light-filled rooms and the large modern canvases provide a striking contrast to the Renaissance paintings which are the main attraction in Venetian churches and museums. A bonus for English speakers is the team of assistants, who are usually arts graduates from Britain.

***Maiastra* by Constant Brancusi**

There are plans by the Guggenheim to acquire the splendidly located customs house at Punta della Dogana. If fulfilled, it will enable many works currently in storage to see the light of day.

Façade of the Palazzo Venier dei Leoni, the home of the Peggy Guggenheim Collection of modern art

Campiello Barbaro ⓲

Map 6 F4. Salute.

AN ENCHANTING little square, the Campiello Barbaro is shaded by trees and flanked on one side by the wisteria-clad walls of Ca' Dario. It is hard to believe the stories of murder, bankruptcy and suicide that have befallen the owners of this Grand Canal palace. The most recent was Raul Gardini, one of Italy's best-known industrialists, who shot himself in 1992.

The ill-fated Ca' Dario, which backs on to Campiello Barbaro

Santa Maria della Salute ⓳

Campo della Salute. **Map** 7 A4.
(041) 522 55 58. *Salute.*
9am–noon, 3–5:30pm daily.

THE GREAT BAROQUE church of Santa Maria della Salute, standing at the entrance of the Grand Canal, is one of the most imposing architectural landmarks of Venice. Henry James likened the church to "some great lady on the threshold of her salon . . . with her domes and scrolls, her scalloped buttresses and statues forming a pompous crown and her wide steps disposed on the ground like the train of a robe". The church was built in thanksgiving for the deliverance of the city from the plague of 1630, hence the name *Salute*, meaning health and salvation. Each November, in celebration, *(see p35)*, worshippers approach across a bridge of boats which span the mouth of the Grand Canal for the occasion. Baldassare Longhena started the church in 1630 at the age of 32, and worked on it for the rest of his life. It was not completed until 1687, some five years after his death.

The Baroque church of Santa Maria della Salute viewed from across the Grand Canal

The interior is comparatively sober. It consists of a large octagonal space below the cupola and six chapels radiating from the ambulatory. The large domed chancel and grandiose high altar dominate the view from the main door. The altar's sculptural group by Giusto Le Corte represents the Virgin and Child giving protection to Venice from the plague. The best paintings inside Santa Mara della Salute are in the sacristy to the left of the altar: Titian's early altarpiece of *St Mark Enthroned with Saints Cosmos, Damian, Roch and Sebastian* (1511–12) and his dramatic ceiling paintings of *Cain and Abel, The Sacrifice of Abraham and Isaac* and *David and Goliath* (1540–49). The *Wedding at Cana* (1551) on the wall opposite the entrance is a major work by Jacopo Tintoretto.

Interior of the Salute showing the octagonal core of the church

Dogana di Mare ⓴

Map 7 A4. *Salute.*

THIS EASTERN promontory of the Dorsoduro provides a panorama which embraces the Riva degli Schiavoni, the island of San Giorgio Maggiore and the eastern section of Giudecca. The *dogana di mare*, or sea customs post, was originally built in the 15th century to inspect the cargo of ships which were intending to enter Venice. The customs house you see today was constructed in the late 17th century and replaced a tower which originally guarded the entrance to the Grand Canal. On the corner tower of the house two bronze Atlases support a striking golden ball with a weathervane figure of Fortuna on the top.

Sights at a Glance

Churches
Gesuiti ❼
Madonna dell'Orto ❶
Santi Apostoli ❿
San Giobbe ⓮
San Giovanni Crisostomo ❾
Santa Maria dei Miracoli ❽
San Marziale ❹
Scalzi ⓭

Streets and Squares
Campo dei Mori ❷
Fondamente Nuove ❺
Fondamenta della Sensa ❸

Historic Buildings
Oratorio dei Crociferi ❻
Palazzo Labia ⓬

Art Gallery
Ca' d'Oro p144 ⓫

Historic Area
The Ghetto ⓯

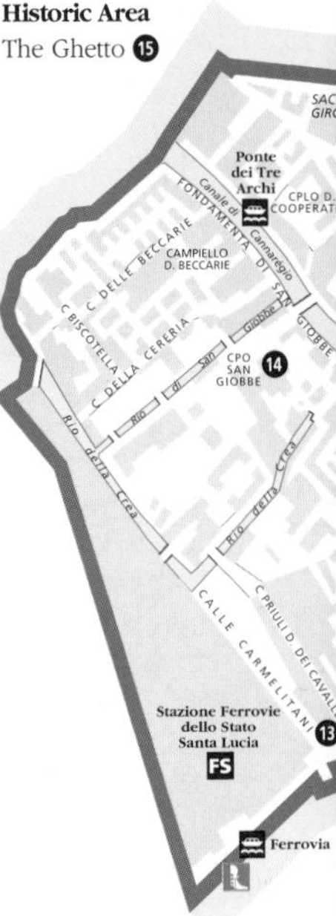

0 metres 250

0 yards 250

◁ **The entrance to Tintoretto's house in Fondamenta dei Mori**

Cannaregio

The city's most northerly *sestiere*, Cannaregio, stretches in a large arc from the 20th-century railway station in the west to one of the oldest quarters of Venice in the east. The northern quays look out towards the islands in the lagoon, while to the south the *sestiere* is bounded by the upper sweep of the Grand Canal.

The name of the quarter derives either from the Italian *canne*, meaning canes or reeds, which grew here centuries ago, or perhaps from "Canal Regio" or Royal Canal – the former name of what is now the Canale di Cannaregio. This waterway was the main entry to Venice before the advent of the rail link with the mainland. Over a third of the city's population lives in Cannaregio. For the most part it is an unspoilt area, divided by wide canals, crisscrossed by alleys and characterized by small stores, simple bars and the artisans' workshops. One of the prettiest and most remote quarters is in the north, near the church of Madonna dell'Orto and around Campo dei Mori.

Hanukah lamp in the Ghetto

Tourism is concentrated along two main thoroughfares: the Lista di Spagna and the wide Strada Nova, both on the well-worn route from the station to the Rialto. Just off this route lies the world's oldest ghetto. Though now largely abandoned by Venetian Jews, this is historically the most fascinating quarter of Cannaregio.

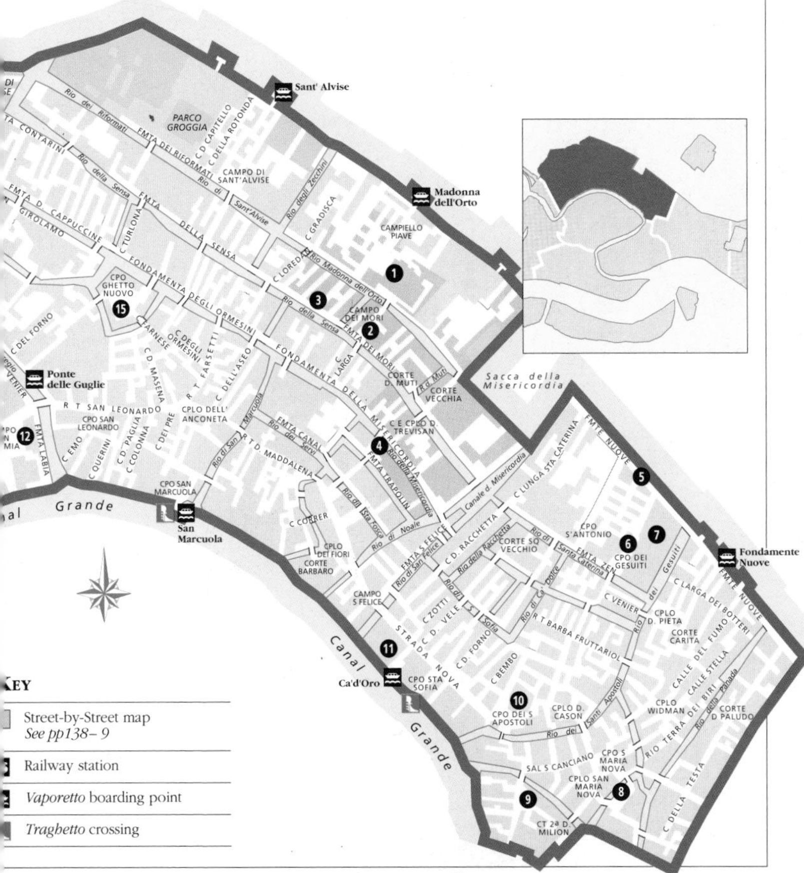

Street-by-Street: Cannaregio

Channel marker in the lagoon

SURPRISINGLY FEW tourists find their way to this unspoilt quarter of northern Cannaregio. This is the more humble, peaceful side of Venice, where clean washing is strung over the waterways and the streets are flanked by the softly crumbling façades of shuttered houses. Along the wide *fondamente*, the little shops and stores stock basic groceries and the bars are always crowded with local Venetians. The quarter's cultural highlight is the lovely Gothic church of Madonna dell'Orto, Tintoretto's parish church.

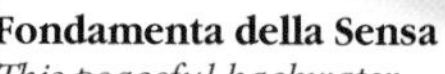

Fondamenta della Sensa
This peaceful backwater, with its peeling façades, is undisturbed by the rigours of tourism ❸

★ Madonna dell'Orto
One of the finest Gothic churches in Venice, Madonna dell'Orto has a richly decorated façade and a wealth of works by Tintoretto ❶

★ Campo dei Mori
This square is named after the stone statues of three Moors (Mori) which are carved on its walls ❷

Tintoretto lived with his family in this house, No. 3399 Fondamenta dei Mori, from 1574 until his death in 1594.

San Marziale
Ceiling paintings by Sebastiano Ricci (1700–25) and a bizarre Baroque altar adorn this Baroque church ❹

KEY

– – – Suggested route

STAR SIGHTS

★ Madonna dell'Orto

★ Campo dei Mori

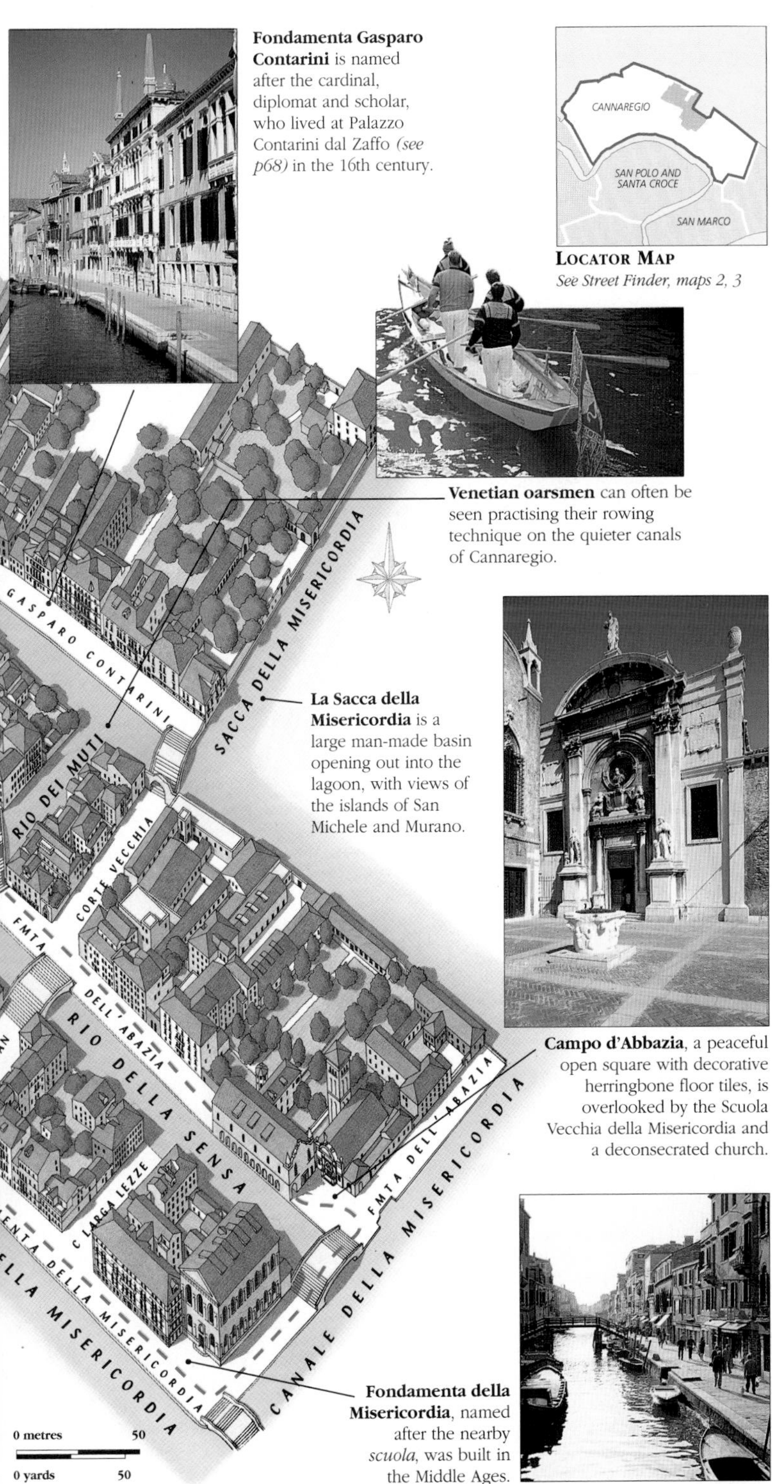

Fondamenta Gasparo Contarini is named after the cardinal, diplomat and scholar, who lived at Palazzo Contarini dal Zaffo *(see p68)* in the 16th century.

Locator Map
See Street Finder, maps 2, 3

Venetian oarsmen can often be seen practising their rowing technique on the quieter canals of Cannaregio.

La Sacca della Misericordia is a large man-made basin opening out into the lagoon, with views of the islands of San Michele and Murano.

Campo d'Abbazia, a peaceful open square with decorative herringbone floor tiles, is overlooked by the Scuola Vecchia della Misericordia and a deconsecrated church.

Fondamenta della Misericordia, named after the nearby *scuola*, was built in the Middle Ages.

The campanile of Madonna dell'Orto, crowned by an onion-shaped cupola

Madonna dell'Orto ❶

Campo Madonna dell'Orto. **Map** 2 F2. *(041) 71 99 33. Madonna dell'Orto. 9:30am–noon, 3:30–5:30pm daily.*

THIS LOVELY Gothic church is frequently referred to as the English Church in Venice for it was British funds that helped restore the building after the 1966 floods *(see p50)*. The original church, founded in the mid 14th century, was dedicated to St Christopher, patron saint of travellers, to protect the boatmen who ferried passengers to the islands in the northern lagoon. The dedication was changed and the church reconstructed in the early 15th century, following the discovery, in a nearby vegetable garden *(orto)*, of a statue of the Virgin Mary said to have miraculous powers. However, a 15th-century statue of St Christopher, newly restored, still stands above the portal.

The interior, faced almost entirely in brick, is large, light and uncluttered. The greatest treasures are the works of art by Tintoretto, who was a parishioner of the church. His tomb, which is marked with a plaque, lies in the chapel to the right of the chancel. The most dramatic of his works are the towering paintings in the chancel (1562–4). On the right wall is *The Last Judgment*, whose turbulent content caused John Ruskin's wife Effie to flee the church. In the painting *The Adoration of the Golden Calf* on the left wall, the figure carrying the calf, fourth from the left, is said to depict Tintoretto himself.

Inside the chapel of San Mauro you can see the radically restored statue of the Madonna which inspired the reconstruction of the church.

To the right of the entrance is Cima da Conegliano's magnificent painting, *St John the Baptist and Other Saints* (c.1493). The vacant space opposite belongs to Giovanni Bellini's *Madonna with Child* (c.1478) which was stolen for the third time in 1993.

TINTORETTO (1518–94)

Jacopo Robusti, nicknamed Tintoretto because of his father's occupation of silk dyer, was born, lived and died in Cannaregio. He left Venice only once in his life. A devout Christian, volatile and unworldly, his was a highly individual and theatrical style, conveyed by vivid exaggeration of light and movement, bold foreshortening and fiery, fluid brushstrokes. His remarkably prolific output has never been ascertained, but scores of his works survive, many still in the places for which they were painted. Examples of his canvases can be seen in the church of Madonna dell'Orto, the Accademia *(see pp130–33)*, and the Doge's Palace *(see pp84–9)* His crowning achievement, however, was the great series of works for the Scuola Grande di San Rocco *(see pp106–7)*.

Campo dei Mori ❷

Map 2 F3. *Madonna dell'Orto.*

ACCORDING TO popular tradition, the "Mori" were the three Mastelli brothers who came from the Morea (the Peloponnese). The brothers, who were silk merchants by trade, took refuge in Venice in 1112 and built the Palazzo Mastelli, recognizable by its camel bas relief, which once backed on to the square. Their weathered stone figures can be seen embedded in the wall of the *campo* on its eastern side. The corner figure with the makeshift rusty metal nose (added in the 19th century) is "Signor Antonio Rioba" who, like the Roman Pasquino, was the focus of malicious fun and satire. A fourth oriental merchant wearing a large turban faces the Rio della Sensa on the façade of Tintoretto's house *(see p138)*.

One of the stone Moors which gave the Campo dei Mori its name

A local trattoria on a quiet quayside by the Rio della Sensa

Fondamenta della Sensa ❸

Map 2 E2. *Madonna dell'Orto.*

When the marshy lands of Cannaregio were drained in the Middle Ages, three long, straight canals were created, running parallel to each other. The middle of these is the Rio della Sensa, which stretches from the Sacca di Sant'Alvise at its western end to the Canale della Misericordia in the east. The Fondamenta cuts through a quiet quarter of Cannaregio, where daily life goes on undisturbed by tourism. With its small grocery shops and simple local bars, the neighbourhood feels far removed from San Marco.

This is one of the poorer areas of the city, though the buildings are interspersed with fine (but neglected) palaces that once belonged to wealthy Venetians. Abbot Onorio Arrigoni lived at No. 3336 with his collection of antiques, and Palazzo Michiel (No. 3218) is an early Renaissance palace which became the French embassy.

San Marziale ❹

Campo San Marziale. **Map** 2 F3. *(041) 71 99 33. San Marcuola. 4–6pm Mon–Sat. Mass 9:30am Sun.*

A baroque church on medieval foundations, San Marziale was rebuilt between 1693 and 1721. The church is mainly visited for the ceiling frescoes by Sebastiano Ricci, a painter of the decorative Rococo style. Executed between 1700 and 1705, relatively early in Ricci's career, these bold, foreshortened frescoes already combine the Venetian tradition with flamboyant Rococo flourishes. Sadly though, the vivid colours for which Ricci was known have been sullied by decades of grime. The central painting shows *The Glory of Saint Martial*, while the side paintings relate to the image of the Virgin.

Fondamente Nuove ❺

Map 3 B3. *Fondamente Nuove.*

The Fondamente Nuove or "New Quays" are actually over 400 years old. This chain of waterside streets borders the northern lagoon for one kilometre (over half a mile), from the solitary Sacca della Misericordia to the Rio di Santa Giustina in Castello on the eastern side.

Altar of San Marziale showing a carving of the Virgin and Child

Before the construction of the quays in the 1580s, this was a desirable residential area where the air was said to be healthy and the houses had gardens sloping down to the lagoon.

One of the residents was Titian, who lived from 1531 to his death in 1576 in a now demolished house at Calle Larga dei Botteri No. 5182–3 (a plaque marks the site).

Today the quaysides are aesthetically uninspiring but they do provide splendid views of the northern lagoon and, on a clear day, the peaks of the Dolomites. The island most visible from the quays is San Michele in Isola *(see p151)*, its dark stately cypress trees rising high above the cemetery walls.

Oratorio dei Crociferi ❻

Campo dei Gesuiti. **Map** 3 B3. *(041) 521 74 11. Fondamente Nuove. Apr–Oct: 10am–1pm Fri–Sun.*

Founded in the 13th century as a hospital for returning Crusaders, the Oratorio dei Crociferi (built for the order of the Bearers of the Cross) was turned into a charitable institution for old people in the 15th century.

Between 1583 and 1591 the artist Palma il Giovane, commissioned by the Crociferi, decorated the chapel with a glowing cycle of paintings, depicting the crucial events in the history of this religious order. The paintings suffered terrible damage in the floods of 1966 *(see p50)*, but thanks to funds from overseas, were successfully restored and the chapel reopened in 1984.

The inscriptions on the walls of some of the surrounding houses in the square are those of art and craft guilds, such as silk weavers and tailors, whose works formerly occupied the buildings.

The sumptuous ceiling frescoes of the Gesuiti church

Gesuiti ❼

Campo dei Gesuiti. **Map** 3 B4. *(041) 528 65 79.* *Fondamente Nuove.* *10am–noon, 5–7pm (4–6pm winter) daily.*

THE JESUITS' close links with the papacy provoked Venetian hostility during the 17th century, and for 50 years they were refused entry to the city. However in 1714 they were given permission to build this church in the north of Venice, on the site of a 12th-century church which had belonged to the Order of the Crociferi. Consecrated as Santa Maria Assunta, the church is always referred to simply as the Gesuiti; thus it is often confused with the Gesuati in Dorsoduro *(see p129)*.

Domenico Rossi's imposing Baroque exterior gives only a hint of the opulence of the interior. The proliferation of green and white marble, carved in parts like great folds of fabric, gives the impression that the church is clothed in damask.

Titian's *Martyrdom of St Lawrence* (c.1555), above the first altar on the left, has been described by the art historian Hugh Honour as "the first successful nocturne in the history of art".

Santa Maria dei Miracoli ❽

Campo dei Miracoli. **Map** 3 B5. *(041) 528 39 03.* *Rialto.* *10am–noon, 3–6pm Mon–Sat.*

AN EXQUISITE masterpiece of the early Renaissance, the Miracoli is the favourite church of many Venetians and the one where they like to get married. Tucked away in a maze of alleys and waterways in eastern Cannaregio, it is small and somewhat elusive, with the opening hours dependent on the mood of the church custodian.

Often likened to a jewel box, the façade is decorated with various shades of marble, with fine bas-reliefs and sculpture. It was built in 1481–9 by the architect Pietro Lombardo and his sons to enshrine *The Virgin and Child* (1408), a painting believed to have miraculous powers. The picture, by Nicolò di Pietro, can still be seen above the altar. The interior of the church, which ideally should be visited when pale shafts of sunlight are streaming in through the windows, is embellished by pink, white and grey marble and crowned by a barrel-vaulted ceiling (1528) which has 50 portraits of saints and prophets. The balustrade, between the nave and the chancel, is decorated by Tullio Lombardo's carved figures of St Francis, Archangel Gabriel, the Virgin and St Clare. The screen around the high altar and the medallions of the Evangelists in the cupola spandrels are also by Lombardo. Above the main door, the choir gallery was used by the nuns from the neighbouring convent, who entered the church through an overhead gallery. The Miracoli has recently undergone a major restoration programme, which was funded by the American Save Venice organization.

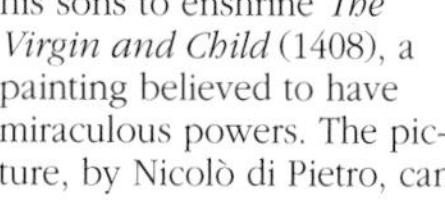

Decorative column, interior of Santa Maria dei Miracoli

SANTA MARIA DEI MIRACOLI

The façade is a harmonious tapestry of decorated panels and multi-coloured polished stone

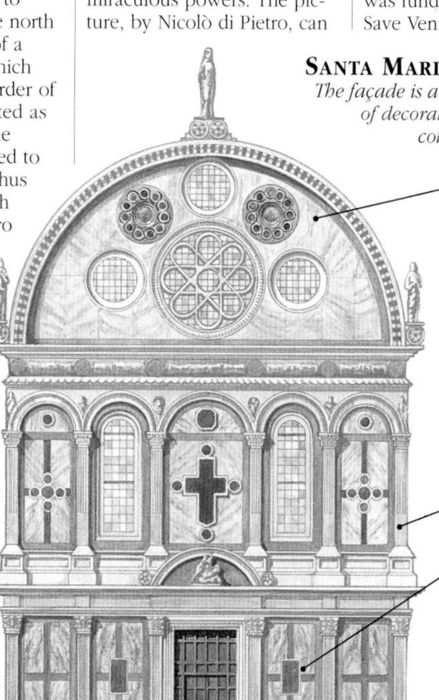

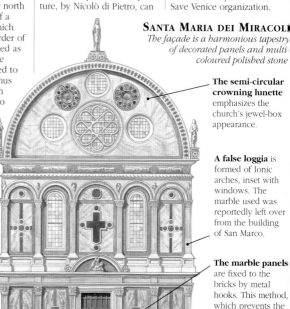

The semi-circular crowning lunette emphasizes the church's jewel-box appearance.

A false loggia is formed of Ionic arches, inset with windows. The marble used was reportedly left over from the building of San Marco.

The marble panels are fixed to the bricks by metal hooks. This method, which prevents the build-up of damp and salt water behind the panels, dates from the Renaissance.

San Giovanni Crisostomo, the last work of Mauro Coducci

San Giovanni Crisostomo ❾

Campo San Giovanni Crisostomo. **Map** 3 B5. *(041) 522 71 55.* *Rialto.* *11am–12:30pm, 3–5pm daily.*

THIS PRETTY little terracotta-coloured church lies in a bustling quarter close to the Rialto. Built between 1479 and 1504, the church was the last work of Mauro Coducci.

The interior, which is built on a Greek-cross plan, is dark and intimate. Giovanni Bellini's *St Jerome with Saints Christopher and Augustine* (1513) hangs above the first altar on the right. Influenced by Giorgione, this was probably Bellini's last painting, executed when he was in his eighties.

Sebastiano del Piombo's *St John Chrysostom and Six Saints* (1509–11), which hangs over the high altar, was also influenced by Giorgione – some believe he actually painted the figures of St John the Baptist and St Liberal.

Marco Polo

Born around 1254 in the quarter of Cannaregio near the Rialto, Marco Polo left Venice at the age of 18 for his four-year voyage to the court of the Emperor Kublai Khan. He impressed the Mongol emperor and stayed for some 20 years, working as a travelling diplomat.

Returning to Venice in 1295, he brought with him a fortune in jewels and a host of spellbinding stories about the Khan's court.

As a prisoner of war in Genoa in 1298 he compiled an account of his travels, with the cooperation of an inmate. Translated into French, this was to become *Le Livre des merveilles.* Despite the fact that many Italians disbelieved his wondrous tales of the east, the book was an instant success. His nickname became Marco Il Milione (of the million lies); hence the name of the two little courtyards where the Polo family lived: Corte Prima del Milion and Corte Seconda del Milion.

Marco Polo leaving on his travels, from a manuscript c.1338

Santi Apostoli ❿

Campo Santi Apostoli. **Map** 3 B5. *(041) 523 82 97.* *Ca' d'Oro.* *7–11:30am, 5–7pm daily (4–6:30pm Sun).*

THE CAMPO Santi Apostoli is a busy crossroads for pedestrians en route to the Rialto or the railway station. Its church is unremarkable architecturally and little remains of the 16th-century building. A notable exception, however, is the enchanting late 15th-century Renaissance Corner Chapel on the right of the nave, believed to have been designed by Mauro Coducci. The chapel contains *The Communion of St Lucy* by Giambattista Tiepolo (1748), the tomb of Marco Corner, probably by Tullio Lombardo (1511), and an inscription to Corner's daughter, Caterina Cornaro, Queen of Cyprus, who was buried here before she was moved to the Church of San Salvatore *(see p94).*

Tomb of Doge Marco Corner in Santi Apostoli (Corner Chapel)

Ca' d'Oro ⓫

See p144.

Palazzo Labia ⓬

Fondamenta Labia (entrance on Campo S Geremia). **Map** 2 D4. *(041) 524 28 12.* *Ponte Guglie.* *by appointment only. Phone 3–4pm Wed–Fri to book.* *public hols.* *concerts.*

THE LABIAS were a wealthy family of merchants from Catalonia who bought their way into the Venetian patriciate in 1646. Towards the end of the century they built their prestigious palace on the wide Cannaregio Canal, close to its junction with the Grand Canal. In 1745–50 the ballroom was frescoed by Giambattista Tiepolo. The scenes are from the life of Cleopatra but the setting is Venice, and the queen's attire is that of a 16th-century noble lady.

Passed from one owner to another the palace gradually lost all trace of its former grandeur and variously served as a religious foundation, a school and a doss-house. Between 1964 and 1992 it was owned by the Italian broadcasting network, RAI, who undertook its restoration. The frescoes can be seen by appointment or by attending a concert *(see p255).*

Ca' d'Oro ⓫

One of the great showpieces of the Grand Canal, the Ca' d'Oro (or House of Gold) is the finest example of Venetian Gothic architecture in the city. The façade, with its finely carved ogee windows, oriental pinnacles and exotic marble tracery, has an unmistakable flavour of the east. But this once gloriously embellished *palazzo* suffered many changes of fortune and there is now little inside to remind you that this was once a 15th-century palace. Since 1984 it has been home to the Giorgio Franchetti Collection.

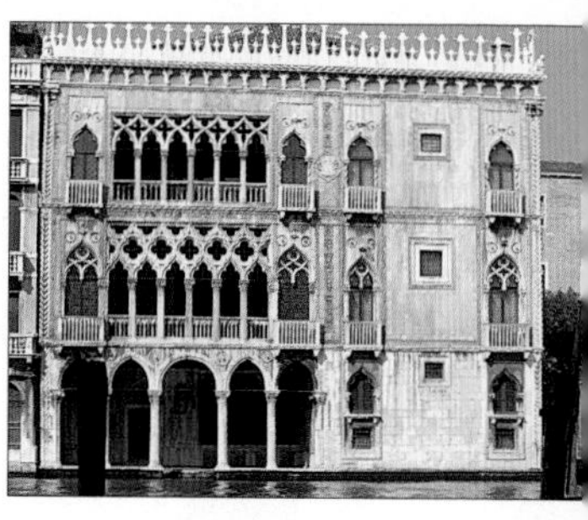

History

In 1420 the wealthy patrician, Marino Contarini, commissioned the building of what he was determined would be the most magnificent palace in the city. The decoration and the intricate carving were executed by a team of Venetian and Lombard craftsmen, and he had the façade adorned in ultramarine, gold leaf and vermilion.

In the course of the 16th century the house was remodelled by a succession of owners, and by the early 18th century was semi-derelict. In 1846 the Russian Prince Troubetzkoy bought it for the famous ballerina Maria Taglioni. Under her direction, the Ca' d'Oro suffered barbaric restoration. The open staircase was ripped out, the wellhead by Bartolomeo Bon (1427–8) was sold and much of the stonework removed. It was finally rescued by Baron Franchetti, a patron of the arts, who restored it to its former glory and bequeathed it to the state in 1915. Another restoration programme was put into action in the 1970s and the façade is once again behind scaffolding. The pretty paved courtyard, which can only be glimpsed through a gateway, contains Bon's beautifully carved wellhead. This was one of the pieces retrieved by Franchetti.

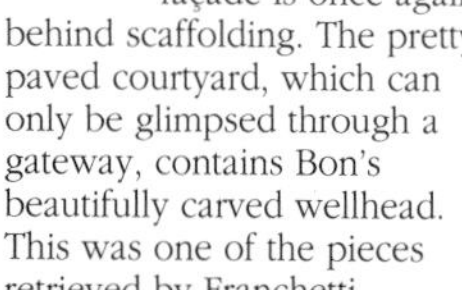

Tullio Lombardo's *Double Portrait*

***The Annunciation* (1504) by Vittore Carpaccio and assistants**

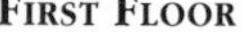

First Floor

Pride of place is given to Andrea Mantegna's *St Sebastian* (1506), the artist's last painting and Franchetti's favourite work of art. The *portego* (gallery) opening on to the Grand Canal is a showroom of sculpture. Among the finest pieces are bronze reliefs by the Paduan sculptor, Il Riccio (1470–1532), Tullio Lombardo's marble *Double Portrait* (c.1493) and Sansovino's lunette of the Madonna and Child (c.1530). Rooms to the right of the *portego* have some fine Renaissance bronzes and, among the paintings, an *Annunciation* and *Death of the Virgin* (both c.1504) by Vittore Carpaccio and assistants. A room to the left of the *portego* is devoted to non-Venetian painting, and includes Luca Signorelli's *Flagellation* (c.1480).

Second Floor

The upper floor, which is often closed for lack of personnel, houses paintings by Venetian masters, including a *Venus* by Titian, two Venetian views by Guardi, and fresco fragments by Giorgione and Titian. Other exhibits include tapestries and ceramics.

Visitors' Checklist

Canal Grande *(entrance on the Calle Ca' d'Oro).* **Map** 3 A4.
(041) 523 87 90. Ca' d'Oro. 9am–1:30pm daily. 1 Jan, 1 May, 25 Dec.

Scalzi ⓭

Fondamenta Scalzi. **Map** 1 C4.
(041) 71 51 15. *Ferrovia.*
7am–noon, 3:30–7pm daily.

BESIDE THE modern railway station *(see p58)* stands the church of Santa Maria di Nazareth, known as the Scalzi. The *scalzi* were "barefooted" Carmelite friars who came to Venice during the 1670s and commissioned their church to be built on the Grand Canal.

Designed by Baldassare Longhena, the huge Baroque interior is an over-elaboration of polychrome marble, gilded woodwork and sculptures. The 1934 ceiling painting, *The Council of Ephesus* by Ettore Tito, replaced Giambattista Tiepolo's fresco of *The Translation of the Holy House to Loreto* (1743–45), which was almost entirely destroyed by the Austrian bombardment of 24 October 1915.

San Giobbe ⓮

Campo San Giobbe. **Map** 1 C3.
(041) 524 18 89. *Ferrovia.*
10am–noon, 2–4pm daily.
If closed, ask at 620 by the church.

THE CHURCH of San Giobbe stands in a remote *campo* full of cats. The early Gothic structure of the church was modified in the 1470s by Pietro Lombardo who added Renaissance elements such as the figures of saints over the portal. The Martini chapel, second on the left, is decorated with Della Robbia-style glazed terracotta. The altarpieces by Giovanni Bellini and Vittore Carpaccio, which were removed when Napoleon suppressed the monastery of San Giobbe, are preserved in the Accademia Gallery *(pp130–33)*.

Saint by Lombardo, San Giobbe portal

The Ghetto ⓯

Map 2 E3. *Ponte Guglie.* **Museo Ebraico** Campo del Ghetto Nuovo.
(041) 71 53 59. *Ponte Guglie.*
10am–7pm (Oct–May: 10am–4:30pm) Sun–Fri. *Jewish hols.*

IN 1516 the Council of Ten *(see p42)* decreed that all Jews in Venice be confined to an islet of Cannaregio. The quarter was cut off by wide canals and the two watergates were manned by Christian guards. The area was named the Ghetto, after a foundry – *geto* in Venetian – which formerly occupied the site. The name was subsequently given to Jewish enclaves throughout the world. By day Jews were allowed out of the Ghetto, but at all times they were made to wear identifying badges and caps. The only occupations they could pursue were trading in textiles, money-lending and medicine.

The rising number of Jews forced the Ghetto to expand.

The wrought iron bridge leading northwards out of the Ghetto

Campo del Ghetto Nuovo, the oldest part of the Ghetto

Buildings rose vertically (the so-called skyscrapers of Venice) and spread into the Ghetto Vecchio (1541) and the neighbouring Ghetto Novissimo (1633). By the mid 17th century the Jewish population numbered over 5,000.

In 1797 Napoleon pulled down the gates, but under the Austrians the Jews were again forced into confinement. It was not until 1866 that they were granted their freedom.

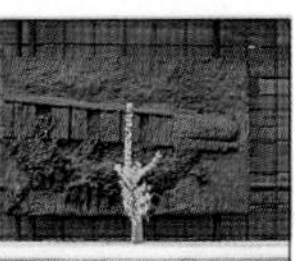

Flowers in front of the Holocaust Memorial

Of the 600 Jews now in Venice, only five families live in the Ghetto. However, the quarter has not lost its ethnic character. There are kosher food shops, a Jewish baker, a Jewish library, and two synagogues where religious ceremonies still take place. There is also a shop on the large, rambling and rather untidy Campo del Ghetto Nuovo, which sells glass rabbis and Hanukah lamps.

Museo Ebraico

The small Jewish Museum in the Ghetto Nuovo houses a collection of artifacts from the 17th–19th centuries. A guided tour of the quarter's synagogues leaves from the Museo Ebraico daily except Friday at 30 minutes past the hour from 10:30am to 5:30pm. Led by English-speaking guides, the tours give a fascinating glimpse into the past life of the Ghetto. A short history of the quarter is followed by a visit to the lavishly decorated German, Spanish and Levantine synagogues.

THE LAGOON ISLANDS

SHROUDED IN MYTH and superstition, the lagoon was once the preserve of fishermen and hunters. But marauders in the 5th and 6th centuries AD drove mainland dwellers to the safety of the marshy lagoon *(see p41)*. Here, they conquered their watery environment, which was protected from the open sea by thin sandbanks *(lidi)*, created from silt washed down by the rivers of the Po delta. In the 13th century the first *murazzi* were built – sea walls of angular stone which safeguard the *lidi* from erosion. Experiments with tidal barriers continue in an effort to combat the ever-present threat of flooding *(see p51)*.

Image of the Madonna, Torcello

The thriving communities that once lived and traded here are long gone. Many of the islands, formerly used as sites for monasteries, hospitals or powder factories, are now abandoned, but each has a tale to tell. The leases of 13 islands are about to be auctioned, with the condition that the lease-holders take on full responsibility for restoration of the buildings.

SIGHTS AT A GLANCE

- Burano 2
- Giudecca 6
- Lazzaretto Vecchio 12
- Lido 10
- Murano 4
- Poveglia 13
- San Clemente 11
- San Francesco del Deserto 3
- San Lazzaro degli Armeni 9
- San Michele 5
- San Servolo 7
- Santa Maria della Grazia 8
- *Torcello pp152–3* 1

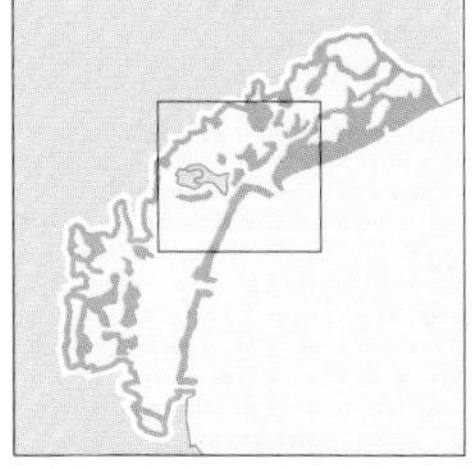

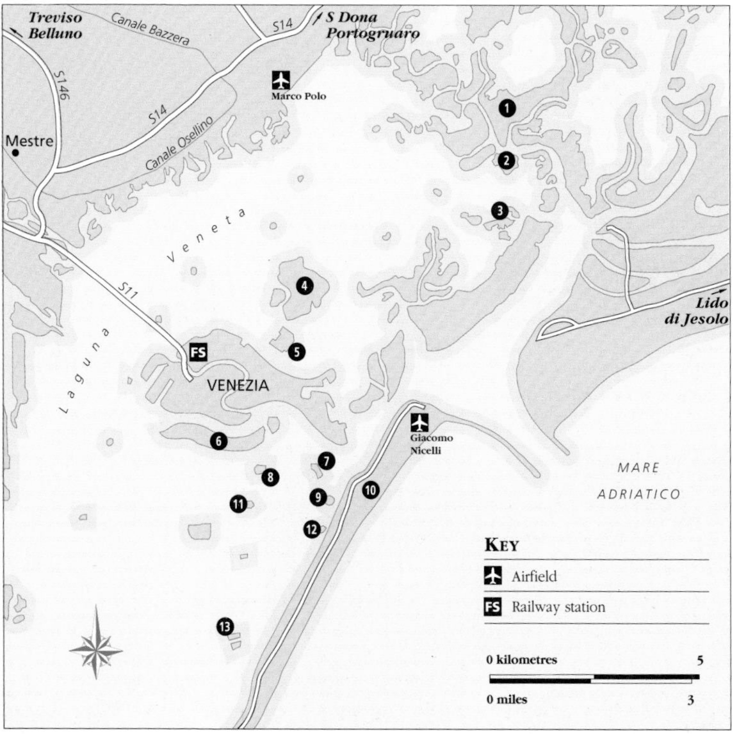

KEY

- Airfield
- FS Railway station

0 kilometres 5

0 miles 3

◁ **The vivid façade of Casa Bepi in Burano**

Exploring the Lagoon

A TRIP TO THE LAGOON ISLANDS makes a welcome break from the densely packed streets of the city. Murano, celebrated for its glass, can be reached in a matter of minutes. Further north, Burano, the "lace island", and ancient Torcello are well worth the longer ride. The Lido, with its sandy beaches, is an easy journey from San Marco. Some of the lesser known islands are worth exploring too, but access can sometimes be difficult.

Murano
Some of Murano's canalside porticoes survive from medieval days 9

Murano and San Michele are clearly visible from the northern quaysides of Venice.

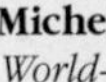

San Michele
World-famous writers and artists are buried alongside Venetians on this island 5

VENEZIA

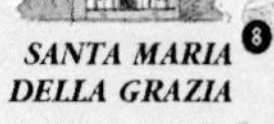

SANTA MARIA DELLA GRAZIA 8

San Giorgio in Alga had its monastery partially destroyed by fire in 1717. It was demolished in the 19th century.

SAN CLEMENTE 11

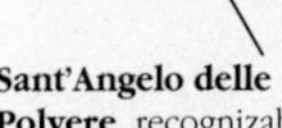

Giudecca
Palladio's great church of the Redentore, on the waterfront, is the island's cultural highlight 6

Sant'Angelo delle Polvere, recognizable by its towers, was formerly a powder factory.

SAN SPIRITO

Sacca Sessola, an artificial island, was the site of a hospital until 1980.

POVEGLIA 13

Lido
Behind the crowded beaches and grand hotels, the Lido has some pleasantly peaceful waterways 10

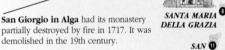

Torcello
The island's cathedral, founded in AD 639, is the oldest building in the lagoon ❶

Sant'Ariano is a former ossuary island where the bones of Venetians were taken.

Le Vignole has market gardens and an ancient fort.

Burano
Gaily painted, shuttered houses are a distinctive feature of the island's streets and quaysides ❷

Sant' Erasmo, once a Roman pleasure ground, is now a vegetable garden.

San Servolo
This is now a centre for artisans learning restoration techniques, such as stucco and plasterwork ❼

Lazzaretto Vecchio
This tiny island with its varied past can be seen from the boat that runs from San Marco to the Lido ⓬

San Lazzaro degli Armeni
Visits to this green and pretty monastery island take in the church, library, museum and printing press ❾

GETTING AROUND

The main islands of Murano, Burano, Torcello, Giudecca, San Michele and the Lido are well-served by the *vaporetto* system *(see pp274–5)*. A few of the smaller islands have a limited public service; others can only be reached by water taxi or by paying a local with a boat.

KEY

Major road

Minor road

0 kilometres 2

0 miles 1

Torcello ❶

See pp152–3.

A stall selling lace and linen in Burano's main street

Burano ❷

No. 12 from Fondamente Nuove, approx. 40–50 minutes (some go direct, others via Torcello). No. 14 from San Zaccaria via the Lido and Punta Sabbioni, approx. 1½ hours.

Burano is the most colourful of the lagoon islands. Lying in a lonely expanse of the northern lagoon, it is distinguished from a distance by the tall, dramatically tilted tower of its church. In contrast to the desolate Torcello, the island is densely populated, its waterways lined by brightly painted houses.

A tour of the island's sights will take an hour or so. The street from the ferry stop takes you to the main thoroughfare, Via Baldassare Galuppi, named after the Burano-born composer (1706–85). The street is lined with lace and linen stalls and open-air trattorias serving fresh fish.

Scuola dei Merletti

Piazza Baldassare Galuppi. *(041) 73 00 34. 9am–7pm Tue–Sat, 10am–4pm Sun. 1 Jan, Christmas.*

The Buranese are fishermen and lacemakers by trade. You can still see the men scraping their boats or mending nets, but lacemakers are rare. In the 16th century the local lace was the most sought after in Europe. It was so delicate it became known as *punto in aria* ("points in the air"). Foreign competition, coupled with the Republic's decline, led to a slump in the 18th century in Burano's industry. However, the need for a new source of income led to a revival of the skill in 1872 and the founding of a lacemaking school, the Scuola dei Merletti.

Today, authentic Burano lace is hard to find. Genuine pieces take weeks of painstaking labour, and are expensive. At the lacemaking school, however, visitors can watch Buranese women stitching busily. Attached to the school is a museum, displaying fine antique lace.

Mazzorbo

Linked to Burano by a footbridge, Mazzorbo is an island of orchards and gardens. Ferries en route to Burano and Torcello pass through its canal. The only surviving church is the Romanesque-Gothic Santa Caterina.

The multicoloured Casa Bepi, in a small square off Burano's main street

San Francesco del Deserto ❸

Access, weather permitting, via *sándolo* (rowing boat) from the quayside near the church in Burano. *Visits to the island: 9–11am, 3–5pm daily.* **Monastery** *(041) 528 68 63.*

This little oasis of greenery, inhabited by nine friars, lies just south of Burano. There is no *vaporetto* service and to get there you must bargain with the boatmen on Burano's quayside, who will row you across the shallow waters and await your return.

One of the multilingual friars will give you a tour of the old church and the enchanting gardens, which have a tree said to have sprouted from the staff of St Francis of Assisi.

A Buranese fisherman about to haul in the day's catch

Murano ❹

No. 52 from San Zaccaria; No. 12, 13 or 23 from Fondamente Nuove.

Like the city of Venice, Murano comprises a cluster of small islands, connected by bridges. It has been the centre of the glassmaking industry since 1291, when the furnaces and glass craftsmen were moved here from the city, prompted by the risk of fire to the buildings and the disagreeable effects of smoke.

Museo Vetrario

Palazzo Giustinian, Fondamenta Giustinian. *(041) 73 95 86. 10am–5pm (10am–4pm winter). Wed.*

Historically Murano owes its prosperity entirely to glass. From the late 13th century, when the population numbered over 30,000, Murano enjoyed self-government, minted its own coins and had its own Golden Book *(see p42)* listing members of the aristocracy. In the 15th and 16th centuries it was the principal glass-producing centre in Europe. Murano's glass artisans were granted unprecedented privileges, but for those who left the island to found businesses elsewhere there were severe penalties – even death.

Although a few of Murano's *palazzi* bear testimony to its former splendour, and its basilica still survives, most tourists visit for glass alone. Some are enticed by offers of free trips from factory touts in San Marco, others go by excursion launch or independently on the public *vaporetti*.

Some of the factories are now derelict, but glass is still produced in vast quantities. Among the plethora of kitsch (including imports from the Far East) are some wonderful pieces, and it pays to seek out the top glass factories *(see p249)*. Many furnaces, however, close at the weekend.

The Museo Vetrario (glass museum) in the huge Palazzo Giustinian houses a splendid collection of antique pieces. The prize exhibit of the collection is the Barovier wedding cup (1470–80), with enamelwork decoration by Angelo Barovier. The section devoted to modern glass, in a separate building at Campo Manin 1c, is closed for restoration.

Basilica dei Santi Maria e Donato

Fondamenta Giustinian. *(041) 73 90 56.* *9am–noon, 4–7:15pm.*

The colonnaded exterior of Murano's Basilica dei Santi Maria e Donato

The island's architectural highlight is the Basilica dei Santi Maria e Donato, whose magnificent colonnaded apse is reflected in the waters of the San Donato canal. Despite some heavy-handed restoration undertaken in the 19th century, this 12th-century church still retains much of its original beauty. Visitors should note the Veneto-Byzantine columns and Gothic ship's keel roof. An enchantingly evocative mosaic portrait of the Madonna, seen standing alone against a gold background, decorates the apse.

The church's floor, or *pavimento*, dating from 1140, is equally beautiful. With its medieval mosaics of geometric figures, exotic birds, mythical creatures and inexplicable symbols, it incorporates fragments of ancient glass from the island's foundries into its imagery.

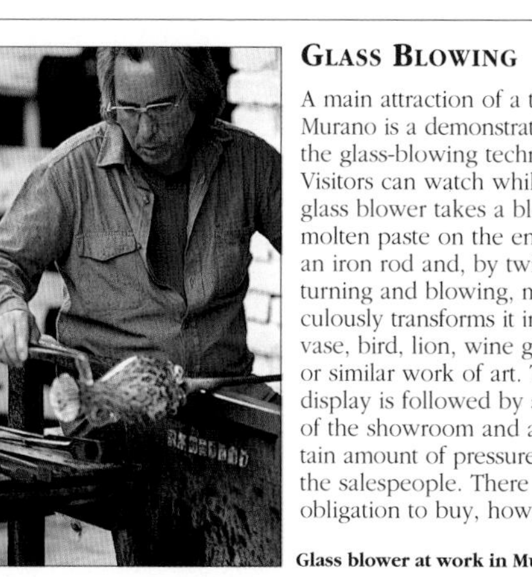

Glass Blowing

A main attraction of a trip to Murano is a demonstration of the glass-blowing technique. Visitors can watch while a glass blower takes a blob of molten paste on the end of an iron rod and, by twisting, turning and blowing, miraculously transforms it into a vase, bird, lion, wine goblet or similar work of art. The display is followed by a tour of the showroom and a certain amount of pressure from the salespeople. There is no obligation to buy, however.

Glass blower at work in Murano

San Michele ❺

No. 23 or 52.

Studded with dark cypresses and enclosed within high terracotta walls, the cemetery island of San Michele lies just across the water from Venice's Fondamente Nuove. The bodies of Venetians were traditionally buried in church graveyards in Venice. But for reasons of hygiene and space, San Michele and its neighbour were designated cemeteries in the 19th century.

Diaghilev's tombstone

The church of San Michele in Isola stands by the landing stage. Designed by Mauro Coducci (c.1469), it was the first church in Venice to be faced in white Istrian stone. The cemetery itself rambles over most of the island. With its carved tombstones and chapels it has a curious fascination. Some graves have suffered neglect, but most are well-tended and enlivened by a riot of flowers.

The most famous graves are those of foreigners: Ezra Pound (1885–1972), in the *Evangelisti* (Protestant) section, and Sergei Diaghilev (1872–1929) and Igor Stravinsky (1882–1971) in the *Greci* or Orthodox section. These bodies have been allowed to rest in peace. Others were dug up after about ten years to make way for new arrivals, and the bones taken to the ossuary island of Sant'Ariano. Today, however, because of increasing lack of space on San Michele, most bodies are buried on the mainland.

Torcello ❶

ESTABLISHED BETWEEN the 5th and 6th centuries, Torcello grew into a thriving colony *(see p40)*, with palaces, churches and a population said to have reached 20,000. But with the rise of Venice the island went into decline. Today, the population is just 60 and all that remains of this once vigorous island is the Byzantine cathedral, the church of Santa Fosca and the memory of its former glory.

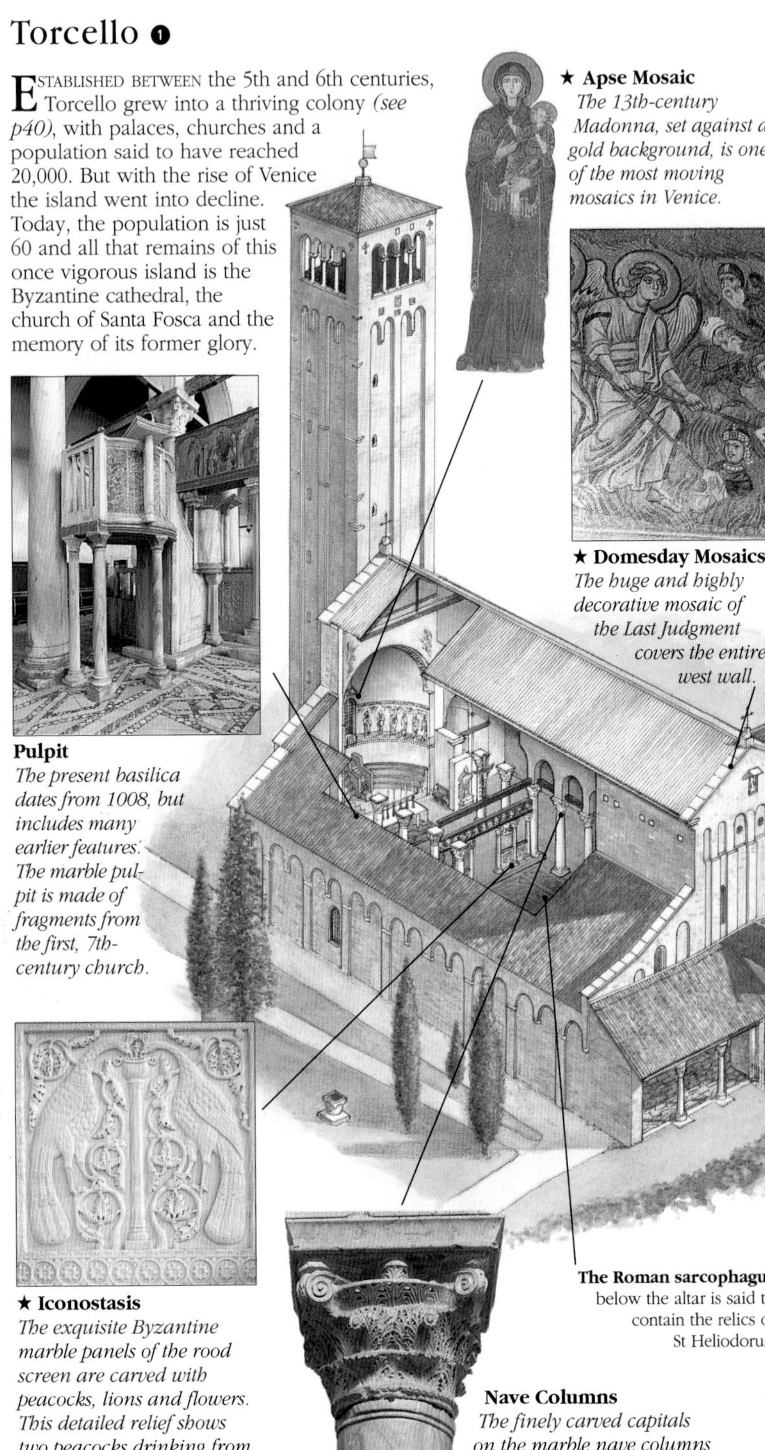

★ Apse Mosaic
The 13th-century Madonna, set against a gold background, is one of the most moving mosaics in Venice.

★ Domesday Mosaics
The huge and highly decorative mosaic of the Last Judgment covers the entire west wall.

Pulpit
The present basilica dates from 1008, but includes many earlier features. The marble pulpit is made of fragments from the first, 7th-century church.

★ Iconostasis
The exquisite Byzantine marble panels of the rood screen are carved with peacocks, lions and flowers. This detailed relief shows two peacocks drinking from the fountain of life.

The Roman sarcophagus below the altar is said to contain the relics of St Heliodorus.

Nave Columns
The finely carved capitals on the marble nave columns date from the 11th century.

Torcello's Last Canals
Silted canals and malaria hastened Torcello's decline. One of the remaining waterways runs from the vaporetto *stop to the basilica.*

VISITORS' CHECKLIST

No. 14 from San Zaccaria.
Basilica di Santa Maria dell' Assunta *(041) 73 00 84.* *10am–12:30pm, 2–6:30pm daily.* **Campanile**
Church of Santa Fosca *(041) 73 00 84.* *10am–12:30pm, 2–6:30pm daily.*
Museo dell'Estuario *(041) 73 07 61.* *10am–12:30pm, 2–5:30pm (4pm winter) Tue–Sun.* *Mon & public hols.*

Santa Fosca
Built in the 11th and 12th centuries on a Greek-cross plan, the church has a lovely portico and a serene Byzantine interior.

The central dome and cross sections are supported by columns of Greek marble with fine Corinthian capitals.

Attila's Throne
It was said that the 5th-century king of the Huns used this marble seat as his throne.

To ***vaporetto*** **boarding point**

Museo dell' Estuario
Old church treasures and archaeological fragments are housed here.

STAR FEATURES

- ★ **Apse Mosaic**
- ★ **Domesday Mosaics**
- ★ **Iconostasis**

Boats moored along the Ponte Lungo on the Giudecca

Giudecca ❻

No. 82 and 52 (to Zitelle only).

In the days of the Republic, the island of Giudecca was a pleasure ground of palaces and gardens. Today it is very much a suburb of the city, its dark narrow alleys flanked by apartments, its squares overgrown and its *palazzi* neglected. However, the long, wide quayside skirting the city side of the island makes a very pleasant promenade and provides stunning views of Venice across the water. The island was originally named Spinalunga (long spine) on account of its shape. The name Giudecca, once thought to have referred to the Jews, or *giudei,* who lived here in the 13th century, is more likely to have originated from the word *giudicati* meaning "the judged". This referred to troublesome aristocrats who, as early as the 9th century, were banished to the island.

The Hotel Cipriani *(see p231)*, among the most luxurious places to stay in Venice, is quietly and discreetly located at the tip of the island. In contrast, at the western end of the island looms the massive Neo-Gothic ruin of the Mulino Stucky. It was built in 1895 as a flour mill by the Swiss entrepreneur Giovanni Stucky, an unpopular employer who was murdered by one of his workers in 1910. The mill ceased functioning in 1954, but in true Venetian form, plans to convert the building have never materialized.

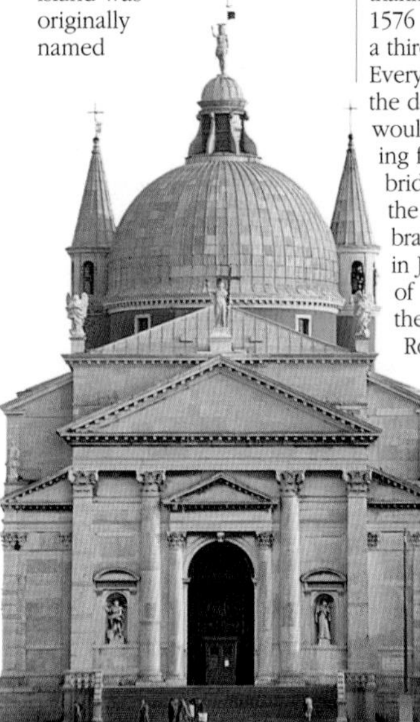

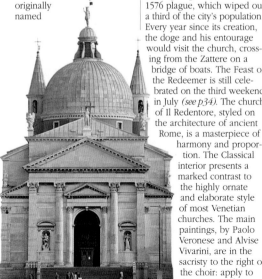

Palladio's Redentore church, Giudecca

Il Redentore

Campo Redentore. *(041) 523 14 15.* *Redentore.* *8am–noon, 4–6pm daily.*

Giudecca's principal monument is Palladio's church of Il Redentore (the Redeemer). It was built in 1577–92 in thanksgiving for the end of the 1576 plague, which wiped out a third of the city's population. Every year since its creation, the doge and his entourage would visit the church, crossing from the Zattere on a bridge of boats. The Feast of the Redeemer is still celebrated on the third weekend in July *(see p34)*. The church of Il Redentore, styled on the architecture of ancient Rome, is a masterpiece of harmony and proportion. The Classical interior presents a marked contrast to the highly ornate and elaborate style of most Venetian churches. The main paintings, by Paolo Veronese and Alvise Vivarini, are in the sacristy to the right of the choir: apply to the sacristan for access to the works. The most rewarding views of the Redentore are from Venice across the water. After dark, when the church is floodlit, it presents a spectacular sight.

Le Zitelle

Fondamenta delle Zitelle. *Zitelle.* *10am for Sun mass only.*

Palladio's church of Le Zitelle is now the site of Venice's most up-to-date congress centre. The church, though restored, is closed, apart from Sunday mass. The building adjoining the church was formerly a hostel for spinsters *(zitelle)*, who occupied themselves by making the fine Venetian *punto in aire* lace.

An artisan at work at the San Servolo training centre

San Servolo ❼

No. 20 from San Zaccaria.
Fondazione Europea pro Venetia Viva *(041) 526 85 46.*

Half-way between San Marco and the Lido is the island of San Servolo. Now a centre for training artists, it started life as one of the original monastery islands of Venice. Benedictine monks established themselves here in the 8th century, and later added a hospital to their monastery.

In 1725 the island became a lunatic asylum and a new hospital was built to house the patients. The Council of Ten *(see p42)* declared that this was to be strictly a shelter for "maniacs of noble family or comfortable circumstances". Poor maniacs were imprisoned or left to their own devices. In 1797 Napoleon scrubbed this discriminatory decree and the asylum became free to all.

In 1980 this rather spartan island was taken over by the

Fondazione Europea pro Venetia Viva (The European Centre for Training Craftsmen in the Conservation of the Architectural Heritage). Within the workshops students experiment with many traditional craft techniques. The centre may be visited, but permission must be granted in advance.

Santa Maria della Grazia ❽

No. 10 from San Zaccaria.

ORIGINALLY CALLED La Cavana or Cavanell, the island lies just a short distance away from San Giorgio Maggiore *(see p95)*. Formerly a shelter for pilgrims on their journey to the Holy Land, it became a monastery island in the 15th century. Its name was changed when a church was constructed to enshrine a miraculous image of the Virgin, brought from Constantinople. The religious buildings, including a Gothic church with some fine paintings, were secularized under Napoleon. The island became a military zone under his rule, but the buildings were subsequently destroyed during the 1848 revolutionary uprising *(see p48)*.

Today, Santa Maria della Grazia is occupied by a hospital for infectious diseases, but rumour has it that this is soon going to close.

Illuminated manuscript, San Lazzarò degli Armeni

San Lazzaro degli Armeni ❾

No. 20 from Riva degli Schiavoni. (041) 526 01 04. 3–5pm daily.

LYING JUST OFF the Lido *(see p156)*, San Lazzaro degli Armeni is a small, very green monastery island, recognizable by the onion-shaped cupola of its white campanile. The buildings are surrounded by well-groomed gardens and dark groves of cypress trees. Since the 18th century it has been an Armenian monastery and centre of learning.

Early history

This small island served as an asylum in the 12th century and later became a hospital island for lepers, named after their patron saint, Lazarus. The lepers were then transferred to the Ospedale di San Lazzaro dei Mendicanti at Santi Giovanni e Paolo *(see pp116–17)*. In 1717 an Armenian monk, Manug di Pietro, known as Mechitar ("the consoler"), was forced to flee his homeland, the Morea, when the Turks invaded. Venetian rulers gave him the island of San Lazzaro in the southern lagoon as a place of shelter. Here, he established a religious order. The Armenians rebuilt the island, setting up a monastery, church, library, study rooms, gardens and orchards. The island became a place of study where monks taught (and still teach) young Armenians their culture.

The garden and cloisters of San Lazzaro degli Armeni

Prince Nehmekhet's sarcophagus (c.1000 BC), San Lazzaro

The island today

Today, multilingual monks give visitors guided tours of the church, the art collection, the library and the museum, which houses Armenian, Greek, Indian and Egyptian artifacts. One of the most famous is an Egyptian sarcophagus complete with mummy, which is one of the best-preserved in the world. The most impressive exhibit is the printing hall where, over 200 years ago, a press produced works in 36 languages. A polyglot press is still in use, producing postcards, maps and prints for visitors.

Lord Byron

In 1816 the poet Byron would often row from Venice to absorb Armenian culture. Full of admiration for the monks, he wrote that the monastery "appears to unite all the advantages of the monastic institution without any of its vices . . . the virtues of the brethren . . . are well fitted to strike a man of the world with the conviction that 'there is another and a better', even in this life." The room where he studied, with mementoes, has been carefully preserved.

The Lido, away from the crowds and glare of the beaches

Lido ⑩

Nos. 1, 82, 52, 6 and 14 to Santa Maria Elisabetta; No. 14 to San Nicolò; No. 17 from Tronchetto to San Nicolò.

THE LIDO is a slender sandbank 12 km (8 miles) long which forms a natural barrier between Venice and the open sea. It is both a residential suburb of the city and – more importantly for tourists – the city's seaside resort. The only island in the lagoon with roads, it is linked to the mainland by car ferry. From Venice, the Lido is served by regular *vaporetti.* The fastest of these (Motonave No. 6) takes little more than ten minutes to reach its destination.

The Lido's main season runs from June to September, the most crowded months being July and August. In winter most hotels are closed.

The elegant bar of the Hôtel des Bains on the Lido

The world's first lido

In the 19th century, before the Lido was developed, the island was a favourite haunt of Shelley, Byron and other literary figures. Byron swam from the Lido to Santa Chiara via the Grand Canal in under four hours.

Bathing establishments were gradually opened and by the turn of the century the Lido had become one of Europe's most fashionable seaside resorts, frequented by royalty, film stars and leading lights of the literati. They stayed in the grand hotels, swam in the sea or sat in deckchairs on the sands by the striped *cabanas.* Life in the Lido's heyday was brilliantly evoked in Thomas Mann's book *Death in Venice* (1912). The Hôtel des Bains, where the melancholic Von Aschenbach stays, features in the novel and in Visconti's 1970 film. It is still a prominent landmark and an elegant place to stay *(see p231).*

The Lido is no longer the prestigious resort it was in the 1930s. Beaches are crowded, the streets busy and the ferries packed with daytrippers. Nevertheless the sands, sea and sporting facilities provide a welcome break from city culture. It is also the summer home for Venice's casino *(see p256)* and, when not overcrowded, the backwaters provide a green respite from the heat of Venice.

Exploring the island

The Lido can be covered by bus but a popular form of transport is the bicycle. You can hire one from the shop almost opposite the *vaporetto* stop at Santa Maria Elisabetta.

The east side of the island is fringed by sandy beaches. For passengers arriving by ferry at the main landing stage, these beaches are reached by bus, taxi or on foot along the Gran Viale Santa Maria Elisabetta. This is the main shopping street of the Lido. At the end of the Gran Viale you can turn left for the beaches of San Nicolò or right along the Lungomare G Marconi, which boasts the grandest hotels and the best beaches. The former control the latter in this area, and levy exorbitant charges (except to hotel residents) for the use of beach facilities.

***Cabanas* on the Lido beaches, hired out to wealthy tourists**

The long straight road parallel to the beach leads southwest to the village of Malamocco. There are some pleasant fish restaurants, but there is little evidence that this was once the 8th-century seat of the lagoon's government. Alberoni, at the southern end of the Lido, is the site of a golf course, a public beach and the landing stage for the ferry across to Pellestrina.

San Nicolò

The Lido's only quarter of cultural interest is San Nicolò in the north. Across the Porto di Lido, you can see the fortress of Sant'Andrea on the island of Le Vignole, built by Michele Sanmicheli between 1435 and 1449 to guard the main entrance of the lagoon. It was to the Porto di Lido that the doge was rowed annually to cast a ring into the sea in symbolic marriage each spring *(see p33)*. After the ceremony he would visit the nearby church and monastery of San Nicolò, founded in 1044 and rebuilt in the 16th century. The nearby Jewish cemetery, open to the public, dates from 1386.

The rest of this northern area is given over to an air-field. The aeroclub organizes flying lessons, parachuting or panoramic flights over Venice and the lagoon.

Aeroclub G Ancillotto

(041) 526 08 08.

San Clemente ⓫

No. 10 from San Zaccaria.

FROM A REFUGE for pilgrims en route to the Holy Land, the island of San Clemente became a hermitage and site of a monastery. During the Republic it was the island where doges frequently met distinguished visitors, but from 1630 when the island was hit by the plague (said to have been brought by the Duke of Mantua) it served as a military depot. In the 19th century the island was turned into a lunatic asylum and most of the existing buildings date from that time.

INTERNATIONAL FILM FESTIVAL

Film fans flock to the Lido every year in late summer for the International Film Festival. The event was inaugurated in 1932 under the auspices of the Biennale *(see p256)* and was so successful that the Palazzo del Cinema was built four years later. During its history the festival has attracted big names in the film world; it has also been plagued by bureaucracy and political in-fighting. There are signs however that the event is making a comeback and the famous names are now returning to the Lido.

The event takes place over two weeks in late August/early September. Films are shown day and night either in the Palazzo del Cinema or the Astra Cinema. Tickets can be hard to come by, but you can normally spot the stars (along with the paparazzi) for the price of a drink on the terrace of the Excelsior Hotel. See also page 255.

Poster advertising the first Lido International Film Festival, 1932

Lazzaretto Vecchio ⓬

No public access.

THIS SMALL ISLAND which lies just west of the Lido has served variously as a hospice for pilgrims travelling to the Holy Land, a home for victims of the plague and an ammunitions depot. The island's church, whose venerated image of the Virgin was taken by the Carmelites to adorn their Scalzi church, sadly no longer exists. Nowadays the island is a home for stray dogs. In summer the casino boat, running from San Marco to the Lido, passes close by.

Poveglia ⓭

No public access.

FORMERLY CALLED Popilia on account of all its poplar trees, the island was once a thriving little community with its own government and monastery. Devastated during the war with Genoa in 1380, it fell into decline, and over the centuries became a refuge for plague victims, an isolation hospital and a home for the aged. Today the land is used for growing crops and vines. Its low hump and distinguishing tower, once part of the city's defence system, can be seen across the lagoon from Malamocco on the Lido.

San Clemente in the southern lagoon, seen through the evening mist

SPORTHOTEL
EUROPA

The Veneto Area by Area

The Veneto at a Glance

THE VENETO'S SHEER VARIETY makes it one of Italy's most fascinating regions to explore. The cities of Verona, Padua and Vicenza are all noted for outstanding architecture, churches and museums. Villas in the rural hinterland are gorgeously frescoed with scenes from ancient mythology. The lagoon has busy fishing ports and beach resorts, while Lake Garda, with its glorious mountain scenery, historic castles and water sports, makes a perfect holiday playground. Northwards lie the majestic Dolomites, Italy's premier region for skiing, which attract visitors in the summer, too, with their alpine beauty and excellent hiking facilities.

Monti Lessini
Scores of scenic villages, such as Giazza (see p191), *nestle in the vineyard-clad valleys of the Lessini mountains.*

Verona
An ancient Roman stronghold, famous as the home of the lovers Romeo and Juliet, Verona today is a city of opera, theatre and art (see pp192–203).

Lake Garda
Most beautiful of all the Italian lakes, Garda is surrounded by Scaligeri castles such as the magnificent Sirmione (see p204).

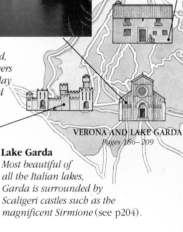

Vicenza
Dominated by the architecture of Palladio, Vicenza (see pp168–73) *is the model Renaissance city.*

Dolomites
Erosion has sculpted the limestone peaks of the Dolomites into bizarre columns and spires, with alpine villages hidden in steep valleys (see p216).

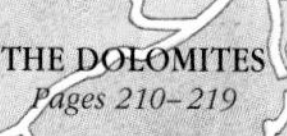

Villa Barbaro
Veronese's lavish frescoes are the perfect complement to one of Palladio's grandest rural villas, surrounded by statue-filled formal gardens, grottoes and pools (see p24).

THE DOLOMITES
Pages 210–219

0 kilometres 30
0 miles 15

THE VENETO PLAIN
Pages 162–185

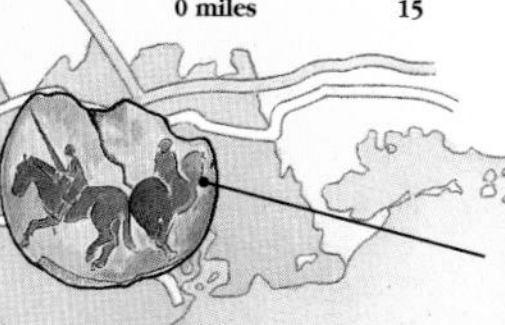

Portogruaro
Roman and early Christian finds fill the museums of this ancient town (see p175).

Padua
The domes and minaret-like spires of St Anthony's basilica (see p182) *lend an Eastern air to this historic university town.*

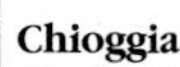

Chioggia
Flocks of wading birds frequent the wild marshland around Chioggia (see p185), *the Venetian lagoon's principal fishing port.*

The Veneto Plain

The great arc of land *that forms the Veneto Plain is one of tremendous contrast, and has much to offer the visitor. Its ancient cities are rich in history and their magnificent architecture is world renowned. The source of the region's wealth is manifest in the industrial landscapes around the towns, but these are never far from beautiful countryside, which includes the green Euganean Hills, calm lagoons and the undulating foothills of the Dolomites.*

The area known as the Veneto Plain sweeps round from the Po river delta in the southwest to the mountains that form the border between Italy and Slovenia. The whole region is crossed by a series of rivers, canals and waterways, all of which converge in the Adriatic sea.

The river-borne silt deposits that created the Venetian Lagoon cover the region, making the land fertile. The Romans established their frontier posts here, and these survive today as the great cities of Vicenza, Padua and Treviso. Their strategic position at the hub of the empire's road network enabled them to prosper under Roman rule, as they continued to do under the benign rule of the Venetian empire more than 1,000 years later.

Wealth from agriculture, commerce and the spoils of war paid for the beautification of these cities through the building of Renaissance palaces and public buildings, many of them designed by the region's great architect, Andrea Palladio. His villas can be seen all over the Veneto, symbols of the idyllic and leisured existence once enjoyed by the region's aristocrats.

The symbols of modern prosperity – factories and scarred landscapes – are encountered frequently, especially around the town of Mestre. Yet there are areas of extraordinary beauty as well. Petrarch *(see p184)*, the great medieval romantic poet, so loved the area that he made his home among the gently wooded Euganean Hills.

Fishing from a breakwater in the lagoon at Chioggia

◁ **Classical figure in the nymphaeum of the Villa Barbaro near Asolo**

Exploring the Veneto Plain

THE LANDSCAPE OF THE VENETO PLAIN is as flat as a board, but it is far from dull. Villagers in the small communities dotted throughout the region used to compete to build the tallest church tower, and these seemingly needle-thin landmarks soaring skywards draw the traveller on. Great stone castles, dating from the 14th century, rise on almost every promontory, each with a backdrop on clear days of the distant Alps.

SIGHTS AT A GLANCE

Asolo 7
Bassano del Grappa 4
Caorle 12
Castelfranco 6
Chioggia 20
Cittadella 5
Conegliano 10
Este 17
Euganean Hills (Colli Euganei) 15
Maròstica 3
Mestre 13
Monsélice 18
Montagnana 16
Padua (Padova) pp176–83 14
Polésine and Rovigo 19
Portogruaro 11
Thiene 2
Treviso 9
Valdobbiadene 8
Vicenza pp168–73 1

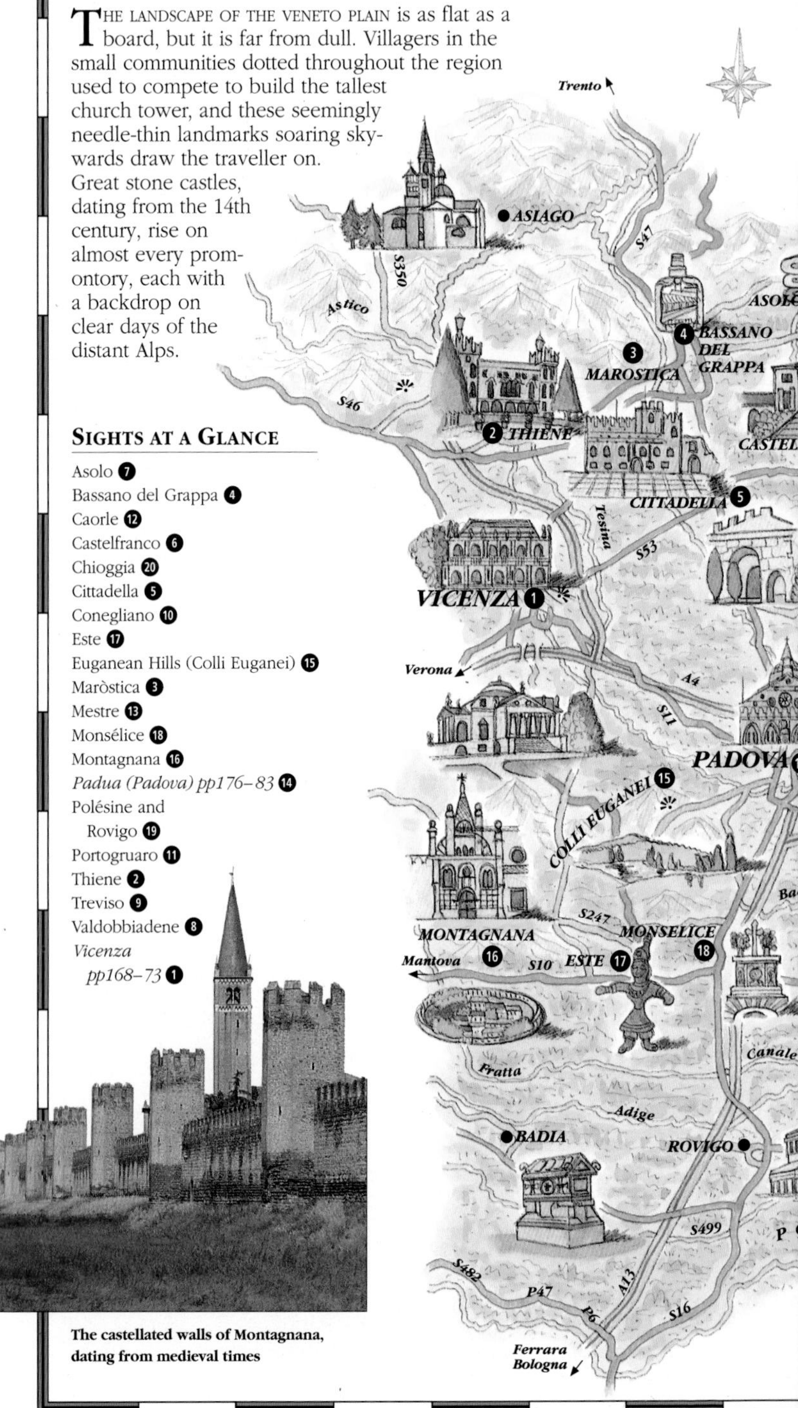

The castellated walls of Montagnana, dating from medieval times

GETTING AROUND

An extensive rail network and good bus services make this region easy to explore by public transport. Roads are heavily used, so avoid cities and *autostrade* during rush hours.

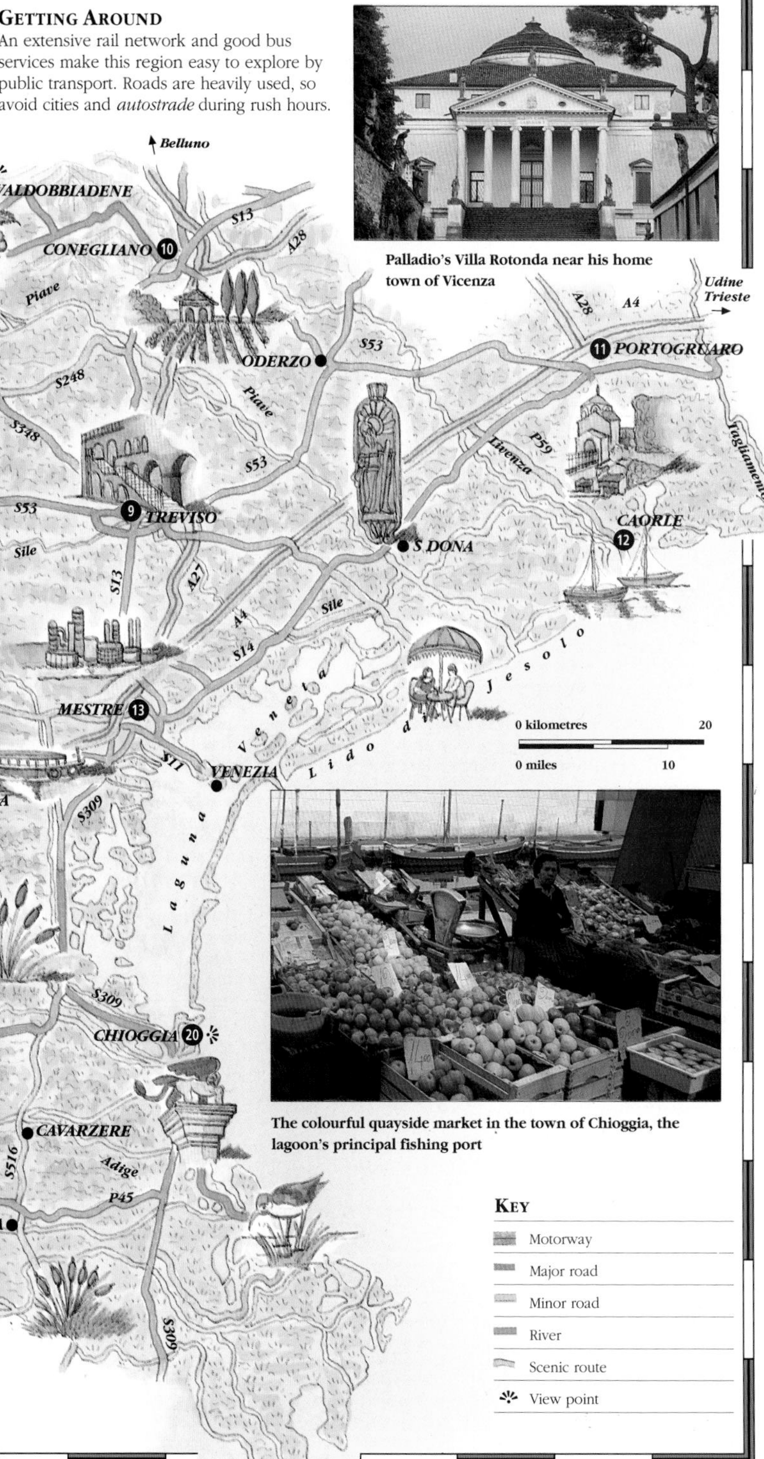

Palladio's Villa Rotonda near his home town of Vicenza

The colourful quayside market in the town of Chioggia, the lagoon's principal fishing port

KEY

- Motorway
- Major road
- Minor road
- River
- Scenic route
- View point

Vicenza ❶

See pp168–73.

Thiene ❷

Road map C3. 20,000. Piazza Ferrarin 20. (0445) 36 95 44. Mon am. ***Shops closed*** *Wed pm.*

THIENE IS ONE of the area's many textile towns, manufacturing jeans and sweatshirts for sale all over Europe. Two villas nearby are worth a visit. The heavily fortified towers and battlemented walls of the **Castello Porto-Colleoni** are offset by pretty Gothic windows. At the time it was built, it stood in open countryside, and the defences were a precaution against bandits and raiders. Inside, 16th-century frescoes by Giambattista Zelotti add a lighter note and many portraits of horses remind the visitor that the villa's owners, the Colleoni family, were employed by the Venetian cavalry.

Zelotti also frescoed the **Villa Godi Malinverni**, the first villa designed by Palladio *(see pp24–5)*. The garden is charming, and the frescoes are magnificent. Inside are works by Italian Impressionists and a lovely portrait by Pietro Annigoni (1910–88) called *La Strega* (the Sorceress).

Castello Porto-Colleoni
Corso Garibaldi 2. (0445) 36 60 15. 15 Mar–15 Nov: Sun & hols, Fri & Sat by arrangement.

Villa Godi Malinverni
Via Palladio 44. (0445) 86 05 61. Mar–Nov: Tue, Sat & Sun.

The human chess game in the town square of Maròstica

Maròstica ❸

Road map C3. 7,000. Piazza Castello 1. (0424) 721 27. Tue.

MAROSTICA is an almost perfect medieval fortified town, surrounded by walls built in 1370 by the Scaligeri *(see p207)*. The rampart walk from the **Castello Inferiore** (lower castle), now the town hall, to the **Castello Superiore** (upper castle) has fine views.

The lower castle exhibits costumes worn by participants in the town's human chess tournament, the *Partita a Scacchi*, held every other September *(see p35)*. Up to 500 people participate in this colourful re-enactment of a game first played here in 1454.

Castello Inferiore
Piazza Castello 1. (0424) 721 27. daily.

Bassano del Grappa ❹

Road map C3. 38,770. FS Largo Corona d'Italia 35. (0424) 52 43 51. Thu & Sat am.

THIS PEACEFUL TOWN, which stands at the foot of Monte Grappa, is synonymous with Italy's favourite after-dinner drink. The crystal clear grappa is not named after the town, although the liquor is produced here. The name is a corruption of *graspa*, the Italian term for the lees left over from wine production and used for the distillation of grappa. Information about the process is given at the **Museo degli Alpini**, reached by crossing the graceful Ponte degli Alpini bridge. Designed in 1569 by Palladio, the structure is built of timber to allow it to flex when hit by spring meltwaters.

Bassano is also famous for majolica wares *(see p252)* on display at **Palazzo Sturm**. The locally born artist Jacopo Bassano (1510–92), and sculptor Antonio Canova (1757–1822) are celebrated in the **Museo Civico**.

Museo degli Alpini
Via Anagarano 2. (0424) 50 36 62. Tue–Sun.

Palazzo Sturm
Via Ferracina. (0424) 52 22 35. Fri–Sun.

Museo Civico
Piazza Garibaldi. (0424) 52 22 35. Tue–Sun.

The Ponte degli Alpini at Bassano del Grappa

Cittadella ❺

Road map C3. 18,000. FS
Piazza Scalco 13. (049) 597 06 27.
Mon am.

THIS ATTRACTIVE TOWN is the twin of Castelfranco. Each was fortified and Cittadella still preserves its 13th-century moated walls. These are interrupted by four gates and by 16 towers. The Torre di Malta near the southern gate was used as a torture chamber by Ezzelino de Romano, who ruled in the mid 13th century. Far more pleasant to contemplate is the *Supper at Emmaus* painting in the **Duomo**, a masterpiece by local Renaissance artist, Bassano.

Fresco from the Villa Emo at Fanzolo, near Castelfranco

Castelfranco ❻

Road map D3. 30,000.
Via Francesco M Preti 39. (0423) 49 50 00. *Tue & Fri am.*

FORTIFIED IN 1199 by rulers of Treviso, as a defence against the neighbouring Paduans, the historic core of this town lies within well-preserved walls. **Casa di Giorgione**, claimed to be the birthplace of artist Giorgione (1478–1511), houses a museum devoted to the life of the man who created such moody and mysterious works as *The Tempest (see p131)*. One of his few directly attributable works is *The Madonna and Child with Saints Liberal and Francis* (1504) displayed in the **Duomo**. This entrancing picture was commissioned by Tuzio Costanza to stand above the tomb of his son, Matteo, killed in battle in 1504.

The pretty town of Asolo in the foothills of the Dolomites

At Fanzolo, 8 km (5 miles) northeast of Castelfranco, is the **Villa Emo**, designed in 1564 by Palladio. Here, Zelotti's sumptuous frescoes reveal the love lives of Greek deities.

Casa di Giorgione
Piazzetta del Duomo. (0423) 49 12 40. Tue–Sun. public hols.

Villa Emo
Fanzolo di Vedelago. (0423) 47 64 14. 5, irregular service. Mar–Oct: daily pm.

Asolo ❼

Road map D3. 2,000.
Piazza d'Annunzio 2. (0423) 52 30 46. Sat. ***Shops closed*** *Mon am & Wed pm.*

ASOLO IS beautifully sited among the cypress-clad foothills of the Dolomites. Queen Caterina Cornaro (1454–1510) once ruled this tiny walled town *(see p43)*, and the poet Cardinal Pietro Bembo coined the verb *asolare* to describe the bittersweet life of enforced idleness she endured. Among others who have fallen in love with the narrow streets and grand houses was poet Robert Browning, who named a volume of poems *Asolanda* (1889) after the town.

Just 10 km (6 miles) east of Asolo is the **Villa Barbaro** at Masèr *(see pp24–5)*, while 10 km (6 miles) north is the village of Passagno, birthplace of Antonio Canova. Canova's remains lie inside the huge temple-like church which he designed himself. Nearby is the family home, the **Casa di Canova**. The Gypsoteca here houses the plaster casts and clay models for many of Canova's sculptures.

Villa Barbaro
Masèr. (0423) 92 30 04. Tue, Sat, Sun & hols.

Casa di Canova
Piazza Canova. (0423) 54 43 23. Tue–Sun. 1 Jan, Easter, Christmas.

Valdobbiadene ❽

Road map D3. 10,700. FS
Viale Vittoria 23. (0423) 97 64 26.
Mon. ***Shops closed*** *Mon pm.*

VALDOBBIADENE, surrounded by beautiful vineyard-planted hills, is a major centre for the sparkling white wine called Cartizze. To the east, the Strada del Vino Bianco (white wine route) stretches 34 km (21 miles) to the town of Conegliano *(see p175)*, passing vineyards offering wine to try and to buy.

ENVIRONS: About 30 km (18 miles) northeast is Follina, where the church of **San Pietro di Foletto** has remarkable Romanesque frescoes. These include a depiction of Christ surrounded by tools connected with various trades.

Vines near Valdobbiadene

Street-by-Street: Vicenza ❶

VICENZA IS KNOWN as the city of Andrea Palladio (1508–80), a man who started out as a humble stone-mason and became the most influential architect of his time. As you walk around the city it is fascinating to study the evolution of his distinctive style. In the centre is the monumental basilica he adapted to serve as the town hall while all around are the palaces he built for Vicenza's wealthy citizens.

Detail on No. 21 Contrà Porti

Loggia del Capitaniato
This covered arcade was designed by Palladio in 1571.

Contrà Porti has some of the most elegant *palazzi* in Vicenza.

Palazzo Valmarana
Palladio's impressive building of 1566 was originally intended to be three times larger. It was not completed until 1680, 100 years after the architect's death.

CORSO ANDREA PALLADIO

CONTRA CA

Duomo
Vicenza's cathedral was rebuilt after bomb damage during World War II left only the façade and choir intact.

KEY

– – – Suggested route

STAR SIGHTS

★ **Piazza dei Signori**

★ **Casa Pigafetta**

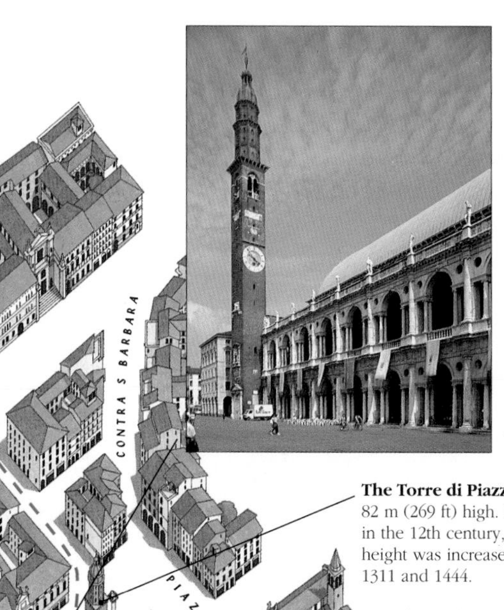

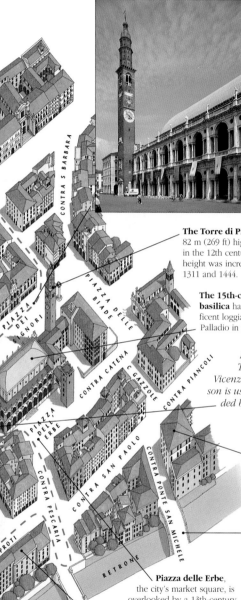

VISITORS' CHECKLIST

Road map C4. *116,000.* FS *Piazza Stazione.* *Piazza Matteotti 12. (0444) 32 08 54.* *Tue & Thu.* ***Shops closed*** *Mon am (clothes), Wed pm (food).* *Epiphany (6 Jan); Theatre season (May, Sep); Fiera di Settembre.*

★ Piazza dei Signori
Encircled by grand 15th-century buildings including the city's green-roofed basilica and slender brick tower, the piazza is a lively spot, with a colourful market and cafés.

The Torre di Piazza is 82 m (269 ft) high. Begun in the 12th century, its height was increased in 1311 and 1444.

The 15th-century basilica has a magnificent loggia built by Palladio in 1549.

Andrea Palladio
This memorial to Vicenza's most famous son is usually surrounded by market stalls.

The Quartiere delle Barche contains numerous attractive palaces built in the 14th-century Venetian Gothic style.

Piazza delle Erbe, the city's market square, is overlooked by a 13th-century torture chamber, the Torre del Tormento.

Ponte San Michele
This elegant stone bridge, built in 1620, provides lovely views of the surrounding town.

★ Casa Pigafetta
This striking house was the birthplace of Antonio Pigafetta, who in 1519 set sail round the world with Magellan.

Exploring Vicenza

VICENZA, THE GREAT PALLADIAN CITY, is celebrated the world over for its architecture. It is also one of the wealthiest cities in the Veneto, with much to offer the visitor, from Roman and Renaissance monuments to elegant shops selling all manner of fine goods.

Statues gazing down from their pillars in the Piazza dei Signori

Piazza dei Signori

At the heart of Vicenza, this square is dominated by the startling bulk of the Palazzo della Ragione, often referred to as the "basilica". Its green, copper-clad roof is shaped like an upturned boat with a balustrade that bristles with the statues of Greek and Roman gods. The colonnades were designed by Palladio in 1549 to support the city's 15th-century town hall, which had begun to subside. This was his first public commission, and his solution ensured the survival of the building.

The astonishingly slender Torre di Piazza alongside has stood since the 12th century. Opposite is the elegant café, the Gran Caffè Garibaldi, which is next to Palladio's Loggia del Capitaniato (1571). The Loggia's upper rooms contain the city's council chamber.

Contrà Porti

Contrà (an abbreviation of *contrada*, or district) is the Vicenza dialect word for street. On the western side is a series of pretty Gothic buildings with painted windows and ornate balconies, including Palazzo Porto-Colleoni (No. 19). These houses reflect the architecture of Venice, a reminder that Vicenza was part of the Venetian empire.

Several fine Palladian *palazzi* stand on this street. The Palazzo Thiene (No. 12) of 1545–50, the Palazzo Porto Barbarano (No. 11) of 1570, and the Palazzo Iseppo da Porto (No. 21) of 1552 all illustrate the sheer variety of Palladio's style – Classical elements are common to all three, but each is unique. The Palazzo Thiene reveals some intriguing details of Palladio's methods: though the building appears to be of stone, close inspection reveals that it is built of cheap lightweight brick, cleverly rendered to look like masonry.

Casa Pigafetta

Contrà Pigafetta. ***No public access.***

This highly decorated Spanish Gothic building of 1481 has clover-leaf balconies, gryphon brackets and Moorish windows. The owner, Antonio Pigafetta, sailed round the world with Magellan in 1519–22, being one of only 20 men who survived the voyage.

Museo Civico

Piazza Matteotti. *(0444) 32 13 48.* *Tue–Sun.*

Vicenza's excellent Museo Civico is housed in Palladio's Palazzo Chiericati, built in 1550. Inside is a fresco by Giulio Carpione of a naked charioteer, representing the Sun, who appears to fly over the ceiling of the entrance hall. The upstairs rooms hold many excellent pictures. Among the Gothic altarpieces from Vicenza's churches is Hans Memling's *Crucifixion* (1468–70), the central panel of a triptych whose side panels are now in New York.

In the later rooms are newly cleaned works by the local artist, Bartolomeo Montagna (c.1450–1523), including his remarkable *Pala di San Bartolo Madonna* and *Madonna with Saints.*

Santa Corona

This impressive Gothic church was built in 1261 to house a thorn from Christ's Crown of Thorns, donated by Louis IX of France. In the Porto Chapel is the tomb of Luigi da Porto (died 1529), author of the novel *Giulietta e Romeo*, upon which Shakespeare based his famous play. Notable paintings

Carpione's ceiling fresco in the large entrance hall of the Museo Civico

nclude Giovanni Bellini's *3aptism of Christ* (c.1500–5) ınd Paolo Veronese's *Adora-ion of the Magi* (1573). In the loister the Museo Naturalistico-Archeologico exhibits natural ıistory and archaeology.

San Lorenzo

The portal of this church is a nagnificent example of Gothic tone carving, richly decorated vith the figures of the Virgin nd Child, and St Francis and ıt Clare. Sadly the frescoes nside are damaged, but there ıre fine tombs. The lovely loister, north of the church, is a flower-filled haven of calm.

he beautiful cloister of the hurch of San Lorenzo

Monte Berico

asilica di Monte Berico. daily.

Monte Berico is the green, ypress-clad hill to the south of Vicenza to which wealthy Vicenzans once escaped in ıe heat of summer to enjoy he cooler air and bucolic harms of their agricultural states. Today, shady *portici*, or colonnades, run alongside he wide avenue linking entral Vicenza to the basilica on top of the hill. The *portici* have numerous hrines along the route, built or the benefit of pilgrims limbing to the domed basilica. The basilica itself vas built in the 15th century nd enlarged in the 18th, nd is dedicated to the Virgin who appeared on this pot during the 1426–8 plague to announce that Vicenza would be spared.

Coachloads of pilgrims till travel to the lovely Baroque church, where Bartolomeo Montagna's moving *Pietà* fresco (1572) makes an impact amid the ornate confectionery of the interior. The other attractions that are worth visiting include a good fossil collection in the cloister, and Veronese's fine painting, *The Supper of St Gregory the Great* (1572). This is located in the refectory; if you would like to see the painting you will have to ask. The large canvas was cut to ribbons by bayonet-wielding soldiers during the revolutionary outbursts of 1848 – a picture on the wall next to the work shows how it was painstakingly restored.

The elegant Villa Rotonda, most famous of all Palladio's works

Villa Valmarana

Via dei Nani 12. *(0444) 54 39 76.* *mid Mar–end Nov.*

The wall alongside the Villa Valmarana (which was built in 1688 by Antonio Muttoni) is topped by the figures of dwarfs which give the building its alternative name – *ai Nani* (of the Dwarfs). Inside, the walls are covered with gravity-defying frescoes by Giambattista Tiepolo, in which the fleshy and pneumatic gods of Mount Olympus float about on clouds watching scenes from the epics of Homer and Virgil. In the separate Foresteria (guest house), the frescoes with themes of peasant life and the seasons, painted by Tiepolo's son, Giandomenico, are equally decorative but more earthily realistic.

The villa can be reached by a short and enjoyable walk from the basilica on Monte Berico. Head downhill along Via M d'Azeglio to the high-walled convent on the right where the road ends, then take the Via San Bastiano. You will reach the villa after some ten minutes.

The Baroque hilltop church, the Basilica di Monte Berico

Villa Rotonda

Via della Rotonda 25. *(0444) 32 17 93.* **Villa** *15 Mar–15 Oct.* **Garden** *all year.*

With its regular, symmetrical forms, this is the epitome of Palladio's architecture, and the most famous of all his villas. The design, a dome on top of a cube, is simple yet aesthetically satisfying, as is the contrast between the green lawns, crisp white walls and terracotta roof tiles. Built between 1550 and 1552, it has inspired lookalikes in cities as far away as Delhi, London and St Petersburg. Fans of *Don Giovanni* will enjoy spotting locations used in Joseph Losey's 1979 film. To reach the Villa Rotonda, follow the path that passes the Villa Valmarana.

Vicenza: Teatro Olimpico

Carved chair in the Odeon

Europe's oldest surviving indoor theatre is a fine and remarkable structure, largely made of wood and plaster and painted to look like marble. Palladio began work on the design in 1579, but he died the following year without finishing it. His pupil, Vincenzo Scamozzi, took over the project, completing the theatre in time for its ambitious opening performance of Sophocles's tragic drama, *Oedipus Rex*, on 3 March 1585.

Bacchantes
Euripides' Greek tragedy is still performed using Scamozzi's versatile scenery.

Main ticket office

★ Odeon Frescoes
The gods of Mount Olympus, after which the theatre is named, decorate the Odeon, a room used for music recitals.

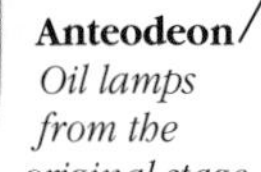

Anteodeon
Oil lamps from the original stage set are now displayed in the theatre's Anteodeon, whose frescoes (1595) depict the theatre's opening performance.

★ Stage Set
Scamozzi's scenery represents the Greek city of Thebes. The streets are cleverly painted in perspective and rise at a steep angle to give the illusion of great length.

Star Features

- ★ Stage Set by Vincenzo Scamozzi
- ★ Odeon Frescoes

Courtyard Sculptures
he courtyard of the former castle is decorated with sculpture donated by members of the Olympic Academy, the learned body that built the theatre.

Visitors' Checklist

Piazza Matteotti. (0444) 32 37 81. to Piazza Matteotti. 9am–12:15pm, 2:15–4:45pm Mon–Sat, 9am–12:15pm (summer: 2:15–6:45pm also) Sun. Regular theatre performances. No general visits during shows.

Armoury Gateway
This stone gateway, with its military-style carvings, leads from Piazza Matteotti into the picturesque theatre courtyard.

The auditorium was designed by Palladio to resemble the outdoor theatres of ancient Greece and Rome, such as the arena at Verona *(see p195)*, with a semi-circle of "stone" benches (actually made of wood) and a ceiling painted to portray the sky.

Costume Designs for Sofonisba
Ancient Greek vases inspired the costumes for this tragedy (1562) by Palladio's patron, GG Trissino.

Façade Statues
The toga-clad figures are portraits of sponsors who paid for the theatre's construction.

The medieval town of Treviso, built around ancient canals

Treviso ❾

Road map D3. *81,655.* FS
Via Toniolo 41. (0422) 54 76 32.
Tue & Sat am.

FULL OF attractive balconied houses overlooking willow-fringed canals, Treviso is a rewarding city for visitors. Comparisons are often made with Venice, but Treviso has its own distinctive character. A good place to explore the architecture is the main street, Calmaggiore, which links the cathedral with the rebuilt 13th-century town hall, the Palazzo dei Trecento. The tradition of painting the exterior of the houses dates back to the medieval period, and this form of decoration, applied to brick and timber, compensated for the lack of suitable building stone. The bustling fish market also dates back to medieval times. It is held on an island in the middle of Treviso's river Sile so that the remains of the day's trading can be flushed away instantly.

Duomo

Treviso's cathedral, founded in the 12th century, was reconstructed in the 15th, 16th and 18th centuries. Inside is Titian's *Annunciation* (1570), but it is upstaged by the striking *Adoration of the Magi* fresco (1520) of Titian's arch rival, Il Pordenone. Other memorable works are *The Adoration of the Shepherds* fresco by Paris Bordone, and the monument to Bishop Zanetti (1501) by Pietro Lombardo and his sons.

Museo Civico

Borgo Cavour 24. *(0422) 51 337.*
Tue–Sun.

The Museo Civico houses an archaeology collection and a picture gallery. The best works are in Room 11 – Lorenzo Lotto's *Portrait of a Dominican* (1526), Titian's *Portrait of Sperone Speroni* (1544) and Bassano's *Crucifixion*. The famous 14th-century frescoes depicting the life of St Ursula, by Tomaso da Modena, have been rehoused in the church of Santa Caterina (for admission ask at the Museo Civico).

San Nicolò

Nestling near the 16th-century town wall is the bulky Dominican church of San Nicolò, full of tombs and frescoes. The piers of the nave bear vivid portraits of saints by Tomaso da Modena who also painted the humorous pictures of monks (1352) on the chapter house walls, which include the first ever depiction of spectacles in art. A magnificent tomb (1500) by Antonio Rizzo is framed by a fresco of page boys (c.1500) by Lorenzo Lotto.

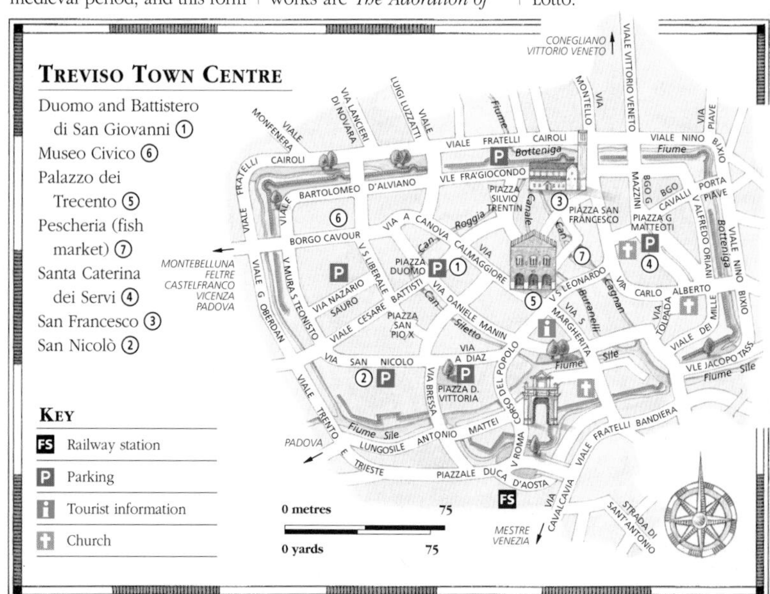

Conegliano ⓾

Road map D3. *35,300.*
Via Colombo 45. (0438) 21 230.
Fri. ***Shops closed*** *Mon am.*

CONEGLIANO lies between the Prosecco-producing vineyards and those which produce fine red wine *(see pp238 –9)*. Wine makers from both areas learn their craft at Conegliano's renowned wine school. The town's winding and arcaded main street, Via XX Settembre, is lined by 15th- to 18th-century *palazzi*, some decorated with external frescoes, some in Venetian Gothic style. The **Duomo** contains the town's one great work of art, a gorgeous altarpiece by Cima da Conegliano (1460–1518) showing the *Virgin and Child with Saints* (1493). This was commissioned by the religious brotherhood whose frescoed headquarters, the Scuola di Santa Maria dei Buttati, stands beside the Duomo.

Reproductions of Cima's famous paintings are displayed in the **Casa di Cima**, the artist's birthplace. His detailed landscapes were based on the hills around the town; they can still be viewed from the gardens surrounding the **Castelvecchio** (old castle). A small museum of local history is housed in the castle.

Casa di Cima
Via Cima. *(0438) 21 660.*
daily (phone to arrange).

Castelvecchio
Piazzale Castelvecchio. *(0438) 228 71.* *Tue–Sun.*

A mythical statue on the theatre in Conegliano's Via XX Settembre

The foundations of Roman buildings in Concordia, near Portogruaro

Portogruaro ⓫

Road map E3. *26,000.* FS
Borgo Sant'Agnese 57. (0421) 27 46 00. *Thu.* ***Shops closed*** *Mon.*

SITUATED ON the main road linking Venice to Trieste, Portogruaro is the medieval successor to the Roman town of Concordia Sagittaria. Finds from Concordia, including statues, mosaics, tomb inscriptions and building materials, are displayed in the town's **Museo Concordiese**. To see where these were unearthed, visit the modern village of Concordia, 2 km (1 mile) south of Portogruaro, where the footings of ruined Roman buildings can be seen all around the church and baptistry.

Museo Concordiese
Via Seminaro 22. *(0421) 726 74.*
daily.

Caorle ⓬

Road map E3. *11,700.*
Riva delle Caoline 1. (0421) 810 85.
Sat am.

LIKE VENICE, Caorle was built among the swamps of the Venetian lagoon by refugees fleeing the Goths in the 5th century. Today it is a fishing village and a busy beach resort perched on the edge of a huge expanse of purpose-built lagoons, carefully managed to encourage fish to enter and spawn. The young are then fed and farmed.

The area is also of great interest to naturalists for the abundant bird life of the reed-fringed waters. The town's 11th-century **Duomo** is worth a visit for its Pala d'Oro, a gilded altarpiece made up of 12th- and 13th-century Byzantine panel reliefs.

Local fishermen at work in the village of Caorle

Mestre ⓭

Road map D4. *183,650.*
Rotatoria Villa Bona Sud, Marghera. (041) 93 77 64. *Wed & Fri.*

MESTRE, the industrial offspring of Venice, is often favoured by visitors as a cheap base for exploring the region, and there are some good restaurants here. Flying into Venice's Marco Polo airport *(see pp270–71)*, you cannot miss the factories and oil terminals that surround Mestre and its neighbour, Marghera, vital to the region's economy.

Street-by-Street: Padua ⓮

The city centre of Padua (Padova) is one of the liveliest in northern Italy, thanks to a large student population and to the two street markets, one specializing in fruit and the other in vegetables. These take place every day except Sunday around the vast Palazzo della Ragione, the town's medieval law court and council chamber. The colonnades round the exterior of the *palazzo* shelter numerous bars, restaurants and shops selling meat, game, cheeses and wine.

Palazzo del Capitanio
Built between 1599 and 1605 for the head of the city's militia, the tower incorporates an astronomical clock made in 1344.

Piazza dei Signori is bordered by attractive arcades which house small speciality shops, interesting cafés and old-fashioned wine bars.

Corte Capitaniato, a 14th-century arts faculty (open for concerts), contains frescoes which include a rare portrait of Petrarch.

Loggia della Gran Guardia
Now used as a conference centre, this fine Renaissance building, dating from 1523, once housed the Council of Nobles.

The Palazzo del Monte di Pietà has 16th-century arcades and statues enclosing a medieval building.

★ Duomo and Baptistry
The 12th-century baptistry of the Duomo contains one of the most complete medieval fresco cycles to survive in Italy, painted by Giusto de' Menabuoi in 1378 and now restored.

Key

— — — Suggested route

0 metres 75

0 yards 75

★ **Caffè Pedrocchi**
Built like a Classical temple, the Caffè Pedrocchi has been a famous meeting place for students and intellectuals since it opened in 1831.

Visitors' Checklist

Road map D4. 250,000. *Piazzale della Stazione. Piazza Boscetti. Piazzale della Stazione. (049) 875 20 77. daily.* ***Shops closed*** *Mon pm (clothes, hardware, gifts), Wed pm (food). concert season (Oct–Apr).*

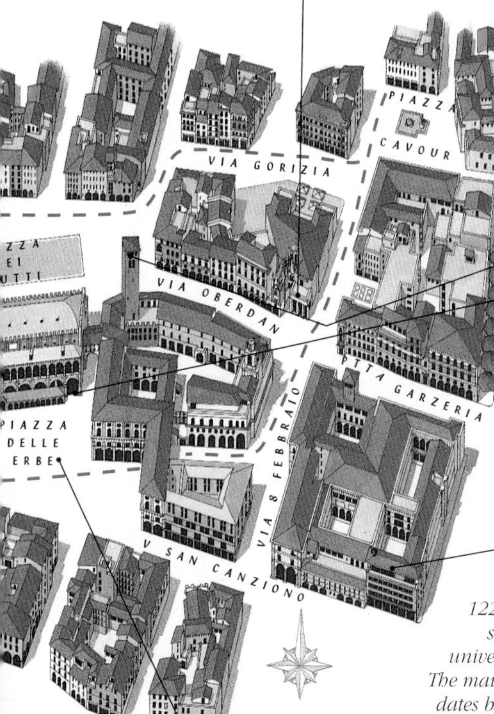

Palazzi Communali
This complex, which houses the city's council offices, has a 13th-century defensive tower.

The Palazzo della Ragione, the "Palace of Reason" was, in medieval times, the city court of justice. Its interior is covered with magnificent astrological frescoes.

Padua University
Founded in 1222, this is the second oldest university in Italy. The main building dates back to the 16th century.

★ **Piazza delle Erbe**
There are good views on to the market place from Palladio's 15th-century loggia, which runs alongside the Palazzo della Ragione.

Star Sights

- ★ **Duomo and Baptistry**
- ★ **Caffè Pedrocchi**
- ★ **Piazza delle Erbe**

Exploring Padua

PADUA IS AN OLD UNIVERSITY TOWN with an illustrious academic history. Rich in art and architecture, it has two particularly outstanding sights. The first is the Scrovegni Chapel *(see pp180–81)*, situated in the north of the city, which is renowned for Giotto's lyrical frescoes. Close to the railway station, it forms part of the Eremitani museums complex. The second is the Basilica di Sant'Antonio, one of Italy's most popular pilgrim shrines, which forms the focal point for a number of sights in the south of the city *(see p182)*.

Detail from the Egyptian room, upper floor of the Caffè Pedrocchi

Caffè Pedrocchi

Via VIII Febbraio 2. *daily.*

Grand cafés have long played an important role in the intellectual life of northern Italy, and many knotty philosophical or political problems have been thrashed out over the tables of the Caffè Pedrocchi since its doors first opened in 1831. Politics superseded philosophy when it became a centre of the Risorgimento movement, dedicated to liberating Italy from Austrian rule; it was the scene of uprisings in 1848, for which several student leaders were executed. Later it became famous throughout Italy as the café that never closed its doors. Today people still come as much to talk, read, play cards or watch the world go by as to eat or drink.

The upstairs rooms, decorated in florid medieval, Moorish, Egyptian, Greek and other styles, are used for lectures, concerts and exhibitions.

Palazzo del Bo (University)

Via Marzolo 8. *(049) 828 31 11.* *10am–1pm, 3–6pm daily.*

Named after a tavern called *Il Bo* (the ox), the historic main university building is mostly used today for graduation ceremonies. Originally the building housed the medical faculty, renowned throughout Europe. Among its famous teachers and students was Gabriele Fallopio (1523–62), after whom the Fallopian tubes are named.

Elena Lucrezia Corner Piscopia became the first woman graduate in 1678 – long before women were allowed to study at many of Europe's other universities. Her statue stands on the staircase leading to the upper gallery of the 16th-century courtyard.

Visitors on the guided tour are shown the pulpit Galileo used when he taught here from 1592 until 1610. They also see the anatomy theatre, built in 1594, the world's oldest surviving medical lecture theatre.

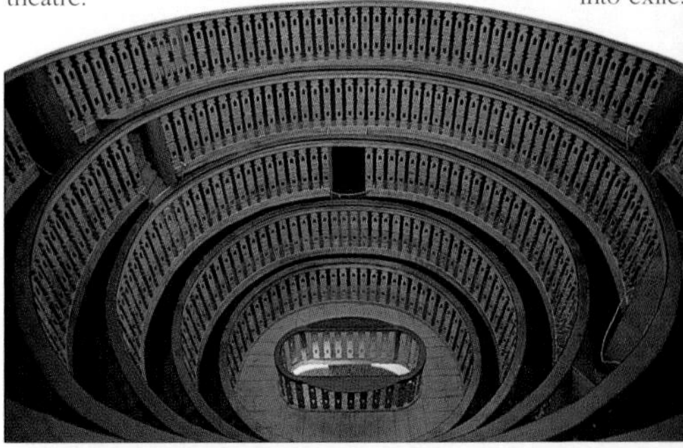

The 16th-century galleried anatomy theatre in the Palazzo del Bo

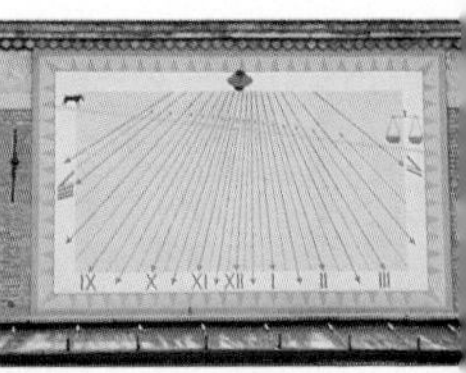

Sundial on the façade of the Palazzo della Ragione

Palazzo della Ragione

Piazza delle Erbe. *(049) 820 50 06.* *daily.*

The "Palace of Reason", also known as the "Salone" by locals, was built to serve as Padua's law court and council chamber in 1218. The vast main hall was originally frescoed by the celebrated artist Giotto, but fire destroyed his work in 1420. The frescoes that survive today are by the relatively unknown Nicola Miretto, though their astrological theme is fascinating.

Despite these changes, the Salone is breathtaking in its sheer size. It is Europe's biggest undivided medieval hall, 80 m (260 ft) long, 27 m (90 ft) wide and 27 m (90 ft) high. The scale is reinforced by the wooden horse displayed at one end – a massive beast, copied from Donatello's Gattamelata statue *(see p183)* in 1466 and originally made to be pulled in procession during Paduan festivities.

The walls are covered in Miretto's frescoes (1420–25), a total of 333 panels depicting the months of the year with appropriate gods, zodiacal signs and seasonal activities.

Also within the *palazzo* is the Stone of Shame, on which bankrupts were exposed to ridicule before they were sent into exile.

Eremitani Museums

THIS MAJOR MUSEUM COMPLEX occupies a group of 14th-century monastic buildings attached to the church of the Eremitani, a reclusive Augustinian order. The admission ticket includes entry to the Scrovegni Chapel *(see pp180–81)*, which stands on the same site, overlooking the city's Roman amphitheatre, and to the Archaeology Museum, the Bottacin Museum of coins and medals, and the Medieval and Modern Art Museum, all of which are housed around the cloisters.

The tomb of the Volumni family in the archaeological collection

Eremitani Church

ALONGSIDE the museum complex is the Eremitani church (1276–1306), with its magnificent roof and wall tombs. Interred here is Marco Benavides (1489–1582), a professor of law at the city university, whose mausoleum was designed by Ammannati, a Renaissance architect from Florence. Sadly missing from the church are Andrea Mantegna's celebrated frescoes of the lives of St James and St Christopher (1454–7), which were destroyed during a bombing raid in 1944. Two scenes from this magnificent work survive in the Ovetari Chapel, south of the sanctuary. *The Martyrdom of St James* was reconstructed from salvaged fragments, and *The Martyrdom of St Christopher* was removed carefully and stored elsewhere before the bombing. Otherwise only photographs on the walls remain to hint at the quality of the lost works.

The Museums

THE HIGHLIGHT of the rich archaeological collection is the temple-like tomb of the Volumni family, dating from the 1st century AD. Among several other Roman tombstones from the Veneto region is one to the young dancer, Claudia Toreuma – sadly, a fairly dull inscribed column rather than a portrait. The collection also includes some fine mosaics, along with several impressive life-size statues depicting muscular Roman deities and toga-clad dignitaries. For most visitors the Renaissance bronzes are likely to be the most appealing feature of the museum, especially the comical *Drinking Satyr* by Il Riccio (1470–1532).

Coin collectors should make a point of visiting the Bottacin Museum. Among the exhibits there is an almost complete set of Venetian coinage and some very rare examples of Roman medallions.

The massive Medieval and Modern Art Museum is in the process of being organized and will eventually cover the entire history of Venetian art, with paintings from Giotto to the present day. Part of the museum that has already been completed looks at Giotto and his influence on local art, using the Crucifix from the Scrovegni Chapel as its centrepiece. The Crucifix is flanked by an army of angels (late 15th century) painted in gorgeous colours by the artist Guariento. Another 15th-century painting worth a look is *Portrait of a Young Senator* by Giovanni Bellini.

***Angels in Armour* (15th century) by Guariento in the Art Museum**

Early 14th-century crucifix on loan from the Scrovegni Chapel

Visitors' Checklist

Piazza Eremitani.
☎ *(049) 820 45 50.* 🚌
○ *Apr–Oct: 9am–7pm Tue–Sun; Nov–Mar: 10am–6pm Tue–Sun.*
Only chapel open *Mon.*
● *1 Jan, 24–6 Dec.*
♿ 🚫

Padua: Scrovegni Chapel

Enrico Scrovegni built this chapel in 1303, hoping thereby to spare his dead father, a usurer, from the eternal damnation wished upon him by the poet Dante in his *Inferno*. The chapel is filled with harmonious frescoes of scenes from the life of Christ, painted by Giotto between 1303 and 1305. As works of great narrative force, they exerted a powerful influence on the development of European art.

The Nativity
The naturalism of the Virgin's pose marks a departure from Byzantine stylization, as does the use of natural blue for the sky, in place of celestial gold

Expulsion of the Merchants
Christ's physical rage, the cowering merchant and the child hiding his face are all typical of Giotto's style.

The Coretti
Giotto painted the two panels known as the Coretti as an exercise in perspective, creating the illusion of an arch with a room beyond.

View towards altar

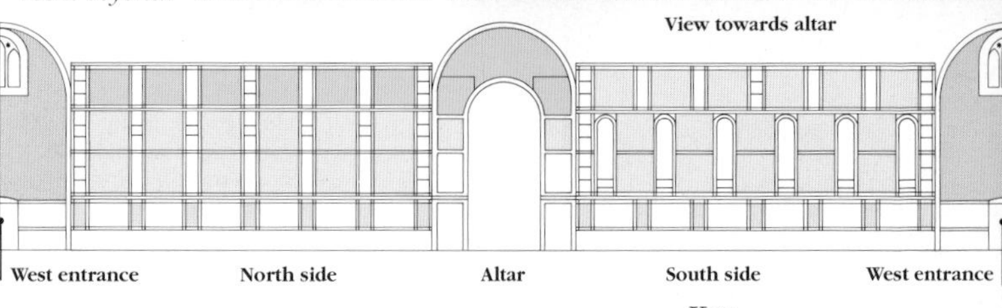

Gallery Guide

Long queues often build up at the entrance to the chapel owing to strict limits on the number of visitors allowed in at any one time. The duration of visits is also timed. To avoid such restrictions, it is best to visit early in the day, before the coach tours arrive, or towards the end of the day. If there is a long queue, try visiting the rest of the Eremitani complex first and return to the chapel later. On entry the custodian will offer you a board, printed in several languages, with a basic numbered key.

Key

- Episodes of Joachim and Anna
- Episodes from the Life of Mary
- Episodes from the Life and Death of Christ
- The Virtues and Vices
- The Last Judgment

The Last Judgment
This scene fills the entire west wall of the chapel. Its formal composition is closer to the Byzantine tradition than some of the other frescoes, with parts probably painted by assistants. A model of the chapel is shown, being offered to the Virgin by Scrovegni.

Visitors' Checklist

Giardini dell'Arena (entrance to chapel in Piazza Eremitani), Padua. *(049) 875 11 53.* *to Piazzale Boschetti.* *9am– 6pm daily.* *1 Jan, 1 May, 15 Aug, 25–6 Dec.* *see also Eremitani Museums p179.*

Mary is Presented at the Temple
Giotto sets many scenes against an architectural background, using the laws of perspective to give a sense of three dimensions.

Injustice
The Virtues and Vices are painted in monochrome. Here Injustice is symbolized by scenes of war, murder and robbery.

View towards entrance

Lament over the Dead Christ
Giotto's figures express their grief in different ways, some huddled, some gesturing wildly.

Giotto

The great Florentine artist Giotto (1266–1337) is regarded as the father of Western art. His work, with its sense of pictorial space, naturalism and narrative drama, marks a decisive break with the Byzantine tradition of the preceding 1,000 years. He is the first Italian master whose name has passed into posterity, and although he was regarded in his lifetime as a great artist, few of the works attributed to him are fully documented. Some may have been painted by others, but his authorship of the frescoes in the Scrovegni Chapel need not be doubted.

The lofty interior of Padua's 16th-century duomo

Duomo and Baptistry
Padua's duomo was commissioned from Michelangelo in 1552, but his designs were sketchy and much altered during the course of the building work. Of the earlier 4th-century cathedral which stood on the site, the domed Romanesque baptistry still survives, with its vibrant frescoes painted by Giusto de' Menabuoi (c.1376). The frescoes cover biblical stories, such as the Creation, the Miracles, Passion, Crucifixion and Resurrection of Christ, and the Last Judgment.

Basilica di Sant'Antonio
This exotic church, with its minaret-like spires and Byzantine domes, is also known as Il Santo. It was begun in 1232 to house the remains of St Anthony of Padua, a preacher who modelled himself on St Francis of Assisi. Although he was a simple man who rejected worldly wealth, the citizens of Padua built one of the most lavish churches in Christendom to serve as his shrine.

The exotic outline reflects the influence of Byzantine architecture – a cone-shaped central dome is surrounded by a further seven domes, rising above a façade that combines Gothic with Romanesque elements. The interior is more conventional, however. Visitors are kept away from the high altar, which features Donatello's magnificent reliefs (1444–5) on the miracles of St Anthony, and his statues of the Crucifixion, the Virgin and several Paduan saints. Access is permitted to the tomb of St Anthony in the north transept, which is hung with offerings and photographs of people who have survived car crashes or serious illness thanks to the saint's

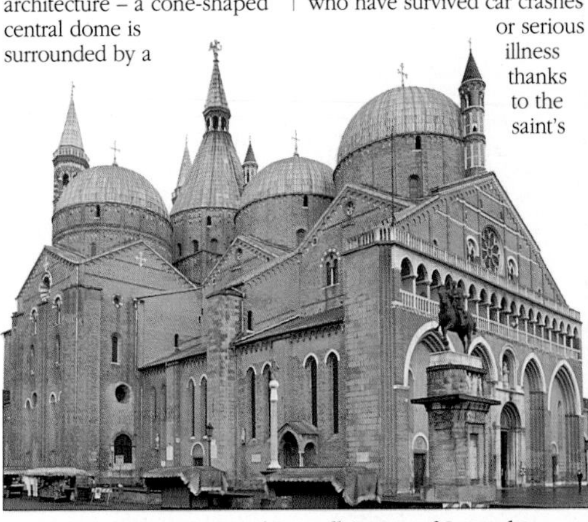

The Basilica di Sant'Antonio and Donatello's statue of Gattamelata

The Brenta Canal

The river Brenta, between Padua and the Venetian Lagoon, was canalized in the 16th century. Flowing for a total of 36 km (22 miles), its potential as a transport route was quickly realized, and fine villas were built along its length. Today, these elegant buildings can still be admired, and three open their doors to the public: the Villa Foscari at Malcontenta, the Villa Widmann-Foscari at Mira, and the Villa Pisani at Stra. All three can be visited on an 8½-hour guided tour from Padua, travelling to Venice (or from Venice to Padua) along the river on board the Burchiello motor launch.

The picturesque town of Mira on the Brenta Canal

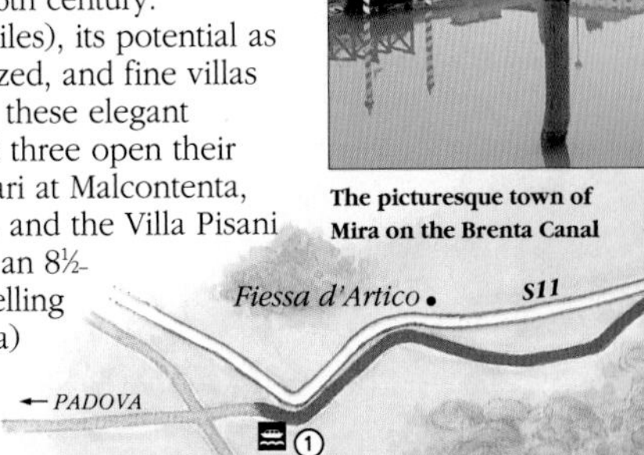

Key

- Tour route
- Roads
- Launch stops

Villa Pisani ①
This 18th-century villa features an extravagant frescoed ceiling by Tiepolo.

assistance. The walls around the shrine are decorated with large marble reliefs depicting St Anthony's life, carved in 1505–77 by various artists, including Jacopo Sansovino and Tullio Lombardo. These are rather cold by comparison with the lively *Crucifixion* fresco (1380s) by Altichiero da Zevio in the opposite transept. This is a pageant-like painting, full of everyday scenes from medieval life and masterly depictions of people, animals and plants.

Statue of Gattamelata

Near the entrance to the basilica stands one of the great Renaissance works. This gritty portrait of the mercenary soldier Gattamelata (whose name means "Honey Cat") was created in 1443–52, honouring a man who in his life did great service to the Venetian Republic. Donatello won fame for the monument, the first equestrian statue made of this size since Roman times.

Scuola del Santo and Oratorio di San Giorgio

Piazza del Santo. *(049) 875 52 35.* *daily.*

These two linked buildings contain five excellent frescoes, including the earliest documented paintings by Titian. These comprise two scenes from the life of St Anthony in the Scuola del Santo, executed in 1511. The delightful saints' lives and scenes from the life of Christ in the San Giorgio oratory are the work of two artists, Altichiero da Zevio and Jacopo Avenzo, who painted them in 1378–84.

One of four stone bridges spanning the canal around Prato della Valle

Orto Botanico

Via Orto Botanico 15. *(049) 65 66 14.* *Apr–Oct: 9am–1pm, 3–6pm daily; Nov–Mar: 9am–1pm Mon–Sat.*

Founded in 1545, Padua's botanical garden is the oldest in Europe, and it retains much of its original appearance; one of the palm trees dates to 1585. Originally intended for the cultivation of medicinal plants, the pathways now spill over with exotic foliage, shaded by ancient trees. The gardens were used to cultivate the first lilacs (1565), sunflowers (1568) and potatoes (1590) grown in Italy.

Prato della Valle

The Prato (field) claims to be the largest public square in Italy, and its elliptical shape reflects the form of the Roman theatre that stood on the site.

St Anthony of Padua used to preach sermons to huge crowds here, but subsequent neglect saw the area turn into a malaria-ridden swamp. The land was drained in 1767 to create the canal that now encircles the Prato. Four stone bridges cross the picturesque channel, which is lined on both sides by statues of 78 eminent citizens of Padua.

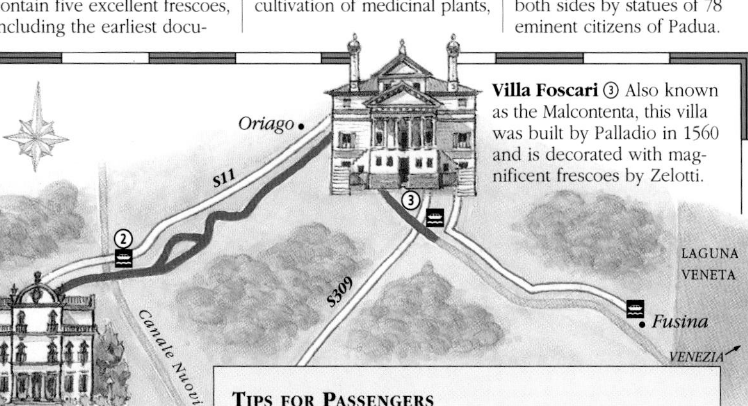

Villa Foscari ③ Also known as the Malcontenta, this villa was built by Palladio in 1560 and is decorated with magnificent frescoes by Zelotti.

lla Widmann-Foscari ② ilt in 1719, but altered in 19th century, the erior is decorated in ench Rococo style.

Tips for Passengers

Padua to Venice.
Wed, Fri and Sun.
Dep *bus station, Piazza Boschetti, 9am.* ***Arr*** *Piazza San Marco 6:55pm.*
Venice to Padua.
Tue, Thu and Sat.
Dep *Piazza San Marco 9am.* ***Arr*** *bus station, Piazza Boschetti 6:55pm.*

Booking necessary *through a local travel agent.*
Ticket includes *bus between Padua and Stra, boat tour and guide, entrance to three villas.*
Return trip *(not included in cost) by train or bus. Travel time approx. 45 mins.*

The Euganean Hills, formed by ancient volcanic activity

Euganean Hills ⓯

Road map C4. *10,000.*
Viale Stazione 60, Montregrotto Terme. *(049) 79 33 84.*

THE CONICAL Euganean Hills, remnants of long-extinct volcanoes, rise abruptly out of the surrounding plain. Hot springs bubble up out of the ground at Abano Terme and Montegrotto Terme where scores of establishments offer thermal treatments, ranging from mud baths to immersion in the hot sulphurated waters. Spa cures such as this date back to Roman times, and you can see extensive remains of the Roman baths and theatre at Montegrotto.

Abbazia di Praglia

Via Abbazia di Praglia, Bresseo di Teòlo. *(049) 990 00 10.*
Tue–Sun pm.
The Benedictine monastery at Praglia, 6 km (4 miles) west of Abano Terme, is a peaceful haven in the tree-clad hills. Here the monks grow herbs commercially and restore ancient books. They also lead guided tours of the dignified Renaissance church (1490–1548), which is renowned for its beautiful cloister.

Casa di Petrarca

Via Valleselle 4, Arquà Petrarca.
(0429) 71 82 94. *Tue–Sun.*
The picturesque town of Arquà Petrarca, on the southern edge of the Euganean hills, was once simply Arquà. Its name changed in 1868 to honour the medieval poet, Francesco Petrarca, known in English as Petrarch (1303–74), who lived here in his old age. He had often sung the praises of the well-tended landscape of olive groves and vineyards, and spent the last few years of his life living in a house frescoed with scenes from his lyrical poems. Though the house has since been altered, it still contains the poet's desk and chair, his bookshelves and his mummified cat, while the setting is as beautiful as ever. Petrarch is buried in a simple sarcophagus located in the piazza in front of the church.

Villa Barbarigo

Valsanzibio. *(049) 913 00 42 .*
Mar–Nov.
To the north of Arquà is the Villa Barbarigo at Valsanzibio, the only one of scores of villas, built by wealthy Paduans, regularly open to the public. The villa itself is of a simple design compared with the Baroque garden. Planted from 1669, it is full of variety, with fountains, statues and lakes.

The house of the poet Petrarch in the town of Arquà Petrarca

Montagnana ⓰

Road map C4. *10,000.*
Piazza Trieste 3. (0429) 813 20.
Thu. ***Shops closed*** *Mon am & Wed pm.*

MEDIEVAL brick walls encircle this town, extending for 1,895 m (2,105 yd), pierced by four gateways and defended by 24 towers. Entering through the eastern Padua Gate, it is impossible to miss Palladio's **Villa Pisani** (c.1560), with its façade featuring the original owner's name (Federico Pisani) in bold letters below the pediment. Just inside the castellated Padua Gate is the town's archaeological museum. The Gothic–Renaissance **Duomo** contains Paolo Veronese's *Transfiguration* (1555).

Antique market in Montagnana

Este ⓱

Road map C4. *17,600.*
Piazza Maggiore 9A. (0429) 36 35. *Wed & Sat am.* ***Shops closed*** *Mon & Wed am.*

EXCAVATIONS at Este have uncovered impressive remains of the ancient Ateste people, who flourished from the 9th century BC until they were conquered by the Romans in the 4th century BC. The archaeological finds, including funerary urns, figurines, bronze vases and jewellery, are on display in the excellent **Museo Nazionale Atestino**, set within the walls of the town's 14th-century castle. The museum also displays examples of Roman and medieval art, and pieces of local pottery, famous since the Renaissance period, and still produced.

Museo Atestino
Palazzo Mocenigo. (0429) 20 85. *daily. 25 Feb, 1 May, 31 Dec.*

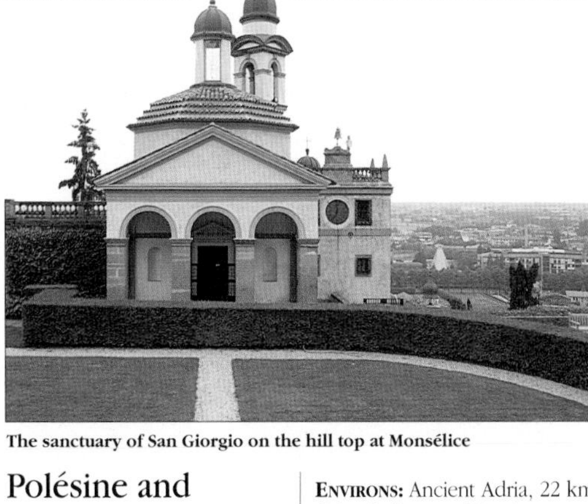

The sanctuary of San Giorgio on the hill top at Monsélice

Monsélice 18

Road map C4. *17,000.* FS
Via Roma 1. (0429) 723 80.
Mon. ***Shops closed*** *Mon am, Wed pm.*

The town of Monsélice stands around the foot of a hill quarried extensively on its western flank for rich deposits of crystalline minerals. The castle-topped hill is now a nature reserve and nesting site for rare birds of prey. The hilltop is therefore out of bounds, but you can climb up Via del Santuario as far as **San Giorgio**, to see its exquisite inlaid marble work.

Returning to the lower town, note the 13th-century cathedral and the statue-filled Baroque gardens of the Villa Nani that can be glimpsed through the villa gates. Nearby is **Ca' Marcello**, a 14th-century castle with wonderful period furnishings, fine frescoes and tapestries, and suits of armour on display.

Marble inlay detail from San Giorgio

Ca' Marcello
Via del Santuario. (0429) 729 31. *Apr–mid-Nov: Tue–Sun.*

Polésine and Rovigo 19

Road map C5. *Piazza Vittorio Emanuele 3, Rovigo. (0425) 42 24 00.*

Polesine is the flat expanse of fertile agricultural land, crisscrossed by canals and subject to flooding, that lies between the river Adige and the Po. The Po delta has a wealth of fascinating birdlife, with visitors such as egrets, herons, curlews and bitterns.

The most scenic areas are around Scardovari and Porto Tolle, on the south side of the Po. Companies in Porto Tolle offer canoe and bicycle hire and half-day boat cruises.

The modern city of Rovigo has one outstanding monument, the splendid octagonal church called **La Rotonda** (1594–1602), decorated with paintings and statues in niches.

Environs: Ancient Adria, 22 km (14 miles) east of Rovigo, was founded as a Greek port, but a deliberate programme of land reclamation and silt deposition has now left Adria dry, apart from a 24-km (15-mile) canal. The exhibits in the **Museo Archeologico** explain the town's changing fortunes and the fascinating display items include a complete iron chariot dating from the 4th century BC.

Museo Archeologico
Via Badini 59, Adria. *daily.*

Chioggia 20

Road map D4. *56,000.* FS
Lungomare Adriatico 101. (041) 40 10 68. *Thu.* ***Shops closed*** *Sun in winter.*

Chioggia is the principal fishing port on the lagoon and the bustling, colourful **fish market** is a good reason to come here early in the day (open every morning except Monday). Many visitors enjoy the gritty character of the port area, with its smells, its vibrantly coloured boats and the tangle of nets and tackle. The town also has numerous inexpensive restaurants which serve fresh fish in almost every variety. Eel, crab and cuttlefish are the local specialities. There is a beach area at Sottomarina, on the western part of the island. Worth seeking out for a special visit is Carpaccio's *St Paul* (1520), the artist's last known work, which is permanently housed in the church of **San Domenico**.

Net mending in the traditional way, Chioggia

PORTO

Verona and Lake Garda

Verona is one of northern Italy's *most alluring cities, its noble palaces, quiet cloisters and ancient streets every bit as romantic as you would expect of Romeo and Juliet's city. On the doorstep are the well-known vineyards of Soave, Bardolino and Valpolicella, set against the rugged slopes of the Little Dolomites. To the west lie the beautiful shores of Lake Garda, a mere 30 minutes' drive from Verona by car, but a world away in atmosphere.*

Set within the curves of the river Adige, Verona has been a prosperous and cosmopolitan city since the Romans colonized it in 89 BC. It stands astride two important trade routes – the Serenissima, connecting the great port cities of Venice and Genoa, and the Brenner Pass, used by commercial travellers crossing the Alps from northern Europe. This helps to explain the Germanic influence in Verona's magnificent San Zeno church, or the realism of the paintings in the Castelvecchio museum, owing more to Dürer than to Raphael.

Verona's passion and panache, however, is purely Italian. Stylish shops and cafés sit amid the impressive remains of Roman monuments. The massive Arena amphitheatre fills with crowds of 20,000 or more, who thrill to opera beneath the stars. All over the city, art galleries and theatres testify to a crowded calendar of cultural activities.

Italy's largest lake, Lake Garda, is renowned for its beautiful scenery. The broad southern end of the lake, with its waterfront promenades, is very popular with Italian and German visitors. Those in search of peace can escape to the heights of the Monte Baldo mountain range, rising above the eastern shore. The ridge marks the western edge of the mountainous region north of Verona. Here is the great plateau of Monti Lessini, with its little river valleys that fan out southwards to join the river Adige.

Giardino Giusti in Verona, one of Italy's finest Renaissance gardens

◁ **The pretty cobbled streets of Sirmione**

Exploring Verona and Lake Garda

VERONA MAKES AN EXCELLENT touring base, with lofty mountains, castles and vineyards all within easy reach of the city. Lake Garda, whose western shore is actually over the border in Lombardy, is a popular destination for excursions from Verona. The many resort towns have excellent hotels, harbourside fish restaurants and lakeside gardens, and the lake is perfect for watersports such as windsurfing or dinghy racing. Less exhausting are the steamer excursions, offering mid-lake views of entrancing beauty.

SIGHTS AT A GLANCE

Bolca 6
Bosco Chiesanuova 8
Garda 10
Gardone Riviera 15
Giazza 7
Grezzana 2
Malcésine 17
Montecchio Maggiore 4
Peschiera 11
Riva del Garda 16
Salò 14
Sant'Anna d'Alfaedo 9
Sirmione Peninsula pp206–7 13
Soave 3
Solferino 12
Valdagno 5
Verona pp192–203 1

Lazise harbour on the eastern shore of Lake Garda

Getting Around

The roads around Verona are heavily used by commercial vehicles and commuter traffic, so expect delays, especially during morning and evening rush hours. Motorways are faster, but are among the oldest in Italy, only two lanes wide and subject to frequent roadworks. There is a good east–west rail service, linking Verona to the southern shores of Lake Garda, but for north–south travel, to the mountains and around the lake, buses are the only public transport option. For ferries across Lake Garda, see p204.

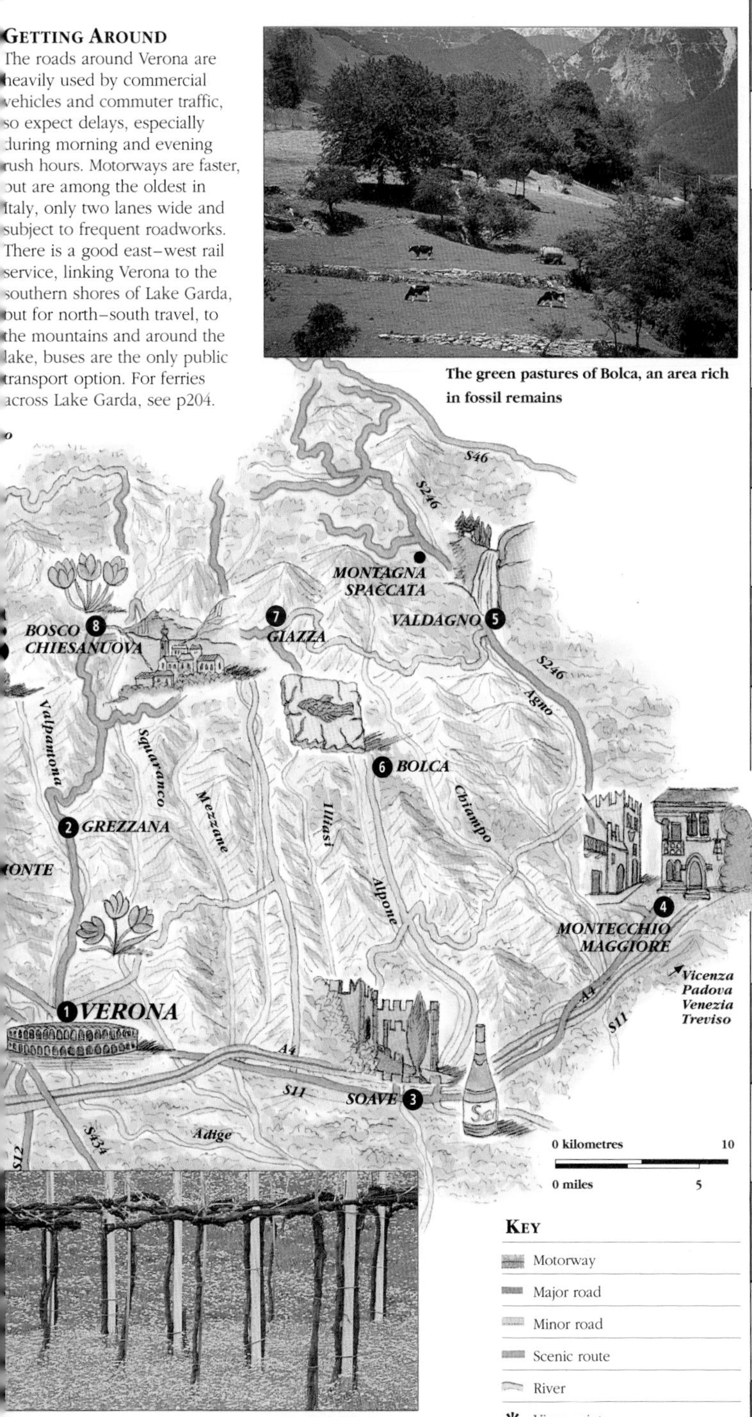

The green pastures of Bolca, an area rich in fossil remains

A vineyard in spring on the hillsides around Verona

The 14th-century Castello Romeo, on a hill overlooking Montecchio

Verona ❶

See pp192–203.

Grezzana ❷

Road map B4. 9,680. *Via Marconi, 14.* *1st Wed and 3rd Fri each month.*

In Grezzana itself, seek out the 13th-century church of Santa Maria which, though frequently rebuilt, retains its robustly carved Romanesque font and its campanile of gold, white and pink limestone.

Environs: Grezzana is in the foothills of the scenic Piccole Dolomiti or Little Dolomites. Close to the town, at nearby Cuzzano, is the 17th-century Baroque **Villa Allegri-Arvedi**. To the south, in Santa Maria in Stelle, is a Roman nymphaeum (a shrine to the nymphs who guard the fresh-water spring) next to the church (known as the Pantheon).

Villa Allegri-Arvedi
Cuzzano di Grezzana.
(045) 90 70 45.
Tue am (phone first).

Soave ❸

Road map B4. 6,820. *Piazza Antenna 2. (045) 768 06 48.* *Tue.*

Soave is a heavily fortified town ringed by 14th-century walls. Its name is familiar all over Europe because of the light and dry white wine that is produced and exported from here in great quantity. Visitors will see few vineyards around the town, since they are mainly located in the hills to the north, but evidence of the industry can be seen in the gleaming factories on the outskirts, where the Garganega grapes are crushed and the fermented wine bottled. Cafés and wine cellars in the town centre provide plenty of opportunity for sampling the local wine.

The city walls rise up the hill to the dramatically sited **Rocca Scaligeri**, an ancient castle enlarged in the 14th century by the Scaligeri rulers of Verona and furnished in period style.

Rocca Scaligeri
Via Castello Scaligero.
(045) 768 00 36. *Tue–Sun.*

Rocca Scaligeri, the ancient castle in Soave

Montecchio Maggiore ❹

Road map C4. *20,000.* *Piazza San Paolo, Alta di Montecchio. (0444) 69 65 46.* *Fri am, Sat pm.*

Visitors to industrialized Montecchio Maggiore come principally to see the two 14th-century castles on the hill above the town. In popular legend these are known as the **Castello di Romeo** and the **Castello di Giulietta**. There is no evidence that they belonged to Verona's rival Capulet and Montague families *(see p199)*, but they look romantic, and provide lovely views over the vineyard-clad hills to the north.

Castello di Romeo
Via Castelli 4 Martiri. *(0444) 49 12 95.* *Sun pm.*

Castello di Giulietta
Via Castelli 4 Martiri. *(0444) 49 12 95.* *Wed pm, Thu–Mon am & pm.*

The dramatic gorge of Montagna Spaccata, north of Valdagno

Valdagno ❺

Road map C3. *28,000.* *Viale Trento. (0445) 40 11 90.* *Tue am, Fri am.*

A scenic drive of 20 km (12 miles) from Montecchio Maggiore leads to Valdagno, a town of woollen mills and 18th-century houses. Just northwest is the Montagna Spaccata, its rocky bulk split by a dramatic 100-m (330-ft) deep gorge and waterfall.

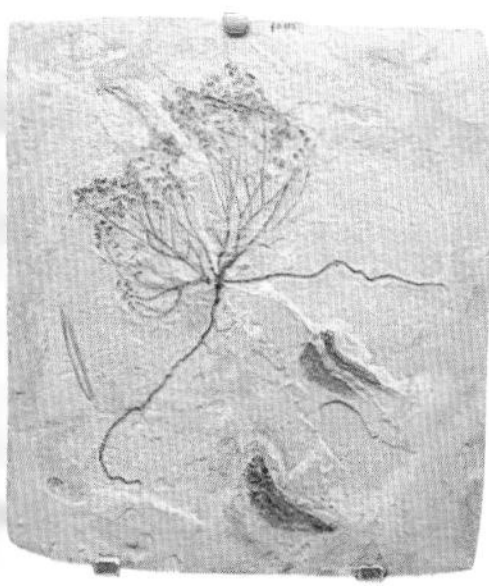

Fossilized plant remains found in the rocks near Bolca

Bolca 6

Road map B3. *500.*
Via San Giovanni Battista. (045) 656 51 11. ***Shops closed*** *Mon am (clothes), Wed pm (food).*

PRETTY BOLCA sits at the centre of the Monti Lessini plateau, looking down the valley of the river Alpone and surrounded by fossil-bearing hills. The most spectacular finds have been transferred to Verona's Museo Civico di Scienze Naturali *(see p203)*, but the local **Museo di Fossili** still has an impressive collection of fish, plants and reptiles preserved in the local basalt stone. A circular walk of 3 km (2 miles) from the town (details available from the museum) takes in the quarries where the fossils were found.

Museo di Fossili
Via San Giovanni Battista. *(045) 656 51 11.* *daily.*

Giazza 7

Road map B3. *220.*
Via Zagar Ruan 24. (045) 784 71 60. ***Shops closed*** *Mon am (clothes), Wed pm (food).*

THE SMALL TOWN of Giazza has an almost Alpine appearance. Its **Museo Etnografico** covers the history of the Tredici Comuni (the Thirteen Communes). In reality there are far more than 13 little hamlets dotted about the plateau, many of them settled by Bavarian farmers who migrated from the German side of the Alps in the 13th century. Cimbro, their German-influenced dialect, has now all but died, but other traditions survive: their huge mountain horns, *tromboni*, are still part of local festivities.

Museo Etnografico
Via dei Boschi, 62. *(045) 784 70 50.* *daily in summer; Sat & Sun in winter.*

Bosco Chiesanuova 8

Road map B3. *2,985.*
Piazza della Chiesa 34. (045) 705 00 88. *Sun.*

ONE OF THE principal ski resorts of the region, Bosco Chiesanuova is well supplied with hotels, ski lifts and cross-country routes. To the east, near Camposilvano, is the **Valle delle Sfingi** (valley of the sphinxes), so called because of its large and impressive rock formations.

Sant'Anna d'Alfaedo 9

Road map B3. *2,458.* *in Bosco Chiesanuova.* *Wed am.*

DISTINCTIVELY Alpine in character, Sant'Anna d'Alfaedo is noted for the stone tiles used to roof local houses. The hamlet of Fosse, immediately to the north, is a popular base for walking excursions up the **Corno d'Aquilio** (1,546 m/5,070 ft), a mountain which boasts one of the world's deepest potholes, the **Spluga della Preta**, 850 m (2,790 ft) deep.

More accessible is another natural wonder, the **Ponte di Veia**, just south of Sant'Anna, a great stone arch bridging the valley. Prehistoric finds have been excavated from the caves at either end. This spectacular natural bridge is one of the largest of its kind in the world.

The town of Giazza, spectacularly situated on the Monti Lessini plateau

Verona ❶

Dragon carving on Duomo façade

VERONA IS A VIBRANT and self-confident city, the second biggest in the Veneto region (after Venice) and one of the most prosperous in northern Italy. Its ancient centre boasts many magnificent Roman remains, second only to those of Rome itself, and *palazzi* built of *rosso di Verona*, the local pink-tinged limestone, by the city's medieval rulers. Verona has two main focal points, the massive 1st-century AD Arena and the Piazza Erbe with its colourful market, separated by a maze of narrow lanes lined with some of Italy's most elegant boutiques.

Verona as seen from the Museo Archeologico

Verona's rulers

In 1263 the Scaligeri began their 127-year rule of Verona. They used ruthless tactics in their rise to power, earning nicknames like Mastino (Mastiff) and Cangrande (Big Dog), but once in power the Scaligeri family brought peace to a city racked by civil strife and inter-family rivalry. They proved to be relatively just and cultured rulers – the poet Dante was welcomed to their court in 1301–4, and he dedicated his *Paradise*, final part of the epic *Divine Comedy*, to Cangrande I.

Verona fell to the Visconti of Milan in 1387, and a succession of outsiders – Venice, France and Austria – followed before the Veneto was united with Italy in 1866.

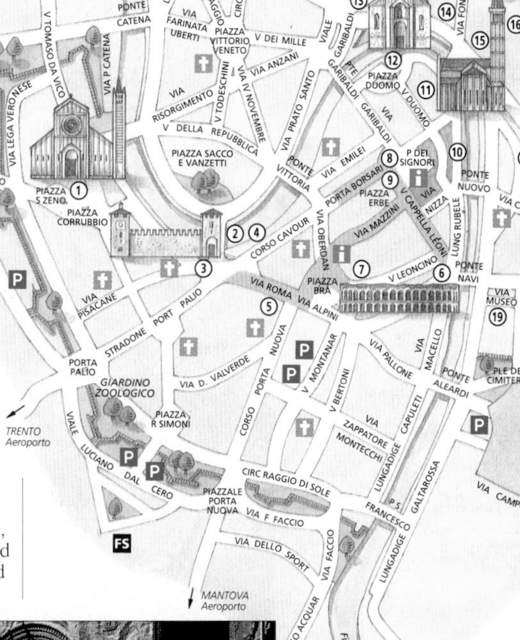

Fruit and vegetable stall in a side street of old Verona

KEY

- Street-by-Street map See *pp196–7*
- Pedestrian area
- FS Railway station
- P Parking
- i Tourist information
- Church

Castelvecchio

Corso Castelvecchio 2. (045) 59 44 34. Tue–Sun.

This spectacular castle, built by Cangrande II between 1355 and 1375, has been transformed into one of the Veneto's finest art galleries. Various parts of the medieval structure have been linked together using aerial walkways and corridors, designed to give striking views of the building itself, as well as the exhibits within. Arranged chronologically, these are excellent and varied.

The first section contains a wealth of late Roman and early Christian material – 7th-century silver plate that shows armoured knights in combat, 5th-century brooches and glass painted with a portrait of Christ the Shepherd in gold. The martyrdom scenes depicted on the carved marble sarcophagus of Saints Sergius and Bacchus (1179) are gruesomely realistic.

The following section, which is devoted to medieval and early Renaissance art, vividly demonstrates the influence of northern art on local painters, suggesting strong links with Verona's neighbours across the Alps. Here, instead of the serene saints and virgins of Tuscan art, the emphasis is on brutal realism. This is summed up in the 14th-century *Crucifixion with Saints*, which depicts the tortured musculature of Christ and the racked faces of the mourners in painful detail. Far more lyrical is a beautiful 15th-century painting by Stefano da Verona called *The Madonna of the Rose Garden*. This contains many allusions to popular medieval fables, including the figure of Fortune with her wheel. In the painting the Virgin sits in a pretty garden alive with decorative birds and angels gathering rosebuds.

Other Madonnas from the 15th century, attributed to Giovanni Bellini, are displayed among the late Renaissance works upstairs. Jewellery, suits of armour, swords and shield bosses feature next, some dating back to the 6th and 7th centuries when Verona was under attack from Teutonic invaders from beyond the Alpine range.

Cangrande I's horse in ceremonial garb

After the armour room, take the walkway that leads out along the river flank of the castle, with its dizzying views of the swirling waters of the river Adige and the Ponte Scaligero *(see p194)*. Next, turning a corner, one finds Cangrande I, his equestrian statue dramatically displayed out of doors on a plinth. This 14th-century statue once graced Cangrande's tomb *(see p198)*, and is remounted here. It is possible to study every detail of the horse and rider draped in their ceremonial garb. Despite Cangrande's cherubic cheeks and inane grin, his face is remarkably forceful and compelling.

Beyond lie some of the museum's celebrated paintings, notably Paolo Veronese's *Deposition* (1565) and a portrait attributed by some to Titian, by others to Lorenzo Lotto.

Visitors' Checklist

Road map B4. 255,700. Villafranca 14 km (9 miles). FS Porta Nuova. (045) 59 06 88. Piazza Cittadella. Piazza Erbe 38. (045) 80 00 065. daily. **Shops closed** Wed pm (food), Mon am (clothing & department stores). Vinitalia – Italy's largest wine fair (Apr); Opera Festival (Jul – Aug); Estate Teatrale Veronese, including Shakespeare Festival (Jun–Aug); Festival Lirico (Jul–Aug).

Sights at a Glance

- Arco dei Gavi ④
- Arena ⑦
- Castelvecchio ③
- Duomo ⑫
- Giardino Giusti ⑱
- Museo Archeologico ⑯
- Museo Civico di Scienze Naturali ⑲
- Museo Lapidario Maffeiano ⑤
- Piazza Erbe ⑨
- Piazza dei Signori ⑧
- Ponte Scaligero ②
- Sant'Anastasia ⑪
- San Fermo Maggiore ⑥
- San Giorgio in Braida ⑬
- Santa Maria Antica ⑩
- Santa Maria in Organo ⑰
- Santo Stefano ⑭
- San Zeno Maggiore ①
- Teatro Romano ⑮

Courtyard of Castelvecchio

Around the Arena

Most visitors to Verona first arrive at Piazza Brà, a large, irregularly shaped square with a public garden. On the north side is an archway known as the Portoni della Brà. Dominating the eastern side of the piazza is the Roman Arena, Verona's most important monument, still in use today for operatic performances. The piazza is ringed with 19th-century buildings that resemble ancient temples and historical landmarks.

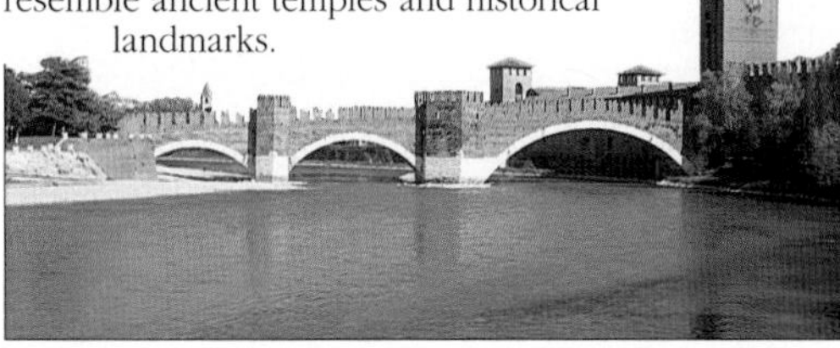

Ponte Scaligero, part of the old defence system of Castelvecchio

Ponte Scaligero

This medieval bridge was built by Cangrande II between 1354 and 1376. The people of Verona love to stroll across it to ponder the river Adige in all its moods, or to admire summer sunsets and distant views of the Alps. Such is their affection for the bridge that it was rebuilt after the retreating Germans blew it up in 1945, an operation that involved dredging the river to salvage the medieval masonry. The bridge leads from Castelvecchio *(see p193)* to the Arsenal on the north bank of the Adige, built by the Austrians between 1840 and 1861 and now fronted by public gardens. Looking back from the gardens it is possible to see how the river was used as a natural moat to defend the castle, with the bridge providing the inhabitants with an escape route.

Arco dei Gavi and Corso Cavour

Dwarfed by the massive brick walls of Castelvecchio, the monumental scale of this Roman triumphal arch is now hard to appreciate. Originally the arch straddled the main Roman road into the city, today's Corso Cavour. But French troops who were occupying Castelvecchio in 1805 damaged the monument so much that a decision was made to move it to its present, less conspicuous position just off the Corso in 1933.

Continuing up Corso Cavour, there are some fine medieval and Renaissance palaces to admire (especially Nos. 10, 11 and 19) before the Roman town gate, the **Porta dei Borsari**, is reached. The gate dates from the 1st century BC, but looking at the pedimented windows and niches it is not difficult to see what influenced the city's Renaissance architects.

The Roman Arco dei Gavi, 1st century AD

Museo Lapidario Maffeiano

Piazza Brà 28. *(045) 59 00 87.* *Tue–Sun.* *Tickets from Castelvecchio booking office.*

This "museum of stone" displays all kinds of architectural fragments hinting at the last splendour of the Roman city. There are numerous carved funerary monuments, and a large part of the collection consists of Greek inscriptions collected by the museum's 18th-century founder, Scipione Maffei.

San Fermo Maggiore

San Fermo Maggiore consists of not one but two churches. This can best be appreciated from the outside, where the eastern end is a jumble of rounded Romanesque arches below with pointed Gothic arches rising above. The lower church, now rather dank because of frequent flooding, dates from 1065, but the upper church of 1313 is more impressive. It has a splendid ship's keel roof, masses of medieval fresco work and some monumental tombs. Frescoes from the 14th century, just inside the main door, are by Stefano de Zevico. They show the fate meted out to four Franciscan missionaries who journeyed to India in the mid 14th century. Nearby is the Brenzoni mausoleum (1439) by Giovanni di Bartolo with Pisanello's *Annunciation* fresco (1426) above. In the south aisle is an unusually ornate pulpit of 1396 with saints in canopied niches above, surrounded by frescoes of the Evangelists and Doctors of the Church.

The apse of the lower church of San Fermo Maggiore

The Arena

VERONA'S ROMAN AMPHITHEATRE, completed in AD 30, is the third largest in the world, after Rome's Colosseum and the amphitheatre at Capua, near Naples. Originally, the Arena could hold almost the entire population of Roman Verona, and visitors came from across the Veneto to watch mock battles and gladiatorial combats. Since then, the Arena has been used for public executions, fairs, theatre performances, bullfighting and opera.

VISITORS' CHECKLIST

Piazza Brà, Verona. *(045) 59 65 17.* *Tue–Sun 8am–6:45pm,* **last adm:** *6:30pm.* *Mon, public hols.* *Access via arcades 5 and 7.* *Classical concerts and operas* (see pp256–7).

Interior
The interior has survived virtually intact, maintained by the Arena Conservators since 1580.

The façade of the Arena seen from Piazza Brà

The elliptical amphitheatre is 139 m (456 ft) long and 110 m (361 ft) wide.

Gladiators and wild beasts entered the arena from both sides.

Stone seats in 44 tiers

Below ground were cages for lions, tigers and other wild beasts, and a maze of passages.

Blood Sports
Prisoners of war, criminals and Christians died in their thousands in the name of entertainment.

Opera in the Arena
Today, performances of Verdi's Aida *and other popular operas can attract a capacity crowd of 25,000.*

Street-by-Street: Verona

SINCE THE DAYS OF THE ROMAN EMPIRE, the Piazza Erbe has been the centre of Verona's commercial and administrative life. Built on the site of the ancient Roman forum, it is an enjoyably chaotic square, bustling with life. Shoppers browse in the colourful market at stalls sheltered from the sun by wide-brimmed umbrellas. The massive towers and *palazzi* of the Scaligeri rulers of Verona have retained their medieval feel, even though they have been altered and adapted many times.

★ Piazza dei Signori
This square is bordered by individual Scaligeri palazzi *linked by Renaissance arcades and carved stone archways.*

Statue of Dante
Dante, the medieval poet, stayed in Verona as a guest of the Scaligeri during his period in exile from his native Florence. His statue (1865) looks down on Piazza dei Signori.

The 17th-century Palazzo Maffei is surmounted by a balustrade supporting statues of gods and goddesses.

Colonna di San Marco (1528) is surmounted by St Mark's Lion, the symbol of Venetian rule.

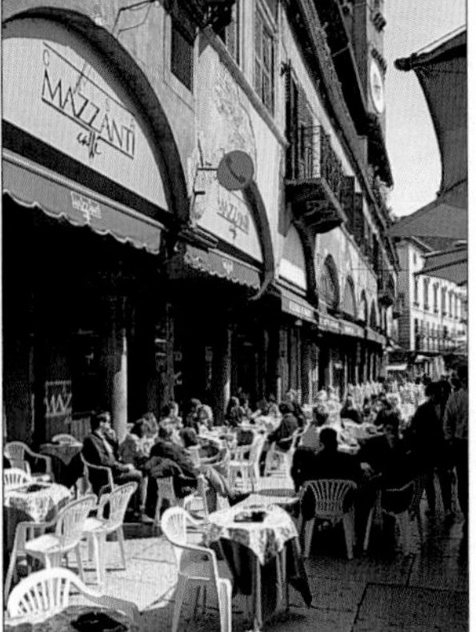

Piazza Erbe
Verona's medieval herb market is now lined with art galleries, upmarket boutiques and inviting pavement cafés.

The fountain of 1368 is topped by a figure known as the Madonna of Verona; in fact, the statue is Roman and probably symbolizes Commerce.

Torre dei Lamberti, 84 m (275 ft) high

Palazzo della Ragione
The medieval Palace of Reason features an elegant Renaissance staircase. It leads from the exterior courtyard into the magistrates' rooms on the upper floor.

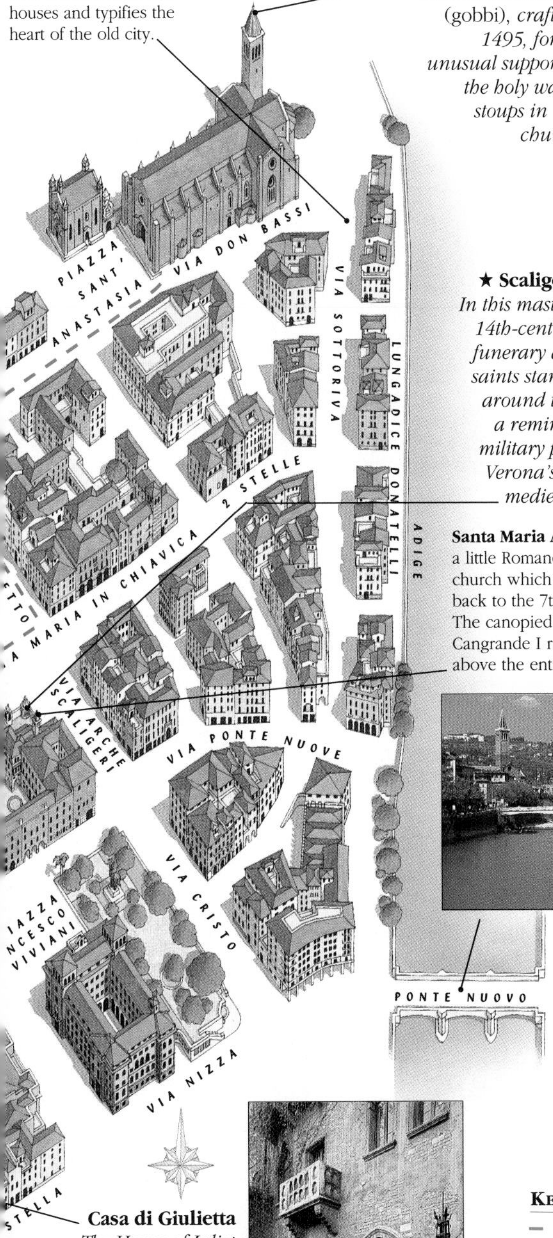

Via Sottoriva is lined with arcaded medieval houses and typifies the heart of the old city.

Sant'Anastasia
Carved hunchbacks (gobbi), *crafted in 1495, form the unusual supports for the holy water stoups in this church.*

★ Scaligeri Tombs
In this masterpiece of 14th-century Gothic funerary art, soldier saints stand guard around the tombs, a reminder of the military prowess of Verona's powerful medieval rulers.

Santa Maria Antica is a little Romanesque church which dates back to the 7th century. The canopied tomb of Cangrande I rises above the entrance.

Ponte Nuovo
The "new bridge" (1540) spans the river Adige, linking the hills on the east bank of the city with Verona's historic centre.

Casa di Giulietta
The House of Juliet looks the part, with its marble balcony and romantic setting, although there is no evidence linking this house with the romantic legend.

0 metres 100

0 yards 100

Key

Suggested route

Star Sights

- **★ Piazza dei Signori**
- **★ Scaligeri Tombs**

Central Verona

THE STREETS OF THIS ANCIENT city centre owe their grid-like layout to the order and precision of the Romans. At the heart is the lively Piazza Erbe, where crowds shop in the ancient market place. The fine *palazzi*, churches and monuments date mostly from the medieval period.

An elegant café in the spacious Piazza dei Signori

Piazza Erbe

Piazza Erbe is named after the city's old herb market. Today's stalls, shaded by huge umbrellas, sell everything from lunchtime snacks of herb-flavoured roast suckling pig in bread rolls to fresh-picked fruit or delicious wild mushrooms.

The **Venetian lion** that stands on top of a column to the north of the square marks Verona's absorption in 1405 into the Venetian empire. The statue-topped building that completes the north end of Piazza Erbe is the baroque **Palazzo Maffei** (1668), now converted to shops and luxury apartments. An assortment of boutiques and cafés line the edge of the square.

The **fountain** that splashes away quietly in the middle of the piazza is often overlooked amid the competing attractions of the market's colourful stalls. Yet the statue at the fountain's centre dates from Roman times, a reminder that this long piazza has been in almost continuous use as a market place for 2,000 years.

Piazza dei Signori

In the centre of Piazza dei Signori is a 19th-century **statue of Dante**, who surveys the surrounding buildings with an appraising eye. His gaze is fixed on the grim **Palazzo del Capitano**, home of Verona's military commander, and the equally intimidating **Palazzo della Ragione**, the palace of Reason, or law court, both built in the 14th century. The Palazzo della Ragione is not quite so grim within. The courtyard has a handsome external stone staircase, added in 1446–50. Fine views of the Alps can be had by climbing the 84-m (275-ft) **Torre dei Lamberti**, which rises from the western side of the courtyard.

Stonework detail, Piazza dei Signori

Behind the statue of Dante is the pretty Renaissance **Loggia del Consiglio**, or council chamber, with its frescoed upper façade (1493) and statues of Roman worthies born in Verona. These include Catullus the poet, Pliny the natural historian and Vitruvius the architectural theorist.

The piazza is linked to Piazza Erbe by the Arco della Costa, or the arch of the rib, whose name refers to the whale rib hung beneath it, put up here as a curiosity in the distant past.

Santa Maria Antica

This tiny Romanesque church is almost swamped by the bizarre Scaligeri tombs built up against its entrance wall. Because Santa Maria Antica was their parish church, the Scaligeri rulers of Verona chose to be buried here, and their tombs speak of their military prowess *(see p207)*.

Over the entrance to the church is the impressive tomb of Cangrande I, or Big Dog (died 1329), topped by his equestrian statue. This statue is a copy; the original is now in the Castelvecchio *(see p193)*. The other Scaligeri tombs are next to the church, surrounded by an intricate wrought-iron fence featuring the ladder motif of the family's original name (*della Scala*, meaning "of the steps"). Towering above the fence are the spire-topped tombs of Mastino II, or Mastiff (died 1351) and Cansignorio, meaning Noble Dog (died 1375).

The fountain in Piazza Erbe, erected in the 14th century

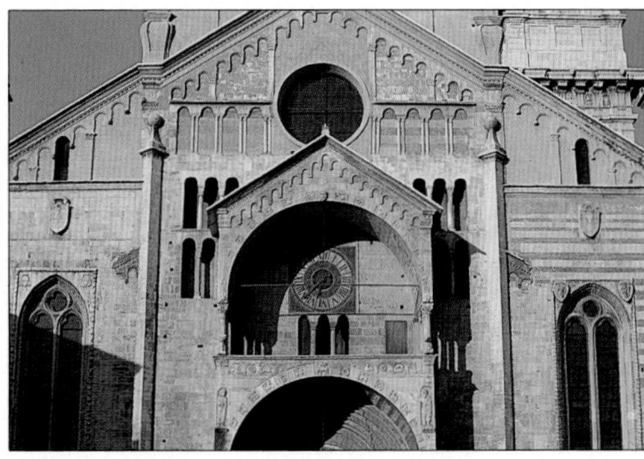
The façade of the Duomo, Santa Maria Matricolare

These two tombs are splendidly decorated with Gothic pinnacles, bristling like lances. In their craftsmanship and design there is nothing else in European funerary architecture quite like these spiky, thrusting monuments.

Plainer tombs nearer the church wall mark the resting place of other members of the Scaligeri family – Mastino (died 1277) who founded the Scaligeri dynasty, having been elected mayor of Verona in 1260, and two who did not have dog-based names: Bartolomeo (died 1304) and Giovanni (died 1359).

Sant'Anastasia

A huge and lofty church, Sant'Anastasia was begun in 1290 and built to hold the massive congregations who came to listen to the rousing sermons typically preached by members of the fundamentalist Dominican order. The most interesting aspect of the church is its Gothic portal, with its faded 15th-century frescoes and carved scenes from the life of St Peter Martyr, and in the two holy water stoups inside. These are supported on delightfully realistic figures of beggars in ragged clothing, known as *i gobbi*, the hunchbacks (the one on the left carved in 1495, the other a century later).

Off the north aisle is the sacristy, home to Antonio Pisanello's fresco, *St George and the Princess* (1433–38). Despite being badly damaged, the fresco still conveys something of the aristocratic grace of the Princess of Trebizond, with her noble brow and her ermine-fringed cloak, as St George prepares to mount his horse in pursuit of the dragon.

The lofty, Romanesque interior of Sant'Anastasia

Duomo

Visitors to Verona's cathedral (begun in 1139) pass through a magnificent Romanesque portal carved by Nicolò, one of the two master masons who carved the façade of San Zeno *(see pp200–201)*. Here he sculpted the swordbearing figures of Oliver and Roland, knights whose exploits in the service of Charlemagne were celebrated in medieval poetry. Alongside them stand saints and evangelists with bold staring eyes and flowing beards. To the south there is a second Romanesque portal carved with Jonah and the Whale (removed for restoration) and comically grotesque caryatids (load-bearing figures).

The highlight of the interior is Titian's *Assumption* (1535–40) in the first chapel on the left. Further down on the left is the entrance to the Romanesque cloister which contains excavated remains of earlier churches on the site. It also leads to the baptistry, known as San Giovanni in Fonte (St John of the Spring). This 8th-century church, built from Roman masonry, features a massive marble font carved in 1200 with scenes from the life of Christ.

Romeo and Juliet

The tragic story of Romeo and Juliet, written by Luigi da Porto of Vicenza in the 1520s, inspired countless poems, films, ballets and dramas. At the **Casa di Giulietta** (Juliet's house), No. 27 Via Cappello, Romeo is said to have climbed to Juliet's balcony: in reality this is a restored 13th-century inn. Crowds throng here, but few visit the rundown house dubbed the **Casa di Romeo** in Via delle Arche Scaligeri. More rewarding for visitors is the so-called **Tomba di Giulietta**, displayed in a crypt below the cloister of San Francesco al Corso on Via del Pontiere. The plain and empty stone sarcophagus is not especially riveting in itself, but the setting is atmospheric.

The lovers *Romeo and Juliet* from a 19th-century illustration

Verona: San Zeno Maggiore

Stone façade detail

BUILT BETWEEN 1123 and 1135 to house the shrine of Verona's patron saint, San Zeno is northern Italy's most ornate Romanesque church. The façade is embellished with marble reliefs of biblical scenes, matched in vitality by bronze door panels showing the miracles of San Zeno. Beneath an impressive rose window, a graceful porch canopy rests on two slim columns. A brick campanile soars to the south, while a squat tower to the north is said to cover the tomb of King Pepin of Italy (777–810).

Nave Ceiling
The nave has a magnificent example of a ship's keel ceiling, so called because it resembles the inside of an upturned boat. This ceiling was constructed in 1386 when the apse was rebuilt.

Striped brickwork is typical of Romanesque buildings in Verona. Courses of local pink brick are alternated with ivory-coloured tufa.

Altarpiece
Andrea Mantegna's three-part altarpiece (1457–59) depicts the Virgin and Child with various saints. The painting served as an inspiration to local artists.

★ Cloister
North of the church the fine, airy cloister (1123) has rounded Romanesque arches on one side and pointed Gothic arches on the other.

STAR FEATURES

- ★ West Doors
- ★ Cloister
- ★ Crypt

Bronze Door Panels

The 48 bronze panels of the west doors are primitive but forceful in their depiction of biblical stories and scenes from the life of San Zeno. Those on the left date from 1030 and survive from an earlier church on the site; those on the right were made 100 years later. Huge staring eyes and Ottoman-style hats, armour and architecture feature prominently, and the meaning of some scenes is not known – the woman suckling two crocodiles, for example.

Descent into limbo | Christ in Glory | Human head

Visitors' Checklist

Piazza San Zeno. (045) 800 61 20. 31, 32, 33 from Castelvecchio. 8:30am–6:30pm Mon–Sat. Sun & after 6pm Mon–Sat. during services.

Nave and Main Altar
The nave of the church is modelled on an ancient Roman basilica, the Hall of Justice. The main altar is situated in the raised sanctuary where the judge's throne would have stood.

The campanile, started in 1045, reached its present height of 72 m (236 ft) in 1178.

Rood Screen
Marble statues of Christ and the Apostles, dating from 1250, are ranged along the sanctuary rood screen.

The rose window symbolizes the Wheel of Fortune: figures around the rim show the rise and fall of human fortunes.

★ West Doors
Each of the wooden doors has 24 bronze plates joined by bronze masks, nailed on to the wood to look like solid metal. A bas relief above the doors depicts San Zeno vanquishing the devil.

Marble side panels, carved in 1140, depict events from the life of Christ to the left of the doors, and scenes from the Book of Genesis to the right.

:rypt
· vaulted crypt contains the ıb of San Zeno, appointed t bishop of Verona in AD 2, who died in AD 380.

Across the Ponte Romano

THE PONTE ROMANO, or Roman Bridge, links Verona's city centre to the eastern bank of the river Adige. This up-market residential district is dotted with fine palaces, gardens and churches, and offers good views back on to the towers and domes of the medieval city.

View from the Teatro Romano across the river Adige

Teatro Romano

Rigaste Redentore 2. (045) 800 03 60. Tue–Sun.

When this theatre was built in the 1st century BC the plays performed would have included satirical dramas by such writers as Terence and Plautus. The tradition continues with open-air performances at the annual Shakespeare festival.

The theatre is built into a bank above the river Adige, and the views over the city must have been every bit as entrancing to the Roman theatregoers as the events being enacted on stage. Certainly it is for the views that the theatre is best visited today, since little survives of the original stage area, though the semi-circular seating area remains largely intact.

In the foreground of the view is one of three Roman bridges that brought traffic into the city. This is the only one to have survived, although it had to be painstakingly reconstructed after World War II. In common with all the city's bridges it was blown up in 1945 by retreating German soldiers who were attempting to delay the advance of Allied troops. Of the five arches, the two nearest to the theatre are least altered.

Augustus Caesar, Museo Archeologico

Museo Archeologico

Rigaste Redentore 2. (045) 800 03 60. Tue–Sun.

A lift carries visitors from the Teatro Romano up through the cliffs to the monastery above. This is now converted into an archaeological museum in which panoramic views over the city vie for attention with the range of exhibits. The first part of the museum displays well-restored mosaics, one of which depicts the kind of gory gladiatorial combat that once went on in Verona's amphitheatre *(see p195)*. Such barbaric performances, seen as a legitimate way of disposing of criminals and prisoners of war, finally came to an end in the early 5th century following a decree from the Christian Emperor Honorius.

In the little monastic cells to the side of this room, visitors can see a bronze bust of the first Roman emperor, the young Augustus Caesar (63 BC–AD 14), who succeeded in outmanoeuvring his opponents, including Mark Antony and Cleopatra, to become the sole ruler of the Roman world in 31 BC. The subject of the female bust in the adjoining cell is unknown. Next comes the tiny cloister, littered with mosaics and ancient masonry fragments, and an extensive warren of ancient rooms which are used to display pottery, glass, inscriptions and tombstones. Labelling stops after a while and visitors are left to puzzle out the age and nature of exhibits for themselves.

Santo Stefano

This is one of the city's oldest churches; the original, long-demolished building was constructed in the 6th century. It served as Verona's cathedral until the 12th century when the new Duomo was built *(see p199)* on the opposite bank of the Adige. Visitors to Santo Stefano are afforded a striking view of the Duomo across the river, taking in the Romanesque apse and the bishop's palace alongside. Santo Stefano itself was rebuilt at the same time by Lombard architects and given its octagonal red brick campanile, but the original apse survives.

Inside the church there is a Byzantine-influenced arrangement of a stone bishop's seat and bench, and a gallery with 8th-century carved capitals. The apse (which is often locked) is even older, dating back to the original 6th-century building. In the crypt there are fragments of 13th-century frescoes and a 14th-century statue of St Peter.

Figure of St Peter, Santo Stefano

Towering above the church to the east is Castel San Pietro, strikingly fronted by flame-shaped cypress trees. The present castle was built in 1854 under Austrian rule, but it stands on the ruins of an earlier castle which was built by the Visconti of Milan when the Milanese captured Verona in 1387.

San Giorgio in Braida

San Giorgio is a rare example in Verona of a domed Renaissance church. It was begun in 1477 by Michele Sanmicheli, an architect best known for his military works. Sanmicheli also designed the classically inspired altar, which is topped by Paolo Veronese's *Martyrdom of St George* (1566). This celebrated painting is outshone by the calm and serene *Virgin Enthroned between St Zeno and St Lawrence* (1526) by Girolamo dai Libri. This work has a beautifully detailed background landscape and a lemon tree growing behind the Virgin's throne.

Marquetry cockerel in Santa Maria in Organo

Santa Maria in Organo

Some of the finest inlaid woodwork to be seen in Italy is in this church. The artist was Fra Giovanni da Verona, an architect and craftsman who worked for nearly 25 years, from 1477 to 1501, on these stunning examples of illusionistic marquetry. The seat backs in the choir and cupboard fronts in the sacristy are full of entertaining detail. By clever interpretation of perspective, Fra Giovanni gave depth to flat landscapes, depicted city views glimpsed through an open window, and created "cupboard interiors" stacked with books, musical instruments, or bowls of fruit. Most charming of all are the little animal pictures – look out for the rabbit on the lectern and the owl and the cockerel in the sacristy.

Fossilized fish from Verona's natural history museum

Museo Civico di Storia Naturali

Lungadige Porta Vittoria 9.
(045) 807 94 00. *daily exc Fri.*

Verona's natural history museum contains an outstanding collection of fossils which can be enjoyed on a purely aesthetic level without any knowledge of palaeontology. Whole fish, trees, fern leaves and dragonflies are captured in extraordinary detail. The fossils were found in rock in the foothills of the Little Dolomites north of the city during quarrying for building stone (*see* Bolca, *p191*).

Human prehistory is represented by finds from ancient settlements round Lake Garda, and there are reconstructions of original lake villages. On the upper floor, cases full of stuffed birds, animals and fish provide an extensive account of today's living world, making this a good museum for visiting with children or on rainy days.

Giardino Giusti

Via Giardino Giusti 2. *(045) 803 40 29.* *daily.*

Hidden among the dusty façades of the Via Giardino Giusti is the entrance to one of Italy's finest Renaissance gardens. They were laid out in 1580 and, as with other gardens of the period, artifice and nature are deliberately juxtaposed. The formal lower garden of clipped box hedges, gravel walks and potted plants is contrasted with an upper area of wilder natural woodland, the two parts linked by stone terracing.

Past visitors have included the English traveller Thomas Coryate who, writing in 1611, called this garden "a second paradise". The diarist John Evelyn, visiting 50 years later, thought it the finest garden in Europe. Today the garden makes an excellent spot for a quiet picnic.

Italianate topiary and statuary in the Giardino Giusti

Around Lake Garda

GARDA, THE LARGEST AND EASTERNMOST of the Italian Lakes, is a favourite summer playground for sports lovers. Strong winds make ideal conditions for windsurfing and sailing, there are numerous yacht harbours and artificial beaches, and luxury hotels offer tennis and horse-riding facilities. The less energetic can explore the lake and shore by steamer, while the magnificent scenery of snow-capped peaks and spectacular sunsets will appeal to every visitor.

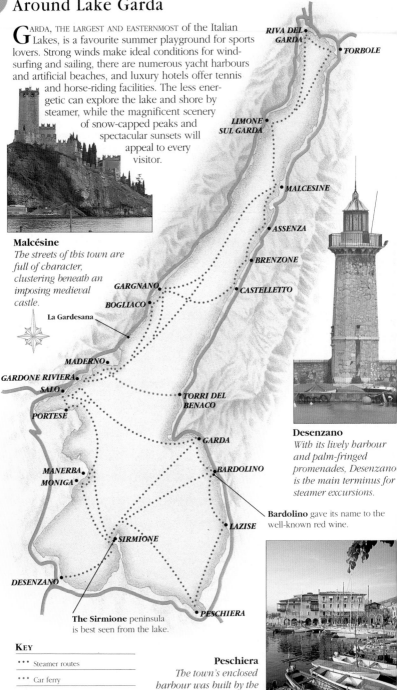

Malcésine
The streets of this town are full of character, clustering beneath an imposing medieval castle.

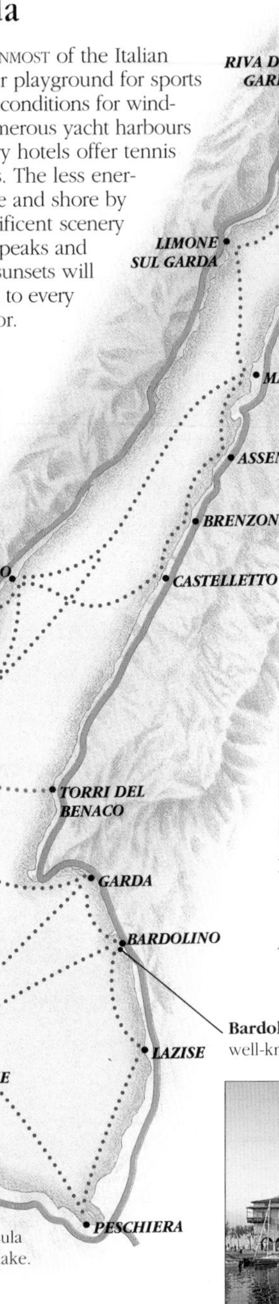

RIVA DEL GARDA
TORBOLE
LIMONE SUL GARDA
MALCESINE
ASSENZA
BRENZONE
GARGNANO
CASTELLETTO
BOGLIACO
La Gardesana
MADERNO
GARDONE RIVIERA
SALO
TORRI DEL BENACO
PORTESE
GARDA
MANERBA
BARDOLINO
MONIGA
LAZISE
SIRMIONE
DESENZANO
PESCHIERA

Desenzano
With its lively harbour and palm-fringed promenades, Desenzano is the main terminus for steamer excursions.

Bardolino gave its name to the well-known red wine.

The Sirmione peninsula is best seen from the lake.

KEY

- Steamer routes
- Car ferry

0 kilometres 10

0 miles 5

Peschiera
The town's enclosed harbour was built by the Austrians in the 1860s, during the Italian Wars of Independence.

.A GARDESANA

his is the name given to the 43-km (89-mile) perimeter road ıat hugs the lake shore. For ıuch of its route the road is cut ırough solid rock, sometimes ɔllowing a narrow ledge in the liff face, sometimes passing ırough tunnels (around 80 in ɔtal). The switchback route ffers spectacular views at every ırn, particularly at Gargnano, nd there are numerous viewing oints. Places of interest along La ardesana include the splendid 8th-century gardens of Palazzo ettoni at Bagliaco and the astle at Riva del Garda.

he scenic road to Limone

LAKE TRIPS

Lake Garda's ferries are still called steamers, even though they are diesel-powered today. The major towns around the southern rim of the lake all have jetties where you can buy a ticket and board the boat for a leisurely cruise. From the water you can see gardens and villas that are otherwise hidden from view. A round trip takes three hours, but allow longer if you want to get off and explore. There is also a regular hydrofoil service around the lake, and catamarans operate around the southern end.

The hydrofoil operating out of Desenzano harbour

Lake Garda steamer at dusk near Peschiera

Garda ⑩

Road map A3. *3,400.*
Lungolago Regina Adelaide 3.
Fri. ***Shops closed*** *Wed pm.*

NUMEROUS PAVEMENT cafés brighten the streets and fill the air with conversation around the central Palazzo dei Capitani, which was built in the 15th century for the use of the Venetian militia. The small town museum has exhibits relating to prehistoric rock engravings found at Punta San Vigilio, 2 km (1 mile) west of the town.

Peschiera ⑪

Road map A4. *8,800.*
Piazzale Betteloni 15. (045) 755 03 81. *Mon am.* ***Shops closed*** *Wed in winter.*

AT PESCHIERA the River Mincio flows out of Lake Garda to join the River Po. The fortress at the entrance was built by Austrians in the 19th century. Named Fortezza del Quadrilatero because of its square shape, it replaced a 15th-century stronghold.

ENVIRONS: Just outside the town is **Gardaland**, a theme park with a replica of the ancient Egyptian Valley of the Kings. Another attraction for children is the **Parco Faunistico**, a zoo with a safari park and models of dinosaurs.

Gardaland
Loc. Ronchi, 37014 Castelnuovo del Garda. *(045) 644 97 77 or 644 95 55.* *Apr–Oct.*

Parco Faunistico
Nr. Bussolengo. *(045) 717 00 80.* *9am–dusk daily.* *Nov–mid Mar: Wed (Safari Park winter).*

Solferino ⑫

Road map A4. *2,118.*
Via Francese. (376) 85 40 68.
Sat pm. ***Shops closed*** *Mon pm.*

THE BATTLE of Solferino (1859) left 40,000 Italian and Austrian troops dead and injured, abandoned without medical care or burial. Shocked by such neglect, a Swiss man named Henri Dunant began a campaign for better treatment. The result was the first Geneva Convention, signed in 1863, and the establishment of the International Red Cross. In the town of Solferino there is a war museum and an ossuary chapel, lined with bones from the battlefield. There is also a memorial to Dunant built by the Red Cross with donations from member nations.

The ossuary chapel at Solferino, lined with skulls

Sirmione Peninsula ⓭

Charming Sirmione is a finger of land extending into the southern end of Lake Garda, connected to the mainland by a bridge. The Roman poet Catullus (born in 848 BC) owned a villa here: the ruins of the Grotte di Catullo lie among ancient olive trees at the northern tip. The Rocca Scaligeri castle stands guard at the base of the peninsula, and beyond, the narrow streets of the village give way to peaceful lakeside walks and elegant spa hotels.

View Towards the Grotto
The high central tower commands views over the castle and the whole of the Sirmione peninsula.

★ Rocca Scaligeri
The castle was built in the 13th century by the Scaligeri of Verona. It is cleverly designed to trap shipborne invaders, leaving them vulnerable to missiles dropped from the castle walls.

The main keep tower was used for bombarding attackers trapped below.

The moat, originally a complex defence system, is today home to carp and waterlilies.

Piazza Castello

Sirmione Old Town
Narrow stone-paved streets are packed with shops selling crafts and souvenirs.

Star Features

- **★ Rocca Scaligeri**
- **★ Grotte di Catullo**

Visiting the Peninsula
Cars must be parked before entering Sirmione, leaving the medieval streets for pedestrians.

Lakeside Walk
Following the eastern shores of the peninsula, this pretty walk links the village to the Grotte di Catullo.

Visitors' Checklist

Road map A4. Viale Marconi 2. (030) 91 61 14.
Rocca Scaligeri and Castle Museum (030) 91 64 68. Summer: 9am–6pm daily. Winter: 9am–1pm Tue–Sun.
Grotte di Catullo (030) 91 61 57. 9am–6pm. Mon except public hols.

San Pietro
Founded in AD 765, on Sirmione's highest point, this church contains a 12th-century fresco of Christ in Majesty.

★ Grotte di Catullo
This complex of villas, baths and shops, built as a resort for wealthy Romans from the 1st century BC, lies ruined here. Finds are displayed in the Antiquarium building.

The inner harbour provided a haven for fishermen during lake storms and an anchorage for the castle fleet.

The drawbridge is heavily fortified, linking the castle to the mainland and offering an escape route to its inhabitants.

The Scaligeri

The Rocca Scaligeri is one of many castles built throughout the Verona and Lake Garda region by the Scaligeri family *(see p192)*. During the turbulent 13th and 14th centuries, powerful military rulers fought each other incessantly in pursuit of riches and power. Despite the autocratic nature of their rule, the Scaligeri brought a period of peace and prosperity to the region, fending off attacks by the predatory Visconti family who ruled neighbouring Lombardy.

The Scaligeri ruler, Cangrande I

Salò ⓮

Road map A3. *10,000.*
Lungolago Zanardelli. *Sat.*

LOCALS PREFER to associate this elegant town with Gaspare da Salò (1540–1609), the inventor of the violin, rather than with Mussolini, the World War II dictator. Mussolini set up the so-called Salò Republic in 1943 and ruled northern Italy from here until 1945, when he was shot by the Italian resistance.

Happier memories are evoked by Salò's buildings, including the cathedral with its unusual wooden altarpiece (1510) by Paolo Veneziano. The main appeal of the town derives from its waterfront buildings, painted in pastel shades, and the lake views. Salò marks the start of the Riviera Bresciana, where the shore is lined with luxurious villas and grand hotels set in semi-tropical gardens.

Gardone Riviera ⓯

Road map A3. *2,500.*
Corso Repubblica 37. *Wed.*

GARDONE'S MOST appealing feature is the terraced public park which cascades down the hillside, planted with noble and exotic trees. Equally exotic are the Mediterranean and African plants in the **Hruska Botanical Gardens**, founded in 1910, which benefit from the town's mild winter climate. Gardone has long been a popular resort – the magnificent 19th-century **Villa Alba** (now a congress centre) was built for the Austrian emperor to escape the bitter winters of his own country. The Art Deco **Grand Hotel** on the waterfront was built for lesser beings.

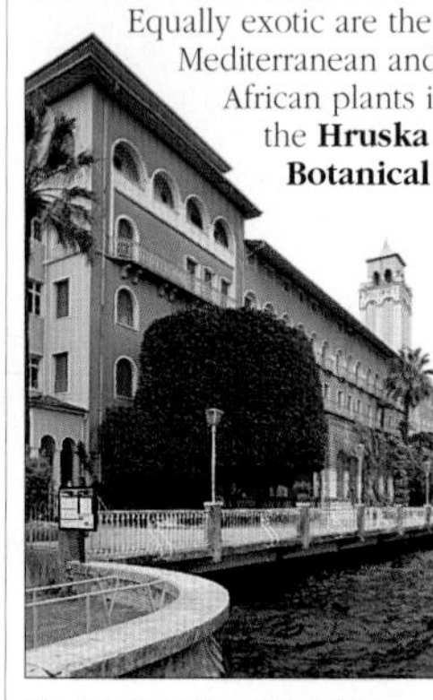

The Art Deco Grand Hotel in Gardone Riviera

High above the town is the **Villa il Vittoriale**, built for the poet, Gabriele d'Annunzio. His Art Deco villa has blacked out windows (he professed to loathe the world) and is full of curiosities, including a coffin-shaped bed. The garden has a landlocked warship, the prow raised high over Lake Garda.

Hruska Botanical Gardens
Via Roma. *daily Mar–Oct.*

Villa il Vittoriale
Via Vittoriale 12. *(0365) 201 30.*
Tue–Sun.

Valpolicella Wine Tour

THIS CIRCULAR TOUR takes in the beautiful, remarkably varied scenery of the wine district that lies between Verona and Lake Garda. On the shores of Lake Garda itself, deep and fertile glacial soils provide sustenance for the grapes that are used to make Bardolino, a wine that is meant to be drunk young *(see pp238–9)*. Inland, the rolling foothills of the Lessini mountains shelter hamlets where lives and working rhythms are tuned to the needs of the vines. These particular vines are grown to produce the equally famed Valpolicella, a red wine that varies from light and fruity to full-bodied.

Affi ④
This wine-producing village is surrounded by vineyards planted in the sheltered basin of the Adige Valley.

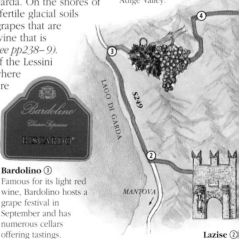

Bardolino ③
Famous for its light red wine, Bardolino hosts a grape festival in September and has numerous cellars offering tastings.

Lazise ②
Lazise has long been the chief port of Garda's eastern shore, its picturesque harbour and medieval church guarded by a 14th-century castle.

TIPS FOR DRIVERS

Starting point: *Verona.*
Length: *45 km (28 miles).*
Approximate driving time: *3 hours.*
Stopping-off points: *The main village of the Valpolicella region, San Pietro in Cariano, has cafés and restaurants.*

KEY

Tour route

Other roads

ooking across Lake Garda from Riva del Garda

iva del Garda ⑯

oad map B3. 13,600.
Giardini di Porta Orientale 8.
)464) 55 44 44. 2nd & 4th Wed
ı summer. ***Shops closed*** *Mon, Sat*
m in winter.

LIVELY RIVA'S waterfront is overlooked by the moated **occa di Riva**, a former caliger fortress. Inside is a good museum, with exhibits from the region's prehistoric lake villages, built by driving huge piles far out into the lake bed to support house platforms. Today, windsurfers use the lake because of the consistent offshore winds.

Rocca di Riva
Piazza Cesare Battista. *(0464) 57 38 69.* *Tue–Sun.*

Malcésine ⑰

Road map B3. *3,500.*
Via Capitanato 6. (045) 740 05 55. *Sat.*

GERMAN VISITORS who come to Malcésine trace the journey taken by the poet Goethe in 1788. His travels were full of mishaps, and at Malcésine he was accused of spying and locked up.

From Malcésine, visitors can take the **cable car** up to the flat summit of Monte Baldo (1,745 m/5,725 ft). The journey takes 15 minutes, and on a clear day it is possible to see the distant peaks of the Dolomites and the Gruppo di Brenta range. Footpaths for walkers are signposted at the top. The lower slopes are designated nature reserves; a good place to see the local flora is the **Riserva Naturale Gardesana Orientale**, just to the north of Malcésine.

ant'Ambrogio di alpolicella ⑤
part from red wine, this village a source of the pink stone used for Verona's palaces.

Gargagnago ⑥
The Alighieri wine estate is owned by a direct descendant of the medieval poet Dante, and set around a 14th-century villa built by Dante's son.

Cloisters of San Giorgio in Valpolicella

Pedemonte ⑦
The Villa Santa Sofia wine estate operates out of a theatrical villa designed by Palladio, but never completed.

TO
San Giorgio
San Floriano
San Pietro in Cariana

S12
Adige

Pescantina
Biffi
PADOVA
S11

Verona ①
The city has numerous old-fashioned bars, called *osterie*, where visitors can go to sample local wine.

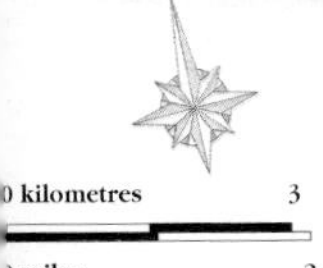

0 kilometres 3
0 miles 2

TRE CIME
TRE CIME
TRE CIME

THE DOLOMITES

THE NAME OF THE DOLOMITES *conjures up a vision of spectacular mountains, as noble and awe-inspiring as the Alps. To the south of the region lie the cities of Feltre, Belluno and Vittorio Veneto. To the north is the renowned ski resort of Cortina d'Ampezzo. In between, travellers will encounter no more cities – just ravishing views, unfolding endlessly, and pretty hamlets tucked into remarkably lush and sunny south-facing valleys.*

The Dolomites cover a substantial portion of the Veneto's land mass, and it is easy to forget, when visiting the cities of the flat Veneto plain, that behind them lies this range of mountains rising to heights of 2,000 m (6,500 ft) and more. Catering for an urban population hungry for fresh air and freedom, the towns and villages of the Dolomites have striven to balance the needs of tourism and nature.

Italian is the region's principal language, although German is also widely spoken, reflecting the region's strong historic links with the Austrian Tyrol. Once ruled by the Hapsburgs, certain areas of the region only became part of Italy in 1918, after the break up of the Austro-Hungarian empire at the end of World War I. Some of that war's fiercest fighting took place in the Dolomites, as both sides tried to wrest control of the strategic valley passes linking Italy and Austria-Hungary. Striking war memorials in many villages and towns provide a sad reminder of that time.

Today the region is renowned for its winter sports facilities. International cross-country ski competitions were held in Cortina d'Ampezzo as early as 1902, and in 1956 the town hosted the Winter Olympics. Today, Cortina is considered to be Italy's most exclusive resort, the winter playground of film stars and royalty.

Outdoor café in the old town of Feltre

◁ **Pleasure boats on Lake Misurina, looking towards the peaks of Tre Cime**

Exploring the Dolomites

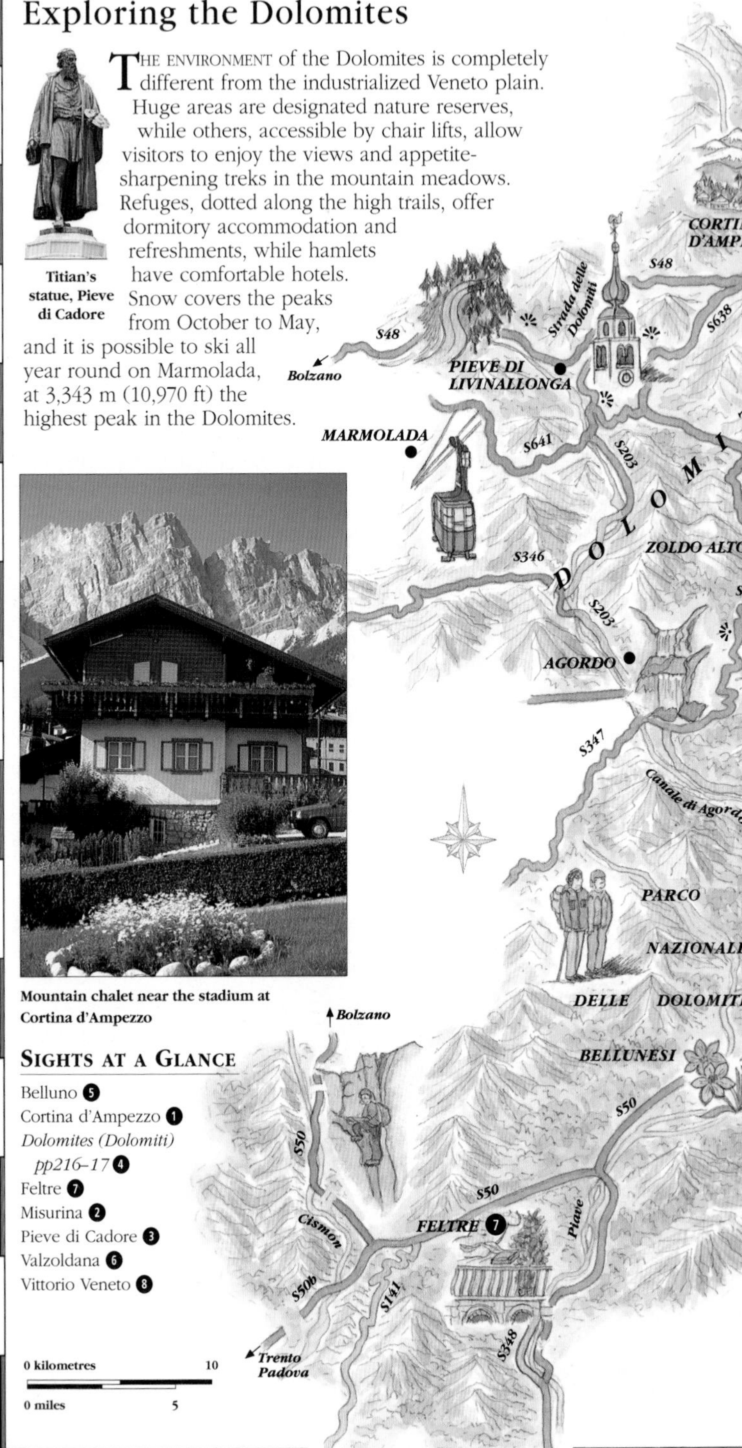

The environment of the Dolomites is completely different from the industrialized Veneto plain. Huge areas are designated nature reserves, while others, accessible by chair lifts, allow visitors to enjoy the views and appetite-sharpening treks in the mountain meadows. Refuges, dotted along the high trails, offer dormitory accommodation and refreshments, while hamlets have comfortable hotels. Snow covers the peaks from October to May, and it is possible to ski all year round on Marmolada, at 3,343 m (10,970 ft) the highest peak in the Dolomites.

Titian's statue, Pieve di Cadore

Mountain chalet near the stadium at Cortina d'Ampezzo

Sights at a Glance

Belluno 5
Cortina d'Ampezzo 1
Dolomites (Dolomiti) pp216–17 4
Feltre 7
Misurina 2
Pieve di Cadore 3
Valzoldana 6
Vittorio Veneto 8

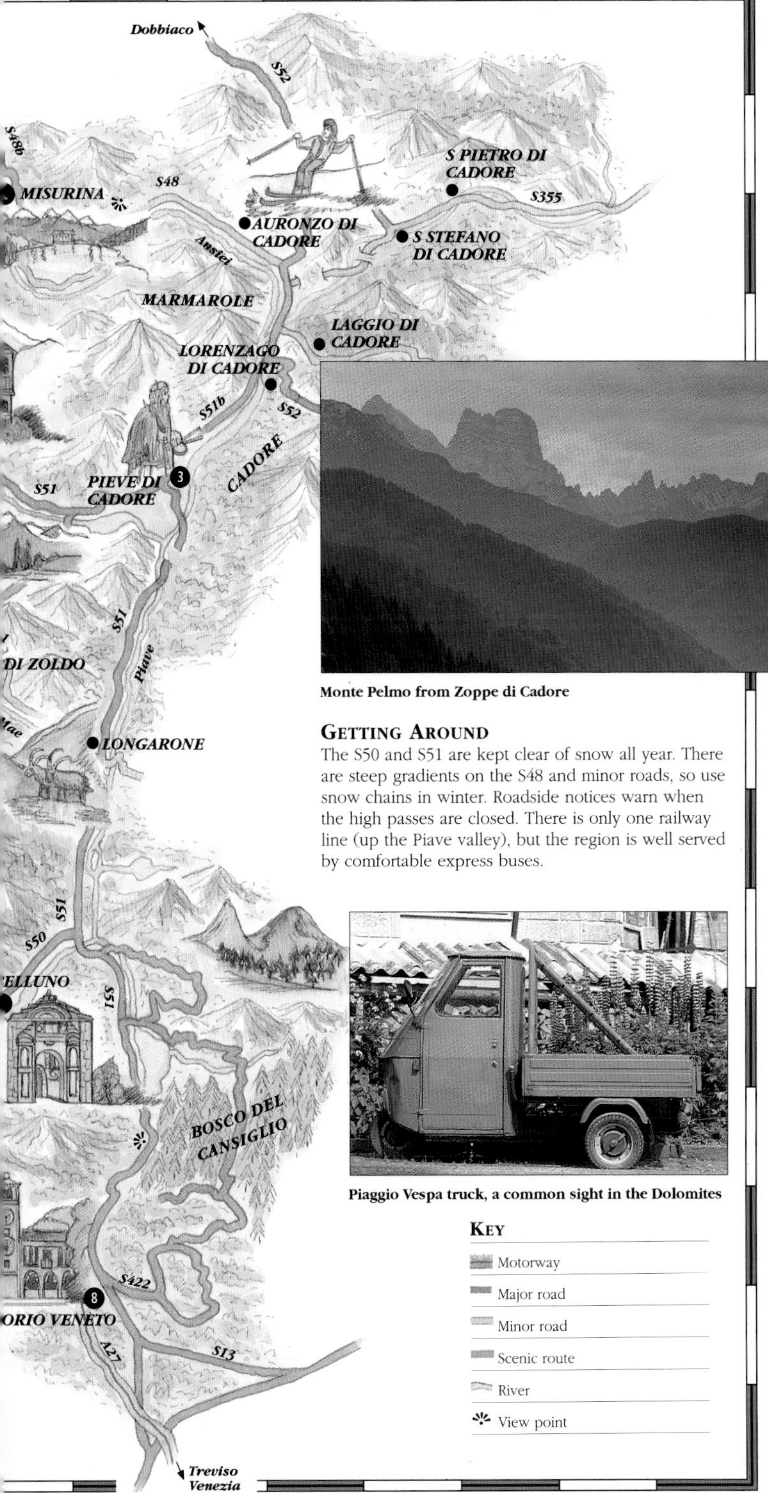

Monte Pelmo from Zoppe di Cadore

Getting Around

The S50 and S51 are kept clear of snow all year. There are steep gradients on the S48 and minor roads, so use snow chains in winter. Roadside notices warn when the high passes are closed. There is only one railway line (up the Piave valley), but the region is well served by comfortable express buses.

Piaggio Vespa truck, a common sight in the Dolomites

Key

- Motorway
- Major road
- Minor road
- Scenic route
- River
- View point

Cortina d'Ampezzo ❶

Road map D1. 7,000. Piazzetta San Francesco 8. (0436) 3231. Tue, Fri.

ITALY'S TOP SKI RESORT, much favoured by the smart set from Turin and Milan, is well supplied with restaurants and bars. The reason for its popularity is the dramatic scenery, which adds an extra dimension to the pleasure of speeding down the slopes.

Skiers at Cortina

Strolling along the Corso Italia in Cortina d'Ampezzo

Wherever you look, crags and spires rise skyward, thrusting their weather-sculpted shapes above the trees.

As a consequence of the resort hosting the 1956 Winter Olympics, Cortina has better than normal sports facilities. If you want something more adventurous than downhill or cross-country skiing, there is a ski jump and a bobsleigh run to test your nerve. The Olympic ice stadium regularly holds skating discotheques, and there are several good swimming pools, tennis courts and riding facilities.

During the summer months Cortina becomes an excellent base for walkers. Information on trails and guided walks is available from the tourist office or, during the summer, from the Club Alpino Italiano next door *(see p224)*.

The Dolomite Road

THE STRADA DELLE DOLOMITI, or Dolomite Road, is one of the most beautiful routes anywhere in the Alps, and is a magnificent feat of highway construction. It starts in the Trentino-Alto Adige region at Bolzano and enters the Veneto region at the Passo del Pordoi, at 2,239 m (7,346 ft) the most scenic of all the Dolomite passes. From here the route follows the winding S48 for another 35 km (22 miles) east to the resort of Cortina d'Ampezzo.

There are plenty of stopping places along the route where you can enjoy spectacular views. In many of the ski resorts, cable cars will carry you up to alpine refuges (some with cafés attached) that are open from mid June to mid September. These refuges mark the start of a series of signposted walks.

Tips for Drivers

Starting point: *Passo del Pordoi.*
Length (within the Veneto): *35 km (22 miles).*
Approximate driving time: *Two hours, but allow a full day to include the return journey and time to stop and enjoy the stunning scenery.*
Stopping-off points: *The small towns of Pieve di Livinallongo and Andraz have good cafés and restaurants.*

Passo del Pordoi ①
To the south is the Gran Vernal at 3,210 m (10,530 ft); to the north the Gruppo di Sella rise to 3,152 m (10,340 ft).

← BOLZANO

S48

Key

Tour route

Other roads

View point

0 kilometres 5

0 miles 2

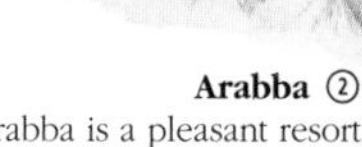

Arabba ②
Arabba is a pleasant resort with a funicular railway to the Col Burz (1,943 m/ 6,370 ft) to the north.

Misurina ❷

Road map D1. 81.
Via Misurina. (0436) 390 16 (summer and Christmas only).

SMALLER AND QUIETER than Cortina, Misurina nestles by the exquisite Lake Misurina. The lake's mirror-like surface reflects the peaks of Monte Cristallo, Cima di Cadini and Tre Cime (Three Peaks) di Lavaredo. Take the road that climbs northeast for 8 km (5 miles) to the Auronzo mountain refuge for a stunning view of the peaks (2,999 m/9,840 ft).

One of the creeks flowing into Lake Misurina

Titian's house at Pieve di Cadore

Pieve di Cadore ❸

Road map D1. *4,000.*
Via XX Septembre 18. (0435) 316 44. Mon. ***Shops closed*** *Wed pm (food), Mon am (clothes).*

FOR CENTURIES the Cadore forests supplied Venice with its timber. The main town of this vast mountainous region is Pieve di Cadore, primarily known as the birthplace of Titian. The humble **Casa di Tiziano** can be visited, and the nearby **Museo Archeologico** has exhibits of finds from the pre-Roman era.

Principally, though, this is a base for touring the scenic delights of the region. North of Pieve the valley narrows to a dramatic ravine, and the road north to Comelica and Sesto is noted for its alpine scenery and its traditional balconied houses. Continuing northeast, you can follow the Piave river to its source, 8 km (5 miles) north of Sappada.

Casa di Tiziano

Via Arsenale 4. *(0435) 322 62.*
Jun–Sep: Tue–Sun.

Museo Archeologico Romano e Preromano

Palazzo della Magnifica Comunità Cadorina, Piazza Tiziano 2. *(0435) 322 62.* *Jun–Sep: Tue–Sun.*

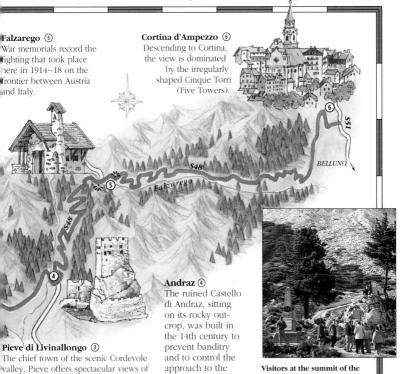

S51
BELLUNO
S48
Falzarego
S48
6
5
4

Falzarego ⑤
War memorials record the fighting that took place here in 1914–18 on the frontier between Austria and Italy.

Cortina d'Ampezzo ⑥
Descending to Cortina, the view is dominated by the irregularly shaped Cinque Torri (Five Towers).

Andraz ④
The ruined Castello di Andraz, sitting on its rocky outcrop, was built in the 14th century to prevent banditry and to control the approach to the Passo di Falzarego.

Pieve di Livinallongo ③
The chief town of the scenic Cordevole valley, Pieve offers spectacular views of dolomitic peaks and cliffs.

Visitors at the summit of the Passo di Falzarego

The Dolomites ❹

IT WAS DR DEODAT DOLOMIEU who, in 1789, first analysed the composition of the mountain range named after him. The Dolomites, formed of mineralized coral laid down beneath the sea during the Triassic era, were uplifted when the European and African continental plates collided 60 million years ago. Unlike the glacier-eroded saddles and ridges of the main body of the Alps, the gleaming white rocks have been sculpted by ice, sun and rain into the characteristic cliffs, spires and "organ pipes" that we see today; an inspiring sight, especially when dawn sunlight turns the rock rose pink.

Outdoor Activities
Chair-lifts provide access to ski-runs, footpaths and picnic sites, for a closer encounter with the grandeur of the peaks.

Climbers and Hikers
Much of the countryside is accessible only on foot, but the Dolomites have an extensive network of waymarked paths for hikers, and mountain huts (rifugi) *for overnight stops.*

Small rural communities live by dairy farming and forestry.

At Lake Misurina and other artificial and natural leisure lakes the outflow is used to supply hydro-electric power

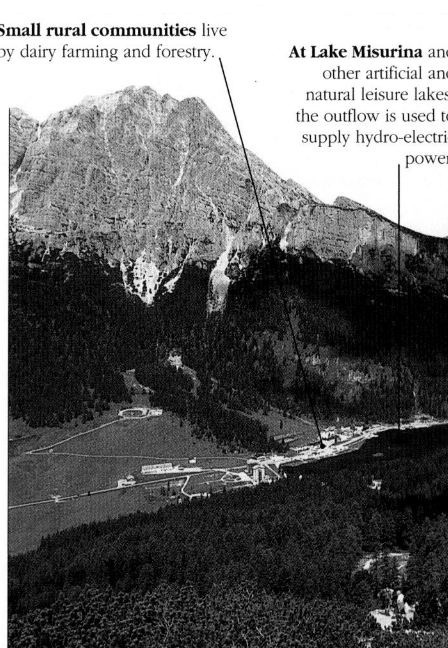

Onion Domes
Church towers in the Dolomites are often capped by onion-shaped domes, a legacy of Austro-Hungarian rule in the region (see p46).

Monte Piana
Reminders of World War I battles still scar the Dolomites. Monte Piana, on the Austro-Italian front, saw much bitter fighting.

Traditional Farms
Large timber barns are used for sheltering animals and storing winter fodder and fuel. Overhanging roof eaves take rain and snow away from the house.

Marked routes now guide walkers along tracks created over several centuries by shepherds and cowherds.

Forests of larch, spruce, pine and fir provide commercial timber, firewood and shelter for wildlife.

The Landscape of the Dolomites

The deep-sided river valleys between the Dolomites' craggy peaks shelter isolated villages which have traditionally survived from forestry and keeping livestock. Tourism has had a major impact on these small communities, some of which have become large resorts. Extensive areas of the Dolomites are now protected by law from development and are accessible only on foot.

Nature in the Dolomites

Forests and meadows support an astonishing richness of wildlife. Alpine plants, in flower between June and September, have evolved their miniature form to survive the harsh winds.

The Flora

***Gentian** seeds are used to make a bitter local liqueur.*

***The red mountain lily** thrives on sun-baked slopes.*

***Potentilla** roots were once used to produce red dye.*

***Saponaria**, a type of soapwort, covers rocks and scree.*

The Fauna

***The ptarmigan's** plumage changes from mottled brown in summer to snow white in winter. This bird feeds on mountain berries and young plant shoots.*

***The chamois**, a shy mountain antelope prized for its soft, pliant skin, is protected in the national parks, where hunting is forbidden.*

***Roe deer** are very common, because their natural predators – wolves and lynx – have died out. Their voracious appetite for tree saplings has become a problem for foresters.*

Belluno ❺

Road map D2. 35,800. FS *Via Rodolfo Psaro 21. (0437) 94 00 83. Sat.* ***Shops closed*** *Mon am & Wed pm.*

Façade and entrance to Palazzo Rettori in Belluno

Picturesque Belluno, capital of Belluno province, serves as a bridge between the two very different parts of the Veneto, with the flat plains to the south and the Dolomite peaks to the north. Both are encapsulated in the picture-postcard views to be seen from the 12th-century **Porta Ruga** at the southern end of Via Mezzaterra, the main street of the old town. Even more spectacular are the views from the campanile of the 16th-century **Duomo** which was designed by Tullio Lombardo, but rebuilt twice after damage by earthquakes.

The nearby baptistry contains a font cover with the figure of John the Baptist carved by Andrea Brustolon (1662–1732), whose elaborate furnishings decorate Ca' Rezzonico in Venice *(see p126)*. Brustolon's works also grace the churches of **San Pietro** (altars and angels) and **Santo Stefano** (candelabra and a Crucifix). On the same square is the **Torre Civica** (12th-century), all that survives of the medieval castle, and the city's most elegant building, the Renaissance **Palazzo dei Rettori** (1491), once home to Belluno's Venetian rulers.

Exterior fresco, Zoppe di Cadore

The **Museo Civico** is worth visiting for the archaeological exhibits, and the paintings by Bartolomeo Montagna (1450–1523) and Sebastiano Ricci (1659–1734). Just to the right of the museum is Belluno's finest square, the **Piazza del Mercato**, which features arcaded Renaissance palaces and a fountain built in 1410.

South of the town are the ski resorts of the Alpe del Nevegal. It is worth taking the chair lift in the summer to the Rifugio Brigata Alpina Cadore (1,600 m/5,250 ft) which has superb views and a botanical garden specializing in alpine plants.

Museo Civico

Via Duomo 16. *(0437) 94 43 36* *Tue–Sun.* *Oct–mid-Apr: Sat*

Valzoldana ❻

Road map D1. *from Longarone* *Via Roma 10, Forno di Zoldo (0437) 78 73 49.*

The wooded Zoldo valley is a popular destination for walking holidays. Its main resort town is Forno di Zoldo and the surrounding villages are noted for their Tyrolean-style alpine chalets and hay-lofts. Examples built in wood

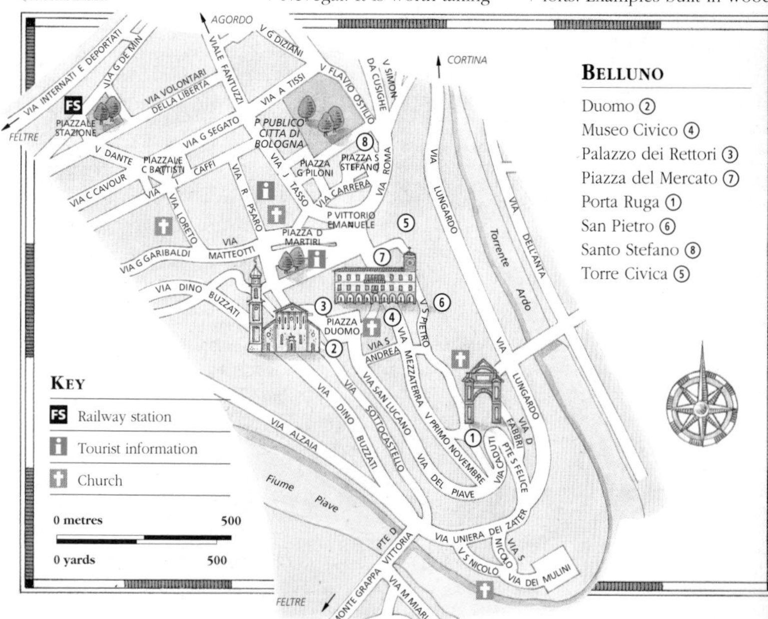

›n stone foundations can be ›een at Fornesighe, 2 km (1 ›nile) northeast of Forno di Zoldo, and on the slopes of ›Ionte Penna at Zoppe di Cadore, 8 km (5 miles) north.

If you have the time, do a ›ircular tour of the area. Drive ›orth on the S251, via Zoldo Alto to Selva di Cadore, then ›vest via Colle di Santa Lucia (a favourite viewpoint for keen ›hotographers). From here ›ake the S203 south through ›he lakeside resort of Alleghe. The route passes through ›vonderful scenery with woodland, flower-filled meadows ›nd pretty mountain hamlets ›which complement the splendour of the rocky crags.

The southernmost town of ›he area is Agordo, nestling in ›he Cordevole Valley. From ›ere, a spectacularly scenic ›oute follows the S34 north-east to the Passo Duran (1,605 m/5,270 ft), descending ›o Dont, close to your starting ›oint. Wayside shrines mark ›he route and it is worth stopping on your way down to ›isit village shops selling local ›voodcarving. Take care when ›riving along this narrow and ›vinding road.

› Renaissance *palazzo* on Via ›uzzo in Feltre

›eltre ❼

›oad map D2. *19,600.* FS 🚌 ℹ *Piazzetta Trento e Trieste 9. (0439) 540.* *Tue & Fri am.* ***Shops closed*** *›Ion am, Wed pm.*

FELTRE OWES its venerable good looks to the vengeful ›Ioly Roman Emperor, Maximilian I. He sacked the town ›wice, in 1509 and in 1510, at ›he outbreak of the war against ›'enice waged by the League ›f Cambrai *(see p44)*. Despite ›he destruction of its buildings ›nd the murder of most of its ›itizens, Feltre remained ›toutly loyal to Venice, and Venice repaid the debt by rebuilding the town after the war. Thus the main street of the old town, Via Mezzaterra, is lined with arcaded early 16th-century houses, most with steeply pitched roofs to keep snow from settling.

Follow the steep main street to the striking Piazza Maggiore, where you can see the remains of Feltre's medieval castle, the church of **San Rocco** and a fountain by Tullio Lombardo (1520).

On the eastern side of the square, Via L Luzzo is lined with fine Renaissance palaces, one of which houses the **Museo Civico**, currently closed for restoration. This displays paintings by the local artist Lorenzo Luzzo, who was known as Il Morto da Feltre (The Dead Man of Feltre), a nickname given to him by his contemporaries because of the deathly pallor of his skin.

🏛 Museo Civico

Palazzo Villabruna, Via L Luzzo 23.
(0439) 88 52 42. ● *for restoration.*

Selva di Cadore from Colle di Santa Lucia, northwest of Valzoldana

Vittorio Veneto ❽

Road map D2. *30,000.* FS 🚌 ℹ *Piazza del Popolo 18. (0438) 572 43.* **Shops closed** *Tue (some Wed pm).* *Mon.*

TWO SEPARATE TOWNS, Ceneda and Serravalle, were merged and renamed Vittorio Veneto in 1866 to honour the unification of Italy under King Vittorio Emanuele II. The town later gave its name to the last decisive battle fought in Italy in World War I. The **Museo della Battaglia** in the Ceneda quarter, the commercial heart of the town, commemorates this. Serravalle is more picturesque, with many fine 15th-century *palazzi*, and pretty arcaded streets. Franco Zeffirelli shot scenes for his film *Romeo and Juliet* in this town that sits at the base of the rocky Meschio gorge. To the east, via Anzano, the S422 climbs up to the Bosco del Cansiglio, a wooded plateau.

🏛 Museo della Battaglia

Piazza Giovanni Paolo I.
(0438) 576 95. ○ *Tue–Sun.*

Vittorio Veneto old town and river

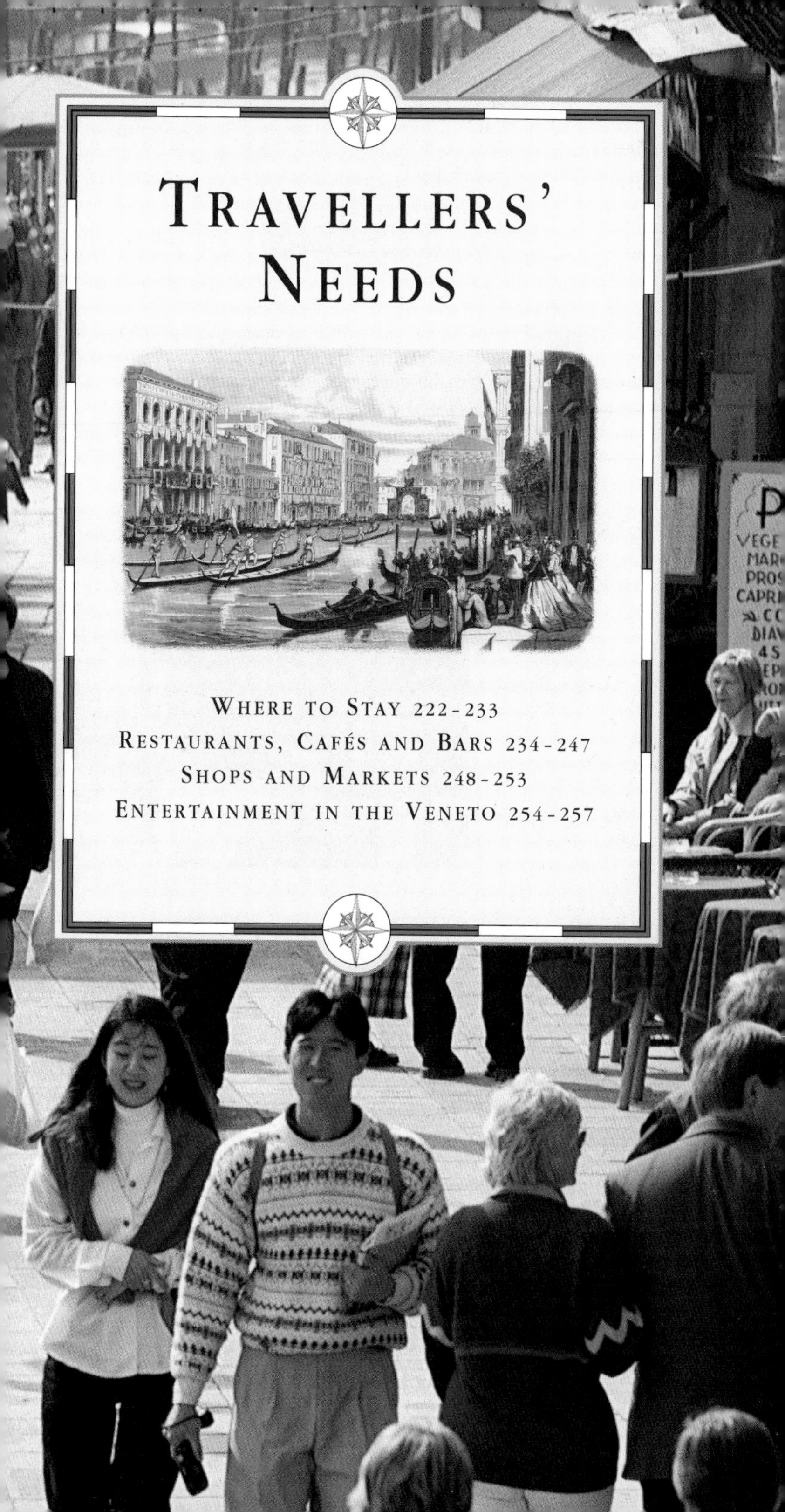

Travellers' Needs

WHERE TO STAY

VENICE'S PERENNIAL attraction to romantics and art lovers means it has an astonishing number of hotels for its size, many of them in former *palazzi*. On the mainland, ancient cities abound with hotels and *pensioni*, often housed in magnificent old buildings and extravagantly decorated. Those in the smaller towns are often run by families who take pride in their reputation.

Sign for a small hotel

Lake Garda is a long-established resort area with many hotels to choose from, and the mountainous north of the region is an all-year-round holiday area with accommodation of all types. Here you can find self-catering in a small farmhouse at very reasonable cost, and there are also numerous idyllically situated and well-equipped campsites. Budget options in the cities include self-catering flats, hostels and dormitory accommodation, and the mountains offer simple refuges for enthusiastic walkers. For more information on hotels in Venice and the Veneto see the listings on pages 228–33.

WHERE TO LOOK

UNLIKE MOST other cities, Venice has hardly any "undesirable" addresses. You will pay considerably more for a hotel in the immediate vicinity of the Piazza San Marco, but in such a compact city even apparently outlying areas such as Cannaregio or Santa Croce *(see pp14–15)* are never far from places of interest. Addresses in Venice are immensely confusing *(see p277)* but a map reference for each hotel is given in the listings. The maps referred to are to be found on pages 282–9 (Venice) and pages 12–13 (the Veneto).

Most visitors feel it is worth splashing out for a few nights' stay in Venice itself, despite the cost, though an increasing number stay in Verona or

Hotel Europa e Regina *(see p229)* overlooking the Grand Canal

Outside the Hotel Marconi *(see p229)* on the busy Riva del Vin

Padua and "commute" into Venice by train. Do not be tempted by the relatively low prices of the Mestre hotels, unless you are prepared to stay in a sprawling industrial town. Remember, too, that if you are travelling by car you will have to pay stiff parking charges at the Piazzale Roma car park or one of its satellites for the duration of your stay in Venice *(see pp278–279)*.

Many of the hotels in minor inland towns of the Veneto cater primarily for business travellers, but if you plan to explore the region you will find some lovely villa hotels in the countryside. Padua and Verona have a number of hotels, but those in Verona are fully booked for months ahead in the summer opera season, so forward planning is essential. Further north there is more choice. The hotels are geared to holiday-makers, with lovely gardens, swimming pools and sports facilities. But bear in mind that Italians as well as foreign tourists flock to the lakes and mountains, so it is always advisable to plan your trip and book in advance.

HOTEL PRICES

HOTEL CHARGES were de-regulated in 1994, so that hotels are free to charge what they feel the market will bear rather than being tied to the tariffs determined by their star rating. Venice is an expensive place to stay and nowadays can hardly be said to have a "low season" with the benefits of lower or negotiable prices. You will be unlikely to find a basic double room for less than L100,000. Occasionally you can find some cheaper ooms from November to 'ebruary, when the weather s often superb. But remem- ber that many hotels close out of season. Some re-open or Carnival – and raise their orices accordingly.

July and August are the nost expensive months at the esorts along Lake Garda. In he Dolomites winter, when kiers flock to the area, is the nigh season and the hotels nay close during the summer.

Single room rates are higher han individual rates for two people sharing a double oom. Prices include tax.

Hidden Extras

If you are travelling on a budget, try to avoid hotels with inclusive breakfast as his is rarely good value for noney. You are expected to ip at least L1,000 for room service and about L2,000 for bellboys, even if service is ncluded in the price of the oom. Laundry services are usually expensive, as are drinks from the minibar and elephone calls from hotel rooms. Check all the rates when you make the booking. Some small hotels in Venice, and most in holiday areas ike Lake Garda, may expect you to take full- or half-board during the high season.

Hotel Gradings and Facilities

Italian hotels are classified by a rating system from one to five stars. However, each province sets its own level for grading, so standards for each category may vary from one area to another. Some hotels may not have a restaurant, but those which do sometimes welcome non-residents who wish to eat.

Air-conditioning is rare in old buildings. Although the thick stone walls provide good insulation against the summer's heat, if you cannot tolerate high temperatures it is well worth choosing air-conditioned accommodation in Venice during the hottest months. Under Italian law, central heating remains off, whatever the temperature outside, until 1 November. This is something that is worth remembering if you plan a late-October trip.

Children are welcome everywhere but smaller hotels have limited facilities. Venice is not an ideal destination for children. If you want to take them with you, it is better to choose a hotel on the Lido where they will have access to the beach and also probably a garden.

Hotel Excelsior *(see p231)*

What to Expect

Hotels are obliged by law to register you with the police, so they will ask for your passport when you arrive. They may need to keep it for a few hours, but make sure you take it back, because you will need identification to change money or

Hotel Do Pozzi *(see p229)*

traveller's cheques. Italian hotel rooms are not "cosy": carpets are rare, the storage space is usually very limited and luxuries such as tea-making facilities are unknown, even in four- and five-star establishments. The decor may be simple and Italian taste can be rather different from what you are used to. However, hotel staff will be friendly and charming, and the standard of cleanliness is high. The bathrooms are, almost without exception, spotlessly clean, even when they are shared. Less expensive hotels are unlikely to have bathtubs; showers are considered more hygienic and more economical on water. Rooms without a bathroom usually have a washbasin and towels are provided.

Breakfast is very light – a cup of coffee and a brioche (a plain or cream- or jam-filled pastry), though hotels generally include fruit juice, bread rolls and jam as well. It is always cheaper to have breakfast in a bar.

With the exception of Venice, where the only sounds are water-borne or human, Italian towns can be very noisy. If you are a light sleeper, ask for a room that is away from the street, or come equipped with earplugs to deaden traffic sounds and church bells.

Check-out time is usually noon in four- and five-star hotels and between ten and noon in small establishments. If you stay longer you will be asked to pay for an extra day.

The elegant Villa Cortine Palace on the Sirmione peninsula *(see p232)*

Booking and Paying

Book at least two months in advance if you want to stay in a particular hotel in the high season; some people book as far as six months or a year ahead in Venice itself. The local tourist office will have listings of all the hotels in the area, and they will be able to advise you on the best hotels in each star category. Hotels above the L100,000 price bracket usually take credit cards, but check which cards are accepted when you make your reservation – Italy is still very much a cash country. You can generally pay the deposit by credit card, or send a Eurocheque or international money order.

Under Italian law, a booking is valid as soon as the deposit is paid and confirmation is received. As in restaurants, you are required by law to keep your hotel receipts until you leave the country.

The conveniently located Hotel Fenice ***(see p229)***

Disabled Travellers

Facilities for the disabled are limited throughout Italy, and Venice poses its own particular problems. A list of tour operators that specialize in holidays for the disabled can be obtained from the **Italian State Tourist Office**. For further advice for the disabled, see p261.

Hotels in Historic Buildings

Many of Venice's hotels are housed in buildings of historical or artistic interest, for example in Gothic *palazzi*. Some of the best are included in the listings here. In the Veneto region there are some attractive villa hotels and a number of these are included in the **International Relais and Châteaux** guide.

Self-Catering

Self-catering flats in Venice proper are hard to come by, owing to the insatiable demand for property in the city. **Italian Chapters** handle some holiday lets within the city, as do **Tailor Made Tours**, or you could try one of the Venetian agents, such as **Serena** or **Sant'Angelo**, though they prefer to deal

Directory

Tourist Office

Italian State Tourist Office (ENIT)
1 Princes Street
London W1R 8AY.
0171-408 1254.

Historic Hotels

Centre d'Information Relais et Châteaux
1 Hay Hill, Berkeley Square, London W1X 7LF.
0171-491 2516.

Self-Catering Agencies

Agriturist Ufficio Regionale
C Monteverdi 15
Venezia Mestre 30170.
(041) 98 74 00.

Alloggi Biasin
Cannaregio 1252.
Map 2 D3.
(041) 71 72 31.

Italian Chapters
102 St John's Wood Terrace,
London NW8 6PL.
0171–722 9560.

Ottolenghi
Cannaregio 180.
Map 2 D3.
(041) 71 52 06.

Sant'Angelo
Campo Sant'Angelo
San Marco 3818.
Map 6 F2.
(041) 522 15 05.

Serena
Ponte Bareteri
San Marco 4931.
Map 7 B1.
(041) 522 46 22.

Tailor Made Tours
22 Church Rise
London SE23 2UD.
0181–291 9736.

Budget Accommodation

Associazione Italiana Alberghi per la Gioventù
Via Cavour 44
00184 Rome.
(06) 487 11 52.

Foresteria Valdese
Campo Santa Maria Formosa,
Castello 5170.
Map 7 C1.
(041) 528 67 97.

Istituto Ciliota
Calle delle Muneghe
San Marco 2976.
Map 6 F2.
(041) 520 48 88.

Ostello Venezia
Fondamenta delle Zitelle
Giudecca 86.
Map 7 B5.
(041) 523 82 11.

Campsites and Mountain Refuges

Club Alpino Italiano
Via Fonseca Pimental 7
20127 Milan.
(02) 26 14 13 78.

Marina di Venezia
Via Montello 6
Litorale del Cavallino.
(041) 96 61 46.

San Nicolò
Riviera San Nicolò 65
Lido. *(041) 76 74 15.*

Touring Club Italiano
Corso Italia 10, 20122
Milan. *(02) 852 61.*

with longer rentals. **Alloggi Biasin** and **Ottolenghi** deal with furnished rooms.

Under the **Agriturismo** scheme there is plenty of self-catering accommodation in the Veneto, usually on working farms. There is an Agriturist office in each region, which will give further information, although you may have to book with the owner. This type of accommodation ranges from simple conversions to luxurious and spacious villas with swimming pools. Prices reflect these variations, and they also fluctuate according to the time of year. Low-season prices for four people start at about L1 million per week.

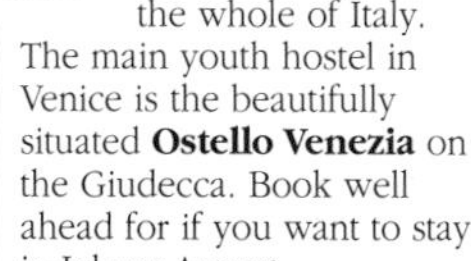

Detail on the Hotel Danieli

Budget Accommodation

One- or two-star budget hotels charging from L40,000 to L80,000 per person per night are generally small, family-run places. These used to be known as *pensioni*, but the term is no longer used very much officially. However, many places retain the name and the personal character that has made them so popular. They rarely offer breakfast and have very few rooms with private bathrooms. You should not expect particularly high standards of service. A variation on these *pensioni* are *affittacamare*, or rented rooms, are even smaller establishments, and they also offer excellent value for money.

Accommodation in hostels and dormitories is sometimes available at convents and religious institutions, and it is often possible to arrange it through the local tourist offices. The **Associazione Italiana Alberghi per la Gioventù** (Italian Youth Hostel Association) in Rome has lists of youth hostels throughout the whole of Italy. The main youth hostel in Venice is the beautifully situated **Ostello Venezia** on the Giudecca. Book well ahead for if you want to stay in July or August.

Lists and booking forms for youth hostels are available through the Italian Tourist Board worldwide, or from local offices. The Venice office also produces a simple typed list of all kinds of hostel accommodation in Venice itself.

Campsites and Mountain Refuges

There are good campsites throughout the region, concentrated mainly on the mainland to the north of Venice, on the shores of Lake Garda and in the northern mountains. A list of campsites and mountain refuges can be obtained from **ENIT** or local tourist offices. Most huts in the mountain districts are owned by the **Club Alpino Italiano**, based in Milan, who can provide full information. The **Touring Club Italiano** publishes annually a list of campsites: *Campeggi e Villaggi Turistici in Italia*.

A suitcase boat transporting visitors' luggage to a hotel

Garden terrace of the Hôtel des Bains on the Lido ***(see p231)***

Using the Listings

Hotels on pages 229–33 are listed according to area and price category. The symbols summarize the facilities available at each hotel.

- all rooms have bath and/or shower unless otherwise indicated
- [1] single-rate rooms available
- rooms for more than two people available, or an extra bed can be put in a double room
- [TV] television in all rooms
- air-conditioning in all rooms
- swimming pool in the hotel
- wheelchair access
- lift
- [P] parking available
- restaurant
- credit cards accepted
- closed out of season *(see p223)*

Price categories for a standard double room per night, including tax and service:

Ⓛ under L100,000
ⓁⓁ L100,000–L160,000
ⓁⓁⓁ L160,000–L240,000
ⓁⓁⓁⓁ L240,000–L340,000
ⓁⓁⓁⓁⓁ over L340,000

Venice's Best Hotels

Hotels in Venice range from the luxurious and renowned, which are mainly clustered along the Grand Canal, to simple, family-run places in the quieter parts of the city. Wherever you stay, you will be within easy reach of the main attractions, with restaurants and shops close at hand. All the hotels shown on this map have something special to recommend them, whether it is the waterside position, a garden or a quiet location away from the crowds. Always book well in advance, and remember that many Venetian hotels are shut at some stage in winter. The hotels shown here are the best in their particular style or price range.

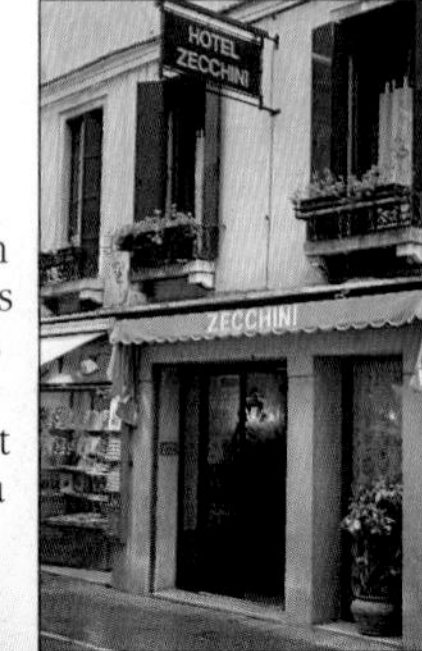

Zecchini
One of the most attractive hotels on the busy Lista di Spagna, the Zecchini is good value. (See p230.)

Cannaregio

Al Sole
Situated beside a tranquil canal, this Gothic palazzo *is away from the main tourist haunts, but within easy reach by foot or water of all the sights.* (See p229.)

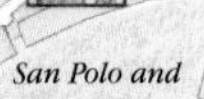

San Polo and Santa Croce

Dorsoduro

0 metres 500

0 yards 500

Agli Alboretti
This charming hotel in a central location has attractive rooms and a garden courtyard. (See p230.)

Gritti Palace
One of Venice's most famous hotels, the Gritti offers rooms and service of impeccable standard in an historic palazzo *on the Grand Canal.* (See p229.)

Giorgione

This high-class, spacious hotel, with its excellent facilities, offers every modern comfort at lower prices than others of similar calibre. (See p231.)

Marconi

This efficiently run hotel, housed in an old palazzo, *has views of the Grand Canal and the Rialto Bridge.* (See p229.)

Castello

Marco

La Residenza

This family-run hotel offers good value for money and is away from the crowds. It has frescoed public rooms and antiques, but the bedrooms are more simple. (See p230.)

Flora

A flower-filled garden is just one of the attractions of this delightful hotel. (See p229.)

Londra Palace

Tchaikovsky once stayed in this grand palazzo *with its views to San Giorgio Maggiore. Today's guests appreciate the welcoming bar and restaurant.* (See p230.)

Hotels in Venice

THIS CHART is a quick reference to recommended hotels in Venice and the lagoon. They are listed within each *sestiere* in price order. More details for these and for hotels in the Veneto are given on the following pages. For information on other types of accommodation, see pages 224–5.

Hotel	Price	Number of Rooms	All Rooms with Bath/Shower	Air Conditioning	Restaurant	Attractive View	Garden/Terrace	Open All Year	Easy Access
SAN MARCO *(see p229)*									
Ai Do Mori	ⓁⓁ	11				■		■	●
Al Gambero	ⓁⓁ	30			●	■		■	●
Do Pozzi	ⓁⓁⓁⓁ	30	●			■	●	■	●
Flora	ⓁⓁⓁⓁ	44	●	■			●	■	●
La Fenice et des Artistes	ⓁⓁⓁⓁ	68	●	■				■	
Santo Stefano	ⓁⓁⓁⓁ	11	●	■		■			
San Moisè	ⓁⓁⓁⓁ	16	●	■		■		■	
Cavalletto e Doge Orseolo	ⓁⓁⓁⓁⓁ	96	●	■	●			■	●
Europa e Regina	ⓁⓁⓁⓁⓁ	192	●	■	●	■	●	■	●
Gritti Palace	ⓁⓁⓁⓁⓁ	93	●	■	●	■	●	■	●
Monaco and Grand Canal	ⓁⓁⓁⓁⓁ	74	●	■	●	■	●	■	●
SAN POLO AND SANTA CROCE *(see pp229–30)*									
Alex	Ⓛ	11						■	
Falier	ⓁⓁ	19	●					■	
Al Sole	ⓁⓁⓁⓁ	80	●	■	●		●	■	
Marconi	ⓁⓁⓁⓁ	26	●	■		■		■	●
Sturion	ⓁⓁⓁⓁ	11	●			■		■	
CASTELLO *(see p230)*									
La Residenza	ⓁⓁⓁ	16	●			■			●
Paganelli	ⓁⓁⓁ	22	●			■	●	■	●
Pensione Wildner	ⓁⓁⓁⓁ	16	●		●	■	●	■	●
Danieli	ⓁⓁⓁⓁⓁ	233	●	■	●	■	●	■	●
Londra Palace	ⓁⓁⓁⓁⓁ	53	●	■	●	■	●	■	●
DORSODURO *(see p230)*									
Montin	Ⓛ	7			●		●	■	
Messner	ⓁⓁ	34	●	■	●		●	■	
Pensione La Calcina	ⓁⓁ	40				■			●
Agli Alboretti	ⓁⓁⓁ	25	●	■	●	■	●	■	●
Pausania	ⓁⓁⓁ	26	●	■			●		
Pensione Seguso	ⓁⓁⓁ	40			●	■	●		●
Pensione Accademia Villa Maravegie	ⓁⓁⓁⓁ	27				■	●	■	
CANNAREGIO *(see pp230–31)*									
Zecchini	ⓁⓁ	27	●					■	●
Abbazia	ⓁⓁⓁ	36					●	■	●
Continental	ⓁⓁⓁⓁ	93	●	■	●	■	●	■	●
Giorgione	ⓁⓁⓁⓁ	70	●	■			●	■	
THE LAGOON ISLANDS *(see p231)*									
Villa Parco (Lido)	ⓁⓁⓁ	23	●	■	●		●		
Quattro Fontane (Lido)	ⓁⓁⓁⓁ	59		■	●		●		
Cipriani (Giudecca)	ⓁⓁⓁⓁⓁ	105	●	■	●	■	●		●
Excelsior Palace (Lido)	ⓁⓁⓁⓁⓁ	197	●	■	●	■	●		●
Hôtel des Bains (Lido)	ⓁⓁⓁⓁⓁ	191	●	■	●	■	●		●

VENICE

San Marco

Ai Do Mori

Calle Larga San Marco, San Marco 658. **Map** 7 B2. *(041) 520 48 17.* ***Rooms:*** *11.* *3.* ⓁⓁ

This friendly little hotel has only a few rooms with their own bath, but it is clean and characterful, and offers extremely good value.

Al Gambero

Calle dei Fabbri, San Marco 4687. **Map** 7 A2. *(041) 522 43 84.* ***Rooms:*** *30.* *4.* ⓁⓁ

Recently modernized, the Al Gambero has simple rooms, many of them singles, and several with a canal view. It is situated a stone's throw from the Piazza on one of Venice's main shopping streets.

Do Pozzi

Corte Do Pozzi, San Marco 2373. **Map** 7 A3. *(041) 520 78 55.* FAX *522 94 13.* ***Rooms:*** *30.* ⓁⓁⓁ

The Do Pozzi *(see p223)* boasts a charming setting in a peaceful green courtyard. Its rooms are quiet, despite its proximity to the tourist-filled Piazza San Marco. The hotel also has a very attractive and welcoming reception area.

Flora

Calle Larga XXII Marzo, San Marco 2283/a. **Map** 7 A3. *(041) 520 58 44.* FAX *522 82 17.* ***Rooms:*** *44.* ⓁⓁⓁ

This quiet hotel in a secluded alley is only a few minutes from the Piazza *(see p227)*. Some rooms are small, although all have their own bathroom. In summer it is possible to enjoy a leisurely breakfast in the fragrant flower-filled garden.

La Fenice et des Artistes

Campiello Fenice, San Marco 1936. **Map** 7 A3. *(041) 523 23 33.* FAX *520 37 21.* ***Rooms:*** *68.*

A pretty hotel, furnished with antiques. On fine days you can sit outside for breakfast, or later in the day enjoy a quiet drink from the bar *(see p224)*.

Santo Stefano

Campo Santo Stefano, San Marco 2957. **Map** 6 F3. *(041) 520 01 66.* FAX *522 44 60.* ***Rooms:*** *11.* ⓁⓁⓁ

The Campo Santo Stefano is one of Venice's most characteristic quarters, and the rooms in this tall, narrow hotel offer a bird's-eye view of the surrounding area. The furnishings are traditionally Venetian, but some of the rooms are rather cramped.

San Moisè

Piscina San Moisè, San Marco 2058. **Map** 7 A3. *(041) 520 37 55.* FAX *521 06 70.* ***Rooms:*** *16.* ⓁⓁⓁ

The San Moisè is a wonderfully quiet but central hotel on a placid back canal. The rooms differ considerably in size, but are in the traditional Venetian style and furnished with antiques.

Cavalletto e Doge Orseolo

Calle Cavalletto, San Marco 1107. **Map** 7 B2. *(041) 520 09 55.* FAX *523 81 84.* ***Rooms:*** *96.*

This excellent hotel has been welcoming visitors for more than 200 years. The rooms are bright and cheerful, with extras ranging from minibars to window boxes. The hotel also has a bar and pleasant restaurant.

Europa e Regina

Calle Larga XXII Marzo, San Marco 2159. **Map** 7 A3. *(041) 520 04 77.* FAX *523 15 33.* ***Rooms:*** *192.* *Private launch service. Access to private beach.* ⓁⓁⓁⓁⓁ

The Europa *(see p222)* offers the best value of all the deluxe hotels. The rooms are large and many of them have views across the Grand Canal. The magnificent public rooms are sumptuously ornate and gilded, in typical Venetian style, and for those who like to eat and drink *al fresco* there is a garden courtyard and canalside terrace with breathtaking views.

Gritti Palace

Santa Maria del Giglio, San Marco 2467. **Map** 7 A3. *(041) 79 46 11.* FAX *520 09 42.* ***Rooms:*** *93.* *Private launch service. Access to private beach.* ⓁⓁⓁⓁⓁ

This renowned deluxe hotel overlooking the Grand Canal is situated in the fine 15th-century *palazzo* that once belonged to the Gritti family *(see p226)*. The hotel is elegant, sumptuous and old-fashioned in the best sense, with superb rooms and meticulous service. The iconic writer Ernest Hemingway was a guest here during his stay in Venice.

Monaco and Grand Canal

Calle Vallaresso, San Marco 1325. **Map** 7 B3. *(041) 520 02 11.* FAX *520 05 01.* ***Rooms:*** *74.* ⓁⓁⓁⓁⓁ

Housed in an 18th-century *palazzo*, this elegant hotel looks across the Grand Canal to the Salute. The public rooms have an intimate feeling and the bedrooms are decorated with lovely furniture and pretty fabrics. The restaurant on the terrace beside the Grand Canal is one of the best in Venice.

San Polo and Santa Croce

Alex

Rio Terrà Frari, San Polo 2606. **Map** 6 E1. *(041) 523 13 41.* ***Rooms:*** *11.* *2.* Ⓛ

It is some time since the rooms of this family-run hotel were overhauled, but the excellent value and good position near the Frari compensate for any shabbiness.

Falier

Salizzada San Pantalon 130, Santa Croce 30135. **Map** 5 C1. *(041) 22 88 82.* FAX *20 65 54.* ***Rooms:*** *19.* ⓁⓁ

Away from the majority of the Venice hotels and on the edge of the student district, the Falier has recently been refurbished.

Al Sole

Fondamenta Minotta, Santa Croce 136. **Map** 5 C1. *(041) 71 08 44.* FAX *71 43 98.* ***Rooms:*** *80.* ⓁⓁⓁ

The station area is not the most attractive part of Venice, but the Al Sole is in a pretty corner. This 14th-century building offers guests marble-floored reception areas and pleasant rooms. There is a shady courtyard where you can have a drink. You may have to accept half-board in the busy season.

Marconi

Riva del Vin, San Polo 729. **Map** 7 A1. *(041) 522 20 68.* FAX *522 97 00.* ***Rooms:*** *26.* ⓁⓁⓁ

Those in search of Venetian opulence will appreciate the reception areas of the Marconi, which was once a 16th-century *palazzo (see p227)*.. The bedrooms are less grand but some overlook the Grand Canal. The Marconi is well situated beside the Rialto Bridge.

Sturion

Calle del Storione, San Polo 679.
Map 7 A1. *(041) 523 62 43.* FAX *52 28 37 80.* ***Rooms:*** *11.* 1 TV ⓁⓁⓁⓁ

Situated just a few yards from the Grand Canal and close to the Rialto, the Sturion's central location has long been part of its appeal. The rooms have recently undergone transformation, making this friendly hotel an attractive base offering exceptional value.

Castello

La Residenza

Campo Bandiera e Moro, Castello 3608.
Map 8 E2. *(041) 528 53 15.* FAX *523 88 59.* ***Rooms:*** *16.* 1 *Jan, Feb, Nov.* ⓁⓁⓁ

The Residenza was once a 14th-century *palazzo (see p227)*. Although the bedrooms are basic, the frescoed ceilings and antique furniture in the public rooms are quite delightfully elegant.

Paganelli

Riva degli Schiavoni, Castello 4182.
Map 8 D2. *(041) 522 43 24.* FAX *523 92 67.* ***Rooms:*** *22.* 1 ⓁⓁⓁ

The cosy and old-fashioned rooms in the Paganelli provide views of St Mark's Basin that would cost three times as much at other hotels along the Riva. Rooms in the *dipendenza* (annexe) on San Zaccaria do not have the view, although they are quieter.

Pensione Wildner

Riva degli Schiavoni, Castello 4161.
Map 8 D2. *(041) 522 74 63.* FAX *526 56 15.* ***Rooms:*** *16.* 1 TV ⓁⓁⓁⓁ

Another family-run hotel on the Riva, the Pensione Wildner offers simple yet immaculate bedrooms with stunning views across to San Giorgio Maggiore. It has a small bar if you prefer not to sit at the tables outside, among the crowds on the busy Riva. In high season you may have to take half-board.

Danieli

Riva degli Schiavoni, Castello 4196.
Map 7 C2. *(041) 522 64 80.* FAX *520 02 08.* ***Rooms:*** *233.* 1 TV *Access to private beach.* ⓁⓁⓁⓁⓁ

One of Venice's top hotels, the Danieli is the epitome of luxury. It was the palace of the Dandolo family and has strong literary and musical connections *(see p113)*. The reception rooms are splendid, lit by resplendent Venetian glass chandeliers, and the service is impeccable. There are two wings, one built in the 1940s, and an older section, obviously the first choice for guests.

Londra Palace

Riva degli Schiavoni, Castello 4171.
Map 8 D2. *(041) 520 05 33.* FAX *522 50 32.* ***Rooms:*** *53.* 1 TV ⓁⓁⓁⓁⓁ

On the long Riva where hotels are cheek by jowl, the monument to King Vittorio Emanuele is a useful landmark outside this lovely stone-faced hotel *(see p227)*. It was once two palaces and has elegant public rooms, a good restaurant and a delightful bar. The rooms are traditional in style and extremely comfortable, though some of the singles are rather small. It was here that Tchaikovsky composed his Fourth Symphony.

Dorsoduro

Montin

Fondamenta Eremite, Dorsoduro 1147. **Map** 6 D3. *(041) 522 71 51.* FAX *520 02 55.* ***Rooms:*** *8.* *7.* 1 Ⓛ

These seven rooms above one of Venice's best-known restaurants are full of charm and character, despite the lack of private bathrooms. It is essential to book.

Messner

Rio Terrà dei Catecumeni, Dorsoduro 216. **Map** 6 F4. *(041) 522 74 43.* FAX *522 72 66.* ***Rooms:*** *34.* *30.* 1 ⓁⓁ

This hotel was renovated a few seasons ago; the rooms are modern and a good size. There is a nice bar and a peaceful courtyard. Children are particularly welcome.

Pensione La Calcina

Zattere ai Gesuati, Dorsoduro 780.
Map 6 E4. *(041) 520 64 66.* FAX *522 70 45.* ***Rooms:*** *40.* *20.* 1 *mid-Jan–mid-Feb.* ⓁⓁ

This simply furnished and airy *pensione*, convenient for the Accademia, is very popular. All the rooms are comfortable and a few have the bonus of splendid views across the Giudecca canal.

Agli Alboretti

Rio Terrà Antonio Foscarini, Dorsoduro 884. **Map** 6 E4. *(041) 523 00 58.* FAX *521 01 58.* ***Rooms:*** *25.* 1 ⓁⓁⓁ

Attractively situated near the Accademia, this hotel has modern, elegant rooms, although they are rather on the small side. The cosy reception area is wood-panelled and there is a garden where you can recuperate after a hard day's sightseeing *(see p226)*. The Agli Alboretti is a great favourite with English and American visitors.

Pausania

Rio di San Barnaba, Dorsoduro 2824.
Map 6 D3. *(041) 522 20 83.* FAX *(041) 52 22 989.* ***Rooms:*** *26.* 1 TV ⓁⓁⓁ

This small, welcoming hotel offers a peaceful escape from the agitation of city life. Most of the bedrooms are situated around a picturesque and secluded garden. With reasonable prices and friendly staff, this is a pleasant place to stay during the busy summer months.

Pensione Seguso

Zattere ai Gesuati, Dorsoduro 779.
Map 6 E4. *(041) 528 68 58.* FAX *522 23 40.* ***Rooms:*** *40.* *28.* 1 *winter.* ⓁⓁⓁ

The rooms of this pleasant *pensione* are traditionally furnished and the reception areas crammed with antiques and books. Nearly all the rooms look out on to water, either the Giudecca canal, or the Rio San Vio. You need to book well ahead, and you may find that in high season half-board is compulsory.

Pensione Accademia Villa Maravegie

Fondamenta Bollani, Dorsoduro 1058–1060. **Map** 6 E3. *(041) 521 01 88.* FAX *523 91 52.* ***Rooms:*** *27.* *26.* 1 ⓁⓁⓁⓁ

Visitors return again and again to this elegant 17th-century villa which was once the Russian Embassy. The rooms are mainly furnished with antiques and some are small, but the garden makes up for these minor inconveniences. Book well ahead.

Cannaregio

Zecchini

Lista di Spagna, Cannaregio 152.
Map 2 D4. *(041) 71 56 11.* FAX *71 50 66.* ***Rooms:*** *27.* *23.* ⓁⓁ

The Zecchini is housed on two floors of an attractive building and is approached via a flight of stairs *(see p226)*. The rooms overlook one of Venice's busiest streets , so it can be a bit noisy. However, the hotel is very good value for money.

Abbazia

Calle Priuli, Cannaregio 66/68. **Map** 1 C4. *(041) 71 73 33.* FAX *71 79 49.* ***Rooms:*** *36.* ⓁⓁⓁ

Quietly situated just off the lively Lista di Spagna, the Abbazia is probably the best hotel in the station area. The rooms are large and comfortable and there is a garden in which you can relax with a drink from the bar.

Continental

Lista di Spagna, Cannaregio 166. **Map** 2 D4. *(041) 71 51 22.* FAX *524 24 32.* ***Rooms:*** *93.*

ⓁⓁⓁⓁ

The Continental is a large, well equipped hotel in the busy station area. It has modern rooms, some of which overlook the Grand Canal. Other rooms look out over the adjacent, tree-shaded *campo* and these are usually quieter.

Giorgione

Santi Apostoli, Cannaregio 4587. **Map** 3 B5. *(041) 522 58 10.* FAX *523 90 92.* ***Rooms:*** *70.* ⓁⓁⓁⓁ

This is a modern, recently refurbished hotel housed in a delightful colour-washed building. The Giorgione has a spacious reception area and bar *(see p 227)*, and a delightful garden. It is conveniently situated for the Rialto area.

THE LAGOON ISLANDS

Villa Parco

Via Rodi 1, Lido di Venezia. *(041) 526 00 15.* FAX *526 76 20.* ***Rooms:*** *23.* ⓁⓁⓁ

With the sea just a few minutes away and set in its own gardens, this family-run hotel is an ideal place to stay if you are holidaying with children. The modern rooms are pleasant and the hotel is quietly situated in a residential area near the summer casino. Parking is available for guests who arrive with their own cars.

Quattro Fontane

Via Quattro Fontane 16, Lido di Venezia. *(041) 526 02 27.* FAX *526 07 26.* ***Rooms:*** *59.* *mid-Nov–mid-Apr.* ⓁⓁⓁⓁ

This is the nicest of the smaller hotels on the Lido. It is set in pretty gardens and has its own tennis court for the use of guests. The reception rooms and bedrooms are all furnished with antiques and the green and white gabled exterior gives the hotel a distinctive alpine look. On warm summer days you can enjoy a meal out of doors in the tranquil creeper-clad courtyard. The Quattro Fontane is only a few minutes from the sea.

Cipriani

Giudecca 10. **Map** 7 C5. *(041) 520 77 44.* FAX *520 39 30.* ***Rooms:*** *105.* *Private launch service.* *winter.* ⓁⓁⓁⓁⓁ

Set in gardens occupying the entire eastern tip of the Giudecca, the Cipriani has been one of the world's great hotels since it opened in 1963. Guests enjoy luxurious surroundings and perfect service. The bedrooms and suites are furnished with taste and opulence, and each one has been decorated in a different style. The renowned restaurant has splendid terraces which are delightful for dining.

Excelsior Palace

Lungomare Marconi 41, Lido di Venezia. *(041) 526 02 01.* FAX *526 72 76.* ***Rooms:*** *197.* *Private launch service. Private beach.* *winter.* ⓁⓁⓁⓁⓁ

When it opened in 1907, the Excelsior was the largest luxury resort hotel in the world *(see pp48–9)*. The exterior is flamboyantly Moorish – even the beach *cabanas* are styled like Arabian tents. The interior, offering every service and comfort, is equally splendid, though it may not appeal to all tastes. The hotel is at its liveliest during the annual Film Festival *(see p255)* – the Palazzo del Cinema and the summer home of the casino are only a few minutes' walk away.

Hôtel des Bains

Lungomare Marconi 17, Lido di Venezia. *(041) 526 59 21.* FAX *526 01 13.* ***Rooms:*** *191.* *Private launch service. Private beach.* *winter.* ⓁⓁⓁⓁⓁ

Film fans will recognize the des Bains as the location for Luchino Visconti's *Death in Venice*, set in the hotel where Thomas Mann wrote the original novel. Built in the early 1900s, its reception rooms have an Art Deco elegance, and there is a verandah for dining *(see p225)*. The bedrooms are spacious. Across the road the hotel's private beach is equipped with bathing huts with a view looking out over the Adriatic.

THE VENETO PLAIN

ASOLO

Villa Cipriani

Road map D3. Via Canova 298, 31011. *(0423) 95 21 66.* FAX *95 20 95.* ***Rooms:*** *31.* ⓁⓁⓁⓁⓁ

Situated in the gentle foothills of the mountains, Asolo is a good base for exploring a wide area. This superbly comfortable hotel is housed in a 16th-century villa where Robert Browning once lived. Standards are very high here, and the beautiful garden, with its fine view, is a major attraction.

BASSANO AREA

Victoria

Road map C3. Viale Diaz 33, 36061 Bassano del Grappa . *(0424) 50 36 20.* FAX *50 31 30.* ***Rooms:*** *23.* Ⓛ

The Victoria is just outside the old city walls and well placed for sightseeing in Bassano. The rooms are comfortable and there is a garden, where you can enjoy a quiet drink.

Belvedere

Road map C3. Piazzale G Giardino 14, 36061 Bassano del Grappa. *(0424) 52 98 45.* FAX *52 98 47.* ***Rooms:*** *91.* ⓁⓁⓁ

One of Bassano's main squares, just outside the old centre, is the setting for the town's most lavishly appointed hotel. This is a large and busy establishment at the top of the orbital road, which can be noisy, but it has comfortable rooms and a high level of service.

CHIOGGIA

Grande Italia

Road map D4. Piazza Vigo 1, 30015. *(041) 40 05 15.* FAX *40 01 85.* ***Rooms:*** *43.* No access. Ⓛ

The Grande Italia stands right at the head of Chioggia's main street and is well situated for the boats running to Venice via Pellestrina and the Lido. A solid, unpretentious and old-fashioned hotel, it offers comfortable rooms at low prices.

For key to symbols *see p225*

Conegliano

Canon d'Oro

Road map D3. Via XX Settembre 129, 31015 . *(0438) 342 46. **Rooms:** 35.* ⓁⓁ

This three-star hotel offers solid comfort and a warm welcome in an old building. The location, on the main street of Conegliano, is ideal for exploring the historic town centre. There is a garden in which to relax, and you can sample a glass of Prosecco in the friendly bar.

Padua

Bellevue

Road map D4. Via L Belludi 11, 35123. *(049) 875 55 47. **Rooms:** 11.* Ⓛ

This is an excellent choice if your budget is tight. The hotel is in the south of the city centre, near the Prato della Valle *(see p183)*. Friendly and family-run, it has simple and immaculate rooms, a bar and a peaceful courtyard.

Leon Bianco

Road map D4. Piazzetta Pedrocchi 12, 35122. *(049) 875 08 14.* FAX *875 61 84. **Rooms:** 22.* ⓁⓁ

It is essential to book ahead to make sure of a room at this central hotel, overlooking the Caffè Pedrocchi *(see p178)*. The rooms are on the small side but it is friendly and welcoming.

Augustus Terme

Road map C4. Viale Stazione 150, 35036 Montegrotto Terme. *(049) 79 32 00.* FAX *79 35 18. **Rooms:** 130.* *8 Jan–28 Feb, 20 Nov–20 Dec.* ⓁⓁⓁ

The spa town of Montegrotto Terme has a number of modern hotels and makes a useful last-minute stopping point for drivers with no pre-booked accommodation. The Augustus Terme is one of the best hotels here: big, comfortable and functional, with a pretty garden and tennis court. You can also sample the thermal hot springs.

Donatello

Road map D4. Via del Santo 102, 35123. *(049) 875 06 34.* FAX *(049) 875 08 29. **Rooms:** 49.* *mid-Dec–mid-Jan.* ⓁⓁⓁ

Named after the sculptor of the great equestrian statue on the piazza outside, the Donatello is opposite the Basilica di Sant' Antonio *(see p182)*. The old shell encloses a modern hotel whose bedrooms are large and comfortable, if lacking in character.

Treviso

Beccherie

Road map D3. Piazza Ancillotto 10, 31100. *(0422) 54 08 71. **Rooms:** 14.* Ⓛ

This is by far the best of Treviso's budget hotels: clean, comfortable and welcoming, and situated right in the middle of the historic town centre. The hotel restaurant offers an excellent range of local dishes.

Valdobbiadene

Diana

Road map D3. Via Roma 49, 31049. *(0423) 97 62 22.* FAX *97 22 37. **Rooms:** 54.* ⓁⓁ

Lying at the northern end of the Strada del Vino Prosecco, the attractive town of Valdobbiadene stands on the slopes of the Venetian pre-Alps overlooking the Piave valley. The Diana has good facilities and comfortable rooms.

Vicenza

Casa San Raffaele

Road map C4. Viale X Giugno 10, 36100 Salita Monte Berico. *(0444) 54 57 67.* FAX *54 22 59. **Rooms:** 24.* Ⓛ

This tranquil hotel lies on the slopes of Monte Berico, with its lovely views over the city. It takes about half an hour to walk into town, but the good rooms make it by far the best choice among the less expensive hotels.

Campo Marzio

Road map C4. Viale Roma 27, 36100. *(0444) 54 57 00.* FAX *32 04 95. **Rooms:** 35.* ⓁⓁⓁⓁ

This stylish, modern hotel, with good facilities, is near the city centre. It is set in a peaceful area not far from the cathedral.

Verona and Lake Garda

Lake Garda South

Peschiera

Road map A4. Via Parini 4, 37010 Peschiera del Garda. *(045) 755 05 26.* FAX *755 04 44. **Rooms:** 30.* *Nov–Mar.* Ⓛ

Just five minutes' drive off the A4 motorway at the southern end of Lake Garda, the little town of Peschiera makes a good overnight stopping place. The Peschiera is a modern building in a traditional style set in its own grounds, with lofty, cool bedrooms and friendly staff. The proprietors will collect train travellers from the station.

Villa Cortine Palace

Road map A4. Via Grotte 6, 25019 Sirmione del Garda. *(030) 990 58 90.* FAX *91 63 90. **Rooms:** 55.* *Nov–Mar.* ⓁⓁⓁⓁⓁ *half-board only.*

The fabulous grounds of this tranquil hotel *(see p223)* cover a third of the Sirmione peninsula *(see pp206–7)*. The huge frescoed rooms of this Classical villa are furnished with antiques, while the bedrooms, standard of comfort, food and service are all you would expect from a hotel of this calibre.

Lake Garda West

Capo Reamol

Road map B3. Via IV Novembre 92, 25010 Limone sul Garda. *(0365) 95 40 40.* FAX *95 42 62. **Rooms:** 48.* ⓁⓁ

This hotel has a stunning location below the main road and right on the lake's edge. All the spacious bedrooms have their own terrace and good meals are served in the waterside dining room. The hotel's facilities are excellent for families, and include a private beach and jetty, a gym, sauna and jacuzzi and good sunbathing terraces.

Hotel du Lac

Road map A3. Via Colletta 21, 25084 Gargnano. *(0365) 711 07.* FAX *725 94. **Rooms:** 12.* *mid-Oct–end Mar.* ⓁⓁ

The dining room of this pretty, stuccoed old building overhangs the water and many bedrooms have balconies facing the lake. This family-run hotel makes a good base for exploring the western shore of Lake Garda.

Lido Palace

Road map B3. Viale Carducci 10, 38066 Riva del Garda. *(0464) 55 26 64.* FAX *55 19 57.* ***Rooms:*** *63.* TV P ⓁⓁⓁⓁ

A spacious country villa houses this immaculately maintained hotel, standing in its own grounds on the outskirts of Riva. Try to book one of the high-ceilinged rooms on the first floor; the upper floors are more cramped. There is a tennis court for visitors' and the lake offers water-sports facilities.

Lake Garda East

Bisesti

Road map A3. Corso Italia 34, 37016 Garda. *(045) 725 57 66.* FAX *725 59 27.* ***Rooms:*** *90.* 1 P *Dec–Feb.* ⓁⓁ

Conveniently placed near the centre of town and only five minutes' walk from Lake Garda, the Bisesti is a well-equipped modern holiday hotel set in its own grounds and with access to a private beach. Many of the rooms have a balcony and the dining room looks out on the garden.

Kriss Internazionale

Road map A3. Lungolago Cipriani 3, 37011 Bardolino. *045) 621 24 33.* FAX *721 02 42.* ***Rooms:*** *33.* 1 TV P ⓁⓁ

Attractively situated on a small promontory jutting into the lake, this modern hotel caters specifically for holiday-makers. Rooms have balconies and lake views, there is a garden, and the hotel has a private beach. The hotel will provide transport from the station for those travellers who arrive without a car.

Sailing Center Hotel

Road map B3. Località Molini Campagnola 3, 37018 Malcésine. *(045) 740 00 55.* FAX *740 03 92.* ***Rooms:*** *32.* TV P ⓁⓁⓁ

Far from the crowds, sited on land jutting into the lake, this modern hotel just outside the main town of Malcésine makes a good base for families for a short stay. The rooms are cool and pleasant, and there is a tennis court and private beach for residents.

Locanda San Vigilio

Road map A3. San Vigilio, 37016 Garda. *(045) 725 66 88.* FAX *(045) 725 65 51.* ***Rooms:*** *7.* TV P *Nov–Feb.* ⓁⓁⓁⓁⓁ

One of the loveliest and most exclusive hotels on Lake Garda, the Locanda lives up to every expectation of comfort, good taste and impeccable service. There is a private beach and a walled garden.

Verona

Il Torcolo

Road map B4. Vicolo Listone 3, 37100. *(045) 800 75 12.* FAX *800 40 58.* ***Rooms:*** *19.* 1 TV P *mid–end Jan.* ⓁⓁ

An extremely popular hotel within a stone's throw of the Arena, the Torcola has some pretty, traditional rooms, and others which are more modern. With its charming interior, breakfast terrace and friendly owners, the hotel is always busy.

Giulietta e Romeo

Road map B4. Vicolo Tre Marchetti 3, 37100. *(045) 800 35 54.* FAX *801 08 62.* ***Rooms:*** *30.* 1 ⓁⓁⓁ

This hotel is situated on a quiet street near the Arena. The spacious bedrooms are comfortable, with modern furnishings Breakfast is served in the bar and there are some good restaurants nearby.

Colomba d'Oro

Road map B4. Via Carlo Cattaneo 10, 37100. *(045) 59 53 00.* FAX *59 49 74.* ***Rooms:*** *49.* 1 TV P ⓁⓁⓁⓁ

An old stone building on a quiet pedestrian street near the Arena houses this central hotel. Rooms are rather plain, but the parking is a bonus and there are good restaurants in the neighbourhood.

Due Torri Hotel Baglioni

Road map B4. Piazza Sant' Anastasia 4, 37100. *(045) 59 50 44.* FAX *800 41 30.* ***Rooms:*** *91.* 1 TV P ⓁⓁⓁⓁⓁ

One of the oldest and perhaps most eccentric Italian hotels, the Due Torri is right in the heart of medieval Verona. The huge bedrooms are each decorated in the style of a different era, with contemporary furnishings, and there are plenty of painted ceilings and frescoed walls. Staying here is an unforgettable experience.

THE DOLOMITES

Belluno

Astor

Road map D2. Piazza dei Martiri 26/e, 32100. *(0437) 94 20 94.* FAX *94 24 93.* ***Rooms:*** *32.* 1 TV ⓁⓁ

This centrally situated hotel is a good choice when visiting the region's main town. The Astor's rooms are comfortable and well designed, and offer very good value. In winter it is popular with skiers and there is a lively atmosphere in the bar.

Cortina d'Ampezzo

Corona

Road map D1. Via Val di Sotto 12, 32043. *(0436) 3251.* FAX *(0436) 86 73 39.* ***Rooms:*** *38.* 1 TV *mid Apr–mid Jun, mid Sep–mid Dec.* ⓁⓁⓁ

Among upmarket Cortina's many stylish hotels, the Corona has much to offer. The bedrooms are pretty and comfortable, the bar and restaurant well-appointed, and there is also a relaxing garden.

Menardi

Road map D1. Via Majon 110, 32043. *(0436) 2400.* FAX *(0436) 86 21 83.* ***Rooms:*** *53.* 1 TV P *mid Apr–mid Jun, mid Sep–mid Dec* ⓁⓁⓁ

The Menardi family have run this hotel on the outskirts of Cortina since 1900. Originally built as a farmhouse, it is furnished with antiques and has a welcoming and homely atmosphere.

San Vito di Cadore

Hotel Ladinia

Road map D1. Via Ladinia 14, 32046. *(0436) 890450.* FAX *992 11.* ***Rooms:*** *46.* 1 TV P *Jun–Sep, Christmas & Easter.* ⓁⓁⓁ

In the scenic Ampezzo valley, the Hotel Ladinia is in a tiny resort village 11 km (7 miles) south of Cortina. Popular in summer as well as during the skiing season, the Ladinia is geared to holiday-makers. There is a tennis court; and you can have a drink in the bar, or outside in the garden.

For key to symbols *see p225*

Restaurants, Cafés and Bars

Restaurants in Venice and the Veneto serve predominantly Italian food from the region, with the emphasis in Venice very much on fish. Wherever you go, you will find the cooking simple, with dishes that make full use of the traditional local ingredients.

Most Venetians eat lunch *(pranzo)* around 12:30pm and dinner *(cena)* from 8pm, though restaurants start serving dinner earlier to cater for the many foreign visitors.

Egyptian detail, Caffè Pedrocchi

Restaurants may be closed for several weeks during the winter and also for two to three weeks during the staff summer holidays. Closing dates are included in the listings, but avoid disappointment by asking your hotel to phone first to confirm that the restaurant is open. Finding restaurants can be confusing in Venice, so use the map references provided. The restaurants listed on pages 240–45 are some of the best across all price ranges.

El Gato restaurant, Chioggia, famous for its fish ***(see p242)***

Types of Restaurants

Italian eating places have a bewildering variety of names, and the differences between them are very subtle. A *trattoria*, an *osteria* and a *ristorante* are pretty similar in terms of price, cooking and ambience, though originally *ristoranti* were a little smarter, while *trattorie* were more homely concerns. Nowadays, these distinctions are blurred so names can be misleading. A *birreria* and a *spaghetteria* are more down-market eating places that sell beer, pasta dishes and snacks; you will mainly find these outside Venice itself. A good *pizzeria* will use wood-fired ovens for the pizza; if this is the case it will normally be open only in the evenings.

If you do not want to eat a full meal at lunchtime you can always stop in a bar or café for a snack. For further information on light meals see page 246.

Opening Times and Closing Days

Opening times are virtually the same throughout Venice and the Veneto: from noon to 2:30pm for lunch, and from 7:30pm to 10:30pm for dinner. Under Italian law all restaurants close one day a week and some close for an additional evening as well; closing days are staggered so there is always somewhere open in the area. Individual restaurants' closing days are given in the listings.

The main bar of the historic Caffè Pedrocchi ***(see p178)***

Vegetarian Food

Italians find it difficult to understand vegetarianism, but if you eat fish you should have no difficulty eating well. If not, there is still a variety of meatless dishes since many starters *(antipasti)*, soups and pasta sauces are vegetable-based. Salads and vegetables are always good, and most places will be happy to serve an omelette *(frittata)* or a selection of cheese.

Fixed-Price Menus

In the days when Italy was building its tourist industry all restaurants had to supply a fixed-price menu. This has largely fallen into abeyance, particularly outside the main tourist centres. Restaurants may often have the so-called *menu turistico* pinned up in the street, but not on offer inside. Such menus, if you do find them, are usually boring and offer no opportunity to sample the wonderful variety of the local cuisine. If money is tight it is far better to have a good pasta dish and some salad, which is acceptable in all but the grandest places.

The *menu gastronomico* is a fixed-price menu consisting of six or seven courses, which allows you to sample the full range of a chef's specialities.

How Much to Pay

Transport charges can add as much as 30 per cent to the price of basic commodities coming into Venice, which partly explains the high cost of eating. In cheaper eating places and *pizzerie* you can have a two-course meal with half a litre of wine for around L20,000–30,000. Three-course meals average about L35,000–50,000, and in up-market restaurants you can easily pay L100,000–140,000. In the Veneto, prices are lower, except for stylish restaurants in Verona and along Lake Garda during the summer.

Nearly all restaurants have a cover charge *(pane e coperto)*, usually L2,000–5,000. Many also add a 10 per cent service charge *(servizio)* to the bill *(il conto)*, so always establish whether or not this is the case. Where leaving a tip is a matter of your own discretion, 12–15 per cent is acceptable.

Restaurants are obliged by law to give you a receipt *(una ricevuta fiscale)*. Scraps of paper with an illegible scrawl are illegal, and you are within your rights to ask for a proper bill. The preferred form of payment is cash, but many restaurants will accept payment by major credit cards. Check which cards are accepted when booking.

Making Reservations

Whatever the price range, Venice's best restaurants are always busy, so it is best to reserve a table, especially if you are making a long boat trip to get there. If restaurants do not accept bookings, try to arrive early to avoid queuing.

Dress Code

Italians like to dress up to dine, though, with rare exceptions, this does not mean that men have to wear a tie, and only occasionally will you feel under-dressed without a jacket. Smart casual clothes are the general rule for men, though ladies in evening attire will not look at all out of place.

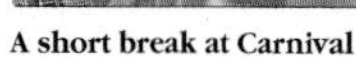

A short break at Carnival

Eating under the loggia of Treviso's Palazzo dei Trecento ***(see p174)***

Reading the Menu

Both lunch and dinner in a restaurant follow the same pattern and usually start with an *antipasto*, or hors d'oeuvres (seafood, olives, beef carpaccio, ham, salami), followed by the *primo* (soup, rice or pasta). The main course, or *secondo*, will be fish or meat, either served alone or accompanied by vegetables *(contorni)* or a salad *(insalata)*. These are never included in the price of the main course.

To finish, there will probably be a choice of fruit *(frutta)*, a pudding *(dolce)* or cheese *(formaggio)*, or a combination of all three. Coffee – Italians always have an *espresso*, never a *cappuccino* – is ordered and served right at the end of the meal, often with a *digestivo*. In cheaper restaurants, the menu *(il menu)* may be chalked up or the waiter may simply recite the day's special dishes at your table.

Choice of Wine

House wines will usually be local *(see pp238–9)*. Cheaper restaurants will have a limited wine list, but at the top of the scale there should be a wide range of Italian and local wines and a selection of foreign vintages.

Children

Children are welcome in restaurants, particularly in simple, family-run ones. Smart places may be less welcoming, particularly in the evenings. Special facilities such as high chairs are not commonly provided. Most restaurants will prepare a half-portion *(mezza porzione)* if requested, and some charge less for these smaller helpings.

Smoking

Italians still smoke more than most Europeans. Virtually no restaurants set aside space for non-smokers.

Wheelchair Access

Very few restaurants make special provision for wheelchairs, though a word when booking should ensure a conveniently situated table and assistance on arrival.

Using the Listings

Key to the symbols in the listings on pp240–45.

fixed-priced menu
jacket and tie required
tables outside
air conditioning
good wine list
★ highly recommended
credit cards accepted. Check which cards are accepted when booking.

Price categories for a three-course meal for one including a half-bottle of house wine, cover charge, tax and service.
Ⓛ under L25,000
ⓁⓁ L25,000–L50,000
ⓁⓁⓁ L50,000–L75,000
ⓁⓁⓁⓁ L75,000–L100,000
ⓁⓁⓁⓁⓁ over L100,000

What to Eat in Venice and the Veneto

Traditional Venetian specialities rely on the freshest of seasonal produce, meat and cheese from the mainland, and a huge variety of fish and seafood. Pasta is eaten here, as all over Italy, but more typical is polenta, made from maize flour, and also the many types of risotto. The mainland produces some renowned salamis. In the mountainous north, game and wild mushrooms provide the basis for some wonderful dishes, while Lake Garda is noted for *coregone*, a firm pink-fleshed fish rather like trout. The long red radicchio from Treviso is famous all over Italy, while Bassano del Grappa and Sant'Erasmo provide wonderful asparagus.

Globe artichoke

Baccalà Mantecata
This is dried salted cod, mixed to a paste with olive oil, parsley and garlic.

Fiori di Zucchini
Courgette flowers stuffed with fish mousse, fried in a light batter, are a seasonal dish.

Antipasto di Frutti di Mare
The seafood selection is dressed with olive oil and lemon juice. It may include exotic shellfish rarely found outside the Lagoon

Risotto alle Seppie
Cuttlefish ink colours the rice in this traditional risotto.

Carpaccio
Wafer-thin slices of raw beef dressed with oil are served here with rocket anc parmesan cheese.

Risi e Bisi
Made with fresh peas and flavoured with bacon, this risotto is soft and liquid.

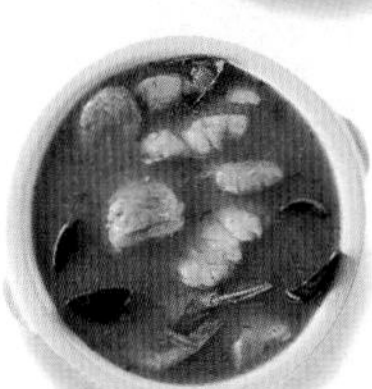

Brodo di Pesce
Fish soup, a classic Venetian dish, is sometimes flavoured with saffron.

Zuppa di Cozze
This is a delicious way of cooking mussels, with white wine, garlic and parsley.

Spaghetti alle Vongole
Fresh clams are served with spaghetti in a piquant sauce made with hot chilli peppers.

Polenta
Made from maize, polenta may be yellow or a more delicate white.

Sardine in Saor
A traditional Venetian way of serving fish is with a sweet and sour sauce.

Anguille in Umido
These eels are cooked in a light tomato sauce with white wine and garlic.

Fegato alla Veneziana
Tender calf's liver is lightly cooked on a bed of onions in this traditional speciality.

Faraona con la Peverada
The sauce for this succulent guinea fowl is based on an ancient Arabian recipe.

Insalata Mista
Wild young leaves of many shapes and colours make up this summer salad.

Radicchio alla Griglia
These red endive leaves from Treviso are grilled over a hot fire.

Tiramisù
This dessert is made with coffee-soaked sponge cake and mascarpone cheese.

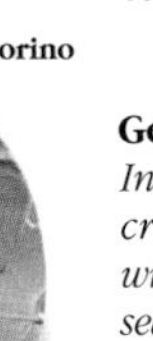

Cheeses
Local cheeses such as asiago, fontina and montasio may be included on the cheese board.

Gelati
In summer, ice cream is made with fresh seasonal fruits.

Amaretti
Small almond biscuits are served with coffee.

What to Drink in Venice and the Veneto

ITALY HAS BEEN MAKING WINE for over 3,000 years, and production in the Veneto reflects this, with the largest output in Italy of superior DOC wines. The area produces an abundance of different wines, which include not only well-known names such as Soave, Valpolicella and Bardolino, but many others which are also excellent value for money. Although Italians tend to drink lighter wines with their food, the area is also noted for some excellent strong wines. Italy's famous *digestivo*, grappa, originated in this corner of the country, and meals are often preceded by an *aperitivo* or a glass of sparkling local Prosecco.

Grapes drying in Valpolicella

Red Venegazzù

Masi's ripasso

Red Wine

RED WINES IN THE VENETO are produced mainly near Bardolino and Valpolicella between Verona and Lake Garda *(see pp208–9)*. Made predominantly from the Corvina grape, they are usually light and fruity, but quality can vary so it is worth looking for reliable names.

Valpolicella comes in several forms. In addition to the normal easy-drinking wine, it is available as a *ripasso*, boosted in colour and strength by macerating the skins of the grapes before pressing. Recioto della Valpolicella is very different, a rich, sweet wine made from selected air-dried grapes. Some Reciotos undergo further fermentation to remove the sweetness, producing the strong, dry Recioto Amarone. These are some of the strongest naturally alcoholic wines in the world and are delicious but expensive.

Excellent red wines are also made by producers such as Venegazzù and Maculan from the Cabernet Sauvignon and Merlot grapes.

Bardolino wine is light, fruity and garnet-red in colour.

Amarone is full-bodied, rich, full of fruit and very alcoholic.

Reading Wine Labels

ITALIAN WINES are classified by four quality levels. Starting at the top, DOCG status *(Denominazione di Origine Controllata e Garantita)* has been awarded to a small number of Italian growing areas, none of which are in the Veneto. Most quality wines – more than 250 in the whole of Italy – are in the DOC category (as above but without the "guarantee") and these can be relied on as good value, quality wines. IGT *(Indicazione Geografica Tipica)* is a category new to Italy, corresponding to the popular French Vin de Pays. The final classification is *vino da tavola*, or table wine, but due to the inflexible Italian wine laws many superb wines appear in this category.

No vintage recommendations are given in the chart because almost all Veneto wines are made for young drinking.

Wine Type	Recommended Producers
White Wine	
Soave	**Anselmi, Bertani, Col Baraca (Masi), Boscaini, CS di Soave, Masi, Pieropan, Scamperle, Tedeschi, Zenato, Zonin**
Bianco di Custoza	**Cavalchina, Le Tende, Le Vigne di San Pietro, Pezzini, San Leone, Tedeschi, Zenato**
Breganze di Breganze	**Maculan**
Gambellara	**CS di Gambellara, Zonin**
Red Wine	
Bardolino	**Alighieri, Bertani, Bolla, Boscaini, Guerrieri-Rizzardi, Masi, Tedeschi**
Valpolicella	**Alighieri, Allegrini, Bertani, Bolla, Boscaini, Guerrieri-Rizzardi, Masi, Tedeschi, Zenato**
Ripasso Valpolicella (non-DOC)	**Serègo Alighieri, Jago (Bolla), Le Cane (Boscaini), Le Sassine (Le Ragose), Campo Fiorin (Masi), Capitel San Rocco (Tedeschi)**
Recioto and Recioto Amarone della Valpolicella	**Serègo Alighieri, Allegrini, Masi, Quintarelli, Le Ragose, Tedeschi**

ʼHITE WINE

anco di ustoza

White Recioto

THE VENETO produces more white wine than red, and most of the region's whites are from vineyards around the hilltop town of Soave *(see p190)*. These wines can be dull, but increasing numbers of producers are trying to raise Soave's image. Bianco di Custoza, a creamy, richer tasting "super Soave" from the eastern shores of Lake Garda, is well worth trying. Breganze is a name to look out for, with Maculan a leader in making fresh, clean, inex-ensive wines and world-class dessert wines. ambellara is made mainly from Soave's Gar-anega grape and is seldom of poor quality. enegazzù is another producer you can trust or good quality white wines.

Pieropan is a top quality producer of Soave. The single-vineyard wines from here are superb.

Venegazzù's Pinot Grigio wine is dry and goes well with Venetian seafood.

White vino da tavola wines range from pale and dry to sweet and golden coloured.

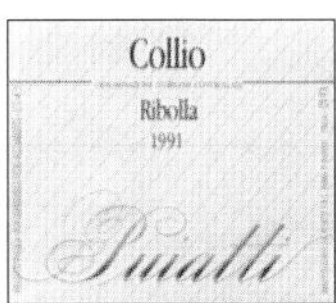

Puiatti's white Ribolla wine is fruity but dry. It is made in neighbouring Friuli.

ʼPERITIFS AND OTHER DRINKS

ITALIAN APERITIFS tend to be wine-based, bitter, herb-flavoured drinks such as Martini and Campari. Less familiar are the herbal Punt e Mes, Cynar (made from artichokes), and the vivid orange Aperol, which is good mixed with white wine and soda. Crodino is a popular non-alcoholic choice.

For settling the stomach after a good meal there are *amari* (bitters) and *digestivi*. Montenegro and Ramazzotti are well worth trying, and grappa, distilled from wine lees *(see Bassano del Grappa p166)*, is another favourite. A local speciality, Trevisana, is mixed with an extract of the long red radicchio from Treviso. Italian brandy can be rather oily, but Vecchia Romagna is a reliable name.

ıppa

Crodino

PROSECCO

The Veneto's own sparkling wine, Prosecco is perfect as either a refreshing light *aperitivo* or with a meal. It originates in Conegliano *(see p175)*, the home of Italy's greatest wine school, and comes in both *secco* (dry) or *amabile* (medium-sweet) forms, and as *frizzante* or *spumante* (semi and fully sparkling). An excellent accompaniment to both fruit and seafood, it is also the traditional base for Bellini, a delicious *aperitivo* of wine mixed with fresh white peach juice *(see p92)*. This drink has bred several variants, such as Mimosa (with orange) and Tiziano (with red grape juice).

Prosecco

Bellini cocktail

ʼOFT DRINKS

ITALIAN BOTTLED fruit juices are . good, and come in delicious avours such as pear, apricot nd peach. Many bars will queeze you a *spremuta* of fresh range *(arancia)* or grapefruit *pompelmo)* juice on the spot. *frullato* is an ice-cold nix of milk and fresh fruit.

Spremuta di arancia

COFFEE

COFFEE IS AN ESSENTIAL part of Italian life. Milky *cappuccino* with chocolate powder is drunk at breakfast time, and tiny cups of strong black *espresso* throughout the day. If you like your coffee with milk, choose a *caffè con latte,* or with just a dash of milk, *caffè macchiato.* Black coffee that is not too strong is *caffè lungo*; a *doppio* has an extra kick and a *corretto* has a good measure of alcohol.

Espresso

Cappuccino

VENICE

SAN MARCO

Al Conte Pescaor

Piscina San Zulian, San Marco 544. **Map** 7 B1. (041) 522 14 83. *Sun.* ⓁⓁⓁ

It is worth seeking out this little restaurant for its superb fish. Despite its position, the clientele are mainly local, which guarantees the quality of the food.

Al Graspo de Ua

Calle Bombaseri, San Marco 5094. **Map** 7 A1. (041) 520 01 50.

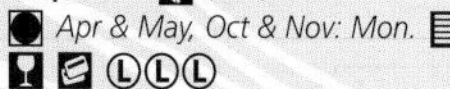

Apr & May, Oct & Nov: Mon. ⓁⓁⓁ

Housed in three small rooms that were once part of the old sacristy of the church of San Bartolomeo, the food here is Venetian with a difference. Worth sampling are the antipasto of prawns, artichokes and rocket, while main courses include fillet of John Dory.

Da Arturo

Calle degli Assassini, San Marco 3656. **Map** 7 A2. (041) 528 69 74. *all Aug & Sun.* ⓁⓁⓁ

This wood-panelled restaurant is unique in Venice in that it serves only meat, and this of the highest quality. Interesting antipasti include aubergines *"in saor"* (a vinegary sweet and sour sauce), and there are red wines of great distinction. Vegetarians are catered for.

Do Forni

Calle dei Specchieri. San Marco 468. **Map** 7 B2. (041) 523 21 48. ⓁⓁⓁⓁ

A large "show business" establishment, the Do Forni has two dining rooms furnished in contrasting styles; one rustic but smart and the other more elegant. The mixed grilled fish is a house speciality which should not be missed.

Harry's Bar

Calle Vallaresso, San Marco 1323. **Map** 7 B3. (041) 528 57 77. ⓁⓁⓁⓁ

People eat at Harry's Bar because of its fame. The food is expensive, but can certainly be good. Prawns with oil and lemon, baked tagliolini with prosciutto and *carpaccio alla Cipriani* (raw marinated beef) are among the more interesting dishes. The wine list is good and the cocktails are justifiably world renowned.

Scala Martini

Campo San Fantin, San Marco 1980. **Map** 7 A2. (041) 522 41 21. *Tue, Wed lunch.* ★ ⓁⓁⓁⓁ

The terrace of this smart restaurant used to overlook the Fenice theatre, and dining inside has its advantages too – meals are served until 1am. The cooking is based on regional specialities; breast of duck with black truffles is recommended. Puddings include an iced mousse with raspberry sauce. The wine list features more than 300 labels.

Tiepolo dell'Europa e Regina

Calle Larga XXII Marzo, San Marco 2159. **Map** 7 A3. (041) 520 04 77. ★ ⓁⓁⓁⓁ

This beautiful hotel restaurant is in a 16th-century palace on the Grand Canal and has a waterside terrace where guests can eat outdoors in summer. The food lives up to the setting, featuring dishes such as seafood salad, porcini mushrooms with parmesan cheese, *fegato alla veneziana*, raspberry crêpes and tiramisù. The wine list is excellent.

Hotel Gritti

Campo Santa Maria del Giglio, San Marco 2467. **Map** 6 F3. (041) 522 60 44. ★ ⓁⓁⓁⓁⓁ

The dining room of this famous hotel is in the Venetian Gothic building. The terrace overlooks the Grand Canal and the church of Santa Maria della Salute. Classic dishes include baby prawns with radicchio, *pasta e fagioli* (pasta and beans), saffron risotto with chicken livers and fried scampi.

La Caravella

Calle Larga XXII Marzo, San Marco 2396. **Map** 7 A3. (041) 520 89 01. ★ ⓁⓁⓁⓁⓁ

One of two restaurants in the Hotel Saturnia, the Caravella is decorated to resemble the interior of a 16th-century Venetian galley. The food is outstanding, with a range of imaginative dishes that includes a smooth lobster soup, bigoli and scampi in champagne.

SAN POLO AND SANTA CROCE

Antica Bessetta

Calle Savio, San Polo 1395. **Map** 2 E5. (041) 72 16 87. *Jan–Mar, May–Aug & Dec: Tue, Wed.* ⓁⓁ

This simple family-run restaurant offers real Venetian home-cooking including *risi e bisi* (fresh spring pea risotto) in season, home-made pasta and good scampi.

Trattoria alla Madonna

Calle della Madonna, San Polo 594. **Map** 7 A1. (041) 522 38 24. *Wed.* ★ ⓁⓁⓁ

This big, lively fish restaurant is a perennial favourite with Venetians and tourists alike. The seafood risotto and fish soup are excellent and the fried and grilled main fish courses all highly recommended.

Antica Trattoria Poste Vecie

Rialto Pescheria, San Polo 1608. **Map** 3 A5. (041) 72 18 22. *Tue.* ★ ⓁⓁⓁⓁ

This stylish restaurant near the fish market claims to be the oldest in Venice. It offers dishes such as home-made ravioli and tagliolini, baked turbot and brill, and there is an impressive pudding trolley.

CASTELLO

Hostaria da Franz

Fondamenta San Giuseppe, Castello 754. **Map** 8 D1. (041) 522 75 05. *Tue.* ★ ⓁⓁⓁ

Situated behind the Public Gardens, the Gasparini's restaurant with its canalside terrace serves some of the best seafood in Venice. Try the gnocchi with prawns and spinach, marinated prawns and peppers, or fried fish with polenta.

Corte Sconta

Calle del Pestrin, Castello 3886. **Map** 7 E2. (041) 522 70 24. *Sun, Mon.* ★ ⓁⓁⓁⓁ

The Proietto family have built up this restaurant from a simple eating house to one of the city's top dining spots. It is hard to find, but worth the effort for the superb fish and home-made pasta, served in summer in the pretty garden.

Les Deux Lions del Hotel Londra Palace

Riva degli Schiavoni, Castello 4171. **Map** 8 D1. (041) 520 05 33. ⓁⓁⓁⓁ

A grand hotel restaurant with a good reputation, Les Deux Lions has a summer terrace right on the Riva. The cooking is Italian and French; the *entrecôte béarnaise* is ecommended, as is the casseroled Verona chicken. The puddings are good, with the chocolate gateau made with plain, milk and white chocolate topping the list.

Arcimboldo

Calle dei Furlani, Castello 3219. **Map** 8 D1. *(041) 528 65 69.* *Tue.* ⓁⓁⓁⓁⓁ

This restaurant takes its name from 16th-century painter Arcimboldo, who used vegetables and fruit to create portraits of people. It offers equally diverse interpretations on the Venetian culinary theme. Try sea-bass with tomatoes or marinated salmon with citrus fruit and herbs.

Danieli Terrace

Riva degli Schiavoni, Castello 4191. **Map** 8 D2. *(041) 522 64 80.* ★ ⓁⓁⓁⓁⓁ

The panorama over St Mark's Basin from this stylish hotel restaurant is magnificent, and the food and service are just what you would expect from an establishment of this calibre. The risotti are very good and the excellent wine list features solely Italian wines.

Dorsoduro

Da Silvio

Calle San Pantalon, Dorsoduro. **Map** 6 D2. *(041) 520 58 33.* *Sun.* Ⓛ

It is a pleasure to find a genuine neighbourhood restaurant in Venice, and this one has a lovely garden for summer eating. The menu has no surprises, but the food is all fresh and home-made.

Taverna San Trovaso

Fondamenta Priuli, Dorsoduro 1016. **Map** 6 E3. *(041) 520 37 03.* *Mon.* ⓁⓁ

This cheerful, bustling restaurant on two floors is between the Accademia and the Zattere. The cooking is straightforward and they also serve pizzas.

Agli Alboretti

Rio Terrà Sant'Agnese, Dorsoduro 882. **Map** 6 E4. *(041) 523 00 58.* *eves only.* *Wed.* ⓁⓁⓁ

Named after the trees outside, Agli Alboretti is welcoming inside and refreshing in summer, when you can eat outside under the pergola. The cooking has innovative touches – try saffron risotto with scampi and mussels, monkfish tails or veal with cherries.

Ai Gondolieri

San Vio, Dorsoduro 366. **Map** 6 F4. *(041) 528 63 96.* *Tue.* ★ ⓁⓁⓁ

Housed in a genuine old inn, sympathetically restored, this restaurant is owned by Giovanni Trevisan, whose wife Marisa does the cooking. She is a dedicated regional cook and her specialities include some excellent vegetable dishes such as the long red radicchio from Treviso and *sformati* of wild leaves.

Linea d'Ombra

Zattere ai Saloni, Dorsoduro 19. **Map** 7 A4. *(041) 520 47 20.* ⓁⓁⓁ

Situated at the end of the Zattere, the Linea comes into its own in summer, when you can dine on the terrace and enjoy breathtaking views across to the Giudecca. The food includes several hard-to-find Venetian dishes such as *baccalà mantecato* and *sarde in saor.*

Locanda Montin

Fondamenta Eremite, Dorsoduro 1147. **Map** 6 D3. *(041) 522 71 51.* *Tue eve.* ★ ⓁⓁⓁ

This famous restaurant, with its artistic and literary connnections, can be variable in quality and service, but the garden is a delight and the atmosphere lively. It also contains a commercial art gallery.

Cannaregio

Ostaria al Bacco

Fondamenta delle Cappuccine, Cannaregio 3054. **Map** 2 D2. *(041) 71 74 93.* *Mon.* ⓁⓁ

The emphasis is on fish in this rustic restaurant with its pretty shaded courtyard. First courses include spaghetti with a black cuttlefish sauce and *frittura di pesce* (lightly fried local fish). *Baccalà*, the classic dried salted cod dish, is also on the menu.

Vini Da Gigio

Fondamenta San Felice, Cannaregio 3628/a. **Map** 3 A4. *(041) 528 51 40.* *Mon, public hols.* ⓁⓁ

Traditional Venetian food with a modern twist is prepared in this cosy but elegant restaurant. The risotto with nettles and prawns is excellent, and puddings include *crema fritta alla veneziana.* The wine list is extensive.

A La Vecchia Cavana

Rio Terrà Vecia Apostoli, Cannaregio 4624. **Map** 3 B4. *(041) 523 86 44.* *Thu.* ⓁⓁⓁ

This well-known establishment is acknowledged to be Cannaregio's smartest eating place. The emphasis is on Venetian specialities and there is a wide choice of simply prepared fish dishes.

Fiaschetteria Toscana

San Crisostomo, Cannaregio 5719. **Map** 3 B5. *(041) 528 52 81.* *Tue.* ⓁⓁⓁ

This restaurant, run by the Busatto family, offers many Venetian dishes as well as Scottish beef. The warm salad of lagoon fish is a delicious antipasto, and main dishes include eels and turbot with black butter and capers.

The Lagoon Islands

Antica Trattoria alla Maddalena

Mazzorbo. *(041) 73 01 51.* *Thu.* ⓁⓁ

Close beside the *vaporetto* stop on Mazzorbo, this modest restaurant originally was, and still is, the local bar and eating house. However, visitors flock here to dine in tranquillity and to savour the renowned wild duck served with fettuccine or polenta.

Do Mori

Eufemia, Giudecca 588. *(041) 522 54 52.* *Sun.* ★ ⓁⓁ

This rustic-style restaurant, right on the Giudecca waterfront, is run by the ex-chef from Harry's Bar, and it shows – though the prices are very attractive. The cooking is homely Venetian, with good antipasti, home-made pasta and several fine fish dishes. Puddings are home made and good.

Ai Pescatori

Via Galuppi 371, Burano. *(041) 73 06 50.* *Wed.* ★ ⓁⓁⓁ

Recently renovated, the Ai Pescatori remains a welcoming restaurant where you can eat outside in summer. Interesting dishes include lobster and tagliolini with cuttlefish. In winter the menu is well-known for its game dishes. The wine list is extensive, with both foreign as well as Italian wines.

For key to symbols *see p235*

Harry's Dolci

Eufemia, Giudecca 773. **Map** 5 C5. *(041) 522 48 44.* *Tue.* ★ ⓁⓁⓁ

What started as a bar and tearoom is now a restaurant serving many of the specialities of the main establishment *(see p240)*. You can enjoy *pasta e fagioli* (pasta and beans), *carpaccio* (marinated raw beef in slivers) and liver, preceded by a genuine Bellini and followed by the delicious house chocolate cake. They also sell the range of Cipriani foodstuffs.

Osteria al Ponte del Diavolo

Torcello. *(041) 73 04 01.* *Thu.* ★ ⓁⓁⓁ

The Osteria gets its name from the graceful bridge across Torcello's now neglected main canal. It was set up by a breakaway trio of staff from Cipriani, so the traditional fish-based menu is good. The large terrace, makes this an attractive spot to stop for lunch.

Da Romano

Piazza Galuppi 221, Burano. *(041) 73 00 30.* *Tue.* ⓁⓁⓁⓁ

Founded in the 1930s, this pretty restaurant with its shady terrace on Burano's main street is now run by the original owner's son. It serves excellent fish at lower prices than you would find in Venice.

Locanda Cipriani

Piazza Santa Fosca 29, Torcello. *(041) 73 01 50.* *Tue & winter.* ⓁⓁⓁⓁ

The most far-flung of the Cipriani establishments was transformed from a fisherman's inn in the 1930s. Dishes include *fritto misto*, and risotto made with fresh greens from the Locanda's picturesque kitchen garden. Lunch here makes a good outing, and the restaurant's launch will collect you from Piazza San Marco.

Hotel Cipriani

Giudecca 10. **Map** 7 C5. *(041) 520 77 44.* ★ ⓁⓁⓁⓁⓁ

The setting and style alone almost justify the high prices of the Hotel Cipriani. The mirror-lined dining room reflects the lagoon and the summer terrace is an oasis of calm. The food is creative and the wine list good. Sample the antipasti buffet before moving on to the other specialities, which include a fish "surprise" in pastry.

THE VENETO PLAIN

Asiago

Ristorante Casa Rossa

Road map C3. Via Kaberlaba 19, 36012. *(0424) 46 20 17.* *Thu. Ann hol 10–25 Jun.* ★ ⓁⓁ

Set in the hills of the Sette Comuni (Seven Communities), this simple restaurant is noted for its regional cooking. Try the savoury cake with wild salad leaves, the duck breast and the wild berry tart. You can eat outside in the garden.

Asolo

Villa Cipriani

Road map D3. Via Canova 298, 31011. *(0423) 95 21 66.* FAX *95 20 95.* ★ ⓁⓁⓁⓁ

This 16th-century villa houses one of the grand hotels of the Veneto *(see p231)*. The restaurant serves innovative and creative food, using local ingredients in such dishes as pappardelle (broad flat pasta) with scampi, tomatoes and basil, and veal medallions with asparagus and artichoke hearts.

Brenta Canal

Alla Posta

Road map D4. Via Ca' Tron 33, Dolo. *(041) 41 07 40.* *Mon. Ann hol 1–15 Jan, 1 week Jul.* ⓁⓁⓁⓁ

The Modena family run this superb fish restaurant in an old Venetian staging posthouse. The anchovy and potato tart is particularly good.

Dall'Amelia

Road map D4. Via Miranesi 113, 30171 Mestre. *(041) 91 39 51.* ⓁⓁⓁⓁ

This restaurant is considered to be one of the finest specializing in classic Venetian cooking. A separate area for those eating fried fish is indicative of the standard here. Specialities include tagliolini with fresh tomatoes and anchovies, and porcini mushrooms with rocket, as well as a wide choice of fish and mouth-watering puddings. The wine list is exceptional.

Castelfranco

Barbesin

Road map D3. Via Montebelluna 41. *(0423) 49 04 46.* *Wed eve and Thu. Ann hol 1–20 Aug, 24 Dec–12 Jan.* ⓁⓁ

This is a good-value restaurant serving regional specialities, including risotto of porcini mushrooms and duck breast with rosemary.

Chioggia

El Gato

Road map D4. Campo San Andrea 653, 30015. *(041) 40 18 06.* *Mon. Ann hol Jan and 1–15 Feb.* ★ ⓁⓁⓁ

Classic cooking in a restrained setting makes this restaurant special. As you would expect in a fishing port, the emphasis is on seafood. The cuttlefish is cooked in the traditional Chioggian way and served with polenta.

Conegliano

Al Salisa

Road map D3. Via XX Settembre 2, 31015. *(0438) 242 88.* *Tue eve, Wed. Ann hol Aug.* ★ ⓁⓁ

This elegant little restaurant in an old house has a pretty veranda where you can eat outside on warm summer evenings. Among the interesting dishes here are home-made fettuccine with a sauce of radicchio and chicken livers, and turbot cooked wrapped in spinach. Try the superlative cold zabaglione for pudding.

Euganean Hills

Trattoria da Piero Ceschi

Road map C4. Piazza Trento 16, 35042 Este. *(0429) 28 55.* *Thu. Ann hol Jul.* ⓁⓁ

Right in the historic centre of Este, this friendly little restaurant serves local prosciutto and genuine porchetta, a whole roast suckling pig with herbs and spices. Pietro, the owner, chef and head waiter, personally selects his excellent wine list from local producers.

La Montanella

Road map C4. Via Costa 33, 35032 Arqua Petrarca. *(0429) 71 82 00.* *Tue eve, Wed. Ann hol Jan, 2 weeks Feb & 2 weeks Aug.* ⓁⓁⓁ

Petrarch spent the last years of his life at Arquà Petrarca, the prettiest of the Euganean spa towns *(see p184)*. Eating at La Montanella is a pleasure, with asparagus tortelloni and guinea-fowl among the treats.

ODERZO

Gambrinus Parco

Road map E3. Località Gambrinus 22, 31020 San Polo di Piave. *(0422) 85 50 43.* *Mon. Ann hol 7–31 Jan.* ★ ⓁⓁⓁ

This restaurant is faithful to the regional culinary tradition. Two outstanding fish dishes are the freshwater prawns alla Gambrinus and the sturgeon cooked with coarse salt.

PADUA

Boccadoro

Road map D4. Via della Resistenza 49, 35100 Noventa Padovana. *(049) 62 50 29.* *Tue eve, Wed. Ann hol 2–12 Jan, 4–27 Aug.* ⓁⓁ

Good Paduan food is served in this family-run restaurant. There is an excellent soup made from radicchio and the goose is worth sampling.

La Braseria

Road map D4. Via N Tommaseo 48, 35100. *(049) 876 09 07.* *Sun. Ann hol 2 weeks Aug.* ⓁⓁ

This is a friendly restaurant where you can enjoy straightforward cooking. The penne with porcini mushrooms and the smoked ham called *speck* are recommended.

Belle Parti El Toulà

Road map D4. Via Belle Parti 11, 35100. *(049) 875 18 22.* *Sun lunch. Ann hol 3 weeks in Aug.* ⓁⓁⓁ

This comfortable, distinguished restaurant is housed in a 16th-century *palazzo*. The vegetable recipes, such as the aubergine *sformato*, are particularly good.

Dotto

Road map D4. Via Soncino 9, 35100. *(049) 875 14 90.* *Sun eve, Mon. Ann hol 10–25 Aug.* ⓁⓁⓁ

This Twenties-style restaurant serves traditional seasonal food. Try the home-made *pasta e fagioli* (pasta and beans) or the *baccalà* (dried salt cod), before moving on to one of the home-made puddings.

Ristorante Marco Polo

Road map C4. Via Forte Marghera, 30173 Rubano. *(041) 98 98 55.* *Sun. Ann hol all Aug.* ⓁⓁⓁⓁ

Familiar dishes are given a personal interpretation here. John Dory with wild fennel, and swordfish with capers are good choices, while meat-eaters will enjoy the veal with porcini mushrooms.

San Clemente

Road map C4. Corso Vittorio Emanuele II 142, 35123. *(049) 880 31 80.* FAX *(049) 880 30 15.* *Sun, Mon lunch. Ann hol Aug.* ★ ⓁⓁⓁⓁ

San Clemente occupies a villa attributed to Palladio, and stands in a tranquil garden. The excellent food and service match the impressive ambience, and dishes such as pigeon with stuffed pear and chicken ravioli are unusual and good. The wine list includes some fabulous Tuscan Brunello wines. If you want lunch it is advisable to ring beforehand.

PORTOGRUARO

Alla Botte

Road map E3. Viale Pordenone 46, 30026. *(0421) 76 01 22.* *Sun in winter.* ★ ⓁⓁ

This cosy hotel restaurant serves good local food, and the menu usually includes some dishes from neighbouring Friuli. The gnocchi flavoured with nettles are well worth trying.

Duilio

Road map E3. Via Strada Nuova 19, 30021 Caorle. *(0421) 810 87.* *Mon in winter.* ★ ⓁⓁ

In this roomy restaurant fish-based regional recipes are given an imaginative modern slant. Turbot is gently poached in champagne or used as a filling for the lightest of savoury pancakes.

ROVIGO

Tre Pini

Road map C5. Viale Porta Po 68, 45100. *(0425) 42 11 11.* *Sun.* ⓁⓁ

The three pines which give the restaurant its name are in the garden of this pretty villa. Specialities of the chef include a delicious onion tart, eels cooked with herbs, and wild duck.

TREVISO

Agnoletti

Road map D3. Via della Vittoria 190, Giavera del Montello. *(0422) 77 60 09.* *Mon, Tue. Ann hol 6–30 Jan, 1–15 Jul.* ⓁⓁ

A pretty hillside location north of Treviso and a lovely garden make this restaurant well worth a visit. Seasonal antipasti and the gnocchi are recommended.

Osteria dalla Pasina

Road map D3. Via Peschiere 15, Dosson di Casier. *(0422) 38 21 12.* *Mon eve, Tue & Sat lunch. Ann hol 1 week Christmas, 3 weeks Aug.* ⓁⓁ

This *osteria* offers tasty dishes which make the most of seasonal produce. The risotto with scampi and asparagus makes a good prelude to tender rabbit cooked with herbs and there are deliciously light mousses and sorbets for pudding.

Toni del Spin

Road map D3. Via Inferiore 7, 31100. *(0422) 54 38 29.* *Sun all day & Mon lunch. Ann hol 3 weeks Aug.* ⓁⓁ

A homely restaurant offering good value, Toni's serves hearty regional fare. Typical house specialities include *pasta e fagioli*, baked tripe and a rich tiramisù.

Ristorante alle Beccherie

Road map D3. Piazza Ancillotto 10, 31100. *(0422) 54 08 71.* *Sun eve, Mon. Ann hol 15–30 Jul.* ⓁⓁⓁ

The Beccherie is one of Treviso's oldest restaurants, housed in a lovely old Venetian-style building with period furniture and discreet service. The guinea fowl with pepper sauce, served with a good risotto of radicchio, is delicious.

For key to symbols *see p235*

Ristorante Enoteca Marchi

Road map D3. Via Castellana 177, Montebelluna. (0423) 238 75. *Tue pm & Wed. Ann hol all Aug.*

This restaurant is worth patronizing for the wines alone – there are over 500 Italian and foreign wines. Try the spaghetti with radicchio and pomegranate, and the lamb cooked in Prosecco.

Vicenza

Al Torresan

Road map C3. Via Zabarella 1, 36042 Breganze. *(0445) 87 32 60. Thu, Fri lunch. Ann hol 15 Jul–15 Aug.* ⓁⓁ

In the autumn locals flock here to enjoy the wild mushroom dishes. The cooking is hearty and is complemented by the local wines.

Antica Trattoria Tre Visi

Road map C4. Contrà Porti 6, 36100. *(0444) 32 48 68. Sun eve, Mon. Ann hol all Jul, 26 Dec–2 Jan.*
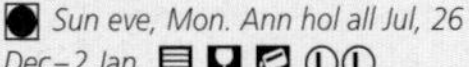
Housed in a building dating from 1483, this restaurant is right in the historic town centre. Diners can see into the busy kitchen, where Luigi cooks good Vicentine dishes.

Leoncino

Road map C4. Via Tavernelle 72, 36077 Altavilla Vicentina. *(0444) 57 20 32. Sun, Mon. Ann hol 27 Dec–10 Jan, Aug.* ⓁⓁ

This rustic restaurant just outside Vicenza is run by the Tecchio family. Good local dishes include the famous *bolliti misti* (boiled meats) served with a delicious herby sauce sharpened with capers.

Taverna Aeolia

Road map C4. Piazza C.Da Schio 1, 36023 Costozza di Longare. *(0444) 55 50 36. Tue. Ann hol 1–15 Nov.* ⓁⓁ

Luca Chemello, chef-patron at this elegant villa, does wonderful things in the kitchen to complement the charm of the frescoed dining room. Tortellini with artichokes and lamb cooked with sesame seeds are two of his delicious specialities.

Ristorante Storione

Road map C4. Strada Pasubio 64, 36100. *(0444) 56 62 44. Sun.* ⓁⓁⓁ

The Zorzo family run this charming restaurant with its pretty garden. Sturgeon *(storione)* is among the many fish specialities. Parents are encouraged to bring children.

Cinzia e Valerio

Road map C4. Piazzetta Porta Padova 65/67, 36100. *(0444) 50 52 13. Sun eve, Mon. Ann hol 1–8 Jan, all Aug.* ★ ⓁⓁⓁⓁ

This stylish restaurant within the old city walls serves nothing but fish. Featured on the menu are marinated raw salmon, prawn risotto, delicate fillets of sole and lobster, mussels and clams.

VERONA AND LAKE GARDA

Lake Garda South

Antica Locanda Mincio

Road map A4. Via Michelangelo 12, 37067 Valeggio sul Mincio. *(045) 795 00 59. Wed eve, Thu. Ann hol 3 weeks Nov, 2 weeks Feb.* ⓁⓁⓁ

The Locanda was once a staging post and is now a delightful dining spot with frescoed walls and open fireplaces. The "rustic" hors d'oeuvres are a good choice for a starter, and many of the main courses have a regional emphasis.

Esplanade

Road map A4. Via Lario 10, 25015 Desenzano del Garda. *(030) 914 33 61. Wed.* ★ ⓁⓁⓁⓁ

This is a waterside restaurant where you can find imaginative food with firmly regional roots. Courgette flowers are stuffed with scampi, there is a light lake-fish terrine and main courses include lamb with onions, potatoes and a rosemary sauce. The extensive wine list has some interesting dessert wines.

Trattoria Vecchia Lugana

Road map A4. Piazzale Vecchia Lugana 1, 25019 Lugana di Sirmione. *(030) 91 90 12. Mon eve, Tue. Ann hol 6 Jan–15 Feb.* ★ ⓁⓁⓁⓁ

Meat is cooked on the old open fire in the kitchen of this charming restaurant. You can sample a good range of local dishes, including lake fish and tender young kid served with polenta. The wine list is strong on Garda labels.

Lake Garda West

Alla Campagnola

Road map A3. Via Brunati 11, 25087 Salò. *(0365) 221 53. Mon, Tues lunch. Ann hol 2 weeks Jan.* Ⓛ

At this mother-and-son establishment Elisa Dal Bom has revived some interesting traditional recipes. She serves an array of vegetables as an antipasto, and pasta stuffed with pumpkin or aubergine.

Capriccio

Road map A3. Piazza San Bernardo 6, Frazione Montinelle, 25080 Manerba del Garda. *(0365) 55 11 24. Tue.* ⓁⓁⓁ

The terrace here has a fine view over the lake and hills and the menu offers a good choice of fish. Local olive oil is extensively used. Puddings include a delicate muscat-grape jelly with a peach sauce.

La Tortuga

Road map A3. Porticciolo di Gargnano, 25084 Gargnano del Garda. *(0365) 712 51. Mon eve, Tue, Jan and Feb.* ★ ⓁⓁⓁⓁ

La Tortuga has a charming lakeside location and light, imaginative cuisine. Lake fish is cooked with care; try the *coregone*, served with tomatoes and capers.

Villa Fiordaliso

Road map A3. Corso Zanardelli 132, 25083 Gardone Riviera. *(0365) 201 58. Mon. Ann hol Jan and Nov.* ★ ⓁⓁⓁⓁ

This stylish restaurant is housed in the villa that was the love-nest of Mussolini and Clara Petacci. In summer you can eat in the garden, which sweeps down to the lake, and enjoy creative cuisine from the best local ingredients.

Lake Garda East

Locanda San Vigilio Regina

Road map A3. Località San Vigilio, 37016 Garda. *(045) 725 66 88. Ann hol Nov–Feb.* eve. ★ ⓁⓁⓁⓁⓁ

This restaurant forms part of one of the loveliest hotels on Lake Garda. You can start with a selection from the antipasto buffet and then choose from the quite astounding range of fish dishes.

VERONA

Al Bersagliere

Road map B4. Dietro Pallone 1, 37100. *(045) 800 48 24.* *Sun.* ★ ⓁⓁ

An old wood-lined dining room makes a pleasant setting for some traditional Veronese cooking: game in season, mixed boiled meats, and polenta with wild mushrooms.

Ciccarelli

Road map B4. Via Mantovana 171, 37100. *(045) 95 39 86.* *Fri eve, Sat. Ann hol 27 Jul–20 Aug.* ★ ⓁⓁ

Ten minutes' drive from Verona, Ciccarelli's has been renowned for over 40 years. There are excellent grilled, roast and boiled meats, home-made pasta, and superb *crème caramel all'amaretto.*

Bottega del Vino

Road map B4. Via Scudo di Francia 3, 37100. *(045) 800 45 35.* *Nov–Jun: Tue.* ⓁⓁ

A haunt of local wine-producers, the Bottega has more than 800 wines on its list. The food is good, with several specialities, including a good selection of antipasti.

Baba-jaga

Road map B4. Via Cabalao 11, 37030 Montecchia di Crosara. *(045) 745 02 22.* *Sun eve, Mon. Ann hol Jan & last 2 weeks Aug.* ⓁⓁⓁ

This is a good choice in the Soave wine-producing area. The lengthy menu includes black truffle risotto, linguine with sturgeon, leeks and tomatoes, and fillet of sea bass.

El Cantinon

Road map B4. Via S Rochetto 11, 37100. *(045) 59 52 91.* *Wed.* ⓁⓁⓁ

Wine-lovers will find a huge range of excellent Italian and foreign wines. Two sample dishes are fillet of pork with porcini mushrooms, and fresh goat's cheese with leeks and scampi. Service in the wood-beamed dining room is excellent.

Arche

Road map B4. Via Arche Scaligere 6, 37100. *(045) 800 74 15.* *Sun, Mon lunch.* ★ ⓁⓁⓁⓁ

The Arche has long enjoyed a good reputation. The accent is on fish: there is a seafood soup with lentils, and a wonderful warm lobster salad with basil and rocket.

Il Desco

Road map B4. Via dietro San Sebastiano 5/7, 37100. *(045) 59 53 58.* *Sun. Ann hol 1–7 Jan, 15–30 Jun.* ★ ⓁⓁⓁⓁⓁ

There is harmony and attention to detail in every aspect of this lovely restaurant in a 16th-century *palazzo.* It is recognized as one of Italy's best. The imaginative menu is full of temptations: potato cake with truffles, lobster risotto, leeks and glazed onions, and a delicious range of puddings.

THE DOLOMITES

BELLUNO

Antica Locanda al Cappello

Road map D2. Piazza Papa Luciani 20, 32026 Mel. *(0437) 75 36 51.* *Tue eve, Wed exc Aug & Christmas.* ★ ⓁⓁ

Antique furniture complements the frescoed rooms of this stylish restaurant, situated on a beautiful late-Renaissance piazza. The chef prepares some interesting dishes, which include poppy-seed gnocchi and ravioli stuffed with radicchio. Game is often on the menu, too.

Locanda San Lorenzo

Road map D2. Via IV Novembre 79, 32015 Puos d'Alpago. *(0437) 45 40 48.* *Wed exc Jul & Aug.* ★ ⓁⓁⓁ

Trout, rabbit, duck and pigeon are served at this hotel restaurant in a hillside village above the lake of Santa Croce. The Dal Farra brothers are the chefs and take justifiable pride in their regional cooking with a modern slant.

Ristorante El Zoco

Road map D2. Via Cademai 18, 32100. *(0436) 86 00 41.* *Mon in low season. Ann hol May–mid-July, Nov–mid-Dec.* ★ ⓁⓁⓁ

Game, particularly chamois meat, features in the culinary repertoire here. The vegetable dishes are of a high standard, especially the *sformato* of Trevisan radicchio and the grilled wild mushrooms.

Dolada

Road map D2. Via Dolada 21, 32010 Pieve d'Alpago. *(0437) 47 91 41.* *Mon, Tue lunch exc Jul & Aug. Ann hol mid Jan–mid Feb.* ★ ⓁⓁⓁⓁ

An obscure village on the slopes of Monte Dolada is home to one of Italy's great restaurants. Enzo and Rossana de Pra's reputation is well-founded, and the cooking is deft and sure. Chicken breast with truffle oil, salmon mousse, and contrefilet of beef with thyme are served with imaginative vegetables and superb wines.

CADORE AREA

Al Capriolo

Road map D1. Via Nazionale 108, 32040 Vodo di Cadore. *(0435) 48 92 07.* *Dec–Apr, Jun–Oct.* *Tue exc Jul & Aug.* ★ ⓁⓁ

You will find good game, including venison and elk, at this restaurant which is situated in stunning surroundings in the heart of the Dolomites. As a bonus, truffles are used for flavouring.

Dal Cavalier

Road map E1. Località Cima Gogna, 32041 Auronzo di Cadore. *(0435) 98 34.* *Wed exc Jul & Aug.* ★ ⓁⓁ

The Munerin family run this mountain restaurant, which has period furniture and a large collection of copper pots. Good dishes include braised venison with wild mushrooms. The wine list is very strong on hard-to-find regional wines.

La Pausa

Road map D1. On SS51, 32044 Pieve di Cadore. *(0435) 300 80.* *Sun eve, Mon in low season.* ★ ⓁⓁ

A wide range of game features on the menu of this little restaurant in the magnificent surroundings of the Piave valley. Smoked hams and horse meat are popular in the area, and you can sample both here.

CORTINA D'AMPEZZO

El Toulà

Road map D1. Via Ronco 123, 32043. *(0436) 33 39.* *Mon. Ann hol Easter–mid-Jul, mid-Sep–mid-Dec.* ★ ⓁⓁⓁⓁ

You can eat on the terrace of this rustic but chic restaurant, which was originally a hay barn. The food is sophisticated and good, and includes dishes such as basil crêpes, veal, duck, and delicious puddings.

For key to symbols *see p235*

Bars and Cafés in Venice

Many bars in Venice draw their trade from tourists and are busy throughout the day, as visitors ease their aching feet and consult their guide-books. Custom is swelled mid-morning and around lunchtime as the Venetians drop in for a drink or snack and to use the telephone. Cafés range from basic one-room bars patronized by local workmen, to opulent coffee houses in old-world style, such as **Quadri's** and **Florian's**. Even the humblest establishment provides a continuous range of refreshments and you can enjoy anything from a morning coffee or lunchtime beer, to an aperitif or a final brandy before bed. Bars also serve snacks throughout their opening hours: freshly baked morning pastries and lunchtime sandwiches, rolls, cakes, biscuits and sometimes home-made ice cream. Wine bars often have a wide range of traditional Venetian snacks, and so make good places to stop for lunch.

Bars

Italians will often stop for breakfast in a bar on their way to work. This normally consists of a *cappuccino* (milky coffee) and a *brioche* (a plain, jam- or cream-filled pastry). **Caffè Poggi**, on the main route from the station to the Rialto, is much favoured by early morning commuters.

A wide range of alcoholic drinks is on offer, and you can ask for a glass of wine or beer on tap. Beer from the keg is called *birra alla spina* and comes in three different sizes: *piccola, media* and *grande*. Italian and imported bottled beers are also available, though the latter can be expensive. All bars serve glasses of mineral water and it is acceptable to request a glass of tap water *(acqua del rubinetto)*, which will be free. Most bars also serve delicious freshly squeezed fruit juices *(una spremuta)* and milk-shakes made with fruit *(un frullato)*. Italian bottled juices are good and are available in delicious flavours such as apricot and pear.

All bars serve a range of sandwiches *(tramezzini)* and filled rolls *(panini)*, and often have toasted sandwiches and pizzas as well. Some double as cake-shops *(pasticceria)*, and these have a tempting range of calorie-filled delights on display to eat in or take away. If you are near the Accademia, seek out the tiny **Pasticceria Vio** for wonderful cakes, or for an expensive treat, go to **Harry's Dolci** on the Giudecca *(see p241)*.

Bear in mind that sitting down to drink in a bar or café can cost a lot more than standing at the bar, as there is a table charge, which can be high. This rises proportionally as you draw nearer San Marco. Some bars, particularly in the less tourist-frequented areas, have a stand-up counter only. All have a lavatory *(il bagno* or *il gabinetto)*, though you may have to ask at the desk for the key. It is also worth noting that bars and cafés tend to shut earlier here than in other parts of Italy, particularly in winter.

The normal procedure is to choose what you want to eat or drink, then ask for it and pay at the cashdesk. You will be given a receipt *(lo scontrino)* which you present at the bar. If they are busy, a L100 coin will usually speed things up. If you decide to sit down, either inside or at an outside table, your order will be taken by a waiter who will bring the bill when he delivers the drinks. You should expect to pay double or more for this, but you can stretch your drink out for as long as you like.

Wine Bars

There is an old tradition in Venice called *cichetti e l'ombra*, meaning "a little bite and the shade". The little bite ranges from a slice of bread and *prosciutto crudo* (raw cured ham), meatballs or fried vegetables, to sardines and *baccalà* (salt cod). The shade is a glass of wine, so called because the gondoliers used to snatch a glass in the shade away from the glare of the sun on the water. Wine bars serving these snacks and a range of wines are numerous and heavily populated by locals. Many, such as **Do Mori**, are in the crowded alleys off the Rialto, but one of the nicest is the **Cantina del Vino già Schiavi** near the Ponte San Trovaso.

Cafés and Ice Cream Parlours

Coffee houses have played their part in the history of the Veneto – notably Padua's Caffè Pedrocchi *(see p178)* – and a visit to Venice would not be complete without a drink at the historic **Florian's** or **Quadri's**. It is a hard decision whether to take a table outside and watch the crowds or to experience the elegant charm of the interior rooms, with their atmosphere of past eras. The prices are sky-high, but you can take your time and be entertained by the resident orchestras.

Harry's Bar *(see p92)*, is another world-famous bar and café. In summer it is crammed with foreigners and the prices are always high, but for a treat, sip a Bellini, a mixture of Prosecco and fresh white peach juice, in the place where it was invented.

The cafés along the Zattere, with their lovely views across the Giudecca Canal, make good places to pause, and the prices are much lower. Many Venetian squares have cafés with tables outside. There are several in the Campo Santo Stefano, or try **Bar Colleoni** in Campo Santi Giovanni e Paolo (San Zanipolo). **Il Caffè** is the nicest in Campo Santa Margherita.

Venetian ice cream is definitely among the best in Italy, with ice cream shops *(gelaterie)* serving a wide selection of seasonal flavours, some unique to Venice. The Venetians eat ice cream all year round, often instead of pudding or as the finale to the evening stroll, or *passeggiata.* It comes as either a cone *(un cono)* or a cup *(una coppa)* and it is normal to have at least three flavours. **Paolin** on Campo Santo Stefano is one of the best ice cream shops. You could also try **Il Doge**, which is in Campo Santa Margherita, and **Nico** on the Zattere, where you will find *gianduiotto,* a rich chocolate-based Venetian speciality. Make certain you buy ice cream made on the premises, *artigianato* or *produzione propria,* and experiment with what is clearly seasonal; the high-summer fruit ices such as melon, peach and apricot are refreshing and mouthwatering.

Directory

San Marco

Bar Gelateria Paolin
Campo Santo Stefano
San Marco 2962.
Map 6 F3.

Florian's
Piazza San Marco,
San Marco 5659.
Map 7 B2.

Harry's Bar
Calle Vallaresso,
San Marco 1323.
Map 7 B3.

Latteria Veneziana
Calle dei Fuseri
San Marco 4359.
Map 7 A2.

Osteria Terrà Assassini
Rio Terrà degli Assassini
San Marco 3695.
Map 7 A2.

Quadri's
Piazza San Marco
San Marco 120–24.
Map 7 B2.

Rosa Salva
Merceria San Salvador
San Marco 951.
Map 7 B1.

Vino Vino
Ponte delle Veste,
San Marco 2007.
Map 7 A3.

San Polo and Santa Croce

Bar Dogale
Campo dei Frari, San Polo 3012. **Map** 6 E1.

Do Mori
Calle Do Mori
San Polo 429. **Map** 3 A5.

Do Spade
Sottoportega delle Do Spade, San Polo 860.
Map 3 A5.

Soto Sopra
Calle San Pantalon
San Polo 3740.
Map 6 D2.

Castello

Bar Colleoni
Campo Santi Giovanni e Paolo, Castello 6811.
Map 3 C5.

Bar Gelateria Orchidea
Campo Santa Maria Formosa, Castello 5840.
Map 7 C1.

Bar Gelateria Riviera
Ponte de la Pietà
Riva degli Schiavoni 4153.
Map 8 D2.

Bar Mio
Via Garibaldi
Castello 1825.
Map 8 F3.

Bar Orologio
Campo Santa Maria Formosa, Castello 6130.
Map 7 C1.

Caffè al Cavallo
Campo Santi Giovanni e Paolo, Castello 6823.
Map 3 C5.

Il Golosone
Salizzada San Lio
Castello 5689.
Map 7 B1.

Dorsoduro

Ai do Draghi
Calle della Chiesa
Dorsoduro 3665.
Map 6 F4.

Ai Pugni
Fondamenta Gherardini
Dorsoduro 2836.
Map 6 D3.

Bar Cucciolo
Zattere ai Gesuati
Dorsoduro 782.
Map 6 E4.

Bar Gelateria Causin
Campo Santa Margherita
Dorsoduro 2996.
Map 6 D2.

Bar Gelateria Il Doge
Campo Santa Margherita
Dorsoduro 3058.
Map 6 D2.

Bar Gelateria Nico
Zattere ai Gesuati
Dorsoduro 922.
Map 6 D4.

Bar Pasticceria Vio
Rio Terrà della Toletta
Dorsoduro 1192.
Map 6 D3.

Cantina del Vino già Schiavi
Ponte San Trovaso
Dorsoduro 992.
Map 6 E4.

Il Caffè
Campo Santa Margherita
Dorsoduro 2963.
Map 6 D3.

Cannaregio

Alla Bomba
Calle dell'Oca,
Cannaregio 4297.
Map 3 A5.

Bar Algiubagio
Fondamenta Case Nuove
Cannaregio 5039.
Map 1 C2.

Bar Gelateria Solda
Campo Santi Apostoli
Cannaregio 4440.
Map 3 B5.

Caffè Pasqualigo
Salizzada Santa Fosca
Cannaregio 2288.
Map 2 F4.

Caffè Poggi
Rio Terrà della Maddalena
Cannaregio 2104.
Map 2 F3.

Cantina Ardenghi
Calle G Gallina
Cannaregio 6369.
Map 3 C5.

Enoteca Boldrin
San Canciano
Cannaregio 5550.
Map 3 B5.

Paradiso Perduto
Fondamente delle Misericordia,
Cannaregio 2540.
Map 2 F4.

The Lagoon Islands

Bar della Maddalena
Mazzorbo.

Bar Ice
Campo San Donato
Murano.
Map 4 F2.

Bar La Palanca
Fondamenta Santa Eufemia, Giudecca 448.
Map 6 D5.

Bar Palmisano
Via Baldassare Galuppi,
Burano.

Bar Trono di Attila
Torcello.

Harry's Dolci
Fondamenta San Biagio
Giudecca 773.
Map 6 D5.

Lo Spuntino
Via Baldassare Galuppi
Burano.

Shops and Markets

The narrow streets of Venice are lined with beautifully arranged windows that cannot fail to tempt shoppers, and the city has the additional bonus of being truly pedestrianized. Few cities of similar size have such a wide variety of goods to browse through as you explore the fascinating and diverse neighbourhoods. There is still a strong artisan tradition in Venice, and alongside glass and lace you will find high-quality fashion and leather goods, antiques and jewellery. In the Veneto which is one of Italy's most prosperous regions, every town boasts a wide range of shops and many have seasonal speciality markets. In country areas you can buy wine and olive oil direct from the producers.

Piece of traditional Murano glass

Display of jewellery in a shop window in the Frezzeria

When to Shop

Generally, shops open around 9 or 9:30am and close for lunch at 12:30 or 1pm, with the exception of food shops and markets, which are in business from 8am. In the afternoon stores are open from 3:30pm to 7:30pm in winter, and 4pm to 8pm in summer. In Venice, many stores aimed directly at tourists are open all day and even on Sundays, as are big out-of-town supermarkets and hypermarkets – useful if you are self-catering in the region.

Monday is usually the traditional closing day in northern Italy though, again, this does not apply to all shops in Venice itself. The smaller towns in the Veneto often have very variable opening hours, with perhaps food shops closing on Mondays but ironmongers and clothes shops closing on Wednesdays. Shops and markets in the Veneto are often closed for two or three weeks during the national holiday time in August.

The best time for finding bargains is during the January and July sales: look out for window signs with the words *saldi* or *sconti*.

Where to Shop in Venice

The glittering Mercerie *(see p95)*, which runs from Piazza San Marco to the Rialto, has been the main shopping street since the Middle Ages and, together with the parallel Calle dei Fabbri, is still a honey pot for the crowds. West of San Marco, the zigzagging Frezzeria is full of interesting and unusual shops. The main route from the Piazza to the Accademia Bridge is lined with up-market speciality stores, while the streets north of Campo Santo Stefano *(see p93)* are another excellent trawling ground for quality souvenirs and gifts.

Across the Grand Canal, the narrow streets from the Rialto southwest towards Campo San Polo *(see p101)* are lined with a wide variety of less expensive stores, while near the station the bustling Lista di Spagna and the route along the Strada Nova towards the Rialto cater for the everyday needs of ordinary Venetians.

A colourful display of T-shirts with the "Venezia" logo

The islands of Murano and Burano *(see pp150–51)* are *the* places to buy traditional glass and lace.

How to Pay

Major credit cards and Eurocheques are usually accepted in the main stores for larger purchases, but cash is preferred for small items, and smaller shops will want cash. Traveller's cheques are also accepted, though the rate that you will get is less favourable than at a bank.

By law, shopkeepers should give you a receipt *(ricevuta fiscale)*, which you should keep until you are some distance away from the store (legally this is 600 m). If a purchased item is defective, most shops will change the article or give you a credit note, as long as you show the till receipt. Cash refunds are not usually given.

VAT Exemption

Visitors from non-European Union countries can reclaim the 19 per cent sales tax (IVA) on goods exceeding L300,000 from the same shop. Ask for an invoice when you buy the goods and inform the shop that you intend to reclaim the tax. The invoice must be stamped at customs as you leave Italy. The shop will reimburse the tax in lire, once they have received the stamped invoice.

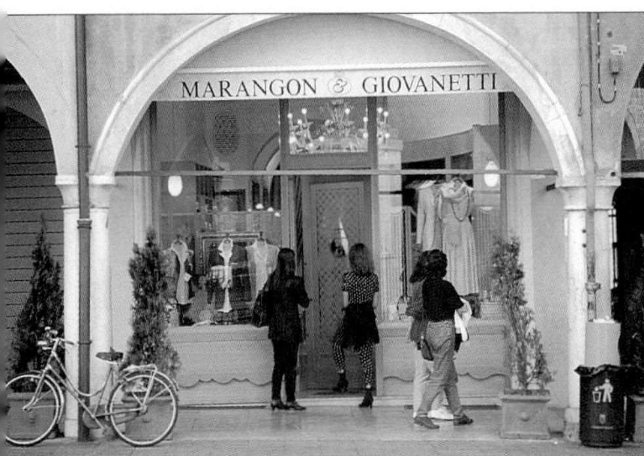

Designer clothes shop in Treviso

Fashion and Accessories

In Venice, the big names in fashion are all found near San Marco. **Armani**, **Laura Biagiotti**, **Missoni** and **Valentino** all have stylish shops just off the Piazza. For really innovative and outrageous designs visit **Fiorella** in Campo Santo Stefano. **La Pantofola**, in the Calle della Mandola, sells a range of good value leather shoes and a wide variety of traditional Venetian slippers in a stunning range of colourful velours. For a genuine gondolier's shirt, take a look at what's on offer in **Emilio Ceccato**.

Fabrics and Interior Design

Venice has long been famed for sumptuous brocades, fine silks and figured velvets. **Trois** sells silks by the metre, including the gossamer-fine pleated silks invented by Fortuny for his Delphos dresses *(see p94)* and **Valli** has wonderful designer silks and other fabrics in its shop in the Mercerie. The famous house of **Rubelli** is based in the Palazzo Corner Spinelli on the Grand Canal. Here you will find rich brocades and velvets. **Color Casa**, in San Polo, has equally lovely textiles at slightly lower prices, and **Jesurum**, not far from the Bridge of Sighs, is the place for beautiful fine linens. They also sell seductive lingerie.

The Lido's Gran Viale has a number of stylish shops that are devoted to modern interior design and beautiful objects for the home.

Masks and Costumes

You can buy cheap, mass-produced masks all over the city, but a genuine one is a good souvenir, and you will be spoilt for choice. **Laboratorio Artigiano Maschere** in Castello revived traditional mask-making and their designs are absolutely stunning. Near Campo San Polo **Tragicomica** sells costumes and masks, as well as Commedia dell'Arte figures. You will find these at **Il Prato** on the Frezzeria too, where they also make string puppets. Dorsoduro has several workshops; **Mondonovo**, just off Campo Santa Margherita, has a marvellous selection of masks and costumes.

A typical Venetian mask

In the weeks leading up to Carnival, maskmakers are, of course, extremely busy, but at other times of the year many workshops welcome visitors and are pleased to show you their craft *(see p31)*.

Glass

The best place to buy glass is on the island of Murano, where it has been made since the 13th century *(see p151)*. All the main manufacturers have their furnaces and showrooms here, catering to mainstream taste. Some manufacturers also have showrooms in Venice itself.

On Murano, **Seguso** and **Barovier e Toso** make glass to traditional designs with good simple lines. You will find similarly attractive pieces in Venice at **Paolo Rossi**, where they specialize in reproductions of antique glass. **Pauly** and **Venini** both have shops near San Marco; they represent the top end of the market and some of their designs are very pleasing.

Jewellery

Venice's smartest jewellers are **Missiaglia** and **Nardi**, both in the arcades of Piazza San Marco. Shops on the Rialto Bridge sell cheaper designs, and this is a good place to find bracelets and chains, whose price is determined by the weight of the gold. For inexpensive, pretty Venetian glass earrings, necklaces and bracelets try **FGB** in Campo Santa Maria Zobenigo.

Wide range of fruit and vegetables for sale in the Rialto market

A typical general food store in the San Marco area

Department Stores

Department stores are not as common in Italy as in many other countries. The main chain store in Venice is Coin, which sells everything from umbrellas to tableware. Standa and Upim are cheaper supermarket-style options. You will find branches of these in other towns in the Veneto.

Treasure trove in one of the art shops on Murano

Books and Gifts

The best general bookshop in Venice is **Goldoni**, in Calle dei Fabbri, which also sells maps. **Filippi Editori Venezia** stocks facsimile editions of old books and books about Venice. **Fantoni** is a specialist art bookshop, and English books are sold at **Libreria Serenissima**.

Hand-made marbled and dragged paper are typically Venetian, and these are used as book covers and made up into writing desk equipment. The **Legatoria Piazzesi** sells hand-blocked papers and stationery items. **Paolo Olbi** constantly adds to their wide range of papers, while **Alberto Valese-Ebru** uses a distinctive marbling technique on fabrics as well as paper.

The San Barnaba area has several art and craft shops where you can buy unusual gifts and souvenirs. **Signor Blum** on the Campo San Barnaba has charming carved and painted wooden objects and toys. Another carver, **Livio de Marchi**, makes large whimsical wooden ornaments. **La Bottega dell'Arte** sells interesting paper objects, as well as masks. For unusual soaps and other toiletries browse in **Il Melograno**, a lovely herbalist in Campo Santa Margherita.

Markets and Food Shops

One of the delights of Venice is a morning spent exploring the food markets and shops around the Rialto. Fruit and vegetable stalls sprawl to the west of the bridge and the Pescheria, or fish market, lies right beside the Grand Canal *(see p100)*. The neighbouring streets are full of unusual and excellent food shops. Olive oil, vinegar and dried pasta, which comes in multifarious colours, shapes and flavours, are all good choices if you are looking for food to take home. **Aliani (Casa del Parmigiano)** is a superlative cheese shop right by the vegetable market, where you can also buy a selection of fresh pasta, salamis and ready-made dishes for a picnic.

Round the corner, on Ruga dei Spezieri, the **Drogheria Mascari** has a fine range of coffees, teas, dried fruits, seeds and nuts. **Pasticceria Marchini** is Venice's best pasticceria, selling traditional Venetian sweetmeats as well as cakes and biscuits.

Viale Santa Maria Elisabetta, the main shopping street of the Lido

Directory

Fashion and Accessories

Emilio Ceccato
Sottoportico di Rialto San Polo 1617.
Map 7 A1.
(041) 522 27 00.

Emporio Armani
Calle dei Fabbri, San Marco 989. **Map** 7 B2.
(041) 522 30 20.

Fiorella
Campo Santo Stefano San Marco 2806.
Map 6 F3.
(041) 520 92 28.

Laura Biagiotti
Calle Larga XXII Marzo San Marco 2400a.
Map 7 A3.
(041) 520 34 01.

Missoni
Calle Vallaresso, San Marco 1312. **Map** 7 B3.
(041) 520 57 33.

La Pantofola
Calle della Mandola, San Marco 3718. **Map** 6 F2.
(041) 522 21 50.

Valentino
Salizzada San Moisè, San Marco 1473. **Map** 7 A3.
(041) 520 57 33.

Fabrics and Interior Design

Annelie
Calle Lunga San Barnaba Dorsoduro 2748.
Map 6 D3.
(041) 520 32 77.

Color Casa
Calle della Madonneta San Polo 1990.
Map 6 F1.
(041) 523 60 71.

Jesurum
Mercerie del Capitello San Marco 4857.
Map 7 B2.
(041) 520 61 77.

Rubelli
Palazzo Corner Spinelli Sant'Angelo, San Marco 3877. **Map** 6 F2.
(041) 521 64 11.

Trois
Campo Santa Maria del Giglio, San Marco 2666.
Map 6 F3.
(041) 522 29 05.

Valli
Merceria San Zulian San Marco 783.
Map 7 B1.
(041) 522 57 18.

Masks and Costumes

Balo Coloc
Calle del Scaleter San Polo 2235.
Map 1 B3.
(041) 524 05 51.

Laboratorio Artigiano Maschere
Barbaria delle Tole Castello 6657.
Map 4 D5.
(041) 522 31 10.

Mondonovo
Rio Terrà Canal Dorsoduro 3063.
Map 6 D3.
(041) 28 73 44.

Il Prato
Frezzeria, San Marco 1770.
Map 7 A2.
(041) 520 33 75.

Tragicomica
Calledei Nomboli San Polo 141.
Map 6 F1.
(041) 72 11 02.

Glass

Barovier e Toso
Fondamenta Vetrai 28 Murano. **Map** 4 E3.
(041) 73 90 49.

Paolo Rossi
Campo San Zaccaria, San Marco 4685. **Map** 8 C2.
(041) 522 70 05.

Pauly
Calle Larga, Ponte dei Consorzi, San Marco.
Map 7 C2.
(041) 520 98 99.

Seguso
Ponte Vivarini 143 Murano.
Map 4 E2.
(041) 73 94 23.

Venini
Piazzetta dei Leoncini San Marco 314. **Map** 7 B2.
(041) 522 40 45.

Jewellery

FGB
Campo Santa Maria Zobenigo, San Marco 2459.
Map 7 C1.
(041) 523 65 56.

Missiaglia
Procuratie Vecchie, San Marco 125. **Map** 7 B2.
(041) 522 44 64.

Nardi
Procuratie Nuove, Piazza San Marco, San Marco 69–71. **Map** 7 B2.
(041) 522 57 33.

Books and Gifts

Alberto Valese-Ebru
Campiello Santo Stefano San Marco 3471.
Map 6 F3.
(041) 520 09 21.

Artigiani Riuniti
Campiello Nomboli, San Polo 2753. **Map** 6 E1.
(041) 523 13 68.

Cartoleria Accademia
Rio Terrà Carità, Dorsoduro 1044. **Map** 6 E3.
(041) 520 70 86.

Cartoleria Testolini
Fondamenta Orseolo San Marco 1744.
Map 7 A2.
(041) 522 30 85.

Erborista Il Melograno
Campo Santa Margherita Dorsoduro 2999.
Map 6 D2.
(041) 528 51 17.

Fantoni
Salizzada San Luca San Marco 4121.
Map 7 A2.
(041) 522 07 00.

Filippi Editori Venezia
Calle Casselleria Castello 5284.
Map 7 A1.
(041) 523 69 16.

Goldoni
Calle dei Fabbri San Marco 4742.
Map 7 A1.
(041) 522 23 84.

La Bottega dell'Arte
Ponte San Barnaba Dorsoduro 2806.
Map 6 D3.

Legatoria Piazzesi
Campiello della Feltrina San Marco 2511.
Map 6 F3.
(041) 522 12 02.

Libreria Serenissima
Merceria San Zulian San Marco 739.
Map 7 B2.
(041) 522 30 50.

Libreria della Toletta
Sacca della Toletta Dorsoduro 1214.
Map 6 D3.
(041) 523 20 34.

Livio de Marchi
Salizzada San Samuele San Marco 3157.
Map 6 E2.
(041) 528 56 94.

Paolo Olbi
Calle della Mandola, San Marco 3653. **Map** 6 F2.
(041) 528 50 25.

Signor Blum
Campo San Barnaba Dorsoduro 2830.
Map 6 D3.
(041) 522 63 67.

Food Shops

Aliani (Casa del Parmigiano)
Campo della Corderia San Polo 214. **Map** 3 A5.
(041) 520 65 25.

Drogheria Mascari
Ruga Rialto, Calle dei Spezieri, San Polo 380.
Map 3 A5.
(041) 522 97 62.

Pasticceria Marchini
Ponte San Maurizio San Marco 2769.
Map 6 F3.
(041) 522 91 09.

What to Buy in the Veneto

Glass is the most popular Venetian souvenir, but there are many other possibilities, ranging from Carnival masks and ceramics to fabrics and lace. For food lovers there is a wide selection of local olive oils, honey, wines and preserves. In the Veneto many food producers sell direct to the public, while different craft and food specialities are found in individual towns and islands.

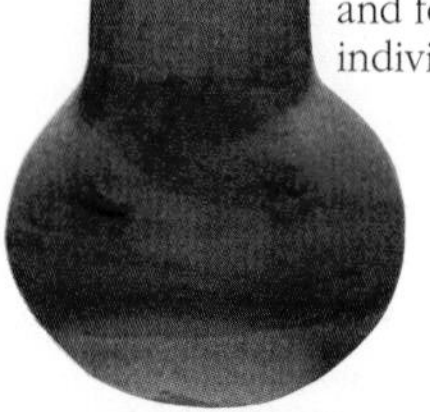

Modern vase of opaque glass

Traditional glass with gold overlay

Two-coloured goblet

Venetian Glass

In traditional rich colours of blue and claret, or in striking modern designs, you will find anything from scent bottles to chandeliers.

Gift box covered in marbled paper

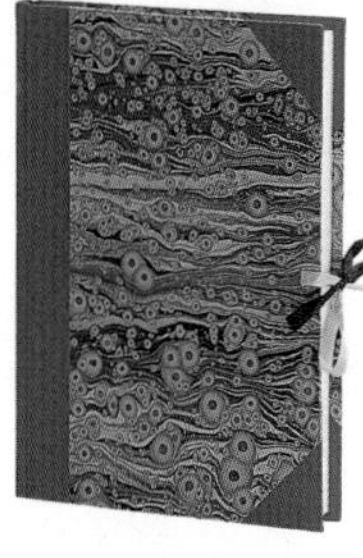

Address book

Venetian Marbled Paper

Marbled paper is a Venetian speciality. The sheets of paper are dipped into liquid gum before adding the paint. You can buy a large range of stationery items covered in the paper, as well as paper by the individual sheet. Each sheet of marbled paper is unique.

Pretty trinket box

Sheets of marbled paper

Decorated ceramic vase from Bassano

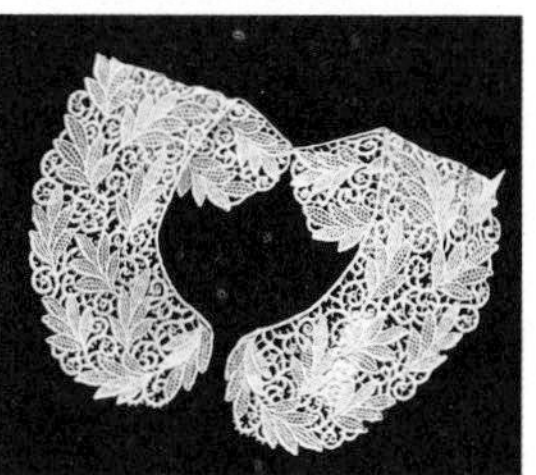

Delicate lace collar from Burano

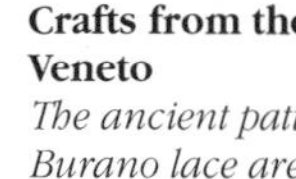

Crafts from the Veneto

The ancient patterns of Burano lace are used to great advantage on table linen and to trim exquisite lingerie. Hand-painted vases, plates and bowls are produced in the picturesque old town of Bassano del Grappa.

Silver spoon with Venetian lion finial

Red and gold mask

Masks *(see pp30–31)*
Mask designs range from Commedia dell'Arte motifs to modern abstracts from young designers, and many are intricate and colourful. They are available all year, but at Carnival time you can buy them from street stalls.

Carnival mask

Velvet slippers

Colourful child's sweater

Clothing
As everywhere in Italy, stylish designer shops abound. Clothes for children are particularly bright and inventive. Velvet slippers, which are made in rich jewel-like colours, are worn at home as well as to dress up in at Carnival time.

Pasta
Attractively packaged dried pasta comes in many colours, shapes and flavours. Tomato, herb and spinach are the most popular varieties, but beetroot, garlic, artichoke, salmon, squid, and even chocolate can also be found in many shops.

Artichoke

Beetroot

Squid

Pasta shapes

Balsamic vinegar and extra virgin olive oil

Panettone

Amaretto biscuits

Delicacies from the Veneto
Panettone is the light yeast cake, flavoured with vanilla and studded with currants and candied peel, that is traditionally eaten at Christmas. Other local delicacies include olive oil from the shores of Lake Garda, vinegars, mountain honey from Belluno, fruit-flavoured liqueurs, grappa from Bassano (see p166)*, and after-dinner Amaretto biscuits.*

Orange liqueur | Lime liqueur | Pear liqueur

Entertainment in the Veneto

Venice was once one of Europe's liveliest night-time cities, and today it still has an impressive range of special events throughout the year. At every season there are some splendid festivals unique to Venice, and in late summer the normal city diet of opera, theatre and concerts is augmented by the International Film Festival and the Biennale, which rank among the best world-class cultural events. The day-to-day evening entertainment in Venice itself now tends to be far less frenetic than in the heyday of the Republic *(see pp46–7)*, but there are a few clubs and discos, and many more across the causeway in Mestre. Or you could have a little flutter at the casino.

Poster advertising the Film Festival

Whatever you choose, your enjoyment will be enhanced by the idyllic backdrop of Venice itself. The ultimate and quintessential Venetian romantic experience is, of course, a gondola ride by moonlight *(see p276)*. However, an evening's entertainment could more usually comprise the traditional stroll, or *passeggiata*, followed by a drink at a bar or café in one of the squares or amid the floodlit splendours of the Piazza San Marco.

Practical Information

Information about what's on in Venice can be found in *Un Ospite di Venezia*, a free Italian and English bilingual booklet which is published weekly during the summer and monthly in winter. You can get a copy at some hotels or from the tourist office, where there are also posters advertising coming cultural events. The Tourist Board also publishes an annual month-by-month list of events: *Venezia Manifestazioni.* The **Department of Culture** may also have useful information, particularly about exhibitions. The Venetian local newspaper, *Il Gazzettino*, lists cinema performances, rock concerts and discos under *Spettacoli.*

For details of events and festivities in the other towns and cities in the Veneto, ask at the local tourist office. The regional press also advertises what's on in advance.

Music and coffee at Florian's café, Piazza San Marco *(see p246)*

Booking Tickets

Booking in advance is not part of the Italian lifestyle, where decisions are made on the spur of the moment. If you want to be certain of a seat you will have to visit the box office in person, as they usually do not take bookings over the telephone. You may also have to pay an advance booking supplement, or *prevendita,* which is usually about 10 per cent of the price of the seat.

The price of a theatre ticket starts at about L15,000, though prices are likely to be five times as much for star-name performances. Tickets for popular music concerts are normally sold through record and music shops whose names are displayed on the publicity posters.

Whereas tickets for classical concerts are sold on the spot for that day's performance, opera tickets are booked weeks ahead. There are very few ticket touts, so it is almost impossible to obtain tickets when the box office has sold out. The **Goldoni** box office is open 9:30am–12:30pm and 4–6pm.

La Fenice opera house before the 1996 fire *(see p93)*

Cinema and the Film Festival

There are five cinemas in Venice, mainly showing dubbed versions of international films. These are known as *prima visione* (first run). The **Accademia** and the **Olimpia** occasionally show more interesting "art-house" films, and you will find these listed in *Un Ospite*.

The annual Film Festival, which takes place in August and September, is one of the major world cinema showcases, and has been running since 1932. Screenings are held in two cinemas on the Lido, the **Palazzo del Cinema** and the **Astra**, and tickets are available to the public direct from the cinema on the day of performance. The programme is available in advance from the tourist office, and you will also see posters advertising the festival around the city.

The courtyard of the Doge's Palace during a Film Festival performance

Gondolier serenading on the Grand Canal

Music and Theatre

Like many Italian cities, Venice makes good use of the most magnificent churches as concert halls. La Pietà was Vivaldi's own church and is still used for concerts, as are the churches of the Frari and Santo Stefano. RAI, the state radio service, uses the **Palazzo Labia** to record classical concerts; tickets are free and you can listen to delightful music while surrounded by Tiepolo frescoes.

Other concerts are held from time to time in Scuola di San Giovanni Evangelista and the Palazzo Prigioni Vecchie, the old prisons attached to the Doge's Palace. In summer, the garden of Ca' Rezzonico and the courtyard of the Doge's Palace are also used as outdoor concert halls.

Unfortunately, La Fenice, one of Italy's most charming opera houses, suffered a disastrous fire in early 1996. Formerly the venue for major operas, its future is now uncertain, although there are plans to rebuild the theatre as soon as money can be raised.

Venice's principal theatre is the **Goldoni** where, not surprisingly, the repertoire is mainly by the eponymous dramatist, though there is a wide range of other productions as well. There are performances from November to June. Venice's other theatre is the **Ridotto**, which also stages plays by Goldoni, as well as 20th-century drama.

At Carnival time in February *(see pp30–31)*, the whole city is invaded by merry-makers in fancy dress and outlandish masks. Numerous theatrical and musical events take place, both in theatres and in the streets and *campi*.

Facilities for the Disabled

Access for disabled people is difficult everywhere in Venice, and theatres are no exception, although concerts are often held in easily accessible churches. For further advice, see page 261.

Masked reveller at Carnival time ***(see pp30–31)***

The Biennale and Other Exhibitions

Venice is without doubt one of the leading art exhibition centres in Europe, offering shows on themes ranging from art history to photography, and frequently playing host to the world's major travelling exhibitions. There are excellent facilities for such exhibitions, and these include the Doge's Palace, the Museo Correr, the Palazzo Grassi, the Querini-Stampalia, the Peggy Guggenheim and the Fondazione Cini. *Un Ospite di Venezia* will give details, as will the tourist office and posters around the city.

One of the best and largest exhibitions is the Biennale, an international display of contemporary and avant-garde art, which is held from June to September in odd-number years (its centenary is 1995). The main site is the Giardini Pubblici *(see p121)*, where the specially built pavilions represent about 40 different countries. Another branch of the exhibition, known as the Aperto and showing the work of less established artists, takes place around the city in venues such as the old rope factory in the Arsenale *(see p119)* and the salt warehouse on the Zattere. The tourist office will have further details.

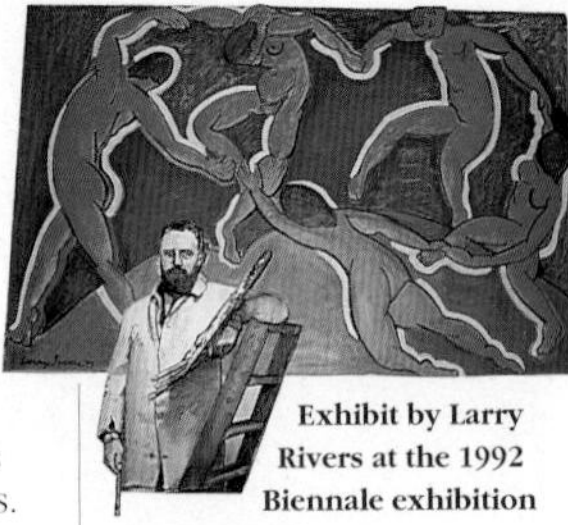

Exhibit by Larry Rivers at the 1992 Biennale exhibition

Casinos, Clubs and Discos

If you want to play roulette on your visit, there is a magnificent casino. In winter, from October to March, the casino is housed in the Palazzo Vendramin-Calergi on the Grand Canal *(see p61)* and you can sweep up to the stately entrance by gondola. In summer the more prosaic *vaporetto* serves the casino's summer quarters on the Lido in the Palazzo del Casino.

The **Scala Martini** is the best-known late-night club. Open until 2am, it has live music in smart surroundings. A few other bars also feature live bands, including the **Paradiso Perduto** in Cannaregio. Discos are few and far between in Venice. You could try **El Souk**, near the Accademia, or go to the mainland, where Mestre has numerous discos. You will find these advertised in the *Spettacoli* listings in *Il Gazzettino*.

Placido Domingo singing at the Verona Festival

Sport and Children

Venetians are very keen on rowing and sailing. There are several clubs in the city, and the tourist office will be able to give you information. Most of the other sporting facilities are on the Lido, where you can ride, swim, cycle, and play golf or tennis.

In the city itself, there are few attractions for young children, but the mainland is more promising. Around Lake Garda there are plenty of watersports and a theme park, Gardaland *(see p205)*.

Music and Theatre in Verona

Verona has two exceptional venues for theatre and music: the superb Arena *(see p95)*, and the 1st-century Teatro Romano *(see p198)* on the far side of the River Adige. Both stage open-air performances during the summer months.

The Arena is a popular site for rock concerts, and is internationally renowned for its summer opera season. The Teatro Romano stages a succession of ballets and drama, including a Shakespeare Festival, in Italian translation. Tickets for the Teatro can be ordered by post; they are also sold at the box office at the Arena. Tickets to some events are free. Information about all the entertainment is given in the Verona newspaper, *L'Arena*.

Opera at the Arena

Almost everyone will enjoy the experience of hearing opera in the magnificent open-air setting of the Arena. Real opera buffs should be aware, however, that Verona performances are very much "opera for all". You should be prepared for less-than-perfect acoustics, noisy audiences, and even small children running about. The opera season runs from the first week in July until the beginning of September, and every year features a lavish production of Verdi's *Aida*. Performances start at 9pm, as dusk is falling, and it is customary to buy one of the little candles that are on sale. Ten

Giant dragon at the Gardaland theme park, Lake Garda *(see p205)*

Aida, **performed annually in Verona's Roman Arena**

minutes before the "curtain goes up", the whole Arena becomes a breathtaking sight, with a sea of flickering lights.

There are normally two intervals, when people eat the picnics they have brought with them, or buy *panini* and ice creams. Glass bottles are not allowed in the Arena, so if you are taking a drink make sure it is in a plastic bottle. Toilets are few and far between and have lengthy queues during the intervals.

Ticket prices are high: an unreserved, un-numbered, backless seat in the *gradinata*, or tiers, is L30,000, while the different types of *poltrone*, literally "armchairs", either on the steps or in the stalls, range from L120,000 to L200,000, with a first-night supplement of L50,000. If you opt for a cheap seat, arrive about two hours before the performance and sit about halfway down the tiers, where the acoustics are better. You can hire an air cushion for about L2,000. Numbered seats are more comfortable, but seats lower in the Arena can be very hot and airless and the view of the stage is limited. You may prefer to sacrifice comfort for fresh air and a bird's-eye view. Unless you have a seat in the best stalls with the glitterati, there is no need to dress up.

Visitors flock to Verona to attend the opera season, so you need to book accommodation months ahead.

DIRECTORY

MUSIC AND THEATRE

Department of Culture
Piazza San Marco
San Marco 52.
Map 7 B2.
(041) 270 76 04.

La Fenice
Cassa di Rispamio,
Campo San Luca.
Map 7 A2.
(041) 521 01 61.

Palazzo Labia
Fondamenta Labia
Cannaregio 275.
Map 2 D4.

Teatro Goldoni
Calle Goldoni,
San Marco 4650.
Map 7 A2.
(041) 520 54 22.

Teatro Ridotto
Calle Vallaresso
San Marco 1335.
Map 7 B3.
(041) 522 29 39.

CINEMAS

Accademia
Calle Corfù
Dorsoduro 1019.
Map 6 E3.
(041) 528 77 06.

Astra
Via Corfù, Lido.
(041) 526 02 89.

Olimpia
Campo San Gallo
San Marco 1094.
Map 7 B2.
(041) 520 54 39.

Palazzo del Cinema
Lungomare G Marconi
Lido. (041) 526 87 00.

NIGHTCLUBS

Scala Martini
Campo San Fantin
San Marco 1980.
Map 7 A2.
(041) 522 41 21.

El Souk
Calle Corfù, Dorsoduro.
Map 6 E3.

Paradiso Perduto
Fondamenta della
Misericordia, Cannaregio
2540. **Map** 1 C4.
(041) 72 05 81.

SPORTS

Cycling
Giorgio Barbieri
Via Zara 5, Lido.
(041) 526 14 90.

Golf
Alberoni
Lido. (041) 73 10 15.

Rowing
Canottieri Bucintoro
Punta Dogana
Dorsoduro 15.
Map 7 B4.
(041) 522 20 55.

Tennis
Tennis Club Venezia
Lungomare G Marconi
41/d.
(041) 526 03 35.

VERONA OPERA

Main box office
ENTE Arena
Piazza Brà 28
37121 Verona.
(045) 59 65 17.

Ticket agent
Cangrande
Via Giovanni della Casa 5
37122 Verona.
(045) 800 49 44.
FAX (045) 801 01 08.

Survival Guide

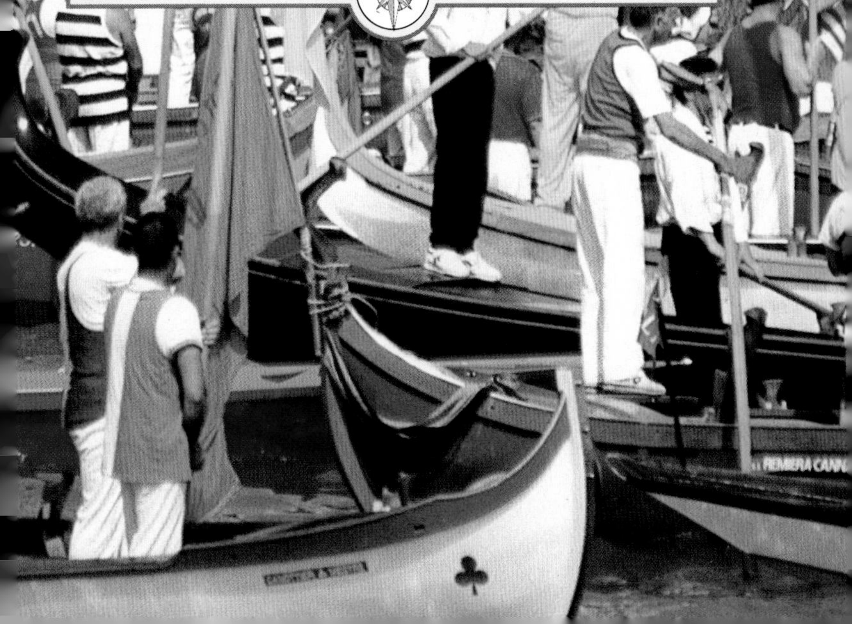

Practical Information

The enormous wealth of art and architecture found in Venice and the historic cities of Padua, Verona and Vicenza can dazzle and overwhelm. The best way to avoid cultural overload is to concentrate on sights in the morning, when they are most likely to be open, relax over your lunch as the Italians do, and leave any shopping or sightseeing of churches until the late afternoon or early evening. The inconsistency of museum opening hours and the fact that some sights or large sections of them are closed for years for restoration can be frustrating. This is particularly true of Venice, where you may often see scaffolding and the signs *chiuso per restauro* (closed for restoration), so it is best to check opening hours with individual museums in advance. However, on the plus side, the principal sights are all within easy walking distance of one another, and exploring by boat and on foot is an exciting experience.

Tourist Board logo

Tourists crossing the white stone Ponte della Paglia

Tourist Information

Most towns in the Veneto have their own tourist offices, and in season Verona and Padua each have three. The offices in smaller towns may not be able to provide much more than a glossy pamphlet. In contrast, the Verona offices publish a useful free booklet, *Passport Verona*, with information on what there is to see and do. Travel agents can supply information on city and other tours on offer. Tourist offices in Venice provide city maps, lists of accommodation, *vaporetto* maps and plenty more printed material on request. The tourist offices and hotels also have leaflets on local entertainment and events *(see pp32–5)*. To obtain information prior to travel contact the **ENIT** (Italian State Tourist Office) in your home country before you leave, or write direct to the **Azienda di Promozione Turistica di Venezia**.

i
AZIENDA PROMOZIONE TURISTICA

A Tourist Information sign

Guided Tours

City tours in Venice with English-speaking guides can be booked through many travel agencies, including **American Express** and **World Vision Travel**. The tours offered take from one to three hours, and are conducted either by foot, motorboat or gondola. In Verona and Padua half-day tours during the tourist season are organized by each town's tourist office. Boat trips along the Brenta Canal *(see pp182–3)* between Venice and Padua, in both directions, are available from April until the end of October.

Museums and Monuments

The opening hours of museums, galleries and palaces change with alarming rapidity. To avoid disappointment ask at the local tourist office for a list of opening times or, if in Venice, consult the current edition of the free booklet *Un Ospite di Venezia*, available from most hotels. Civic museums are shut on Mondays, otherwise there is no real pattern to opening and closing times. Many places shut at 1pm or 2pm and do not re-open in the afternoon. The majority of museums charge an admission fee, but there are usually concessions for children, students and, in a few cases, senior citizens. Churches are normally open in the mornings from around 9am until noon, then again from mid-afternoon until 6 or 7pm.

A selection of local tourist information publications

A mechanically operated wheelchair ramp across a bridge

Venice for the Disabled

The stepped bridges of Venice make it almost impossible for the disabled to get around the city – it is difficult enough for a parent with a pushchair. A further problem are the *vaporetti* – particularly the smaller, sleeker *motoscafi* which are especially hazardous for those confined to wheelchairs.

One of the few aids for the disabled is the *Veneziapertutti* plan of the city, indicating places of interest which can be reached by *vaporetto*, or by streets avoiding any bridges. This is produced by the University of Architecture in Venice, Santa Croce 191. Few of the sights have special facilities for the disabled, but one or two of the bridges in the city have now been adapted for wheelchairs by the installation of a mechanically operated ramp.

The Venice tourist office brochure that lists available accommodation indicates which hotels are suitable for disabled guests.

Etiquette

Any attempt by visitors to speak Italian is always appreciated by local people. Few people speak English in the Veneto, but hotel receptionists are usually helpful and will readily offer to make any enquiries and reservations on your behalf.

To avoid offence always dress decently, particularly for churches, and make sure you are never drunk in a public place. Smoking is common in bars and restaurants but banned on the *vaporetti*.

Visiting Churches

Bare shoulders and shorts are frowned upon in Italian churches and those unsuitably dressed may well be refused entry. Church interiors tend to be very dark but there are usually coin-operated light meters to illuminate works of art. Make sure you take plenty of L200 and L500 coins. Machines that provide recorded information on the church and its artifacts are available to hire, but the commentaries are sometimes inaudible. Although most churches do not charge an entry fee, you are encouraged to leave a few coins for the upkeep.

Pedestrians shopping in Verona

Tipping

Always keep L1,000 and L2,000 notes handy for porters, chambermaids, restaurant staff and custodians of churches. Italian taxi drivers do not expect a tip and there is no need to tip a gondolier.

WCs

There is a dearth of public toilets in the Veneto. The few in Venice are very inconspicuous. You will need L500 or L1,000 for those at the station, or you can use toilets in cafés and bars, or the facilities in museums. Ask for *il bagno* or *il gabinetto*. The toilets are always short of paper, so it is a good idea to carry tissues with you.

Useful Addresses

American Express
Salizzada San Moisè
San Marco 1471, Venice.
Map 7 A3.
☎ *(041) 520 08 44.*

Azienda di Promozione Turistica di Venezia
Castello 4421, Venice.
Map 7 C1.
☎ *(041) 529 87 01.*

ENIT UK
1 Princes Street
London WIR 8AY.
☎ *0171-408 1254.*

Uffici Informazioni (Tourist Information)
Palazzetto del Selva,
Giardinetti Reale, San Marco.
Map 7 B3.
☎ *(041) 522 63 56.*

Ferrovia Santa Lucia, Venice.
Map 1 B4.
☎ *(041) 71 90 78.*

Gran Viale 6, Lido, Venice.
☎ *(041) 526 57 21.*

Piazza Erbe 38, Verona.
☎ *(045) 800 00 65.*

Riviera Mugnai 8, Padua.
☎ *(049) 875 06 55.*

Piazza Duomo 5, Vicenza.
☎ *(0444) 54 41 22.*

World Vision Travel
Campiello de la Feltrina
San Marco 2513/a,
Venice.
Map 6 F3.
☎ *(041) 523 09 33.*

Gondola with passengers

Students relaxing in the sun in Verona

Immigration and Customs

European union (EU) residents and visitors from the US, Canada, Australia and New Zealand do not need visas for stays of up to three months. However, all non-EU visitors need to bring a full passport. British citizens may use a Visitor's Card. A visa is needed for stays longer than three months. Vaccination certificates are not necessary.

All visitors to Italy should, by law, register with the police within three days of arrival. Most hotels will register visitors when they check in. If in doubt, contact a local police department or phone the **Questura**.

Duty-free allowances are as follows: non-EU residents can bring in either 400 cigarettes, 100 cigars, 200 cigarillos or 500 grams of tobacco; 1 litre of spirits and 2 litres of wine; 50 grams of perfume. Goods such as watches and cameras may be imported as long as they are for personal or professional use. EU residents no longer have to declare goods, but random checks are often made to guard against drugs traffickers.

The refund system for Value Added Tax (IVA in Italy) for non-EU residents is complicated and is worth reclaiming only if you have spent at least L300,000 in a single establishment.

Student Information

An international Student Identity Card (ISIC) or a Youth International Educational Exchange Card (YIEE) will usually get reductions on museum admissions and other charges. Venice Municipality's Rolling Venice card for 14- to 30-year-olds offers, for a small fee, a package of useful books and information on the city; this includes alternative itineraries, haunts for the young, and hotels, theatres, shops and restaurants offering discounts to card holders.

Discount rail tickets for students are available from the Transalpino office at Venice railway station.

Educational Courses

The societa Dante Alighieri in Venice organizes monthly Italian courses for foreigners. The **European Centre for Training Craftsmen in the Conservation of the Architectural Heritage**, on the island of San Servolo, offers standard three-month or intensive two-week courses to Italian and foreign craftsmen.

The **Scuola Internazionale di Grâfica** in Venice specializes in short summer courses in painting, printing and sketching while **Zambler** offers Italian courses in Venice, with accommodation arranged if required. In Padua the **Istituto Linguistico Bertrand Russell** organizes courses in the Italian language, which run throughout the year.

Newspapers, Radio and TV

The local newspapers are the *Gazzettino* and the *Nuova Venezia*. European and American newspapers and magazines are available at the main news kiosks, normally a day or two after publication. The state TV channels are RAI Uno, RAI Due and RAI Tre. Satellite and cable TV transmit European channels in many languages, as well as CNN news in English. BBC World Service is broadcast on radio on 15.070 MHz (short wave) in the mornings and 648 KHz (medium wave) at night.

Newspaper stall selling national and international publications

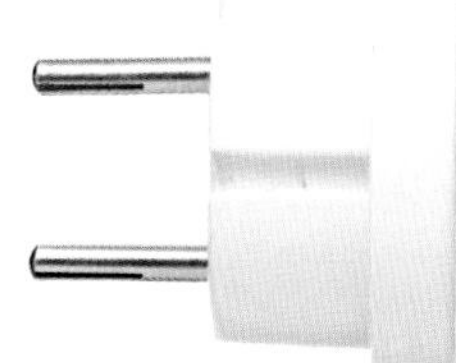
Standard Italian plug

Embassies and Consulates

If you lose your passport or need other help, contact your national embassy or consulate as listed in the directory below.

Electrical Adaptors

Electrical current in Italy is 220V AC, with two-pin, round-pronged plugs. It is probably better to purchase an adaptor before leaving for Italy. Most hotels that are graded above three star have electrical points for shavers and hairdryers in all bedrooms.

Italian Time

Italy is one hour ahead of Greenwich Mean Time (GMT). The time difference between Venice and other cities is: London: –1 hour; New York: –6 hours; Perth: +7 hours; Auckland: +11 hours; Tokyo: +8 hours. These figures may vary for brief periods in summer with local changes. For all official purposes the Italians use the 24-hour clock.

The clock of San Giacomo di Rialto in San Polo, Venice

Conversion Table

Imperial to Metric
1 inch = 2.54 centimetres
1 foot = 30 centimetres
1 mile = 1.6 kilometres
1 ounce = 28 grams
1 pound = 454 grams
1 pint = 0.6 litres
1 gallon = 4.6 litres

Metric to Imperial
1 centimetre = 0.4 inches
1 metre = 3 feet, 3 inches
1 kilometre = 0.6 miles
1 gram = 0.04 ounces
1 kilogram = 2.2 pounds
1 litre = 1.8 pints

Directory

Police (Questura)

Venice
Campo San Polo.
Map 6 F1.
(041) 271 55 93.

Vicenza
Viale Mazzini 213.
(0444) 33 75 11.

Padua
Via Santa Chiara 11.
(049) 66 16 00.

Verona
Lungadige Porta Vittoria.
(045) 809 06 11.

Student Information

Comune di Venezia Assessorato alla Gioventù
San Marco 1529, Venice.
Map 7 A3.
(041) 274 76 50.

Transalpino
Ferrovia Santa Lucia
Venice.
Map 1 B4.
(041) 71 66 00.

Ufficio Informativo Rolling Venice
Ferrovia Santa Lucia
Venice. **Map** 1 B4.
(041) 270 76 45.

Educational Courses

European Centre for Training Craftsmen in the Conservation of the Architectural Heritage
Isola di San Servolo
Casella Postale 676
30100 Venice.
(041) 526 85 46.

Istituto Linguistico Bertrand Russell
Via Filiberto 6
35122 Padua
(049) 65 40 51.

Istituto Zambler
Cannaregio 3764
30121 Venice. **Map** 3 A4.
(041) 522 43 31.

Società Dante Alighieri
Ponte del Purgatorio
Arsenale, Castello
Venice. **Map** 8 F2.
(041) 528 91 27.

Scuola Internazionale di Gràfica
Calle della Regina
Santa Croce 2213
30135 Venice.
Map 2 F5.
(041) 72 19 50.

Embassies and Consulates

Australia
Via Reno 9, Rome.
(06) 85 43 05.

Canada
Via Vittor Pisani 19
Milan.
(02) 675 81.

New Zealand
Via Zara 28
Rome.
(06) 440 29 28.

United Kingdom
Palazzo Querini
Accademia
Dorsoduro 1051
Venice. **Map** 6 E3.
(041) 522 72 07.

US
Via Principe Amedeo 2/10
Milan.
(02) 29 03 51.

Religious Services

Anglican
St George's Anglican Church, Campo San Vio
Dorsoduro
Venice. **Map** 6 F4.
(041) 520 05 71.

Greek Orthodox
Greek Orthodox Church
Ponte dei Greci
Castello 3412
Venice. **Map** 8 D2.
(041) 522 54 46.

Jewish
Sinagoga, Ghetto Vecchio
Cannaregio
Venice. **Map** 2 D3.
(041) 71 50 12.

Lutheran
Chiesa Evangelica Luterana
Campo Santi Apostoli
Cannaregio 4443
Venice. **Map** 3 B5.

Methodist
Chiesa Evangelica Valdese e Metodista
Campo S Maria Formosa
Castello 5170
Venice. **Map** 7 C1.
(041) 522 54 46.

Personal Security and Health

VENICE IS ONE of the safest cities in Europe. Violent crime is very rare and petty crime minimal in comparison with other main cities. Nevertheless, it is wise to take a few simple precautions, particularly against pickpockets, both in Venice and elsewhere in the Veneto. Leave valuables and any important documents in the hotel safe and carry only the minimum amount of money necessary for the day.

Make sure you take out adequate travel insurance before leaving for Italy, as it is very difficult to obtain once you are in the country.

Venice by night, not always well-lit but safe

Looking After Your Property

TRAVELLER'S CHEQUES or Eurocheques are the safest way to carry large sums of money. Try to carry your cheque receipts and Eurocheque card separately to be on the safe side, and keep a photocopy of all vital documents such as your passport.

Safeguard against attracting the attention of pickpockets and bagsnatchers, particularly at railway stations, markets and on the buses. In Venice take extra care while waiting at the *vaporetto* landing stages; be especially vigilant when crowds are jostling to get on to the boats.

If you drive while in the Veneto, always remember to lock the car before you leave it and never leave valuables on display inside. Hired cars or those with foreign number plates are favourite targets of car thieves.

Personal Safety

VENICE IS UNEVENTFUL by night and you can stroll through the streets without any threat. There is no red light quarter or any area that could be described as unsavoury. Women alone in Venice are unlikely to encounter anything more troublesome than the usual Latin roving eye. Elsewhere in the Veneto, in the less touristy towns particularly, unescorted females are likely to attract more attention.

Avoid unauthorized taxi drivers, who may not be insured and almost invariably overcharge. Airports are their favourite haunts. Make sure you take only official taxis which have the licence number clearly displayed *(see p278)*.

Police

THE VIGILI URBANI, or municipal police, are most often seen in the streets regulating traffic and enforcing local laws. They wear blue uniforms in winter and white during the summer. The *carabinieri*, with red striped trousers, are the armed military police, responsible for public law and order. *La polizia*, or state police, wear blue uniforms with white belts and berets. They specialize in serious crimes. Any of these should be able to help you.

In the event of theft go straight to the nearest police station *(polizia* or *carabinieri)* to make a statement. If there is a language problem, you should consult your nearest consulate *(see p263)*, which you should also do in the case of a lost passport.

Emergency Numbers

Ambulance
Venice
(041) 118.
Verona
(045) 118.
Padua
(049) 75 55 55.
Vicenza
(0444) 118.

Automobile Club d'Italia
116.
Car accident and breakdown.

Fire
115.

General SOS
113.

Medical Emergencies
118.

Police
112.

Traffic Police
Venice
(041) 270 70 70.
Verona
(045) 807 84 11.
Padua
(049) 880 55 00.
Vicenza
(0444) 51 29 78.

Medical Precautions

VISITORS FROM the European Union (EU) are entitled to reciprocal state medical care in Italy. Before you travel, pick up form E111 from the post

A group of Venetian policemen on the Riva degli Schiavoni

office, which covers you for emergency medical treatment. You may wish to take out additional medical insurance, as E111 does not cover repatriation costs or additional expenses such as accommodation, food and flights for anyone travelling with you. Visitors from outside the EU should take out a comprehensive insurance policy covering emergency medical treatment. If you are taking prescribed medication, take supplies or prescriptions with you.

Inoculations are not needed for the Veneto, but take sunscreen and mosquito repellent in the summer. Because of the canals, mosquitoes can be irksome in Venice. An electric gadget, from pharmacies or department stores, will repel insects in your room for up to 12 hours. Tap water is safe to drink, but locals often prefer mineral water, either fizzy *(con gas)* or still *(naturale)*.

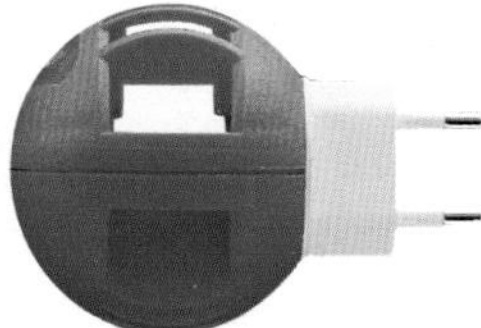

Electric mosquito deterrent

Medical Treatment

If you are in need of urgent medical attention, go to the *Pronto Soccorso* (First Aid) department of the nearest main hospital. Standards of health care are generally better than those in the south of Italy, although not as high as in Britain or the US. There are usually queues at the emergency departments and many hospitals expect the patient's family or friends to help with his or her nursing.

Should you require a consultation with a doctor, ask the advice of your hotel or look in the yellow pages of the telephone directory, under *medici*. (A *dottore* is not necessarily a doctor of medicine.) If you have a serious medical complaint or allergy you should bring a letter, preferably translated, from your doctor at home.

Pharmacy sign

Many doctors in the region speak at least a little English. There are first aid facilities with the services of a doctor at airports and at most railway stations.

Dentists are expensive in Italy. You can find the nearest one in the yellow pages of the telephone directory, listed under *dentisti medici chirurghi*, or ask your hotel receptionist for his or her recommendation.

For insurance claims, make sure you keep all receipts for medical treatment and any medicines prescribed.

Pharmacies are open during the summer months from 8:30am to 12:30pm and 4pm to 8pm Monday to Friday, and from 9am to noon on Saturday. Winter hours are slightly shorter. All towns offer a 24-hour pharmacy service, with a night-time and Sunday rota. You will find the rota posted on the doors of pharmacies. Opening times can also be found in the local newspapers or, if you are in Venice, the booklet *Un Ospite di Venezia*.

Italian pharmacists are well-trained to deal with minor ailments and can prescribe many drugs without needing a doctor's prescription. The majority of pharmacies do not stock quantities of foreign medicines but can usually supply the local equivalent. Many of the words for minor complaints and remedies are similar in Italian, for example *lassativo* (laxative), *aspirina* (aspirin) and *tranquillante* (tranquillizer).

Useful Information

Venice Hospital
Ospedale Civile, Campo Santi Giovanni e Paolo. **Map** 3 C5.
(041) 529 41 11.

Verona Hospital
Ospedale Borgo Trento, ULSS No.25 Verona Piazzale Stefani 1.
(045) 807 11 11.

Padua Hospital
Ospedale Civile, Via Giustiniani 2.
(049) 821 11 11.

Vicenza Hospital
Ospedale di Vicenza
Via Rodolfi 37.
(0444) 99 31 11.

Red Cross
San Marco 52. **Map** 1 B4.
(041) 528 63 46.

Lost Property Offices
Ferrovia Santa Lucia, Venice.
Map 1 B4.
(041) 78 52 38.
Railway Station, Verona.
(045) 809 38 27.
Railway Station, Padua.
(049) 822 41 11 (freephone).

Missing Credit Cards
American Express
(06) 722 82.
Diners Club
167 86 40 64 (freephone).
Eurocard
167 82 80 47 (freephone).
MasterCard (Access)
172 10 11 then ask for 314 275 66 90.
VISA
167 87 72 32 (freephone).

Missing Traveller's Cheques
American Express
167 87 20 00 (freephone).
Thomas Cook
167 87 20 50 (freephone).
VISA
167 87 41 55 (freephone).

An ambulance boat on the Grand Canal

Banking and Local Currency

Visitors to the Veneto have a number of options available to them for changing money. Banks tend to give more favourable rates than bureaux de change, hotels and travel agents, but the paperwork is usually more time consuming. Alternatively, credit cards or Eurocheques can be used for purchasing goods. When changing traveller's cheques you will need to show some form of identification.

Cash dispenser which also accepts VISA and MasterCard (Access)

Changing Money

Banking hours can be erratic, especially the day before a public holiday, so it is safest to acquire some local currency before you arrive in Italy. Exchange rates will vary from place to place.

A more convenient way to change money is to use the electronic exchange machines. These are found at Marco Polo airport, Venice railway station and at several banks in the city. All major towns in the Veneto have foreign exchange machines. These machines have multilingual instructions and the exchange rate is displayed on the screen. You simply feed in up to ten notes of the same foreign currency, and you will get lire in return.

Traveller's Cheques

Traveller's cheques are probably the safest way to carry large sums of money. Choose a name that is well known such as Thomas Cook, American Express or cheques issued through a major bank. There is usually a minimum commission charge, which may make changing small sums of money uneconomical. Some establishments will charge you for each cheque.

Check the exchange rates before you travel and decide whether sterling, dollar or lire traveller's cheques are more appropriate for your trip. Bear in mind that it may be more difficult to cash lire traveller's cheques, especially in hotels, because they are not very profitable for the exchanger.

Eurocheques

Eurocheque cards can be used in any cashpoint machine displaying the Eurocheque logo. Eurocheques can also be used as direct payment for a wide variety of goods and services. Most up-market shops, hotels and restaurants in the popular tourist areas will accept Eurocheques, but it is worthwhile checking before you try. Alternatively, you can cash cheques at any bank displaying the Eurocheque logo. A Eurocheque card guarantees cheques up to a maximum amount of L300,000.

Eurocheque logo

Credit Cards

Credit cards are not as widely used in Italy as they are in the rest of Europe. However, most establishments in the cities will accept them. VISA and Access (MasterCard) are the most popular, followed by American Express and Diners Club. Some banks and cash dispensers accept VISA or Access for cash advances, although interest is payable once the money is withdrawn. In Venice, cash dispensers accepting credit cards can be found at the Banca d'America e d'Italia in the Calle Larga XXII Marzo (**map** 7 A3), and at the Cassa di Risparmio in Campo Manin (**map** 7 A2).

Banking Hours

Banks are usually open between 8:30am and 1:30pm, Monday to Friday. Most also open for an hour in the afternoon from about 2:35pm until 3:35pm. They close at weekends and for public holidays, and they also close early the day before a major holiday. Exchange offices stay open longer, but the rates are less favourable. The exchange offices at Venice airport and railway station stay open until the evening and at weekends.

Plaque of the Cassa di Risparmio in Campo Manin

Using Banks

Changing money at a bank can at times be a frustrating process, as it inevitably involves endless form-filling and queuing. You must apply first at the window displaying the *cambio* sign, then move to the *cassa* to obtain your Italian money.

For security, most branches have electronic double doors. Press the button to open the outer door, then wait for it to close behind you. The inner door then opens automatically.

Currency

Italy's currency is the lira (plural lire) and is usually written as L or, confusingly, £. Lira means pound, so the English pound is referred to as the *lira sterlina*.

Initially, the thousands of units of currency may seem very confusing, but the distinctive colours of the notes make distinguishing between them easy. From time to time, however, familiar coins and notes are redesigned or even discontinued, although plans to alter the basic unit have not yet been implemented. You will find that the last three noughts are regularly ignored in spoken Italian: *cinquanta* will usually mean L50,000 and not, as you may think, L50.

Shop and bar owners are not keen to accept high-value notes for small purchases, so try to get some small denomination notes when changing money. Always keep a few coins in reserve for telephones, tips and for coin-operated lights which illuminate works of art in churches *(see p261)*. *Gettoni* (telephone tokens) have a value of L200 and can be used as ordinary coins. There is a dearth of small-denomination coins; you will sometimes be offered change in *gettoni* or even sweets.

Bank Notes

All bank notes are issued by the Banca d'Italia. Each denomination is a different colour and carries a portrait of a historic person. As a further aid to identification, the notes also increase in size according to their value. The lira itself is never divided up into smaller units.

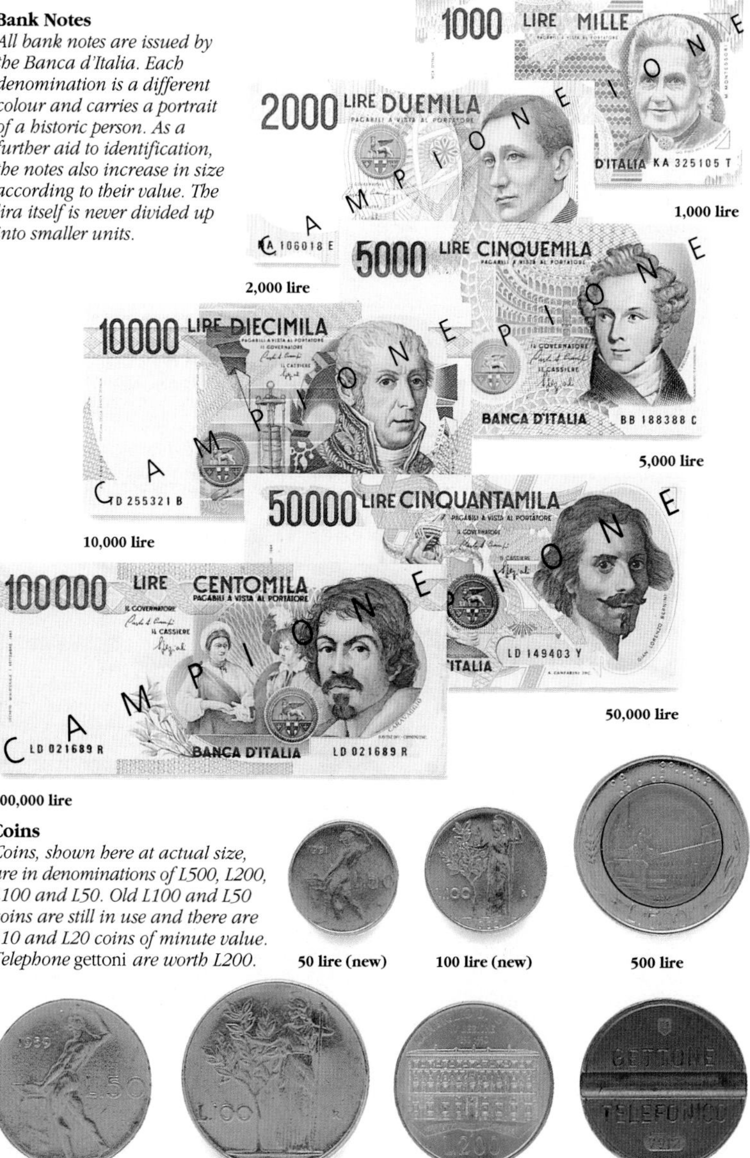

1,000 lire

2,000 lire

5,000 lire

10,000 lire

50,000 lire

100,000 lire

Coins

Coins, shown here at actual size, are in denominations of L500, L200, L100 and L50. Old L100 and L50 coins are still in use and there are L10 and L20 coins of minute value. Telephone gettoni *are worth L200.*

50 lire (new)

100 lire (new)

500 lire

50 lire (old)

100 lire (old)

200 lire

Gettone (200 lire)

Using the Telephone

You can find public telephones on the streets of all the main towns in the Veneto, as well as at bars and post offices. In Venice there are public phones in most of the main squares and at virtually every *vaporetto* landing stage. About half of these now accept prepaid phonecards, so it is far easier and more convenient for visitors to make long-distance calls.

Telephone company logo

Telephone Offices

Telephone offices *(Telefoni)* are run by the Italian telecom companies Telecom Italia (previously called SIP) and ASST. When making international or long-distance calls you may find it easier to get a connection using one of these offices. Each *Telefono* has several soundproof booths with a metered phone. You will be assigned a booth by an assistant and your call will be metered once you are connected to the number. No premium is charged, but you will find that the opening hours of *Telefoni* rarely coincide with Italy's cheap-rate calling hours. The calls are measured in *scatti* (units) costing L200 per unit. All major post offices and **Telecom Italia** offices have *Telefoni*. The *Telefono* at Venice's main post office *(see page 269)* is open every day until 6:45pm. If you need to send a telegram abroad or to anywhere in Italy, you can either go to any post office or call 186 for assistance.

Telephone sign

Call Charges

Calls within Italy are cheapest between 10pm and 8am from Monday to Saturday and all day Sunday. It is also relatively inexpensive to call between 6:30pm and 10pm weekdays and after 1pm on Saturdays.

International calls within Europe are cheapest between 10pm and 8am and all day Sunday. Calls to Canada and the United States are cheapest from 11pm to 8am weekdays, and 11pm to 2pm at weekends. For calls to Australia, phone between 11pm and 8am Monday to Saturday and all day Sunday. There are no cheap rates for Japan.

Hotels will charge more for calls made from your room. Calls from Italy cost more than the equivalent call from the USA or the UK.

Using Public Telephones

You can dial long-distance and international calls from SIP public telephones. When making long-distance

Using a Telecom Italia Coin and Card Telephone

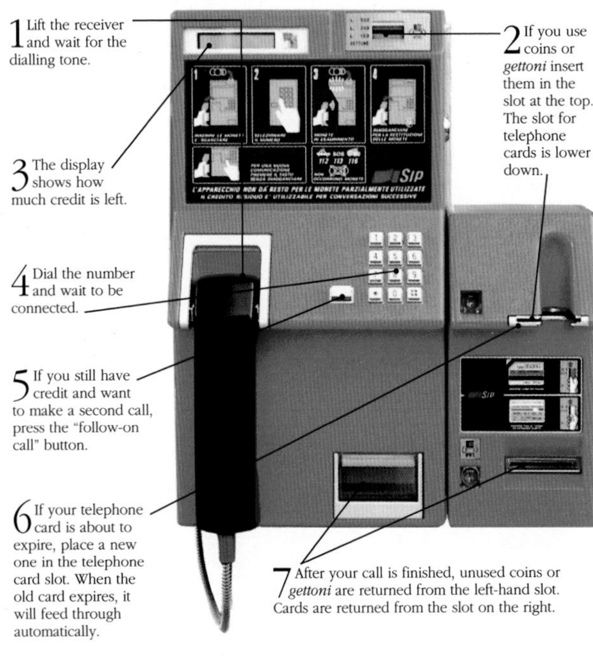

1 Lift the receiver and wait for the dialling tone.
2 If you use coins or *gettoni* insert them in the slot at the top. The slot for telephone cards is lower down.
3 The display shows how much credit is left.
4 Dial the number and wait to be connected.
5 If you still have credit and want to make a second call, press the "follow-on call" button.
6 If your telephone card is about to expire, place a new one in the telephone card slot. When the old card expires, it will feed through automatically.
7 After your call is finished, unused coins or *gettoni* are returned from the left-hand slot. Cards are returned from the slot on the right.
SIP

Telephone cards have values of L5,000 or L10,000.

To use a card, break off the marked corner and insert, arrow first.

L200 **L500**

L100 ***Gettone***

calls, always have at least L2,000 in change ready. If you don't put enough coins in to start with, the telephone disconnects you and retains your money. The most up-to-date payphones also take SIP telephone cards *(carta* or *scheda telefonica)*. You can buy these from post offices, newspaper kiosks and tobacconists *(tabacchi)* that display the black and white T sign. However, the older-style phones which are found in more remote villages accept only tokens *(gettoni)*. The *gettoni* cost L200 each and can be bought from bars, newsagents and all post offices. If you want to make a long-distance call on an older style telephone it is best to find a metered phone. Ask a bar owner if you can use the phone and the meter will be set. You pay the bar owner when you have finished making your call.

Tabacchi sign

Using a public phone

Main Telefoni

Venice
Ferrovia Santa Lucia. **Map** 1 B4.
Piazzale Roma. **Map** 5 B1.

Verona
Ferrovia Porta Nuova.

Padua
Piazzetta A Sartori 17a.

Vicenza
Via Napoli 11.

Reaching the Right Number

To ring Italy from the UK, dial 0039 then the number (omit the first 0). From Australia, dial 00 11 39.

- Dialling code for
  - Venice 041
  - Verona 045
  - Vicenza 0444
  - Padua 049
  - Treviso 0422
- European directory enquiries 176
- European operator assistance 15
- Intercontinental operator assistance 170
- General telephone information 1800
- Telegrams and cables in Italy and abroad 186

With Telecom Italia, you can dial directly to an operator in your own country to place a reverse charge or credit card call. Dial 172 followed by: 0044 for UK; 1061 for Australia.

- *See also* Emergency Numbers, *p264*.

Sending Letters

The Italian postal service is notoriously inefficient. Expect anything sent abroad to take some time, especially during the August holiday season. Postcards to the UK can take up to a month if sent during the summer, and letters sent within Italy can take up to a week to reach their destination. For urgent or important communications, it is better to use the more expensive express system.

You can buy stamps *(francobolli)* from any tobacconist with the black and white T sign, as well as from post offices. Sub-post office hours are usually 8:30am to 2pm, Monday to Friday and 8:30am to noon on Saturday and the last day of the month. Main offices stay open until early evening.

Italian post box

Post Office sign

Sending Parcels

Sending parcels from Italy can be extremely difficult. Unless certain rules are adhered to, it is unlikely that your package will be sent. The package must be placed in a rigid box, wrapped in brown paper and bound with string and a lead seal. You will also need to fill in a simple customs declaration form. Often a stationery or gift shop in the major towns will, for a fee, wrap your package. Very few post offices offer this service.

Poste Restante

Letters and parcels should be sent care of (c/o) Fermo Posta, Ufficio Postale Principale, followed by the name of the town in which you wish to pick them up. Print the surname clearly in block capitals to make sure the letters are filed correctly. To collect your post, you need to show some form of identification and pay a small fee.

Main Post Offices

Fondaco dei Tedeschi 929, Venice. **Map** 7 B1.
☎ *(041) 271 71 11.*

Piazza Viviani, Verona.
☎ *(045) 59 09 55.*

Piazza Garibaldi, Vicenza.
☎ *(0444) 32 24 88.*

Corso Garibaldi, Padua.
☎ *(049) 820 85 11.*

Travel Information

The easiest way to reach the Veneto is by air. Direct flights link Venice to major European cities, but there are no direct intercontinental flights. Visitors from outside Europe have to transfer at Milan or Rome. Venice's Marco Polo airport, 10 km (6.5 miles) north of the city, receives both domestic and European flights as well as some charter flights. The airport is small and not well equipped for the volume of traffic. Treviso and Verona have their own small airports, both of which receive flights from the UK. Car drivers who plan to tour the Veneto must bear in mind that toll charges are expensive on European motorways. Visitors to Venice itself will have to leave their cars in one of the large car parks on the outskirts of the city because there are no streets for cars in the centre. Parking fees are heavy, and owners run the risk of leaving their cars unattended for the length of their stay. However, the region's rail network is good, and Venice railway station links the city to towns of the Veneto, and major European cities.

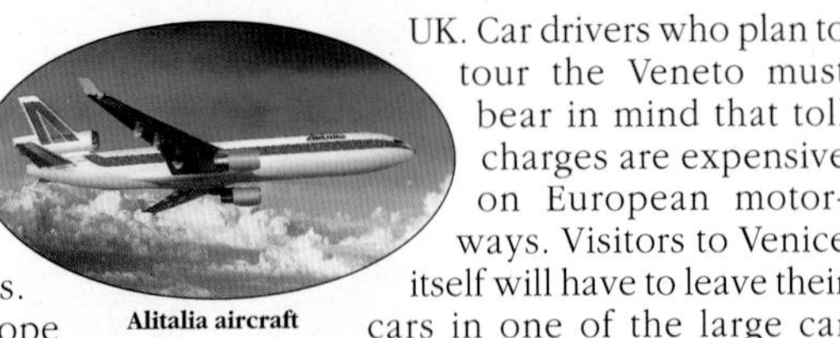

Alitalia aircraft

The quayside at Venice's Marco Polo airport

Arriving by Air

Venice is served by two airports: Marco Polo for scheduled flights and Treviso for charter flights. The city is linked to London, Paris and all other major European cities by direct flights. Visitors from outside Europe must fly to Rome or Milan then take a connecting flight or a train to Venice. The alternative, which is usually cheaper, is to take a budget flight to London, Paris, Amsterdam or Frankfurt, and then a connecting flight to Venice from there.

Daily scheduled flights to Venice are operated by **British Airways** and **Alitalia** from London. PEX or SuperPEX generally offer the best deals in scheduled flights to Venice (Marco Polo), but these are subject to booking restrictions and need to be reserved in advance. They may also be subject to penalty clauses, so it is advisable to take out suitable insurance cover against unforeseen cancellation.

Many charter flights operate to Venice (Treviso). Specialists such as **Sky Shuttle** in the UK offer charter flights and reduced prices on scheduled services flying into Venice (Treviso) and Verona.

If you wish to book flights during your stay, travel agents such as **American Express** or **World Vision** in Venice offer a good service.

Package Holidays

Taking a package holiday to the Veneto is more convenient but not always cheaper than going independently of a tour operator. It is always worth comparing the costs, particularly if you are intending to travel off-season when charter flights are at their cheapest. For visitors who prefer the convenience of a package holiday, Venice is offered as a single destination or as part of a two- or three-centre holiday with Florence and Rome. Transfer from the airport on arrival is usually included in the holiday price. Most tour operators tend to concentrate on Venice as a centre, though some offer packages to Verona or touring trips of the Veneto taking in the popular villas, museums and art galleries.

Useful Numbers

Alitalia
Venice
(147) 865 643.

British Airways
Venice
(041) 541 56 29.

Sky Shuttle
Air Travel Group (Holidays) Ltd
227 Shepherds Bush Road
London W6 7AS.
0181-748 1333.

American Express
See p261.

World Vision Travel
See p273.

Airport Information
Venice
(041) 260 92 60.
Verona
(045) 809 56 66.
Treviso
(0422) 203 93.

Marco Polo Airport (Venice)

Facilities are limited at the airport, but there is a hotel reservations office and a currency exchange office which is open all day. The only restaurant is 3 minutes' walk from the airport building.

Verona airport check-in desk

The most dramatic entry from the airport into Venice is by boat. The Cooperativa San Marco public water launch to San Marco and the Lido departs at hourly or two-hourly intervals depending on the time of day. Tickets are available from the office close to the exit of the arrivals hall. The journey to Venice takes about 50 minutes and costs around L15,000 per person one way. The boat stops only a short walk away from the San Marco *vaporetto* landing stage. Water taxis operating from the airport to San Marco take only about 20 minutes but a journey will cost around six times as much as the public launch. Beware of water taxi touts who will charge you a good deal more than the official fare.

The less spectacular but quicker and cheaper alternative to the lagoon crossing is the ATVO bus to Piazzale Roma. The service meets all scheduled flights and costs around L5,000. Cheaper still, but stopping along the way, is the public bus to Piazzale Roma, which departs every 30 minutes. There is also a land taxi rank at the front of the airport. The journey takes 15 minutes and the drop-off point is Piazzale Roma.

Treviso Airport

This is a small airport which receives charter flights from London (Gatwick) twice a week and from Amsterdam daily. An exchange office is open when flights are in operation, mainly in the afternoon. The coach service to Piazzale Roma in Venice, which connects with flights, costs about L7,000 and takes approximately 45 minutes. Alternatively, you can take the public bus No. 6 which runs to Treviso station, where there is a regular rail service to Venice. For those on package tours the transport to Venice is pre-arranged and normally included in the overall price of the holiday.

Verona Airport

Verona airport receives flights from London (Gatwick), Frankfurt and Paris. The currency exchange office opens on weekdays only. The bus service from the airport to Verona, which links up with scheduled flights, costs about L6,000.

Porters in Venice

Unless you are staying very close to your arrival point, you will have to take a *vaporetto* to the landing stage nearest to your hotel. Porters are very expensive; you will have to pay for the porter's boat fare, as well as for each piece of luggage – each piece costs the same as an adult. The cost of a porter handling two suitcases, including your *vaporetto* fare, could amount to nearly L20,000. If there are no porters available, which is often the case, you can call your hotel and ask for a porter to meet you.

Arriving by Car

To drive your own car in Italy you will need an international Green Card (for insurance purposes) and your vehicle registration documentation. EU nationals who do not have the standard pink licence will need an Italian translation of their licence, available from most motoring organizations and Italian tourist offices. Requirements vary for visitors from non-EU countries, and drivers should check with their insurance companies before leaving for Italy. Insurance can always be bought at the border. This is also required to hire a car. *(See also p278.)*

Car Hire Numbers

In addition to those below, each major car hire company has an office at Venice, Treviso and Verona airports.

Piazzale Roma, Venice
Avis *(041) 522 58 25.*
Hertz *(041) 528 40 91.*
Thrifty *(041) 541 52 99.*

Padua Railway Station
Maggiore *(049) 875 28 52.*

Verona Railway Station
Avis *(045) 800 66 36.*
Hertz *(045) 800 08 32.*
Maggiore *(045) 800 48 08.*

Vicenza Railway Station
Maggiore *(0444) 54 59 62.*

Boat from Venice's Marco Polo airport into the city

Travelling by Train

ITALY'S STATE RAILWAY (Ferrovie dello Stato or FS) runs an extensive and efficient rail network throughout the Veneto. Services are regular, trains tend to be punctual and the cost of travel is very reasonable. The variety of trains ranges from the painstakingly slow *locale*, which stops at almost every station, through various levels of fast intercity services to the high-speed *pendolino*, which links Venice with Rome.

A *pendolino* – Italy's fastest train

ARRIVING BY TRAIN

SANTA LUCIA railway station in Venice is the terminus for trains from Paris, Munich, Innsbruck, Vienna, Geneva, Zürich and other European cities. Passengers travelling from London have to change in Paris or Ostend. Fast inter-city trains link Venice with Verona, Bologna, Milan, Rome and other major Italian cities.

Europe-wide train passes such as Eurail (US) or Inter-Rail for those under 26 (Europe) are accepted on the FS network. You may have to pay a supplement, however, to travel on fast trains.

SANTA LUCIA STATION, VENICE

STANDING AT the west end of the Grand Canal, Ferrovia Santa Lucia is a modern, well-equipped station. There are *vaporetti* landing stages below the steps of the station with boats going to San Marco and all stops en route. There is also a water taxi and gondola service. Porters are not so easy to find, however. The bus and coach terminal and the only land taxi rank in Venice are in Piazzale Roma nearby (follow yellow signs).

Automatic ticket machines in the station are easy to use and display instructions in six languages. Notes in every denomination up to L50,000, coins and some credit cards are accepted. Tickets can be booked free of charge in advance through travel agents.

Multilingual display screens give information on arrivals, departures, costs of travel and details about city services and tours. A tourist office offers to make hotel reservations, but queues are long in summer. There is also a Rolling Venice office *(see p262)*, a bank and bureau de change, an international telephone office, a left luggage facility, a cafeteria and bar, and a shop that sells international newspapers and magazines. Another useful facility is the *albergo diurno*. This is a daytime hotel where you can rest in a private room with an en-suite shower.

Santa Lucia station in Venice – gateway to the Veneto

ORIENT EXPRESS

FROM MARCH to November the *Venice Simplon-Orient-Express* runs between London and Venice with stops at Paris, Düsseldorf, Cologne, Frankfurt, Zürich, St Anton, Innsbruck and Verona. A one-way journey with cabin from London to Venice costs four to five times the price of a charter flight, but if you opt for the return trip, the outward flight to Venice is thrown in.

FS train in Verona station

The Orient-Express logo

VERONA STATIONS

VERONA LIES at the intersection of the main railway lines from Venice to Milan and Bologna to Munich. The main station, Porta Nuova, lies south of the centre, connected to it by frequent bus services. Train information is available at the ticket office and automatic help points give information in English. Other facilities include a left luggage office, a bar, an automatic exchange machine and a newspaper shop which sells bus tickets.

The small Porta Vescova station, serving local stations to the east of Verona, is used mainly by locals.

PADUA STATION

PADUA IS ONLY 30 minutes by train from Venice. The station is in the north of the town, and local buses for the centre leave from outside the station. The main bus terminal, with services to Venice and other towns of the Veneto is at Piazzale Boschetti, 10 minutes' walk from the station. Padua's tourist office is within the station building. There is a left luggage office, restaurant, a tobacconist selling bus tickets and a currency exchange office open 9am to noon and 3pm to 5:30pm Monday to Saturday; 9am to noon Sunday.

Vicenza Station

Vicenza, 55 minutes from Venice, is on the main railway line between Verona and Padua. The station is south of the city centre. Facilities include a bar and offices for left luggage, tickets, information and currency exchange open 8am to 12:30pm and 2pm to 6:30pm Monday to Saturday, 8:30am to noon and 2pm to 6pm Sunday.

Train Travel

If you plan to travel around, there are passes allowing unlimited travel on the FS network. These include a ticket for unlimited travel *(biglietto turistico libera circolazione)* and the *biglietto chilometrico*, which allows 20 trips totalling no more than 3,000 km (1,865 miles) for up to four people. These are available from international and Italian **CIT** offices and from some travel agents. There are facilities for disabled travellers on some services.

Tickets for Local Journeys

Ask at newsstands on the station for a biglietto a fasce chilometriche *to the destination you require.*

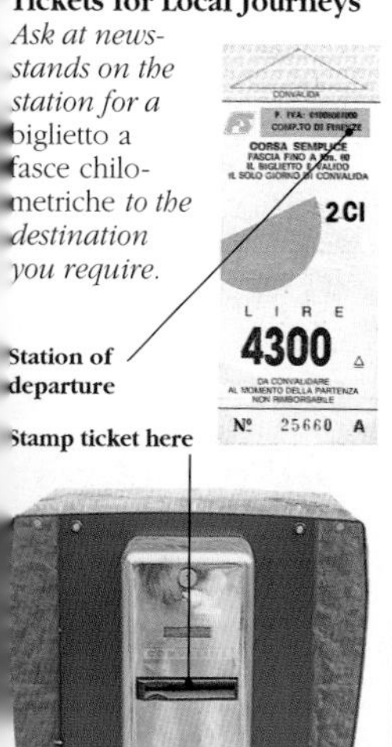

Machine for validating tickets

Machines for FS Rail Tickets

These machines are easy to use, and most have instructions in six languages on a printed panel.

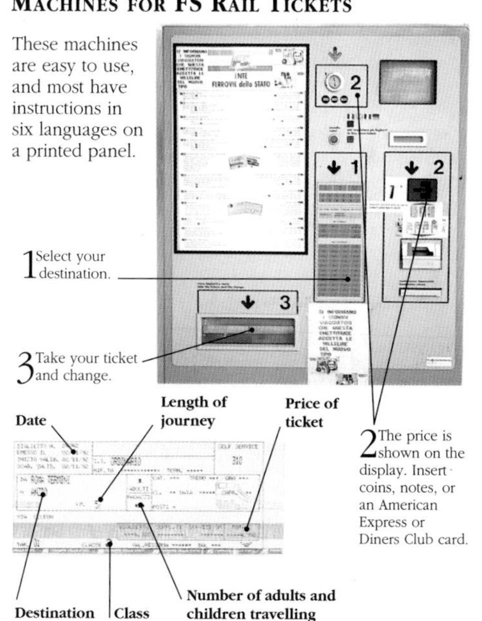

Tickets

On all intercity trains a supplement is charged, even if you have a rail card. Booking is obligatory on the *pendolino* and some other intercity services, and it is also advisable on other trains if you wish to travel at busy times. Buying your international city ticket at least five hours before travelling entitles you to a free seat reservation.

If you are travelling less than 200 km (125 miles), a short-range ticket *(biglietto a fasce chilometriche)* is available. The ticket is stamped with the destination you require and it must then be validated in one of the machines at the entrance to the platforms. Both outward and return portions of a ticket must be used within three days of purchase. Tickets can be bought on the train but these are liable to a flat rate surcharge *and* a supplement based on the ticket price.

Railway Information Offices

Venice *(041) 71 55 55.*
Verona *(045) 59 06 88.*
Padua *(049) 875 18 00.*
Vicenza *(0444) 32 50 46.*

Booking Agents

CIT
Via Giacomo Matteotti 12, Padua
(049) 66 33 33.

London *0171-258 8000.*
Sydney *(02) 299 4754.*

World Vision Travel
Corso Porta Nuova 7, Verona.
(045) 59 09 77.
See also p270.

Multilingual information board showing train departures

Getting Around Venice by Boat

FOR VISITORS TO VENICE, the *vaporetti* or waterbuses provide an entertaining form of public transport, although most journeys within the city can usually be covered more quickly on foot. The main route through the city for the *vaporetti* is the Grand Canal, and these waterbuses also supply a useful service connecting outlying points on the periphery of Venice and linking the city with the islands in the lagoon. The best value service from a visitor's point of view is the No. 1. This operates from one end of the Grand Canal to the other and travels sufficiently slowly for you to admire the parade of palaces at the waterside *(see pp56–71)*.

A *vaporetto* pulling into San Marco

The smaller, sleeker *motoscafo*

A two-tier *motonave* on its way to Torcello

The Boats

THE ORIGINAL *vaporetti* were steam-powered (*vaporetto* means little steamer); today they are diesel-run motor boats. Although all the boats tend to be called *vaporetti*, strictly speaking the word applies only to the large wide boats used on the slow routes, such as No. 1. These boats provide the best views. The *motoscafi* are the slimmer, smaller and faster boats, such as No. 52. Some of them might look old and rusty, but they go at quite a pace. The two-tier *motonavi*, which look huge in comparison to the *vaporetti* or *motoscafi*, are used on routes to outlying islands and the Lido.

Types of Ticket

THE PRICE OF A TICKET depends not on the length of your journey but on the line you are taking. The faster *diretto* boats cost about half as much again as those which stop at every landing stage.

Tickets that are purchased on board, rather than at the kiosk found at each stop, are more expensive. If you buy a book of 10 or 20 tickets, the cost is no cheaper than single tickets, but does avoid the inconvenience of queuing at a ticket office for each journey.

You can save money, however, by buying a 24-hour or 72-hour ticket, which entitles the holder to unlimited travel on most lines.

Vaporetto* stop at the Giardini Pubblici *(see pp120–21)

If you are staying more than a few days, you can also save money by purchasing a weekly or monthly season ticket *(abbonamento)* which is available from ticket offices. Holders of Rolling Venice cards *(see p262)* can buy a *Tre Giorni Giovane*, or three-day youth pass, for L16,000.

The new "Isles Ticket" enables you to travel on No. 12 for a *one-way* run and stay for the day on the islands of Murano, Mazzorbo, Burano and Torcello. (You will, however, need to buy a ticket for the return trip.)

Hours of Service

THE MAIN ROUTES run every 10 to 20 minutes until the early evening. Services are reduced at night, particularly after 1am. From June to September the services are more frequent and certain routes are extended. Details of main lines are also given in the booklet *Un Ospite di Venezia (see p260)*. From May to September the main routes and island boats are very crowded.

Sightseeing from a *vaporetto* on the Grand Canal

The Main Routes

(1) Confusingly called the *Accelerato*, this is the slow boat down the Grand Canal, stopping at every landing stage. The route starts at Piazzale Roma, travels the length of the Grand Canal, then from San Marco it heads east to the Lido.

(82) A relative newcomer to the canals, the No. 82 is the faster route down the Grand Canal, making only six stops. It goes westwards as far as Tronchetto (the car park island), then eastwards along the Giudecca Canal to San Zaccaria. In season the line is extended to the Lido.

(52) Taking the place of the old No. 5, the 52 skirts the periphery of Venice and takes in the island of Murano. It has also been extended to the Lido. The circular route provides a scenic tour of Venice, though to do the whole circuit you have to change boats at Fondamente Nuove.

(3) A new seasonal line, this goes from Tronchetto, down the Grand Canal to San Zaccarie and returns via the Giudecca Canal.

(4) Reverse route of No. 3, starting at San Zaccarie.

(23) Seasonal line to Murano from San Zaccarie.

(12) Departing from the Fondamente Nuove, this line serves the main islands in the northern lagoon: Murano, Mazzorbo, Burano and Torcello. The service only runs about once an hour.

(14) A new line, this goes to Torcello and Burano from San Zaccaria (near San Marco). It is a much longer route than that from the Fondamente Nuove, going via the Lido, Punta Sabbioni and Treporti.

Vaporetto Information

ACTV (Information Office)
Ramo dei Fuseri, San Marco 1810, Venice. **Map** 7 A2.
(041) 272 23 10 & 272 23 11.

Using the Vaporetti

The service is run by **ACTV** *(Azienda del Consorzio Trasporti Veneziano).* In 1993 the waterbus system was given an overhaul; the map on the inside back cover of this guide reflects the changes made to the lines. Many maps on sale in Venice, therefore, will now be out of date. If you are not sure which boat to take to reach your destination, check with the boatman – the *vaporetti* crew tend to be very helpful.

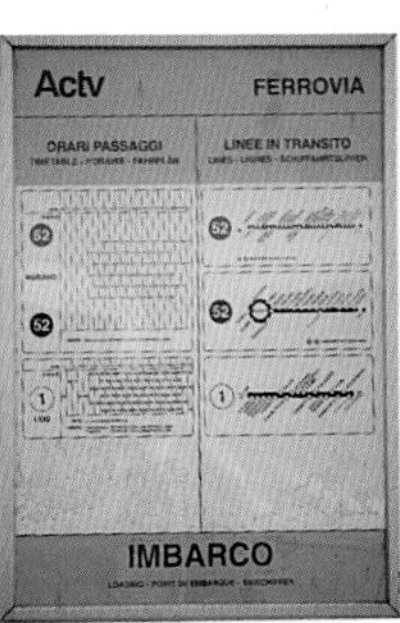

Timetable and routes at a *vaporetto* boarding point

1 Tickets are available at most landing stages, some bars, shops and tobacconists displaying the ACTV sign. The price of a ticket remains the same whether you are going one stop or doing the whole circuit, but some routes are more expensive than others and a variety of season tickets is available *(see* Types of Ticket*).*

2 Signs on the landing stage tell you at which end you should board the boat.

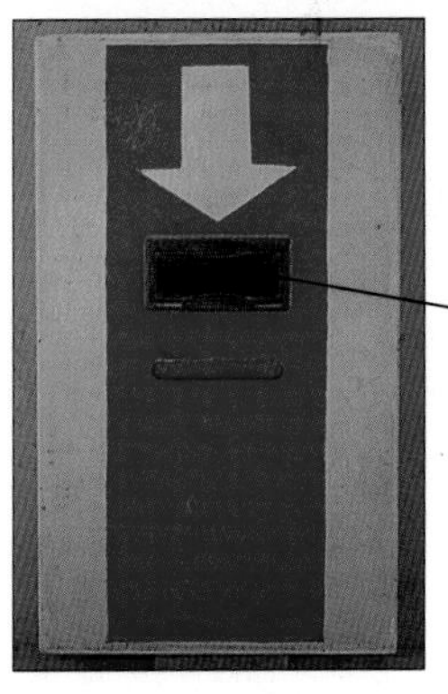

3 Tickets should be punched at the automatic machines on the landing stages before each journey. Inspectors rarely board the boats and this makes it surprisingly easy for tourists (and Venetians) to hop on and off the boats without a validated ticket. However, there are steep fines for passengers without tickets, and there are notices in English to this effect in all the boats.

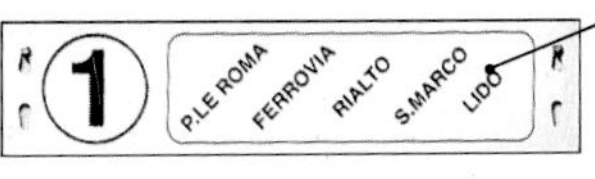

4 An indicator board at the front of each boat gives the line number and main stops. (Ignore the large black numbers on the side of the boat.)

5 Each landing stage has its name clearly marked on a yellow board. Most stops have two landing stages and it is quite easy, particularly if it is crowded and you can't see which way the boat is facing, to board a boat travelling in the wrong direction. It is helpful to watch which direction the boat is approaching from; if in doubt, check with the boatman on board.

Finding Your Way in Venice

VENICE IS SURPRISINGLY SMALL and most of the sights can be covered comfortably on foot. However, to avoid losing your way in the maze of little alleys, make sure you have the Street Finder *(see pp280–93)* handy. The gondola is the most romantic way to see the city, but prices are high, while the water taxi is the fastest means of travelling through the city and out to the islands.

Gondolas

GONDOLAS ARE a luxury form of transport used only by tourists (apart from Venetians on their wedding day). There are a number of gondola ranks throughout the city and plenty of gondoliers waiting for business.

Before boarding, check the official tariffs and agree a price with the gondolier. Prices are in the booklet *Un Ospite di Venezia* *(see p260)* and should also be available at gondola ranks. Official costs are around L70,000 for 50 minutes, rising to L90,000 after 8pm. However, gondoliers are notorious for overcharging, sometimes by double the official price: try bargaining, whatever the cost quoted. During the low season, or when business looks slack, you may be able to negotiate a fee below the official rate and a journey shorter than the minimum of 50 minutes. Another way of cutting costs is to share a gondola – five is the maximum number of passengers.

Gondoliers all speak a smattering of English and have taken basic exams in Venetian history and art. Do not expect your gondolier to burst into *O Sole Mio*, however; the most you are likely to hear are the low melodious cries of *Oe*, *Premi* and *Stai* – the warning calls that have been echoing down the canals of Venice for centuries. If you want to go on a serenaded tour, join an evening flotilla with accompanying musicians, organized regularly from May to October. Hiring a gondola independently is more romantic, but will cost considerably more. Details are available from any local travel agent.

The romance of an early evening gondola ride

Sign for one of the *traghetti* across the Grand Canal

Traghetti

TRAGHETTI are gondola ferries that cross the Grand Canal at seven different points, providing an invaluable service for pedestrians. Surprisingly, few tourists make use of this cheap, constant service. The points where the *traghetti* cross the Grand Canal are marked on the Street Finder maps *(see pp280–93)*. Yellow street signs show the way to the *traghetti*, illustrated with a little gondola symbol.

Crossing the Grand Canal by *traghetto*

Water Taxis

FOR THOSE with little time and sufficient funds, the fastest and most practical means of getting from A to B is by water taxi. The craft are sleek, white or polished wood motorboats, all equipped with a cabin. They zip to and from the airport in only 20 minutes. There are 16 water taxi ranks, including one at the airport and at the Lido. Telephone numbers and official tariffs are listed in the booklet *Un Ospite di Venezia* *(see p260)*. Extra is charged for luggage, waiting, night service and for calling out a taxi. If the *vaporetti* are on strike, which is not uncommon, water taxis are like gold dust.

A water taxi

Gondola Stands

San Marco (Molo)
(041) 520 06 85.

Rialto (Riva Carbon)
(041) 522 49 04.

Railway Station (San Simeone Piccolo)
(041) 71 85 43.

Water Taxi Stands

Radio Taxi (all of Venice)
(041) 522 23 03.

Ferrovia Santa Lucia
(041) 71 62 86.

Piazzale Roma
(041) 71 69 22.

San Marco
(041) 522 97 50.

Walking in Venice

One of the great pleasures of exploring Venice is walking. In the absence of traffic you soon get used to crossing streets and squares without so much as a glance to left or right. What you do have to contend with is the constant flow of tourists. The narrow alleys, particularly in the *sestiere* of San Marco, become extremely congested. However, the vast majority of tourists never venture beyond San Marco and it will be probably little more than a matter of minutes before you find yourself with only a few locals for company.

When sightseeing in Venice you will inevitably do a lot of walking, and a day taking in the sights can be extremely tiring. You need to allow only 35 minutes to cross the city from north to south on foot – provided you do not lose your way. Most visitors do, and this is, of course, part of the fun of exploring, but sensible shoes are a must, however short you think your journey may be.

Venice is so compact that you are never very far from the yellow signs that give directions to the key points of the city. Another useful landmark is the Grand Canal, which sweeps through the heart of the city in an inverted S-shape *(see p57)*.

The city has countless *campi*, or squares, which open out

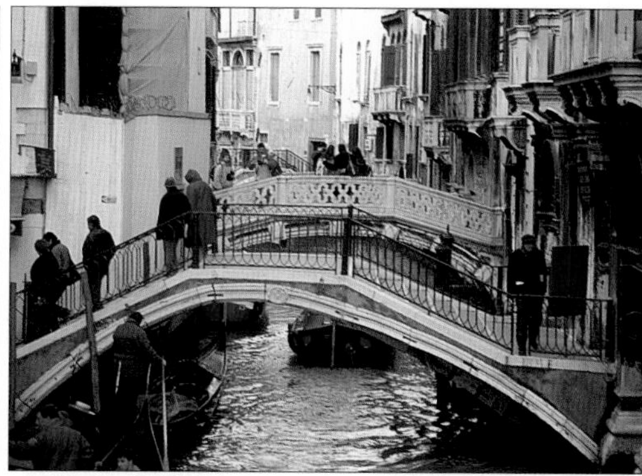

Venice, city of bridges

from the narrow alleys. Very few of these are equipped with public benches, but footsore tourists can sit at an open-air café. However, you will be expected to buy a drink, and drinks are expensive. Many of the main sights are concentrated in the *sestiere* of San Marco, either in Piazza San Marco *(see pp74–5)* or close by. Even the main sights in the *sestieri* beyond San Marco are, for most people, within comfortable walking distance of the main square. Apart from the Accademia and San Marco, the signposting is not very impressive, but the maps in this guide will help you find your way around.

An ornate Venetian door knocker

In July and August, when temperatures are at their highest, it is wise to avoid walking around midday. You should also be prepared, particularly at this time of year, for nasty smells which waft from some of the canals.

From October there is always the risk of high tides *(acqua alta)*, which cause flooding in the city. The first area to flood is Piazza San Marco. Duckboards are laid out in the square, however, and along main thoroughfares. If you are not equipped with wellington boots you can always buy cheap knee-high plastic shoe covers from local shops.

Addresses in Venice

For any newcomer to the city of Venice, the system of addresses is initially very confusing. All buildings are numbered by the *sestiere* (administrative district) in which they fall rather than by the street. Hence a typical address would merely give the name of the *sestiere* followed by the number of the building, for example San Marco 2517 or Cannaregio 3499.

To locate an address it is, therefore, essential for you to establish the name of the street or square or, failing that, the nearest landmark. The Venetians resort to a book called *Indicatore Anagrafico*, which lists all the numbers in Venice and their corresponding streets. Better still is the more recent publication, *Calli, Campielli e Canali*, which also provides detailed maps of the city and islands.

You will find translations of Venetian words commonly used in place names in the Street Finder *(see p280)*.

A plethora of confusing signs in Cannaregio

Getting Around the Veneto

Taxi in Verona

Day trips can be made from Venice by train or bus, and the city centres can be covered easily on foot or by local bus. Although the train and bus networks are excellent, the most practical and pleasant means of travel is by car, allowing total independence to explore the countryside. However, some of the roads between towns tend to be very congested in the summer, city centres are banned to tourist traffic, and the cost of petrol is high.

A city bus in central Verona

On Foot and by Bus

All the cities of the Veneto are small enough to get around reasonably comfortably on foot. Limited traffic zones means that walking is pleasant and there are plenty of squares where you can sit and watch the world go by.

City buses are cheap and regular. Tickets, which must be bought prior to travel, are available from news-stands, bars, tobacconists and shops which display the bus company sign. There are also ticket vending machines in the streets, usually near stops, which take coins and L1,000 and L5,000 notes. A flat fee is charged for rides within the city and the suburbs. The ticket becomes valid only when you time-stamp it in the machine at the front or rear of the bus.

It is normally cheaper and quicker to travel between towns by train. In some cases the bus will take twice as long as the train, but there are a few towns such as Asolo where your only choice is a bus. In most cases the bus departure point is near the train station. You can usually buy a ticket valid for one, two or more hours of travel. Daily passes, or *tesserini*, are also available.

The city of Venice has excellent rail connections, but a limited bus service. The most popular routes connect nearby towns such as Mestre, Mira, Marghera and Stra.

Taxis

Travelling by taxi in the Veneto is not cheap. Meters show a fixed starting charge, then clock up every kilometre. There are extra charges for luggage, trips to the airports and journeys taken between 10pm and 7am, on Sundays and public holidays. Taxi drivers do not necessarily expect a tip – Italians give small tips or none at all.

Take taxis only from the official ranks, not from touts at railway stations and airports. In Venice the taxi rank is in Piazzale Roma; in the Veneto towns of Verona, Vicenza and Padua, taxis can be found at the main piazzas.

Car Hire

If you book your car within the Veneto, local Italian firms such as Maggiore *(see p271)* tend to be cheaper than the international ones. Whichever company you choose, make sure that quoted prices include collision damage waiver, theft breakdown service and IVA (Italian VAT, currently 19 per cent). Insurance against theft is usually an extra. To hire a car you must be over 21 and have held a licence for at least a year. Visitors from outside the EU need an international licence, though in practice hire firms may not insist on this.

Driving and Parking

Cities in the Veneto have limited traffic zones and normally only residents and taxis can drive into the centre. Visitors can unload at their hotel, but must then park on the outside of town and come in by foot or bus. Some hotels have a limited number of parking permits, but this is no guarantee of a space. Your

Rules of the Road

Drive on the right and, generally, give way to the right. Seat belts are compulsory in the front and back, and children should be properly restrained. You must also carry a warning triangle in case of breakdown. In town centres, the speed limit is 50 km/h (30 mph); on ordinary roads 90 km/h (55 mph); and on motorways 110 km/h (70 mph) for cars up to 1099cc, and 130 km/h (80 mph) for more powerful cars. Penalties for speeding include spot fines and licence points, and there are drink-driving laws as elsewhere in the EU.

One-way street

Speed limit (on minor road)

End of speed restriction

Pedestrianized street – no traffic

Give way to oncoming traffic

Give way 320 m (350 yd) ahead

Danger (often with description)

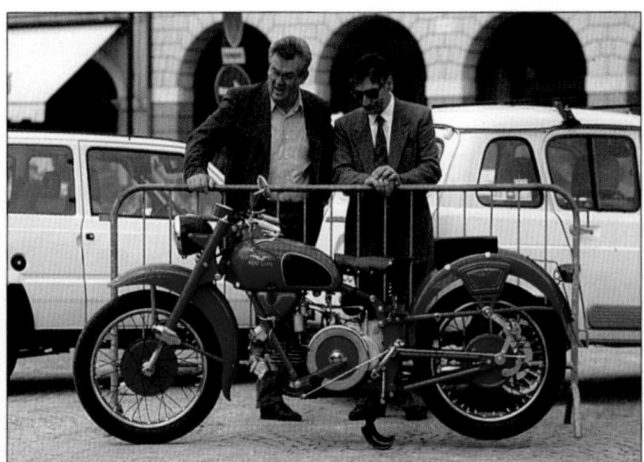

Moto Guzzi's classic Gambalunga in the market place at Montagnana

best bet is to telephone in advance and warn the hotel of your arrival.

Official parking areas are marked by blue lines, usually with meters or an attendant nearby. The *disco orario* system allows free parking for a limited period in certain areas. The cardboard discs, which you place on your windscreen, are provided by car hire companies, or can be purchased at petrol stations and also at supermarkets in the cities of the Veneto.

***Disco orario* parking disc**

If your car is towed away, phone the **Polizia Municipale**, or Municipal Police.

In Venice parking is prohibitively expensive. The closest car parks to the centre are at Piazzale Roma, where space is at a premium. There is a huge car park on the Isola del Tronchetto, linked to Venice by *vaporetto*. The cheaper parks at Fusina and San Giuliano in Mestre are open only in summer, for Carnival and Easter.

Many of the main roads are surprisingly old, with only a couple of lanes, and traffic is often very heavy. What looks like just a short trip on the map may take much longer than you would expect.

Autostrada tolls, which are levied on the motorways, are expensive. Payment can be made in cash or by pre-paid magnetic "swipe" cards called Viacards. These are available from tobacconists and all ACI offices.

The picturesque but hair-raising Gardesana ***(see p204)***

Petrol

Motorway service stations are open 24 hours a day. Petrol stations are scarce in the countryside and the majority do not accept credit cards. Many are closed all afternoon, all day Sunday and the whole of August. However, you will find that some of the self-service petrol stations have automatic machines which accept L10,000 notes.

Breakdowns

The ACI (**Automobile Club d'Italia**) provides an efficient 24-hour service which is also available to foreign visitors. The organization has basic reciprocal arrangements with affiliated associations in other countries such as the AA or RAC in Britain.

Directory

Bus Information

Belluno
Dolomiti Bus, Via Col da Ren
(0473) 94 11 67.

Padua
Autobus
ACAP Office
Ferrovia.
(049) 66 20 55.

Verona
Azienda Municipalizzata Trasporti
Via Torbido 1.
(045) 887 11 11.

Vicenza
Aziende Industriali
Municipalizzate
Via San Biagio
(0444) 39 49 09.

Breakdown

Automobile Club d'Italia
Emergencies
116.

Via Ca' Marcello 67
Mestre.
(041) 531 03 62.

Via della Valverde 34
Verona.
(045) 59 53 33.

Via degli Scrovegni 19
Padua.
(049) 61 68 01.

Viale della Pace 260
Vicenza.
(0444) 51 33 71.

Towing Away

Polizia Municipale (Municipal Police)
Venice
(041) 533 92 98, or (041) 522 45 76.

Padua
(049) 807 23 99.

Verona
(045) 807 84 11.

Vicenza
(0444) 39 49 09.

Venice Street Finder

All the sights, hotels, restaurants, shops and entertainment venues in Venice have map references which refer you to this section of the book. The key map below indicates the areas of the city covered by the Street Finder, and includes the colour coding specific to each area. Following the map section is a complete index of street names *(see pp290–93)*. The standard Italian spelling has been used on the maps throughout this book, but when exploring the city you will find that the street signs are often printed in Venetian dialect. Sometimes this means only a slight variation in the spelling (see the word Sotoportico/Sotoportego below), but some names look completely different. For example, Santi Giovanni e Paolo *(see Map 3)* is often signposted as "San Zanipolo". Major sights are labelled in Italian.

Recognizing Street Names

The signs for street *(calle)*, canal *(rio)* and square *(campo)* will soon become familiar, but the Venetians have a colourful vocabulary for the maze of alleys which makes up the city. When exploring, the following may help.

FONDAMENTA S.SEVERO

Fondamenta A street that runs alongside a canal, often named after the canal it follows.

RIO TERRA GESUATI

Rio Terrà A filled-in canal. Similar to a *rio terrà* is a *piscina*, which often forms a square.

SOTOPORTEGO E PONTE S.CRISTOFORO

Sotoportico or Sotoportego A covered passageway.

SALIZADA PIO X

Salizzada A main street (formerly a paved street).

RIVA DEI PARTIGIANI

Riva A wide *fondamenta*, often facing the lagoon.

RUGAGIUFFA

Ruga A street lined with shops.

CORTE DEI DO POZZI

Corte A courtyard.

RIO MENUO O DE LA VERONA

Many streets and canals in Venice often have more than one name: *o* means "or".

0 metres 500

0 yards 500

Key to Street Finder

- Major sight
- Place of interest
- Railway station
- Ferry boarding point
- *Vaporetto* boarding point
- *Traghetto* crossing
- Gondola waiting point
- Coach station
- Tourist information office
- Hospital with casualty unit
- Parking
- Police station
- Church
- Synagogue
- Post office
- Railway line

Scale of Map Pages

0 metres 200

0 yards 200

Scale of Murano Inset

0 metres 500

0 yards 500

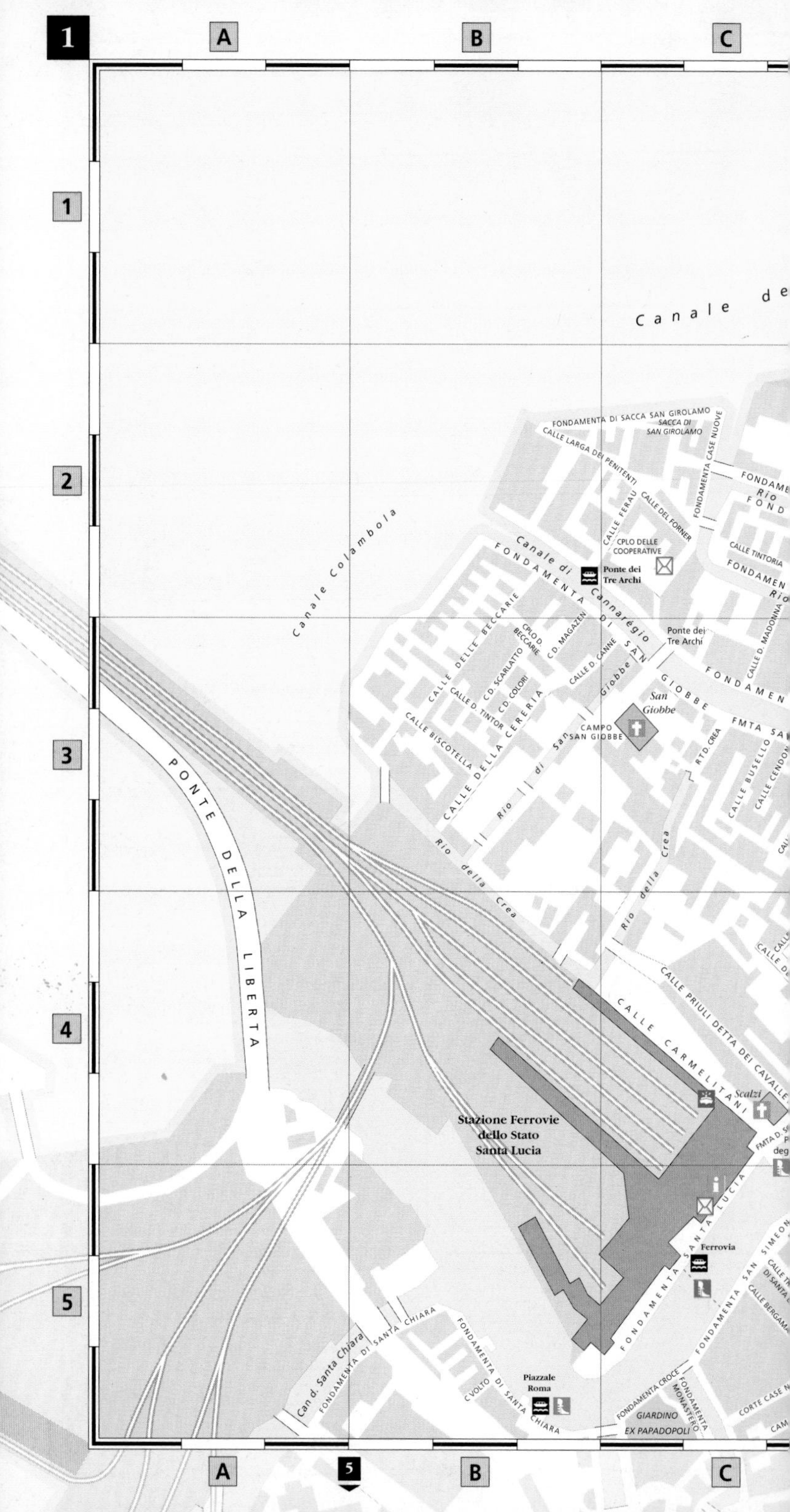
A
B
C
1
2
3
4
5
Canale de
FONDAMENTA DI SACCA SAN GIROLAMO
SACCA DI SAN GIROLAMO
CALLE LARGA DEI PENITENTI
FONDAMENTA CASE NUOVE
CALLE FERAU
CALLE DEL FORNER
CPLO DELLE COOPERATIVE
CALLE TINTORIA
Canale Colambola
Canale di Cannaregio
FONDAMENTA DI SAN Giobbe
Ponte dei Tre Archi
CALLE DELLE BECCARIE
CPLO D. BECCARIE
C.D. MAGAZEN
CALLE D. CANNE
C.D. SCARLATTO
C.D. COLORI
CALLE D. TINTOR
CALLE BISCOTELLA
CALLE DELLA CERERIA
CAMPO SAN GIOBBE
San Giobbe
GIOBBE
FMTA SA
RTD. CREA
CALLE BUSELLO
CALLE CENDON
CALLE D. MADONNA
Rio di San Giobbe
Rio della Crea
PONTE DELLA LIBERTÀ
CALLE PRIULI DETTA DEI CAVALLETTI
CALLE CARMELITANI
Scalzi
Stazione Ferrovie dello Stato Santa Lucia
FONDAMENTA SANTA LUCIA
Ferrovia
FONDAMENTA SAN SIMEON
CALLE BERGAMA
Can d. Santa Chiara
FONDAMENTA DI SANTA CHIARA
C.VOLTO
Piazzale Roma
FONDAMENTA CROCE
FONDAMENTA MONASTERO
GIARDINO EX PAPADOPOLI
CORTE CASE N
5

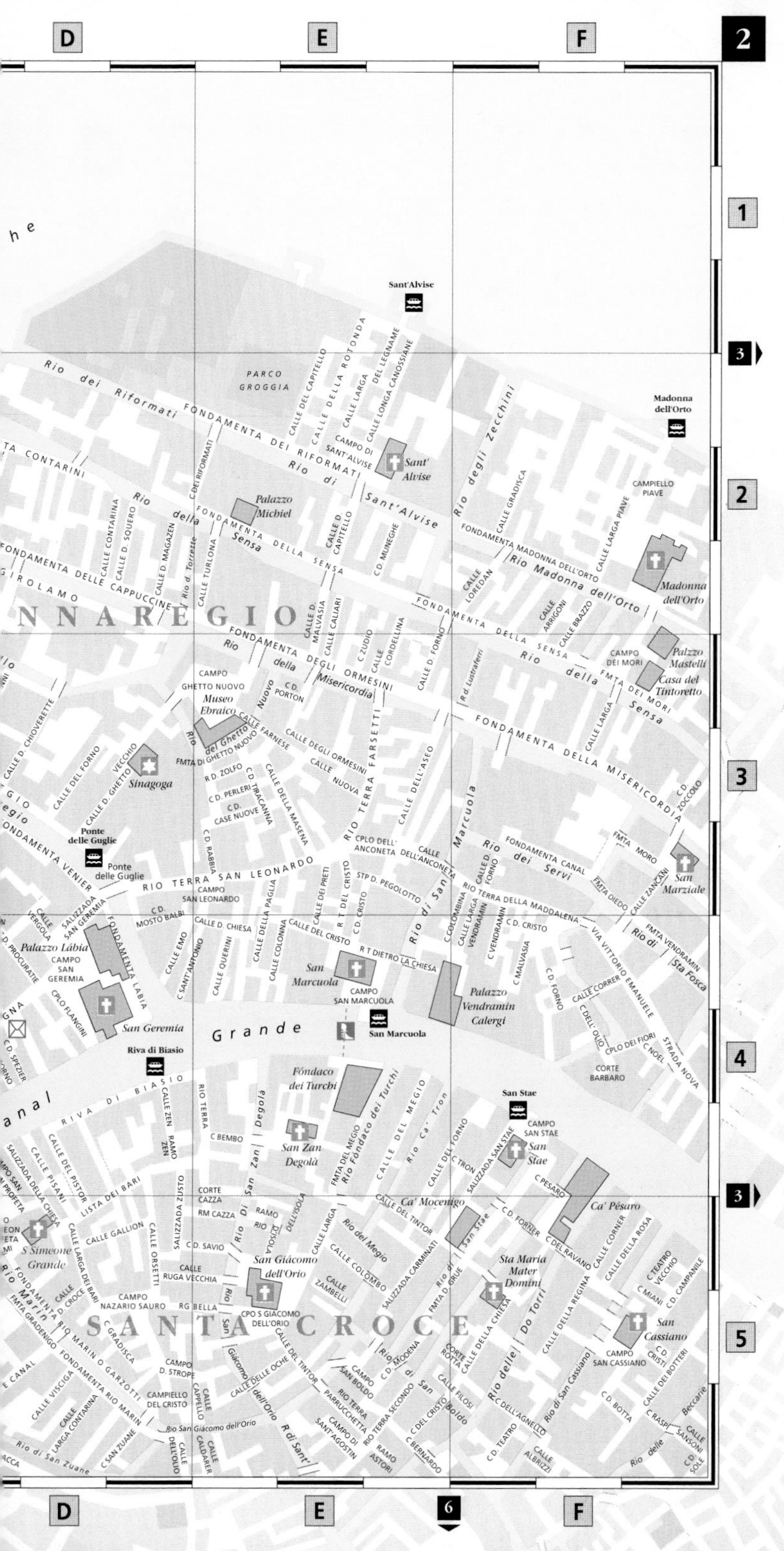

D
E
F
2
1
3
4
5
6
CANNAREGIO
SANTA CROCE
Canal Grande
Sant'Alvise
Madonna dell'Orto
Ponte delle Guglie
Riva di Biasio
San Marcuola
San Stae
PARCO GROGGIA
Sant' Alvise
Palazzo Michiel
Madonna dell'Orto
Palzzo Mastelli
Casa del Tintoretto
Museo Ebraico
Sinagoga
San Marziale
Palazzo Lábia
San Geremia
San Marcuola
Palazzo Vendramin Calergi
Fóndaco dei Turchi
San Zan Degolà
San Stae
Ca' Mocenigo
Ca' Pésaro
San Giácomo dell'Orio
Sta Maria Mater Domini
San Cassiano
S Simeone Grande
Rio dei Riformati
FONDAMENTA DEI RIFORMATI
Rio di Sant'Alvise
Rio della Sensa
FONDAMENTA DELLA SENSA
FONDAMENTA DELLE CAPPUCCINE
FONDAMENTA DEGLI ORMESINI
Rio della Misericordia
FONDAMENTA DELLA MISERICORDIA
Rio degli Zecchini
Rio Madonna dell'Orto
FONDAMENTA MADONNA DELL'ORTO
FONDAMENTA DELLA SENSA
Rio della Sensa
FMTA DEI MORI
CAMPO DEI MORI
CAMPIELLO PIAVE
CAMPO GHETTO NUOVO
Rio del Ghetto
FMTA DI GHETTO NUOVO
RIO TERRA SAN LEONARDO
CAMPO SAN LEONARDO
RIO TERRA FARSETTI
Rio di San Marcuola
Rio dei Servi
FONDAMENTA CANAL
RIO TERRA DELLA MADDALENA
VIA VITTORIO EMANUELE
STRADA NOVA
Rio di Sta Fosca
FONDAMENTA VENIER
FONDAMENTA LABIA
CAMPO SAN GEREMIA
CAMPO SAN MARCUOLA
CORTE BARBARO
RIVA DI BIASIO
CAMPO SAN STAE
Rio Fondaco dei Turchi
Rio Ca' Tron
Rio del Megio
Rio di San Boldo
Rio delle Do Torri
Rio di San Cassiano
Rio Marin
Rio San Giácomo dell'Orio
Rio di San Zuane
CAMPO NAZARIO SAURO
CPO S GIACOMO DELL'ORIO
CAMPO SAN CASSIANO
CAMPO SAN BOLDO
CAMPO DI SANT'AGOSTIN
CAMPIELLO DEL CRISTO
CAMPO D. STROPE
FONDAMENTA RIO MARIN
FONDAMENTA RIO MARIN O GARZOTTI
LISTA DEI BARI
CALLE LARGA DEI BARI
RUGA VECCHIA
CORTE CAZZA

3
A
B
C
1
2
3
4
5
2
2
7
Canale delle Fondamente Nuove
Canale de
Sacca della Misericordia
Santa Maria della Misericordia
FONDAMENTA GASPARO CONTARINI
CORTE D. MUTI
R d. Muti
CORTE VECCHIA
C E CPLO D. TREVISAN
FONDAMENTA DELL'ABAZIA
CALLE LARGA LEZZE
Canale della Misericordia
Rio della Misericordia
FMTA TRAPOLIN
Rio di Noale
CALLE SALAMON
FMTA SAN FELICE
Rio di San Felice
CALLE DELLA RACCHETTA
Rio della Racchetta
CORTE SQ VECCHIO
FMTA SANT'ANDREA
CALLE CORRENTE
C DEGLI ALBANESI
RAMO ALBANESI
RUGA DO POZZI
CALLE LUNGA SANTA CATERINA
C DELLA MASENA
FONDAMENTA SANTA CATERINA
Rio di Santa Caterina
CALLE ZANARDI
Rio di Ca' Dolce
FONDAMENTE NUOVE
CALLE MARCO FOSCARINI
CALLE D. CADENE
C DEI LEGNAMI
SAL DEI SPECCHIERI
Oratorio dei Crociferi
CAMPO SANT'ANTONIO
Gesuiti
CAMPO DEI GESUITI
FONDAMENTA ZEN
Fondamente Nuove
RAMO DONA
C.D. CROCI
CALLE D. VIDA
CALLE LARGA DEI BOTTERI
Rio dei Gesuiti
CALLE D. MAGAZEN
C DELLA PIETA
CALLE DEI VOLTI
CALLE VENIER
CPLO D. PIETA
CORTE CARITA
CALLE DEI CORDONI
CALLE DEL FUMO
C LARGA BERLENDIS
C.D. SARTORI
FMTA DEI SARTORI
SAL SERIMAN
CALLE D. SQUERO
SAL L. BORGATO
CALLE DEL REMER
RIO TERRA BARBA FRUTTARIOL
C.D. VOLTO
CALLE STELLA
C D. STUA
FMTA D. CHIESA
CALLE LARGA DOGE PRIULI
CALLE D. FORNO
Rio di Santa Sofia
CAMPO S FELICE
C S FELICE
CALLE ZOTTI
CALLE DELLE VELE
CALLE PRIULI
CALLE DEL FORNO
CALLE D. PISTOR
CALLE DEI TAGLIAPIETRA
RIO TERRA D. SANTI APOSTOLI
CALLE D. FORNO
C.D. MADONNA
C MORO
C.D. POSTA
CALLE VARISCO
Rio della Panada
CORTE D PALUDO
CALLE DELLO SQUERO
STRADA NOVA
Ca' d'Oro
C.D. CA D'ORO
C DEL DUCA
C DEL VERDE
CALLE BEMBO
RT D. FRANCESCHI
C LARGA D. PROVERBI
CALLE DEL TRAGHETTO
Santi Apostoli
CAMPIELLO WIDMAN
CALLE BONDI
RIO TERRA DEI BIRI
Rio dei Mendicanti
FONDAMENTA DEI MENDICANTI
CALLE GABRIELLA
C.D. OCA
CALLE DRAGAN
SALIZZADA DEL PISTOR
CPLO D. CASON
C COMELLO
CALLE WIDMAN
Ca' d'Oro
CAMPO SANTA SOFIA
Canal Grande
CAMPO DEI SANTI APOSTOLI
Santi Apostoli
Rio dei Santi Apostoli
FMTA DELL'OLIO
Pescheria
CAMPO D. PESCHERIA
Rio delle Beccarie
C BECCARIE
C PRIMO D. DONZELLA
RUGA D. SPEZIALI
Fabbriche Nuove
Erboria
CALLE D. MAGAZEN
RIO TERRA BAGATIN
SALIZZADA SAN CANCIANO
CAMPO S MARIA NOVA
CPLO SANTA MARIA NOVA
FMTA PIOVAN
C LARGA G GALLINA
CALLE DELLA TESTA
Scuola di S Marco
Rio San Giovanni Crisostomo
C.D. FORNO
Santa Maria dei Miracoli
San Giovanni Crisostomo
C CASTELLI
Bartolomeo Colleoni
CALLE DELLE ERBE
C.D. CAVALLO
CALLE BRESSANA
C.D. MADONNA
FMTA FELZI
C MODENA
CORTE 2ª D. MILION
C ASEO
SAL SAN GIOVANNI CRISOSTOMO
Rio di San Lio
C SCALETTA
CALLE D. DOSE
CAMPO SAN MARINA
Rio di San Marina
CALLE SAN MATTIO
CALLE ARCO
RG VECCHIA S GIOVANNI
RUGA DEGLI OREFICI
CAMPO RIALTO NUOVO
San Giácomo di Rialto
CALLE DEL BO

D
E
F
1
2
3
4
5
MURANO
CAMPO SPORTIVO
Canale di San Donato
FONDAMENTA SEBASTIANO SANTI
FMTA SAN LORENZO
CALLE DEL CONVENTO
CALLE VOLPI
FONDAMENTA LORENZO RADI
CALLE DELL'ARTIGIANO
CALLE SAN BERNARDO
COM.D. CIMITERO
STRADA
FONDAMENTA SEBASTIANO VENIER
CALLE BRUSSA
CAMPO SAN SALVADOR
CAMPO SAN BERNARDO
VIA BAROVIER
CALLE DEL CRISTO
CALLE CONTERIE
SS Maria e Donato
GIUSTINIAN
FONDAMENTA
Ponte San Donato
Museo dell'Arte Vetraria
Venier
Canale degli Angeli
CALLE DEL MISTRO
FONDAMENTA CAVOUR
Museo
Canal Grande di Murano
FONDAMENTA
NAVAGERO
CALLE S GIACOMO
FONDAMENTA ANDREA NAVAGERO
C PARADISO
Navagero
Canale Ondello
Canale Serenella
CALLE VIVARINI
RAMO DA MULA
FMTA DA MULA
Palazzo da Mula
Ponte Vivarini
FMTA ANTONIO COLLEONI
CPO STO STEFANO
San Pietro Martire
MANIN
CPLO TURELLA
FMTA SAN GIOVANNI DEI BATTUTI
STRADA VICINALE ORTI
CALLE SAN CIPRIANO
C BERTOLINI
Rio dei Vetrai
FONDAMENTA DEI VETRAI
DANIELE
CPLO BIGAGLIA
VIALE GARIBALDI
CALLE DELL'OLIO
VIA BRIATI
FMTA PIAVE
Faro
CALLE MIOTTI
CALLE BERTOLINI
FONDAMENTA SERENELLA
Colonna
Canale dei Marani
Canale delle Fondamente Nuove
FONDAMENTE NUOVE
Ospedale Civile
CALLE DELLE CAPPUCCINE
Rio di Santa Giustina
FONDAMENTA DI SANTA GIUSTINA
Celestia
FMTA CASE NUOVE
CALLE D. MOSCHETTE
C.D. CAFFETTIER
CALLE SAN FRANCESCO
San Francesco della Vigna
CALLE DEGLI ORTI
C SAGREDO

8

A
B
C
1
2
3
4
5
SANTA CROCE
DORSO
GIUDECCA
PIAZZALE ROMA
GIARDINO EX PAPADOPOLI
CAMPO SANT'ANDREA
Canale di Santa Chiara
Canale Scomenzera
Canale di Santa Maria Maggiore
FONDAMENTA SANT'ANDREA
CAMPAZZO TRE PONTI
Rio Nuovo
Rio delle Burchielle
FONDAMENTA DELLE BURCHIELLE
FONDAMENTA FABBRICA TABACCHI
RIO TERRA DEI PENSIERI
Rio della Cazziola
FONDAMENTA RIZZI
FONDAMENTA DELLE PROCURATIE
FMTA DI SANTA MARIA MAGGIORE
Rio del Tintor
Rio Briati
FONDAMENTA ROSSA
CALLE RAGUSEI
CALLE LARGA RAGUSEI
CORTE CONTARINI
Rio dell'Arziere
FONDAMENTA DELL'ARZIERE
RIO TERRA DEI SECCHI
FONDAMENTA SANTA MARTA
Rio delle Terese
FONDAMENTA DELLE TERESE
Rio dei Carmini
Santa Maria dei Carmini
Collegio Armeno
Rio dell'Angelo Raffaele
FONDAMENTA BARBARIGO
FONDAMENTA DI PESCHERIA
Rio di San Nicolò
San Nicolò dei Mendicoli
CAMPO SAN NICOLO
CPLO D. STENDARDO
C DIETRO AI MEGAZZINI
Angelo Raffaele
CPO ANGELO RAFFAELE
Rio di San Sebastiano
San Sebastiano
CALLE DELL'AVOGARIA
Rio dell'Avogaria
CALLE BALASTRO
SALIZZADA SAN BASEGIO
Stazione Marittima
BANCHINA DEL PORTO COMMERCIALE
BANCHINA DI SAN BASEGIO
FMTA ZATTERE
San Basilio
Canale di Fusina
Sacca Fisola
FONDAMENTA BEATA GIULIANA
CALLE DELLA SCUOLA
CALLE DEI FIGHERI
CALLE DEL VAPORETTO
CALLE FISOLA
CALLE DELLA SACCA
CALLE RIMINI
CAMPO DELLA CHIESA
CALLE LARGA DEI LAVRANERI
Canale dei Lavraneri
Mulino Stucky
Rio di San Biagio
CAMPIELLO PRIULI
Rio delle Convertite
FMTA DELLE CONVERTITE
FMTA SAN GERARDO SAGREDO
Can Sacca Fisola San Biagio

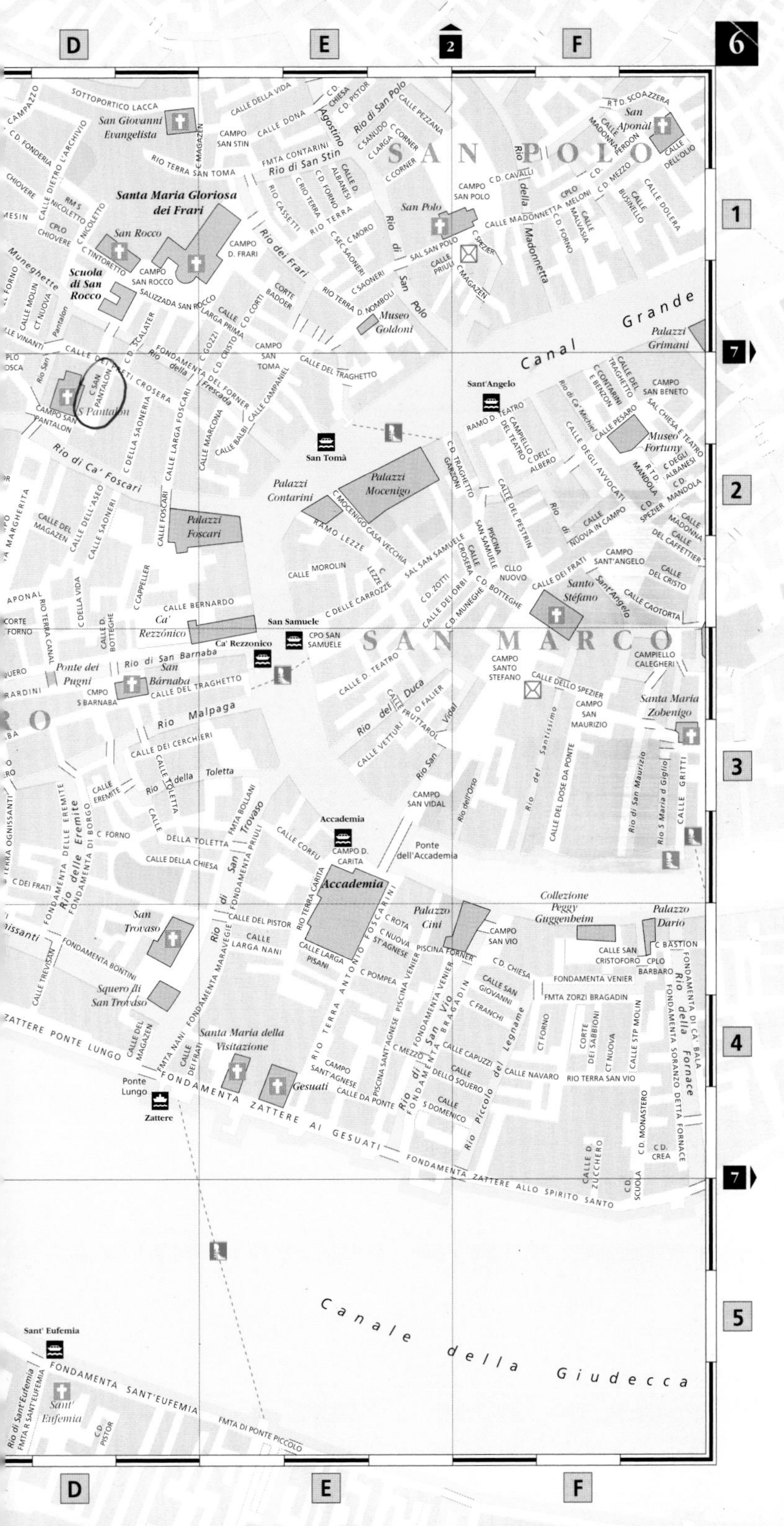
D
E
2
F
1
2
3
4
5
7
SAN POLO
SAN MARCO
Canal Grande
Canale della Giudecca
Santa Maria Gloriosa dei Frari
San Giovanni Evangelista
San Rocco
Scuola di San Rocco
S Pantalon
San Polo
Museo Goldoni
San Aponal
Palazzi Grimani
Sant'Angelo
Museo Fortuny
San Tomà
Palazzi Mocenigo
Palazzi Contarini
Palazzi Foscari
Ca' Rezzónico
Ca' Rezzonico
San Samuele
Santo Stéfano
San Bárnaba
Ponte dei Pugni
Santa Maria Zobenigo
Accademia
Ponte dell'Accademia
Palazzo Cini
Collezione Peggy Guggenheim
Palazzo Dario
San Trovaso
Squero di San Trovaso
Santa Maria della Visitazione
Gesuati
Ponte Lungo
Zattere
Sant' Eufemia
Sant' Eufemia
Rio di Ca' Foscari
Rio Malpaga
Rio di San Barnaba
Rio della Toletta
Rio delle Eremite
Rio di San Trovaso
Rio del Duca
Rio dei Frari
Rio di San Polo
Rio della Madonnetta
Rio di San Stin
Rio della Frescada
Rio di Santo Stefano
Rio del Santissimo
Rio della Fornace
Rio di San Vio
Rio Piccolo del Legname
FONDAMENTA ZATTERE AI GESUATI
FONDAMENTA ZATTERE ALLO SPIRITO SANTO
ZATTERE PONTE LUNGO
FONDAMENTA SANT'EUFEMIA
FMTA DI PONTE PICCOLO
CAMPO SAN STIN
CAMPO SAN POLO
CAMPO D. FRARI
CAMPO SAN ROCCO
CAMPO SAN TOMA
CAMPO SAN BENETO
CAMPO SANT'ANGELO
CAMPO SANTO STEFANO
CAMPO SAN MAURIZIO
CAMPO SAN VIDAL
CAMPO SAN VIO
CAMPO SANT'AGNESE
CAMPO D. CARITA
CAMPIELLO CALEGHERI
CAMPO SAN PANTALON
CMPO S BARNABA
CPO SAN SAMUELE
CALLE DEI FRATI
CALLE BERNARDO
CALLE DEL TRAGHETTO
CALLE DELLA TOLETTA
CALLE CORFU
CALLE DEL PISTOR
CALLE NAVARO
RIO TERRA SAN VIO
RIO TERRA SAN TOMA
RIO TERRA ANTONIO FOSCARINI
SALIZZADA SAN ROCCO
SOTTOPORTICO LACCA
CALLE LARGA FOSCARI
CALLE DEI CERCHIERI
FONDAMENTA VENIER
FMTA ZORZI BRAGADIN
CALLE CAOTORTA
CALLE DELLO SPEZIER
CALLE GRITTI

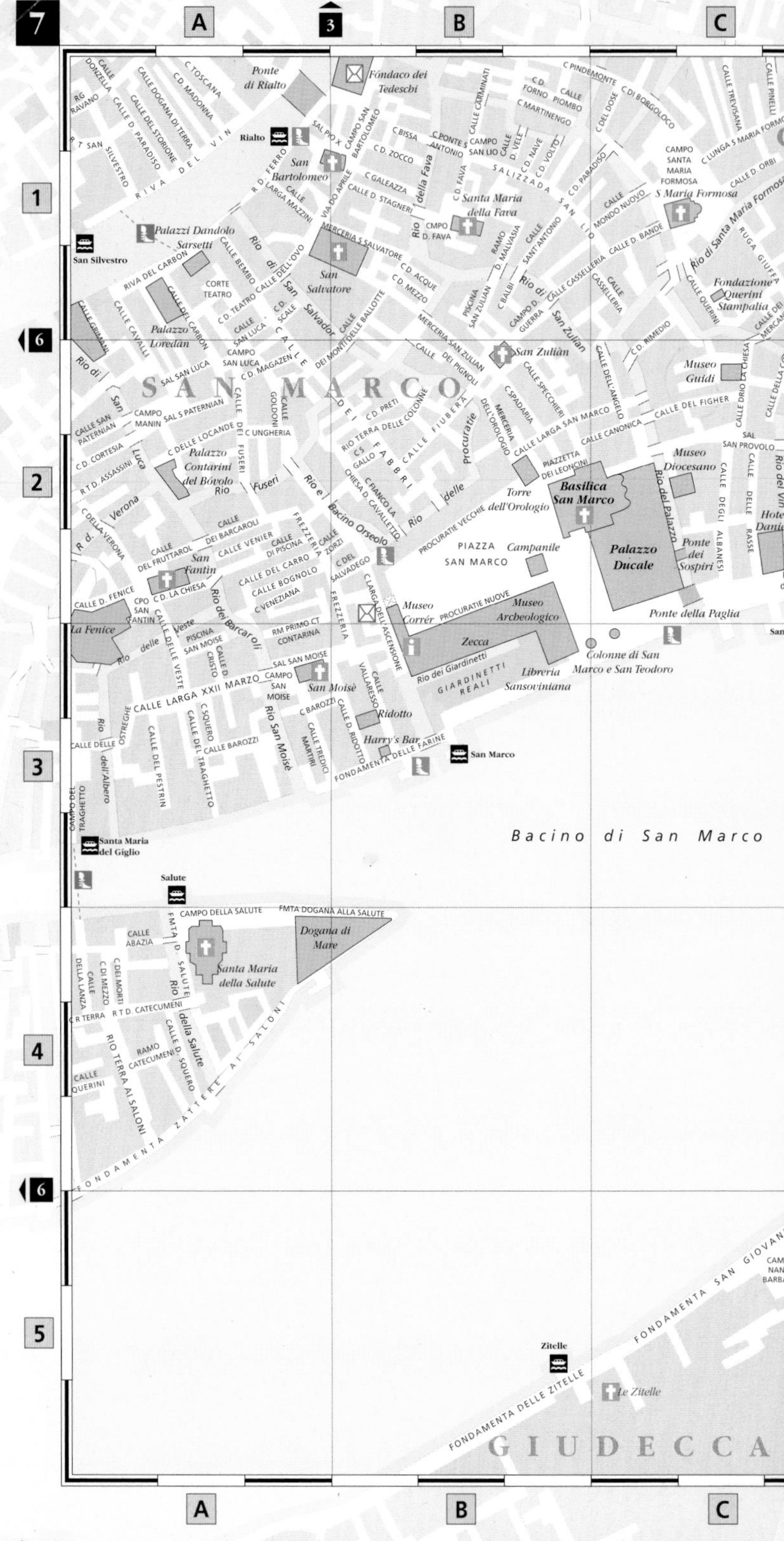

7
A
B
C
1
2
3
4
5
6
Ponte di Rialto
Fóndaco dei Tedeschi
Rialto
San Bartolomeo
San Silvestro
Palazzi Dandolo
Sarsetti
San Salvatore
Santa Maria della Fava
S Maria Formosa
Fondazione Querini Stampalia
Palazzo Loredan
San Zulian
Museo Guidi
SAN MARCO
Palazzo Contarini del Bóvolo
Torre dell'Orologio
Basilica San Marco
Museo Diocesano
Palazzo Ducale
Ponte dei Sospiri
Hotel Daniel
PIAZZA SAN MARCO
Campanile
San Fantin
La Fenice
Museo Corrér
Museo Archeologico
Zecca
Ponte della Paglia
Colonne di San Marco e San Teodoro
Libreria Sansoviniana
GIARDINETTI REALI
San Moisè
Ridotto
Harry's Bar
San Marco
Bacino di San Marco
Santa Maria del Giglio
Salute
Dogana di Mare
Santa Maria della Salute
CAMPO DELLA SALUTE
FMTA DOGANA ALLA SALUTE
FONDAMENTA ZATTERE AI SALONI
RIO TERRA AI SALONI
Rio della Salute
CALLE LARGA XXII MARZO
RIVA DEL VIN
RIVA DEL CARBON
MERCERIA S SALVATORE
MERCERIA SAN ZULIAN
CALLE LARGA SAN MARCO
CALLE CANONICA
PROCURATIE VECCHIE
PROCURATIE NUOVE
FONDAMENTA DELLE FARINE
Rio dei Giardinetti
FONDAMENTA SAN GIOVANNI
Zitelle
Le Zitelle
FONDAMENTA DELLE ZITELLE
GIUDECCA

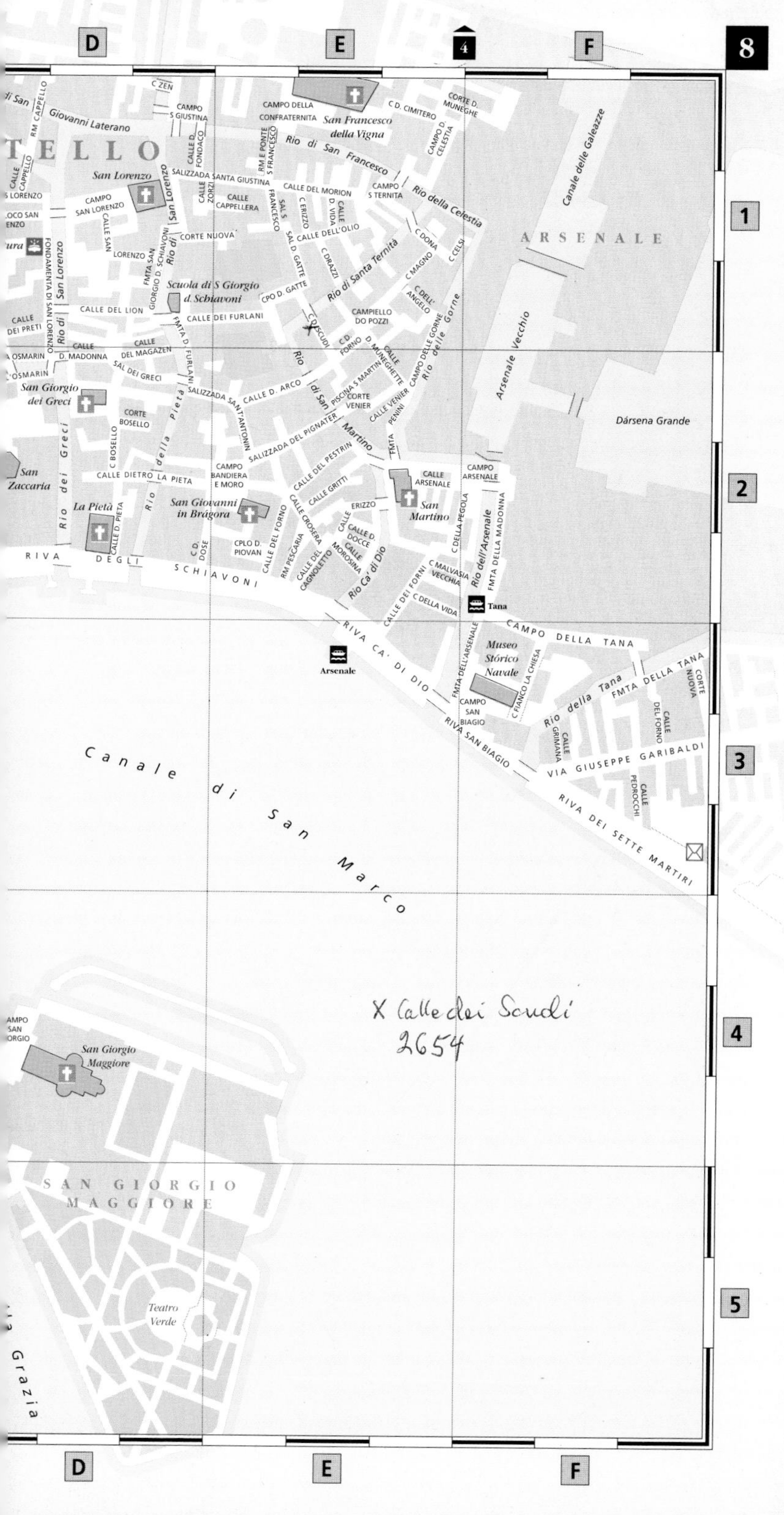

D
E
4
F
8
C ZEN
CAMPO S GIUSTINA
Giovanni Laterano
RM CAPPELLO
CAMPO DELLA CONFRATERNITA
San Francesco della Vigna
C D. CIMITERO
CORTE D. MUNEGHE
CAMPO D. CELESTIA
Canale delle Galeazze
TELLO
San Lorenzo
CALLE D. FONDACO
RM E PONTE S FRANCESCO
Rio di San Francesco
SALIZZADA SANTA GIUSTINA
CALLE DEL MORION
CAMPO S TERNITA
Rio della Celestia
CAMPO SAN LORENZO
CALLE CAPPELLERA
CALLE ZORZI
C ERIZZO
CALLE D. VIDA
CALLE DELL'OLIO
1
ARSENALE
CORTE NUOVA
C DONA
C CELSI
Rio di San Lorenzo
FONDAMENTA DI SAN LORENZO
C DRAZZI
Rio di Santa Ternità
C MAGNO
Scuola di S Giorgio d. Schiavoni
CPO D. GATTE
SAL D. GATTE
C DELL' ANGELO
CALLE DEL LION
CALLE DEI FURLANI
CAMPIELLO DO POZZI
CALLE DEI PRETI
C D. SCUDI
C D. FORNO
CALLE D. MUNEGHETTE
CAMPO DELLE GORNE
Rio delle Gorne
Arsenale Vecchio
D. MADONNA
CALLE DEL MAGAZEN
FMTA D. FURLANI
SAL DEI GRECI
Rio di San Martino
San Giorgio dei Greci
SALIZZADA SANT'ANTONIN
CALLE D. ARCO
PISCINA S MARTIN
CORTE VENIER
CALLE VENIER
Dársena Grande
CORTE BOSELLO
C BOSELLO
SALIZZADA DEL PIGNATER
FMTA PENINI
San Zaccaria
Rio dei Greci
CALLE DIETRO LA PIETA
Rio della Pietà
CAMPO BANDIERA E MORO
CALLE DEL PESTRIN
CALLE ARSENALE
CAMPO ARSENALE
2
La Pietà
CALLE D. PIETA
San Giovanni in Brágora
CALLE GRITTI
ERIZZO
San Martino
C DELLA PEGOLA
Rio dell'Arsenale
FMTA DELLA MADONNA
C.D. DOSE
CPLO D. PIOVAN
CALLE DEL FORNO
CALLE CROSERA
RM PESCARIA
CALLE DEL CAGNOLETTO
MOROSINA
CALLE D. DOCCE
Rio Ca' di Dio
CALLE DEI FORNI
C MALVASIA VECCHIA
C DELLA VIDA
RIVA DEGLI SCHIAVONI
Tana
CAMPO DELLA TANA
RIVA CA' DI DIO
Arsenale
Museo Stórico Navale
FMTA DELL'ARSENALE
C FIANCO LA CHIESA
Rio della Tana
FMTA DELLA TANA
CORTE NUOVA
CALLE DEL FORNO
CAMPO SAN BIAGIO
RIVA SAN BIAGIO
CALLE GRIMANA
VIA GIUSEPPE GARIBALDI
CALLE PEDROCCHI
3
Canale di San Marco
RIVA DEI SETTE MARTIRI
X Calle dei Scudi
2654
4
San Giorgio Maggiore
SAN GIORGIO MAGGIORE
Teatro Verde
Grazia
5
D
E
F

Street Finder Index

Key to Abbreviations Used in the Street Finder

C	Calle	**d.**	di, del, dell', dello, della, dei, delle, degli	**R**	Rio	**Sta**	Santa
Can	Canale			**R T**	Rio Terrà	**Sto**	Santo'
Cpo	Campo			**Rg**	Ruga/Rughetta	**SS**	Santi/Santissimo
Cplo	Campiello	**Fmta**	Fondamenta	**Sal**	Salizzada	**Stp**	Sottoportico
Ct	Corte	**Rm**	Ramo	**S**	San/Sant'		

Crea, Rio Terrà della **1 C3**
Cremonese, Calle **5 C1**
Cristi, Calle dei **2 F5**
Cristo, Calle de **2 E4**
Cristo, Calle del **2 E3**
Cristo, Calle del **2 E5**
Cristo, Calle del **2 F4**
Cristo, Calle del **4 E2**
Cristo, Calle del **5 B3**
Cristo, Calle del **6 E2**
Cristo, Calle del **6 F2**
Cristo, Calle del **7 A3**
Cristo, Campiello del **2 D5**
Cristo, Rio Terrà del **2 E3**
Croce, Calle della **2 D5**
Croce, Fondamenta **1 C5**
Croci, Calle delle **3 C4**
Crociferi, Oratorio dei **3 B4**
Crosera, Calle **6 F2**
Crosera, Calle **8 E2**
Crotta, Fondamenta **2 D4**

D

Dario, Sottoportico **2 D5**
Diedo, Fondamenta **2 F3**
Do Aprile, Via **7 A1**
Do Pozzi, Campiello **8 E1**
Do Pozzi, Ruga **3 A4**
Do Torri, Rio delle **2 F5**
Docce, Calle delle **8 E2**
Dogana alla Salute, Fondamenta **7 A4**
Doge Priuli, Calle Larga **3 A4**
Dogolin, Calle **5 C3**
Dolera, Calle **6 F1**
Donà, Calle **6 E1**
Donà, Calle **8 E1**
Donà, Ramo **3 B4**
Donzella, Calle **7 A1**
Dose, Calle del **3 B5**
Dose, Calle del **7 C1**
Dose, Calle del **8 D2**
Dose da Ponte, Calle del **6 F3**
Dragan, Calle **3 A5**
Drazzi, Calle **8 E1**
Duca, Calle del **3 A4**
Duca, Rio del **6 E3**

E

Emo, Calle **2 D4**
Erbe, Calle delle **3 C5**
Eremite, Calle **6 D3**
Eremite, Fondamenta delle **6 D3**
Eremite, Rio delle **6 D3**
Erizzo, Calle **8 E1**
Erizzo, Calle **8 E2**

F

Fabbri, Calle dei **7 B2**
Fabbrica Tabacchi, Fondamenta **5 B1**
Fabbriche Nuove **3 A5**
Falier, Calle **6 D1**
Farine, Fondamenta delle **7 B3**
Farnese, Calle **2 E3**
Farsetti, Rio Terrà **2 E3**
Fava, Calle della **7 B1**
Fava, Campo della **7 B1**
Fava, Rio della **7 B1**
Felzi, Fondamenta **3 C5**
Fenice, Calle della **7 A2**
Fenice, Fondamenta **7 A3**
Ferau, Calle **1 C2**
Ferro, Riva del **7 A1**
Figher, Calle del **7 C2**
Figheri, Calle dei **5 A5**
Filosi, Calle **2 F5**
Fiori, Campiello dei **2 F4**
Fisola, Calle **5 B5**
Fiubera, Calle **7 B2**
Flangini, Campiello **2 D4**
Fóndaco, Calle del **8 D1**
Fonderia, Calle della **6 D1**
Fóndaco dei Turchi, Rio **2 E4**
Fornace, Rio della **6 F4**
Forner, Calle del **1 C2**
Forner, Calle del **2 F5**
Forner, Fondamenta del **6 D1**
Forner, Piscina **6 E4**
Forni, Calle dei **8 E2**
Forno, Calle **6 D3**
Forno, Calle del **2 D3**
Forno, Calle del **2 D4**
Forno, Calle del **2 E3**
Forno, Calle del **2 E4**
Forno, Calle del **2 F3**
Forno, Calle del **2 F4**
Forno, Calle del **3 A4**
Forno, Calle del **3 A4**
Forno, Calle del **3 B4**
Forno, Calle del **3 B5**
Forno, Calle del **6 D1**
Forno, Calle del **6 D2**
Forno, Calle del **6 E1**
Forno, Calle del **6 F1**
Forno, Calle del **7 B1**
Forno, Calle del **8 E1**
Forno, Calle del **8 E2**
Forno, Calle del **8 F3**
Forno, Corte **6 F4**
Forno, Corte del **6 D2**
Foscari, Calle **6 D2**
Foscari, Calle Larga **6 D2**
Foscarini, Calle Marco **3 B3**
Foscarini, Fondamenta **5 C2**
Foscarini, Rio Terrà Antonio **6 E4**
Franceschi, Rio Terrà dei **3 B4**
Franchi, Calle **6 F4**
Frari, Campo dei **6 E1**
Frari, Rio dei **6 E1**
Frati, Calle dei **5 B3**
Frati, Calle dei **6 D3**
Frati, Calle dei **6 D4**
Frati, Calle dei **6 F2**
Frescada, Rio della **6 D2**
Frezzeria **7 A2**
Fruttarol, Calle **6 E3**
Fruttarol, Calle del **7 A2**
Fumo, Calle del **3 C4**
Furlani, Calle dei **8 E1**
Furlani, Fondamenta dei **8 D2**
Fuseri, Calle dei **7 A2**
Fuseri, Rio **7 A2**
Fusina, Canale di **5 A4**

G

Gabriella, Calle **3 C5**
Gaffaro, Fondamenta del **5 C1**
Galeazza, Calle **7 B1**
Galeazze, Canale delle **8 F1**
Gallina, Calle Larga Giacinto **3 C5**
Gallion, Calle **2 D5**
Gallo, Corte **5 C1**
Garibaldi, Via Giuseppe **8 F3**
Garibaldi, Viale **4 E3**
Gatte, Campo delle **8 E1**
Gatte, Salizzada delle **8 E1**
Gesuiti, Campo dei **3 B4**
Gesuiti, Rio dei **3 B4**
Gherardini, Fondamenta **6 D3**
Ghetto Nuovo, Campo **2 E3**
Ghetto Nuovo, Fondamenta di **2 E3**
Ghetto Nuovo, Rio del **2 E3**
Ghetto Vecchio, Calle del **2 D2**
Giardinetti Reali **7 B3**
Giardinetti, Rio dei **7 B3**
Giardino Ex Papadopoli **1 C5**
continues **5 C1**
Giochin, Calle **2 D4**
Giudecca, Canale della **6 E5**
Giuffa, Ruga **7 C1**
Giustinian, Fondamenta **4 E2**
Goldoni, Calle **7 A2**
Gorne, Campo delle **8 E1**
Gorne, Rio delle **8 E1**
Gozzi, Calle **6 E2**
Gradenigo, Fondamenta **2 D5**
Gradisca, Calle **2 D5**
Gradisca, Calle **2 F2**
Grazia, Canale della **8 D5**
Greci, Rio dei **8 D2**
Greci, Salizzada dei **8 D2**
Grimana, Calle **8 F3**
Grimani, Calle **7 A1**
Gritti, Calle **6 F3**
Gritti, Calle **8 E2**
Groggia, Parco **2 E2**
Grue, Fondamenta delle **2 E5**
Guardiani, Calle dei **5 B2**
Guerra, Campo della **7 B1**

I

Isola, Ramo dell' **2 E5**
Isola, Rio dell' **2 E5**

L

Labia, Fondamenta **2 D4**
Lacca, Calle della **2 D5**
Lacca, Sottoportico **6 D1**
Lana, Campo della **1 C5**
Lanza, Calle della **7 A4**
Lardona, Corte **5 B3**
Larga, Calle **2 E5**
Larga, Calle **2 F3**
Larga, Calle **6 E1**
Lavraneri, Calle Larga dei **5 A5**
Lavraneri, Canale dei **5 B5**
Le Zitelle **7 C5**
Legname, Calle Larga del **2 E2**
Legnami, Calle dei **3 A3**
Leoncini, Piazzetta dei **7 B2**
Lezze, Calle **6 E2**
Lezze, Calle Larga **3 A3**
Lezze, Ramo **6 E2**
Lion, Calle del **8 D1**
Lista di Spagna, Rio Terrà **2 D4**
Lizza, Fondamenta **5 B3**
Locande, Calle delle **7 A2**
Loredan, Calle **2 F2**
Lustraferri, Rio dei **2 F3**

M

Maddalena, Calle **5 B3**
Maddalena, Rio Terrà della **2 F3**
Madonna, Calle **6 F1**
Madonna, Calle **6 F2**
Madonna, Calle della **1 C3**
Madonna, Calle della **3 B4**
Madonna, Calle della **3 C5**
Madonna, Calle della **7 A1**
Madonna, Calle della **8 D1**
Madonna, Fondamenta della **5 B2**
Madonna, Fondamenta della **8 F2**
Madonna dell'Orto, Fondamenta **2 F2**
Madonna dell'Orto, Rio **2 F2**
Madonnetta, Calle **6 F1**
Madonnetta, Rio della **6 F1**
Magazen, Calle **6 D1**
Magazen, Calle **6 F1**
Magazen, Calle del **1 B3**
Magazen, Calle del **2 D2**
Magazen, Calle del **3 B4**
Magazen, Calle del **3 B5**
Magazen, Calle del **6 D2**
Magazen, Calle del **6 D4**
Magazen, Calle del **7 A2**
Magazen, Calle del **8 D1**
Magazen, Fondamenta **5 C1**
Magazzini, Calle dietro ai **5 A3**
Maggiore, Corte **5 B3**
Magno, Calle **8 E1**
Malcanton, Fondamenta **5 C1**
continues **5 C2**
Malcanton, Rio del **5 C1**
Malpaga, Rio **5 C3**
Malvasia, Calle **2 F4**
Malvasia, Calle **6 F1**
Malvasia, Calle della **2 E2**
Malvasia, Ramo della **7 B1**
Malvasia Vecchia, Calle **8 E2**
Mandola, Calle della **6 F2**
Mandola, Rio Terrà della **6 F2**
Manin, Campo **7 A2**
Manin, Fondamenta Daniele **4 D3**
Marani, Canale dei **4 E3**
Maravegie, Fondamenta **6 D4**
Marcona, Calle **6 E2**
Marin, Rio **2 D5**
Martinengo, Calle **7 B1**
Masena, Calle della **2 E3**
Masena, Calle della **3 B3**
Masena, Calle della **5 C4**
Mazzini, Calle Larga **7 A1**
Megio, Calle del **2 E4**
Megio, Fondamenta del **2 E4**
Megio, Rio del **2 E5**
Meloni, Campiello **6 F1**
Mendicanti, Fondamenta dei **3 C5**
Mendicanti, Rio dei **3 C5**
Mercanti, Calle dei **7 C1**
Mercerie **7 B2**
Mezzo, Calle **6 E4**
Mezzo, Calle di **6 D1**
Mezzo, Calle di **6 F1**
Mezzo, Calle di **7 A4**
Mezzo, Calle di **7 B1**
Miani, Calle **2 F5**
Minotto, Fondamenta **5 C1**
Miotti, Calle **4 D3**
Misericordia, Calle della **1 C4**
Misericordia, Calle della **5 C1**
Misericordia, Canale della **3 A3**
Misericordia, Fondamenta della **2 F3**
Misericordia, Ramo della **1 C4**
Misericordia, Rio della **2 E3**
continues **3 A3**
Misericordia, Sacca della **3 A3**
Mistro, Calle del **4 E2**
Mocenigo Casa Vecchia, Calle **6 E2**
Modena, Calle **3 B5**
Modena, Calle della **2 E5**
Molin, Calle **6 D1**
Monastero, Calle del **6 F4**
Monastero, Fondamenta **1 C5**
Mondo Nuovo, Calle **7 C1**
Monfalcone, Calle **5 B5**
Monti delle Ballotte, Calle dei **7 B1**
Mori, Campo dei **2 F3**
Mori, Fondamenta dei **2 F3**
Morion, Calle del **8 E1**
Moro, Calle **3 B4**
Moro, Calle **6 E1**
Moro, Fondamenta **2 F3**
Morolin, Calle **6 E2**
Morosina, Calle **8 E2**
Morti, Calle dei **5 C4**
Morti, Calle dei **7 A4**
Mosca, Campiello **6 D2**
Moschette, Calle delle **4 D5**
Mosto Balbi, Calle da **2 D4**
Mula, Fondamenta da **4 E2**
Mula, Ramo da **4 E2**
Muneghe, Calle delle **2 E2**
Muneghe, Calle delle **6 E3**
Muneghe, Campiello **2 D5**
Muneghe, Corte delle **8 E1**
Muneghette, Calle delle **8 E2**
Muneghette, Rio delle **6 D1**
Murano, Canal Grande di **4 E2**
Muti, Corte dei **3 A3**
Muti, Rio dei **3 A3**

N

Nani e Barbaro, Campo **7 C5**
Nani, Calle Larga **6 E4**
Nani, Fondamenta **6 D4**
Navagero, Fondamenta Andrea **4 E3**
Navaro, Calle **6 F4**
Nave, Calle **5 B3**
Nave, Calle della **7 B1**

O

P

Q

R

S

T

U

V

W

Z

General Index

D

H

I

J

K

L

M

N

O

P

Q

R

Acknowledgments

DORLING KINDERSLEY would like to thank the many people whose help and assistance contributed to the preparation of this book.

MAIN CONTRIBUTORS
Susie Boulton studied languages and history of art at the University of Cambridge, where she still has her home. She has been visiting Venice for over 20 years and is the author of several guide books on the city. Although a specialist on Venice, she has also written travel guides and articles on various other cities and regions of Europe.

Christopher Catling has been visiting Italy for over 22 years since his first archaeological dig there while he was a student at Cambridge University. He is the author of several guide books on Italian cities and regions, including *The Eyewitness Travel Guide to Florence and Tuscany.*

ADDITIONAL CONTRIBUTOR
Sally Roy first got to know Venice while she was at school in Rome and has been returning to the country ever since. She read medieval history at St Andrew's University, Edinburgh and has contributed to several books on Italy where she now spends six months of every year.

ADDITIONAL ILLUSTRATIONS
Annabelle Brend, Dawn Brend, Neil Bulpitt, Richard Draper, Nick Gibbard, Kevin Jones Associates, John Lawrence, The Maltings Partnership, Simon Roulstone, Sue Sharples, Derrick Stone, Paul Weston, John Woodcock.

DESIGN AND EDITORIAL ASSISTANCE
Michael Blacker, Dawn Brend, Michael Ellis, Irena Hoare, Annette Jacobs, Stephen Knowlden, Erika Lang, Steve Rowling, Janis Utton, Fiona Wild.

RESEARCH ASSISTANCE
Hans Erlacher, Paolo Frullini, Oscar Gates, Marinella Laini, Elizabetta Lovato, Fabiola Perer, Sarah Sole.

INDEX
Indexing Specialists, 202 Church Road, Hove, East Sussex, UK.

SPECIAL ASSISTANCE
Comune di Vicenza; Arch. Gianfranco Martinoni at the Assessorato Beni Culturali Comune di Padova; Ca' Macana; Cesare Battisti at the Media Tourist Office, Venice; Curia Patriarcale Venezia; D.ssa Foscarina Caletti at the Giunta Regionale di Venezia; Jane Groom, Brian Jordan; Alexandra Kennedy; Joy Parker; Frances Hawkins, Lady Frances Clarke and John Millerchip of the Venice in Peril Fund; the staff of the APT offices throughout the Veneto, in particular Anna Rita Bisaggio in Montegrotto Terme, Stephano Marchioro in Padua; Anna Maria Carlotto, Virna Scarduelli and Christina Erlacher in Verona, Anselmo Centomo in Vicenza; Heidi Wenyon.

PHOTOGRAPHY PERMISSIONS
DORLING KINDERSLEY would like to thank the following for their kind permission to photograph at their establishments:
VENICE: Amministrazione Provinciale di Venezia (Museo dell'Estuario, Torcello); Ca' Mocenigo; Ca' Pesaro; Ca' Rezzonico; Caffè Quadri; Collegio Armeni; Fondazione Europea Pro Venetia Viva, San Servolo; Fondazione Giorgio Cini (San Giorgio Maggiore); Peggy Guggenheim Museum; Hôtel des Bains; Libreria Sansoviniana; Museo Archeologico; Museo Correr; Museo Diocesano d'Arte Sacra; Museo Fortuny; Museo Storia Navale; Museo Storico Naturale; Museo Vetrario, Murano; Arch. Umberto Franzoi and staff at the Palazzo Ducale; Procuratie di San Marco (Basilica San Marco); Santi Giovanni e Paolo; San Lazzaro degli Armeni; Santa Maria Gloriosa dei Frari; Scuola Grande dei Carmini; Scuola Grande di San Rocco.
VENETO: Arena Romano, Verona; Basilica, Vicenza; Caffè Pedrocchi, Padua; Duomo, Padua; Duomo, Vicenza; Giardini Giusti, Verona; Museo Archeologico, Verona; Museo di Castelvecchio, Verona; Museo Civico, Malcésine; Museo Civico, Vicenza; Museo Concordiese, Portogruaro; Museo dei Eremitani, Padua; Museo Lapidario Maffeiano, Verona; Museo dei Storia Naturale, Verona; Ossuario di San Pietro, Solferino; Sant'Anastasia, Verona; San Fermo Maggiore, Verona; San Giorgio, Monsélice; San Giorgio in Braida, Verona; San Lorenzo, Vicenza; Santa Maria in Organo, Verona; San Pietro in Malvino, Sirmione; San Severo, Bardolino; San Stefano, Verona; San Zeno Maggiore, Verona; Santuario di Monte Berico, Vicenza; Teatro Olimpico, Vicenza; Università di Padova; Contessa Diamante Luling-Buschette, Villa Barbaro, Masèr; Conte Marco Emo, Villa Emo, Fanzolo di Vedelago.

PICTURE CREDITS
t = top; tc = top centre; tr = top right; cla = centre left above; ca = centre above; cra = centre right above; cl = centre left; c = centre; cr = centre right; clb = centre left below; cb = centre below; crb = centre right below; bl = bottom left; bc = bottom centre; br = bottom right.

Every effort has been made to trace the copyright holders, and we apologize in advance for any unintentional omissions. We would be pleased to insert the appropriate acknowledgments in any subsequent edition of this publication.

Works of art have been reproduced with the permission of the following copyright holders:: © ADAGP, Paris and DACS, London 1995: Intérieur Hollandais II (1928) by Joan Miró,

Maiastra by Constantin Brancusi, 134cr.

The publishers are grateful to the following museums, companies, and picture libraries for permission to reproduce their phograph: Accademia Olimpica, Vicenza: 172tr, 173crb; Ace Photo/Agency Tore Gill: 254b; Ace/Mauritius: 257t; Ancient Art & Architecture Collection: 38c, 40 br, 40 tl; APT del Bresciano: 35b; Archiv für Kunst und Geschichte: 26tl/c/cr, 30bl, 36, 43br, 44crb, 45bl, 45t, 46tl, 46tr, 48cla, 49clb, 50b, 54br, 130b, 131b, 131c, 132b, 133t, 30/31c; Archivio Raimondo Zago 49cl. Archivio Veneziano: Sarah Quill 29br, 51crb, 67cl, 106tr, 144t.

Benetton: 51tl; Biblioteca Civica di Trieste (Foto Halupca): 119b; Bridgeman Art Library, London: *Madonna and Child and Saints* (triptych altarpiece) by Giovanni Bellini (c.1431-1516), Santa Maria dei Frari, Venice 27tl/tr; *The Siege of Antioch* 1098 by William of Tyre. Bibliothèque Nationale, Paris 40cla; *Marco Polo with Elephants and Camels* from Livre des Merveilles, Bibliothèque Nationale, Paris, 42cb; *View of Venice* by Bernardo von Breitenbach, from Opusculum Sanctarum Peregrinationum in Terram Sanctam, Bibliothèque Nationale, Paris, 8-9; *Family Tree of the Cornaro Family*, Italian School (18th century), Palazzo Corner Ca' Grande, Venice, 69tr; *Salome* by Gustav Klimt (1862-1918) Museo d'Arte Moderna, Venice, 105br; *The Nuns' Visiting Day* by Francesco Guardi (1712-93), Museo Ca' Rezzonico, Venice, 112bl; *St George Killing the Dragon* by Vittore Carpaccio (c.1460/5-1523/6), Scuola di San Giorgio degli Schiavoni, Venice, 118tl; *The Stealing of the Body of St Mark* by Tintoretto (1518-94), Accademia, Venice, 131t; *The Rape of Europa* by Francesco Zuccarelli, Accademia, Venice, 133c; *Marco Polo dressed in Tartar costume* (Italian, c.1700) Museo Correr, Venice/Giraudon, 4tr/115br; Osvaldo Böhm: 20b, 20cl, 29bl, 41crb, 63tl, 79tl, 132c, 132t, 144c.

Cephas Picture Library: Mick Rock, 35cra, Mick Rock, 238tr; Ciga Hotels: 70tl; Claire Calman: 55cr; Giancarlo Costa: 47clb; Joe Cornish: 15b, 209c, 216bl; Stephanie Colasanti: 56, 195br.

Chris Donaghue The Oxford Photo Library: 3 (inset), 5tl, 76t, 156br, 280t; Michael Dent: 82b, 252tl.

E.T. Archive: Sala dei Prior Siena, 41tl; Baroque Hall of Mirrors, Palazzo Papadopoli, Venice, 64tl; *The Apothecary's Shop* by Pietro Longhi, Accademia, Venice, 130c; Electa, Milan: 180cr, 181t/cl; Erizzo Editrice srl: 105clb; Mary Evans Picture Library: 7 (inset), 9 (inset), 24tl, 39clb, 44bl, 47br, 47t, 53 (inset), 58cra, 58tr, 63c, 64c, 69ca, 143tr, 159 (inset), 221 (inset), 259 (inset), 32c.

Grazia Neri: 41b, 49tl, 50crb, 50tr, 65br, 92tl, 157b, 254t; Marco Bruzzo, 5tr, 32t, 35clb, 40tr, 166t; Cameraphoto 71tr, 115bl, 42/43c; Peggy Guggenheim Museum, Venice: 134b; The Ronald Grant Archive: 50cla.

Hotel Excelsior, Venice Lido: 48crb; Robert Harding Picture Library: 216cl; The Hulton Deutsch Collection: 38br, 38tr, 39br, 43clb, 46br, 48bc, 49br, 61tr, 66b, 66cra, 69cl, 101c, 140bl, 181br, 199b

Image Bank: Guido A. Rossi, 11 (inset); Image Select: Ann Ronan, 44cl.

Hugh McKnight Photography: 71tl; Magnum Photos/David Seymour: 48tr; Marka: 37b; Moro Roma: 30cl, 48/49c; Museo Archeologico, Verona: 38bl; Museo Civico agli Eremitani: 179t/c/b; Museo Civico di Oderzo: 39t; The Mansell Collection: 46crb, 47bl.

NHPA/Gerard Lacz: 217b; NHPA/Laurie Campbell: 217c; NHPA/Silvestris Fotoservice: 217crb; The National Gallery, London: 29t.

Olympia/Select: 34b, 51br, 51clb, 157t, 255t; Olympia/Select/Larry Rivers: 256t.

Performing Arts Library/Gianfranco Fainello: 256c.

The Royal Collection ©1994 Her Majesty Queen Elizabeth 11: 46/47c.

John Ferro Sims: 167t; Scala, Firenze: *Madonna di Ca' Pesaro* by Titian (1477/89-1576), S. Maria Gloriosa dei Frari, Venice 27, c/cr; *Ultimi Momenti del Doge Marin Faliero* by Francesco Hayez (1791-1881), Pinacoteca di Brera, Milan, 43t; *Banquet of Antony and Cleopatra* by Giambattista Tiepolo (1692-1770), Palazzo Labia, Venice, 60t; *Ultimi Momenti del Doge Marin Faliero* by Francesco Hayez (1791-1881), Pinacoteca di Brera, Milan, 68cra; *Crocifissione* by Tintoretto (1518-1594), Scuola Grande di S. Rocco, Venice,106c/cb; *S Michele* by Giambono (15th century), Accademia, Venice, 123b; *Il Trasporto della Santa Casa di Loreto* by Giambattista Tiepolo (1692-1770), Accademia, Venice, 130tr; *Presentazione al tempio* by Titian (1477/89-1576), Accademia, Venice,133b; *Intérieur Hollandais II* by Joan Miró (1928), Museo Guggenheim, Venice, 134c; *Madonna col Bambino* by Filippo Lippi (1406-69), Cini Collection, Venice 134t; *Annunciazione* by Vittore Carpaccio (1460 c.-1526), Ca' d'Oro, Venice,144b; Pala di San Zeno by Andrea Mantegna (1431-1506), San Zeno, Verona, 200cl; Science Photo Library: Earth Satellite Corporation,10 (inset); Settore Beni Culturali, Padua: 180t/cla/clb; Spectrum Colour Library: 28b; Studio Pizzi: 45cla, 47crb; Tony Stone Images: 214cla, 216t, 50/51c.

Phrase Book

In Emergency

Help!	**Aiuto!**	eye-**yoo**-toh
Stop!	**Fermate!**	fair-**mah**-teh
Call a doctor.	**Chiama un medico**	kee-**ah**-mah oon **meh**-dee-koh
Call an ambulance.	**Chiama un' ambulanza**	kee-**ah**-mah oon am-boo-**lan**-tsa
Call the police.	**Chiama la polizia**	kee-**ah**-mah lah pol-ee-**tsee**-ah
Call the fire brigade.	**Chiama i pompieri**	kee-**ah**-mah ee pom-pee-**air**-ee
Where is the telephone?	**Dov'è il telefono?**	dov-**eh** eel teh-**leh**-foh-noh?
The nearest hospital?	**L'ospedale più vicino?**	loss-peh-**dah**-leh pee-**oo**vee-**chee**-noh?

Communication Essentials

Yes/No	**Sì/No**	**see**/ **noh**
Please	**Per favore**	pair fah-**vor**-eh
Thank you	**Grazie**	**grah**-tsee-eh
Excuse me	**Mi scusi**	mee **skoo**-zee
Hello	**Buon giorno**	bwon **jor**-noh
Goodbye	**Arrivederci**	ah-ree-veh-**dair**-chee
Good evening	**Buona sera**	**bwon**-ah **sair**-ah
morning	**la mattina**	lah mah-**tee**-nah
afternoon	**il pomeriggio**	eel poh-meh-**ree**-joh
evening	**la sera**	lah **sair**-ah
yesterday	**ieri**	ee-**air**-ee
today	**oggi**	**oh**-jee
tomorrow	**domani**	doh-**mah**-nee
here	**qui**	**kwee**
there	**la**	**lah**
What?	**Quale?**	**kwah**-leh?
When?	**Quando?**	**kwan**-doh?
Why?	**Perchè?**	pair-**keh**?
Where?	**Dove?**	**doh**-veh

Useful Phrases

How are you?	**Come sta?**	**koh**-meh stah?
Very well, thank you.	**Molto bene, grazie.**	**moll**-toh **beh**-neh **grah**-tsee-eh
Pleased to meet you.	**Piacere di conoscerla.**	pee-ah-**chair**-eh dee coh-**noh**-shair-lah
See you soon.	**A più tardi.**	ah pee-**oo tar**-dee
That's fine.	**Va bene.**	va **beh**-neh
Where is/are ...?	**Dov'è/Dove sono ...?**	dov-**eh**/doveh **soh**noh?
How long does it take to get to ...?	**Quanto tempo ci vuole per andare a ...?**	**kwan**-toh **tem**-poh chee voo-**oh**-leh pair an-**dar**-eh ah ...?
How do I get to ...?	**Come faccio per arrivare a ...?**	koh-meh **fah**-choh pair arri-**var**-eh ah..?
Do you speak English?	**Parla inglese?**	**par**-lah een-**gleh**-zeh?
I don't understand.	**Non capisco.**	non ka-**pee**-skoh
Could you speak more slowly, please?	**Può parlare più lentamente, per favore?**	pwoh par-**lah**-reh pee-**oo** len-ta-**men**-teh pair fah-**vor**-eh
I'm sorry.	**Mi dispiace.**	mee dee-spee-**ah**-cheh

Useful Words

big	**grande**	**gran**-deh
small	**piccolo**	**pee**-koh-loh
hot	**caldo**	**kal**-doh
cold	**freddo**	**fred**-doh
good	**buono**	**bwoh**-noh
bad	**cattivo**	kat-**tee**-voh
enough	**basta**	**bas**-tah
well	**bene**	**beh**-neh
open	**aperto**	ah-**pair**-toh
closed	**chiuso**	kee-**oo**-zoh
left	**a sinistra**	ah see-**nee**-strah
right	**a destra**	ah **dess**-trah
straight on	**sempre dritto**	**sem**-preh **dree**-toh
near	**vicino**	vee-**chee**-noh
far	**lontano**	lon-**tah**-noh
up	**su**	**soo**
down	**giù**	**joo**
early	**presto**	**press**-toh
late	**tardi**	**tar**-dee
entrance	**entrata**	en-**trah**-tah
exit	**uscita**	oo-**shee**-ta
toilet	**il gabinetto**	eel gah-bee-**net**-toh
free, unoccupied	**libero**	**lee**-bair-oh
free, no charge	**gratuito**	grah-**too**-ee-toh

Making a Telephone Call

I'd like to place a long-distance call.	**Vorrei fare una interurbana.**	vor-**ray far**-eh oona in-tair-oor-**bah**-nah
I'd like to make a reverse-charge call.	**Vorrei fare una telefonata a carico del destinatario.**	vor-**ray far**-eh oona teh-leh-fon-**ah**-tah ah **kar**-ee-koh dell dess-tee-nah-**tar**-ree-oh
I'll try again later.	**Ritelefono più tardi.**	ree-teh-**leh**-foh-noh pee-oo **tar**-dee
Can I leave a message?	**Posso lasciare un messaggio?**	**poss**-oh lash-**ah**-reh oon mess-**sah**-joh?
Hold on.	**Un attimo, per favore**	oon **ah**-tee-moh, pair fah-**vor**-eh
Could you speak up a little please?	**Può parlare più forte, per favore?**	pwoh par-**lah**-reh pee-**oo for**-teh, pair fah-**vor**-eh?
local call	**la telefonata locale**	lah teh-leh-fon-**ah**-ta loh-**kah**-leh

Shopping

How much does this cost?	**Quant'è, per favore?**	kwan-**teh** pair fah-**vor**-eh?
I would like ...	**Vorrei ...**	vor-**ray**
Do you have ...?	**Avete ...?**	ah-**veh**-teh.. ?
I'm just looking.	**Sto soltanto guardando.**	stoh sol-**tan**-toh gwar-**dan**-doh
Do you take credit cards?	**Accettate carte di credito?**	ah-chet-**tah**teh **kar**-teh dee **creh**-dee-toh?
What time do you open/close?	**A che ora apre/ chiude?**	ah keh **or**-ah **ah**-preh/kee-**oo**-deh?
this one	**questo**	**kweh**-stoh
that one	**quello**	**kwell**-oh
expensive	**caro**	**kar**-oh
cheap	**a buon prezzo**	ah bwon **pret**-soh
size, clothes	**la taglia**	lah **tah**-lee-ah
size, shoes	**il numero**	eel **noo**-mair-oh
white	**bianco**	bee-**ang**-koh
black	**nero**	**neh**-roh
red	**rosso**	**ross**-oh
yellow	**giallo**	**jal**-loh
green	**verde**	**vair**-deh
blue	**blu**	bloo
brown	**marrone**	mar-**roh**-neh

Types of Shop

antique dealer	**l'antiquario**	lan-tee-**kwah**-ree-oh
bakery	**la panetteria**	lah pah-net-tair-**ree**-ah
bank	**la banca**	lah **bang**-kah
bookshop	**la libreria**	lah lee-breh-**ree**-ah
butcher's	**la macelleria**	lah mah-chell-eh-**ree**-ah
cake shop	**la pasticceria**	lah pas-tee-chair-**ee**-ah
chemist's	**la farmacia**	lah far-mah-**chee**-ah
delicatessen	**la salumeria**	lah sah-loo-meh-**ree**-ah
department store	**il grande magazzino**	eel **gran**-deh mag-gad-**zee**-noh
fishmonger's	**la pescheria**	lah pess-keh-**ree**-ah
florist	**il fioraio**	eel fee-or-**eye**-oh
greengrocer	**il fruttivendolo**	eel froo-tee-**ven**-doh-loh
grocery	**alimentari**	ah-lee-men-**tah**-ree
hairdresser	**il parrucchiere**	eel par-oo-kee-**air**-eh
ice cream parlour	**la gelateria**	lah jel-lah-tair-**ree**-ah
market	**il mercato**	eel mair-**kah**-toh
news-stand	**l'edicola**	leh-**dee**-koh-lah
post office	**l'ufficio postale**	loo-**fee**-choh pos-**tah**-leh
shoe shop	**il negozio di scarpe**	eel neh-**goh**-tsioh dee **skar**-peh
supermarket	**il supermercato**	su-pair-mair-**kah**-toh
tobacconist	**il tabaccaio**	eel tah-bak-**eye**-oh
travel agency	**l'agenzia di viaggi**	lah-jen-**tsee**-ah dee vee-**ad**-jee

Sightseeing

art gallery	**la pinacoteca**	lah peena-koh-**teh**-kah
bus stop	**la fermata dell'autobus**	lah fair-**mah**-tah dell **ow**-toh-booss
church	**la chiesa**	lah kee-**eh**-zah
	la basilica	lah bah-**seel**-i-kah
closed for the public holiday	**chiuso per la festa**	kee-**oo**-zoh pair lah **fess**-tah
garden	**il giardino**	eel jar-**dee**-no
library	**la biblioteca**	lah beeb-lee-oh-**teh**-kah
museum	**il museo**	eel moo-**zeh**-oh
railway station	**la stazione**	lah stah-tsee-**oh**-neh
tourist information	**l'ufficio turistico**	loo-**fee**-choh too-**ree**-stee-koh

Staying in a Hotel

Do you have any vacant rooms?	**Avete camere libere?**	ah-**veh**-teh **kah**-mair-eh **lee**-bair-eh?
double room	**una camera doppia**	oona **kah**-mair-ah **doh**-pee-ah
with double bed	**con letto matrimoniale**	kon **let**-toh mah-tree-moh-nee-**ah**-leh
twin room	**una camera con due letti**	oona **kah**-mair-ah kon **doo**-eh **let**-tee
single room	**una camera singola**	oona **kah**-mair-ah **sing**-goh-lah
room with a bath, shower	**una camera con bagno, con doccia**	oona **kah**-mair-ah kon **ban**-yoh, kon **dot**-chah
porter	**il facchino**	eel fah-**kee**-noh
key	**la chiave**	lah kee-**ah**-veh
I have a reservation.	**Ho fatto una prenotazione.**	oh **fat**-toh oona preh-noh-tah-tsee-**oh**-neh

Eating Out

Have you got a table for ...?	**Avete una tavola per ... ?**	ah-**veh**-teh oona **tah**-voh-lah pair ...?
I'd like to reserve a table.	**Vorrei riservare una tavola.**	vor-**ray** ree-sair-**vah**-reh oona **tah**-voh-lah
breakfast	**colazione**	koh-lah-tsee-**oh**-neh
lunch	**pranzo**	**pran**-tsoh
dinner	**cena**	**cheh**-nah
The bill, please.	**Il conto, per favore.**	eel **kon**-toh pair fah-**vor**-eh
I am a vegetarian.	**Sono vegetariano/a.**	**soh**-noh **veh**-jeh-tar ee-**ah**-noh/nah
waitress	**cameriera**	kah-mair-ee-**air**-ah
waiter	**cameriere**	kah-mair-ee-**air**-eh
fixed price menu	**il menù a prezzo fisso**	eel meh-**noo** ah **pret**-soh **fee**-soh
dish of the day	**piatto del giorno**	pee-**ah**-toh dell **jor**-no
starter	**antipasto**	an-tee-**pass**-toh
first course	**il primo**	eel **pree**-moh
main course	**il secondo**	eel seh-**kon**-doh
vegetables	**il contorno**	eel kon-**tor**-noh
dessert	**il dolce**	eel **doll**-cheh
cover charge	**il coperto**	eel koh-**pair**-toh
wine list	**la lista dei vini**	lah **lee**-stah day **vee**-nee
rare	**al sangue**	al **sang**-gweh
medium	**al puntino**	al poon-**tee**-noh
well done	**ben cotto**	ben **kot**-toh
glass	**il bicchiere**	eel bee-kee-**air**-eh
bottle	**la bottiglia**	lah bot-**teel**-yah
knife	**il coltello**	eel kol-**tell**-oh
fork	**la forchetta**	lah for-**ket**-tah
spoon	**il cucchiaio**	eel koo-kee-**eye**-oh

Menu Decoder

l'acqua minerale gasata/naturale	**lah**-kwah mee-nair-**ah**-leh gah-**zah**-tah/nah-too-**rah**-leh	mineral water fizzy/still
l'agnello	lahn-**yell**-oh	lamb
al forno	al **for**-noh	baked
alla griglia	ah-lah **greel**-yah	grilled
l'anguilla	lahng-**gwee**-lah	eel
l'aragosta	lah-rah-**goss**-tah	lobster
arrosto	ar-**ross**-toh	roast
il baccalà	eel bahk-kah-**lah**	dried salted cod
la birra	lah **beer**-rah	beer
la bistecca	lah bee-**stek**-kah	steak
il brodetto	eel-broh-**det**-toh	fish soup
il burro	eel **boor**-oh	butter
il caffè	eel kah-**feh**	coffee
i calamari	ee kah-lah-**mah**-ree	squid
il carciofo	eel kar-**choff**-oh	artichoke
la carne	la **kar**-neh	meat
carne di maiale	**kar**-neh dee mah-**yah**-leh	pork
i fagioli	ee fah-**joh**-lee	beans
il fegato	eel **fay**-gah-toh	liver
il formaggio	eel for-**mad**-joh	cheese
le fragole	leh **frah**-goh-leh	strawberries
il fritto misto	eel free-toh **mees**-toh	mixed fried fish
la frutta	la **froot**-tah	fruit
frutti di mare	**froo**-tee dee **mah**-reh	seafood
i funghi	ee **foon**-ghee	mushrooms
i gamberi	ee **gam**-bair-ee	prawns
il gelato	eel jel-**lah**-toh	ice cream
l'insalata mista	leen-sah-lah-tah **mees**-tah	mixed salad
l'insalata verde	leen-sah-lah-tah **vehr**-day	green salad
il latte	eel **laht**-teh	milk
i legumi OR i contorni	ee leh-**goo**-mee ee kon-**tor**-nee	vegetables
il manzo	eel **man**-tsoh	beef
la melanzana	lah meh-lan-**tsah**-nah	aubergine
la minestra	lah mee-**ness**-trah	soup
il pane	eel **pah**-neh	bread
il panino	eel pah-**nee**-noh	bread roll
le patate	leh pah-**tah**-teh	potatoes
le patatine fritte	leh pah-tah-**teen**-eh **free**-teh	chips
il pepe	eel **peh**-peh	pepper
la pesca	lah **pess**-kah	peach
il pesce	eel **pesh**-eh	fish
il pollo	eel **poll**-oh	chicken
il prosciutto cotto/crudo	eel pro-**shoo**-toh **kot**-toh/**kroo**-doh	ham cooked/cured
il riso	eel **ree**-zoh	rice
il sale	eel **sah**-leh	salt
la salsiccia	lah sal-**see**-chah	sausage
le seppie	leh **sep**-pee-eh	cuttlefish
secco	**sek**-koh	dry
la sogliola	lah **soll**-voh-lah	sole
i spinaci	ee spee-**nah**-chee	spinach
succo d'arancia/ di limone	**soo**-koh dah-**ran**-chah/ dee lee-**moh**-neh	orange/lemon juice
il tè	eel **teh**	tea
la tisana	lah tee-**zah**-nah	herbal tea
il tonno	eel **ton**-noh	tuna
la torta	lah **tor**-tah	cake/tart
la trippa	lah **treep**-pah	tripe
vino bianco	**vee**-noh bee-**ang**-koh	white wine
vino rosso	**vee**-noh **ross**-oh	red wine
il vitello	eel vee-**tell**-oh	veal
le vongole	leh **von**-goh-leh	clams
lo zucchero	loh **zoo**-kair-oh	sugar
gli zucchini	lyee dzu-**kee**-nee	courgettes
la zuppa	lah **tsoo**-pah	soup

Numbers

1	**uno**	**oo**-noh
2	**due**	**doo**-eh
3	**tre**	treh
4	**quattro**	**kwat**-roh
5	**cinque**	**ching**-kweh
6	**sei**	**say**-ee
7	**sette**	**set**-teh
8	**otto**	**ot**-toh
9	**nove**	**noh**-veh
10	**dieci**	dee-**eh**-chee
11	**undici**	**oon**-dee-chee
12	**dodici**	**doh**-dee-chee
13	**tredici**	**tray**-dee-chee
14	**quattordici**	kwat-**tor**-dee-chee
15	**quindici**	**kwin**-dee-chee
16	**sedici**	**say**-dee-chee
17	**diciassette**	dee-chah-**set**-teh
18	**diciotto**	dee-**chot**-toh
19	**diciannove**	dee-chah-**noh**-veh
20	**venti**	**ven**-tee
30	**trenta**	**tren**-tah
40	**quaranta**	kwah-**ran**-tah
50	**cinquanta**	ching-**kwan**-tah
60	**sessanta**	sess-**an**-tah
70	**settanta**	set-**tan**-tah
80	**ottanta**	ot-**tan**-tah
90	**novanta**	noh-**van**-tah
100	**cento**	**chen**-toh
1,000	**mille**	**mee**-leh
2,000	**duemila**	**doo**-eh **mee**-lah
5,000	**cinquemila**	**ching**-kweh **mee**-lah
1,000,000	**un milione**	oon meel-**yoh**-neh

Time

one minute	**un minuto**	oon mee-**noo**-toh
one hour	**un'ora**	oon **or**-ah
half an hour	**mezz'ora**	medz-**or**-ah
a day	**un giorno**	oon **jor**-noh
a week	**una settimana**	oona set-tee-**mah**-na
Monday	**lunedì**	loo-neh-**dee**
Tuesday	**martedì**	mar-teh-**dee**
Wednesday	**mercoledì**	mair-koh-leh-**dee**
Thursday	**giovedì**	joh-veh-**dee**
Friday	**venerdì**	ven-air-**dee**
Saturday	**sabato**	**sah**-bah-toh
Sunday	**domenica**	doh-**meh**-nee-kah

Vaporetto Routes Around Venice

The Vaporetto Routes

The ACTV network runs regula services around the city and ou to most of the islands. Some services are circular for part of their route; others extend their routes during the high season. Full details of the different type of vaporetti *and how to use the are given on pages 274–5.*

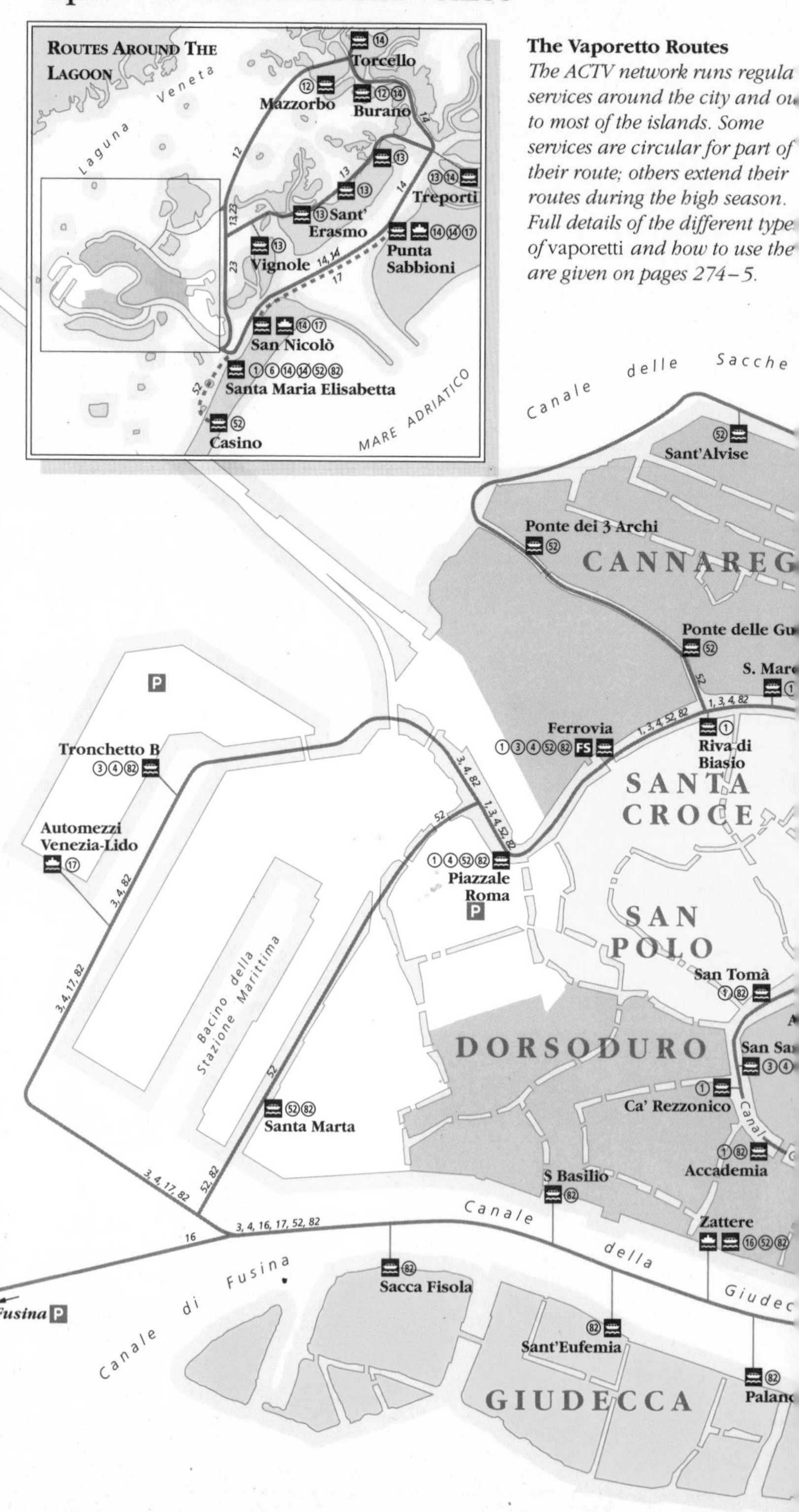